T0207176

Communications
in Computer and Information Science 1881

Rationale

The CCIS series is devoted to the publication of proceedings of computer science conferences. Its aim is to efficiently disseminate original research results in informatics in printed and electronic form. While the focus is on publication of peer-reviewed full papers presenting mature work, inclusion of reviewed short papers reporting on work in progress is welcome, too. Besides globally relevant meetings with internationally representative program committees guaranteeing a strict peer-reviewing and paper selection process, conferences run by societies or of high regional or national relevance are also considered for publication.

Topics

The topical scope of CCIS spans the entire spectrum of informatics ranging from foundational topics in the theory of computing to information and communications science and technology and a broad variety of interdisciplinary application fields.

Information for Volume Editors and Authors

Publication in CCIS is free of charge. No royalties are paid, however, we offer registered conference participants temporary free access to the online version of the conference proceedings on SpringerLink (http://link.springer.com) by means of an http referrer from the conference website and/or a number of complimentary printed copies, as specified in the official acceptance email of the event.

CCIS proceedings can be published in time for distribution at conferences or as post-proceedings, and delivered in the form of printed books and/or electronically as USBs and/or e-content licenses for accessing proceedings at SpringerLink. Furthermore, CCIS proceedings are included in the CCIS electronic book series hosted in the SpringerLink digital library at http://link.springer.com/bookseries/7899. Conferences publishing in CCIS are allowed to use Online Conference Service (OCS) for managing the whole proceedings lifecycle (from submission and reviewing to preparing for publication) free of charge.

Publication process

The language of publication is exclusively English. Authors publishing in CCIS have to sign the Springer CCIS copyright transfer form, however, they are free to use their material published in CCIS for substantially changed, more elaborate subsequent publications elsewhere. For the preparation of the camera-ready papers/files, authors have to strictly adhere to the Springer CCIS Authors' Instructions and are strongly encouraged to use the CCIS LaTeX style files or templates.

Abstracting/Indexing

CCIS is abstracted/indexed in DBLP, Google Scholar, EI-Compendex, Mathematical Reviews, SCImago, Scopus. CCIS volumes are also submitted for the inclusion in ISI Proceedings.

How to start

To start the evaluation of your proposal for inclusion in the CCIS series, please send an e-mail to ccis@springer.com.

Michael Khachay · Yury Kochetov ·
Anton Eremeev · Oleg Khamisov ·
Vladimir Mazalov · Panos Pardalos
Editors

Mathematical Optimization Theory and Operations Research

Recent Trends

22nd International Conference, MOTOR 2023
Ekaterinburg, Russia, July 2–8, 2023
Revised Selected Papers

 Springer

Editors
Michael Khachay ⓘ
Krasovsky Institute of Mathematics
and Mechanics
Ekaterinburg, Russia

Anton Eremeev ⓘ
Sobolev Institute of Mathematics
Omsk, Russia

Vladimir Mazalov ⓘ
Institute of Applied Mathematical Research
Petrozavodsk, Russia

Yury Kochetov ⓘ
Sobolev Institute of Mathematics
Novosibirsk, Russia

Oleg Khamisov ⓘ
Melentiev Energy Systems Institute
Irkutsk, Russia

Panos Pardalos
University of Florida
Gainesville, FL, USA

ISSN 1865-0929   ISSN 1865-0937 (electronic)
Communications in Computer and Information Science
ISBN 978-3-031-43256-9   ISBN 978-3-031-43257-6 (eBook)
https://doi.org/10.1007/978-3-031-43257-6

This Springer imprint is published by the registered company Springer Nature Switzerland AG
The registered company address is: Gewerbestrasse 11, 6330 Cham, Switzerland

Paper in this product is recyclable.

Preface

This volume contains selected papers presented at the 22nd International Conference on Mathematical Optimization Theory and Operations Research (MOTOR 2023)[1] held during July 2–8, 2023, in Ekaterinburg, capital of the Urals, Russia.

This year, we celebrated the 90th anniversary of academician Ivan Ivanovich Eremin (1933–2013). Academician I. I. Eremin was a widely known Soviet and Russian mathematician, a specialist in mathematical programming and operations research. He introduced the well-known Eremin-Zangwill exact penalty functions, established a brilliant theory of improper (singular) linear and convex programs having valuable applications in mathematical economics, and put forward a family of highly efficient iterative algorithms, called by him Fejér methods. In the early 1960s, he established the Ural scientific school on optimization theory and methods. MOTOR 2023 is devoted to his blessed memory.

MOTOR 2023 was the fifth joint scientific event unifying a number of well-known conferences held in Ural, Siberia, and the Far East of Russia for a long time

– The Baikal International Triennial School Seminar on Methods of Optimization and Their Applications (BITSS MOPT), established in 1969 by academician N. N. Moiseev, with 17 events held up to 2017,
– The All-Russian Conference on Mathematical Programming and Applications (MPA), established in 1972 by I. I. Eremin, with 15 events held up to 2015,
– The International Conference on Discrete Optimization and Operations Research (DOOR), which was organized nine times between 1996 and 2016,
– The International Conference on Optimization Problems and Their Applications (OPTA), which was organized seven times in Omsk between 1997 and 2018.

The first four events of this series, MOTOR 2019, MOTOR 2020, MOTOR 2021, and MOTOR 2022 were held in Ekaterinburg, Novosibirsk, Irkutsk, and Petrozavodsk,

---

[1] http://motor2023.uran.ru

Russia, respectively. As per tradition, the main conference scope included, but was not limited to, mathematical programming, bi-level and global optimization, integer programming and combinatorial optimization, approximation algorithms with theoretical guarantees and approximation schemes, heuristics and meta-heuristics, game theory, optimal control, optimization in machine learning and data analysis, and their valuable applications in operations research and economics.

In response to the call for papers, MOTOR 2023 received 189 submissions. Out of 89 full papers considered for review (100 abstracts and short communications were excluded for formal reasons) only 29 papers were selected by the Program Committee for publication in the first volume of conference proceedings, LNCS, vol. 13930. Each submission was reviewed by at least three PC members or invited reviewers, experts in their fields, in order to supply detailed and helpful comments. In this volume, the PC recommended for publication 29 papers, after their presentation and discussion during the conference and subsequent revision with respect to the reviewers' comments.

The conference featured ten invited lectures:

- Kamil Aida-Zade (Institute of Control Systems, Azerbaijan), "Feedback control on the class of zonal control actions"
- Mario R. Guarracino (Higher School of Economics, Russia), "Semi-supervised Learning with Depth Functions"
- Milojica Jaćimović (University of Montenegro, Montenegro), "Strong convergence of extragradient-like methods for solving quasi-variational inequalities"
- Pinyan Lu (Shanghai University of Finance and Economics, China), "Algorithms for Solvers: Ideas from CS and OR"
- Panos Pardalos (University of Florida, USA), "Artificial Intelligence, Smart Energy Systems, and Sustainability"
- Eugene Semenkin (Reshetnev Siberian State University of Science and Technology, Russia), "Hybrid evolutionary optimization: how self-adapted algorithms can automatically generate applied AI tools"
- Yaroslav D. Sergeev (University of Calabria, Italy), "Numerical Infinities and Infinitesimals in Optimization"
- Alexander A. Shananin (Lomonosov Moscow State University, Russia), "General equilibrium models in production networks with substitution of inputs"
- Predrag S. Stanimirović (University of Niš, Serbia), "Optimization methods in gradient and zeroing neural networks"
- Vladimir V. Vasin (Krasovsky Institute of Mathematics and Mechanics, Russia), "Fejér type iterative processes for quadratic minimization problems".

We thank the authors for their submissions, members of the Program Committee, and all the external reviewers for their efforts in providing exhaustive reviews. We thank our sponsors and partners: Krasovsky Institute of Mathematics and Mechanics, the Ural Mathematical Center, the Ural Branch of the Russian Academy of Sciences, the Sobolev Institute of Mathematics and Mathematical Center in Akademgorodok, the Center for Research and Education in Mathematics, and the Higher School of Economics (Nizhny

Novgorod). We are grateful to the colleagues from the Springer LNCS and CCIS editorial boards for their kind and helpful support.

August 2023

Michael Khachay
Yury Kochetov
Anton Eremeev
Oleg Khamisov
Vladimir Mazalov
Panos Pardalos

Organization

Honorary Chairs

Yury Evtushenko   Dorodnicyn Computing Centre, Russia
Panos Pardalos   University of Florida Gainesville, USA

General Chair

Michael Khachay   Krasovsky Institute of Mathematics and Mechanics, Russia

Program Committee Chairs

Anton Eremeev   Sobolev Institute of Mathematics, Russia
Oleg Khamisov   Melentiev Energy Systems Institute, Russia
Yury Kochetov   Sobolev Institute of Mathematics, Russia
Vladimir Mazalov   Institute of Applied Mathematical Research, Russia

Program Committee Members

Alexander Afanasiev   Institute for Information Transmission Problems, Russia
Vladimir Beresnev   Sobolev Institute of Mathematics, Russia
George Bolotashvili   Georgian Technical University, Georgia
Igor Bykadorov   Sobolev Institute of Mathematics, Russia
Tatiana Davidović   Mathematical Institute of the Serbian Academy of Sciences and Arts, Serbia
Alexander Dolgui   IMT Atlantique, France
Adil Erzin   Sobolev Institute of Mathematics, Russia
Alexander Gasnikov   Moscow Institute of Physics and Technology, Russia
Milojica Jaćimović   University of Montenegro, Montenegro
Valeriy Kalyagin   National Research University Higher School of Economics, Russia

Vadim Kartak	Ufa State Aviation Technical University, Russia
Alexander Kazakov	Matrosov Institute for System Dynamics and Control Theory, Russia
Andrey Kibzun	Moscow Aviation Institute, Russia
Alexnader Kononov	Sobolev Institute of Mathematics, Russia
Dmitri Kvasov	University of Calabria, Italy
Bertrand Lin	National Yang Ming Chiao Tung University, Taiwan
Vittorio Maniezzo	University of Bologna, Italy
Nevena Mijajlovic	University of Montenegro, Montenegro
Evgeni Nurminski	Fast East Federal University, Russia
Nicholas Olenev	Doronicyn Computing Centre, Russia
Mikhail Posypkin	Federal Research Center on Computer Science and Control, Russia
Oleg Prokopyev	University of Pittsburg, USA
Artem Pyatkin	Sobolev Institute of Mathematics, Russia
Soumyendu Raha	Indian Institute of Science, Bengaluru, India
Anna Rettieva	Institute of Applied Mathematical Research, Russia
Eugene Semenkin	Siberian State Aerospace University, Russia
Yaroslav Sergeyev	Università della Calabria, Italy
Angelo Sifaleras	University of Macedonia, Greece
Alexander Strekalovsky	Matrosov Institute for System Dynamics and Control Theory, Russia
Tatiana Tchemisova	University of Aveiro, Portugal
Raca Todosijević	Université Polytechnique Hauts-de-France, France
Igor Vasilyev	Matrosov Institute for System Dynamics and Control Theory, Russia
Alexander Vasin	Lomonosov Moscow State University, Russia
Vladimir Vasin	Krasovsky Institute of Mathematics and Mechanics, Russia

Additional Reviewers

Natalia Aizenberg	Rentsen Enkhbat	Igor Izmest'ev
Anatoly Antipin	Vladimir Erokhin	Lev Kazakovtsev
Artem Aroslonkin	Majid Forghani	Daniil Khachai
Vladimir Berikov	Mikhail Gomoyunov	Vladimir Khandeev
René van Bevern	Tatiana Gruzdeva	Dmitry Khlopin
Ivan Davydov	Victor Iliev	Elena Khoroshilova
Zhang Dong	Sergey Ivanov	Vladimir Kotov

Sergey Kumkov
Igor Kulachenko
Sergey Lavlinskii
Anna Lempert
Tatiana Makarovskikh
Andrey Melnikov
Katherine Neznakhina
Natalia Obrosova
Yuri Ogorodnikov
Andrey Orlov

Anna Panasenko
Artem Panin
Elena Parilina
Alexander Plyasunov
Leonid Popov
Alexey Ratushnyi
Valeriy Rozenberg
Vladimir Servakh
Ruslan Simanchev
Margarita Sotnikova

Fedor Stonyakin
Oleg Sukhoroslov
Alexander Tarasyev
Anton Ushakov
Rashid Yarullin
Igor Zabotin
Yulia Zakharova
Vladimir Zubov
Alexander Zyryanov

Industry Session Chair

Igor Vasilyev   Matrosov Institute for System Dynamics and
                             Control Theory, Russia

Organizing Committee

Alexey Makarov (Chair)   Ural Branch of RAS, Ekaterinburg, Russia
Igor Kandoba (Co-chair)   Krasovsky Institute of Mathematics and
                             Mechanics, Ekaterinburg, Russia
Alexander Petunin (Co-chair) Ural Federal University, Ekaterinburg, Russia
Boris Digas   Krasovsky Institute of Mathematics and
                             Mechanics, Ekaterinburg, Russia
Artem Firstkov   Krasovsky Institute of Mathematics and
                             Mechanics, Ekaterinburg, Russia
Majid Forghani   Krasovsky Institute of Mathematics and
                             Mechanics, Ekaterinburg, Russia
Nina Kochetova   Sobolev Institute of Mathematics, Novosibirsk,
                             Russia
Polina Kononova   Sobolev Institute of Mathematics, Novosibirsk,
                             Russia
Timur Medvedev   Higher School of Economics, Nizhny Novgorod,
                             Russia
Ekaterina Neznakhina   Krasovsky Institute of Mathematics and
                             Mechanics, Ekaterinburg, Russia
Yuri Ogorodnikov   Krasovsky Institute of Mathematics and
                             Mechanics, Ekaterinburg, Russia
Roman Rudakov   Krasovsky Institute of Mathematics and
                             Mechanics, Ekaterinburg, Russia
Ksenia Ryzhenko   Krasovsky Institute of Mathematics and
                             Mechanics, Ekaterinburg, Russia

Organizers

Krasovsky Institute of Mathematics and Mechanics, Russia
Ural Mathematical Center, Russia
Ural Branch of RAS, Russia
Sobolev Institute of Mathematics, Russia
Mathematical Center in Akademgorodok, Novosibirsk, Russia
Higher School of Economics, Nizhny Novgorod, Russia
Melentiev Energy Systems Institute, Russia
Ural Federal University, Russia

Contents

Invited Papers

Towards Subderivative-Based Zeroing Neural Networks 3
  Predrag S. Stanimirović, Dimitrios Gerontitis, Vladimir N. Krutikov,
  and Lev A. Kazakovtsev

Mathematical Programming

An Algorithm for Decentralized Multi-agent Feasibility Problems 19
  Olga Pinyagina

Online Optimization Problems with Functional Constraints Under Relative
Lipschitz Continuity and Relative Strong Convexity Conditions 29
  Oleg Savchuk, Fedor Stonyakin, Mohammad Alkousa, Rida Zabirova,
  Alexander Titov, and Alexander Gasnikov

A Cutting Method with Successive Use of Constraint Functions
in Constructing Approximating Sets 44
  Rashid Yarullin and Igor Zabotin

Implementing One Variant of the Successive Concessions Method
for the Multi-objective Optimization Problem 54
  Igor Zabotin, Oksana Shulgina, and Rashid Yarullin

Stochastic Optimization

UCB Strategy for Gaussian and Bernoulli Multi-armed Bandits 67
  M. A. Ershov and A. S. Voroshilov

Estimation of Both Unknown Parameters in Gaussian Multi-armed Bandit
for Batch Processing Scenario 79
  Sergey Garbar

Zero-Order Stochastic Conditional Gradient Sliding Method
for Non-smooth Convex Optimization 92
  Aleksandr Lobanov, Anton Anikin, Alexander Gasnikov,
  Alexander Gornov, and Sergey Chukanov

Discrete and Combinatorial Optimization

Tabu Search Metaheuristic for the Penalty Minimization Personnel Task
Scheduling Problem 109
    Ivan Davydov, Igor Vasilyev, and Anton V. Ushakov

An $O(n \log n)$-Time Algorithm for Linearly Ordered Packing of 2-Bar
Charts into $OPT + 1$ Bins 122
    Adil Erzin, Alexander Kononov, Stepan Nazarenko,
    and Konstantin Sharankhaev

Approximation Algorithms for Graph Cluster Editing Problems
with Cluster Size at Most 3 and 4 134
    Victor Il'ev and Svetlana Il'eva

On Cone Partitions for the Min-Cut and Max-Cut Problems
with Non-negative Edges 146
    Andrei V. Nikolaev and Alexander V. Korostil

A Pattern-Based Heuristic for a Temporal Bin Packing Problem
with Conflicts 161
    A. Ratushnyi

Integer Models for the Total Weighted Tardiness Problem on a Single
Machine 176
    R. Yu. Simanchev and I.V. Urazova

Solving Maximin Location Problems on Networks with Different Metrics
and Restrictions 188
    Gennady G. Zabudsky

Operations Research

On Probability Shaping for 5G MIMO Wireless Channel with Realistic
LDPC Codes 203
    Evgeny Bobrov and Adyan Dordzhiev

Additive Routing Problem for a System of High-Priority Tasks 218
    Alexandr G. Chentsov and Pavel A. Chentsov

Public-Private Partnership Model with a Consortium 231
    Sergey Lavlinskii, Artem Panin, and Alexander Plyasunov

Variable Neighborhood Search Approach for the Bi-criteria Competitive
Location and Design Problem with Elastic Demand 243
Tatiana Levanova, Alexander Gnusarev, Ekaterina Rubtsova,
and Sigaev Vyatcheslav

Decomposition Approach for Simulation-Based Optimization of Inventory
Management 259
Alexander Yuskov, Igor Kulachenko, Andrey Melnikov, and Yury Kochetov

Optimal Control and Mathematical Economics

The Algorithm for the Construction of a Symbolic Family of Regulators
for Nonlinear Discrete Control Systems with Two Small Parameters 277
Yulia Danik and Mikhail Dmitriev

Analytical Construction of the Singular Set in One Class of Time-Optimal
Control Problems in the Presence of Linear Segments of the Boundary
of the Target 292
Lebedev Pavel and Uspenskii Alexander

On the Existence of Fuzzy Contractual Allocations, Fuzzy Core
and Perfect Competition in an Exchange Economy 308
Valeriy Marakulin

Linear Interpolation of Program Control with Respect
to a Multidimensional Parameter in the Convergence Problem 324
Vladimir Nikolaevich Ushakov, Aleksandr Anatol'evich Ershov,
Anna Aleksandrovna Ershova, and Aleksandr Vladimirovich Alekseev

Behavior of Stabilized Trajectories of a Two Factor Economic Growth
Model Under the Changes of a Production Function Parameter 338
Anastasiia A. Usova and Alexander M. Tarasyev

Optimization in Machine Learning

Uncertainty of Graph Clustering in Correlation Block Model 353
Artem Aroslankin and Valeriy Kalyagin

Multi-target Weakly Supervised Regression Using Manifold
Regularization and Wasserstein Metric 364
Kirill Kalmutskiy, Lyailya Cherikbayeva, Alexander Litvinenko,
and Vladimir Berikov

Using General Least Deviations Method for Forecasting of Crops Yields 376
Tatiana Makarovskikh, Anatoly Panyukov, and Mostafa Abotaleb

Searching for Distance Graph Embeddings and Optimal Partitions
of Compact Sets in Euclidean Space 391
V. A. Voronov, A. D. Tolmachev, D. S. Protasov, and A. M. Neopryatnaya

Author Index 405

Invited Papers

Towards Subderivative-Based Zeroing Neural Networks

Predrag S. Stanimirović[1,4(✉)]📝, Dimitrios Gerontitis[2]📝,
Vladimir N. Krutikov[3,4], and Lev A. Kazakovtsev[4]📝

[1] Faculty of Sciences and Mathematics, University of Niš,
Višegradska 33, 18000 Niš, Serbia
pecko@pmf.ni.ac.rs
[2] International Hellenic University, Thessaloniki, Greece
dimitrios.gerontitis@yahoo.gr, dimger@iee.ihu.gr
[3] Kemerovo State University, 6 Krasnaya street, Kemerovo 650043, Russia
[4] Laboratory "Hybrid Methods of Modelling and Optimization in Complex Systems",
Siberian Federal University, Prosp. Svobodny 79, 660041 Krasnoyarsk, Russia
levk@bk.ru

Abstract. Zeroing Neural Networks (ZNN) are dynamic systems suitable for studying and solving time-varying problems. The advantage of this particular type of recurrent neural networks (RNNs) is their global and exponential convergence property, which can be accelerated to a finite-time convergence. The dynamic flow of the ZNN requires the use of an appropriate error (Zhang) function $E(t)$, which can be in matrix, vector, or scalar form, and the element-wise time derivative of $\dot{E}(t)$ at each time instant t. A possible difficulty arises in all cases where the time derivative $\dot{E}(t)$ does not exist, for any element of $E(t)$ and any time instant t_0 from a predefined time interval $[0, T]$.

In this research, we propose improvements to the ZNN formula for the case where the time-derivative of the Zhang function does not exist at some points. The non-differentiability occurs in various forms in several cases and occurs frequently. One possible solution in convex and non-differentiable environments is based on the use of subderivatives instead of the time derivative. Another solution is applicable in nonconvex cases and situations with discontinuity, and it is based on shifting in singular points to avoid the division by zero (DBZ) problem that often occurs in division with time-varying expressions.

Keywords: Zeroing neural network · subderivative · Gradient descent methods · Division by zero

Supported by the Laboratory "Hybrid Methods of Modelling and Optimization in Complex Systems", Siberian Federal University, Prosp. Svobodny 79, 660041 Krasnoyarsk, Russia.

M. Khachay et al. (Eds.): MOTOR 2023, CCIS 1881, pp. 3–15, 2023.
https://doi.org/10.1007/978-3-031-43257-6_1

1 Introduction

Nowadays, it is considered normal in science and industry to tackle intractable problems and solve complex computational problems by using neural networks. These problems can be manifested in the form of equations and systems of equations in matrix, vector or scalar form. Zeroing neural networks (ZNNs) are a class of recurrent neural networks specifically designed to approximate zeros of these equations. In recent years, ZNNs have played an essential part in finding online solutions of time-varying (TV) problems.

Zhang Neural Network or Zeroing Neural Network (ZNN) dynamics was proposed by Zhang *et al.* in 2011 for solving TV problems, mainly in computing the matrix inversion and pseudoinversion, solving linear and matrix equations or systems of such equations, solving convex quadratic optimization, calculating various TV functions in matrix, vector, or scalar form, solving some practical optimization problems in robotics.

A ZNN model is often designed in implicit dynamics instead of explicit dynamics assumed in Gradient Neural Networks (GNNs). Each ZNN dynamics is developed in two global steps.

Step 1ZNN. The first step requires the construction of a suitable error function, also known as *Zhang function* (ZF) (or Zhangian), termed as $E(t)$. A ZF $E(t)$ is stated appropriately if the solution to the equation $E(t) = 0$ coincides with the exact solution of the problem under consideration. Zhang and Guo, in [22,24], gave a comprehensive overview of the various Zhang functions and classified them into four categories: scalar-valued ZFs in the real and complex domain, vector-valued ZFs in the real and complex domain, and matrix-valued ZFs in the real and complex domain.

Step 2ZNN. The second step is based on the exploitation of the defined ZF in the ZNN dynamics, which is given by the evolution law

$$\dot{E}(t) = \frac{\mathrm{d}E(t)}{\mathrm{d}t} = -k\mathcal{F}\left(E(t)\right), \tag{1}$$

where $\dot{E}(t)$ denotes the time derivative of $E(t)$, $k \in \mathbb{R}^+$ is a positive real quantity necessary for scaling the convergence and $\mathcal{F}(\cdot) : \mathbb{R}^{n \times n} \to \mathbb{R}^{n \times n}$ element-wise application of a suitable odd and monotonically-increasing activation function (AF). The essence of the AF $\mathcal{F}$ is to derive an output based on a set of input values to a node. The purpose of an activation function is to include non-linearity to the RNN. Different kinds of nonlinear activation functions improve behaviour of the ZNN models in practice. Moreover, particular activation functions lead to a finite-time convergence. In general $\mathcal{F}$ is an odd and monotonically increasing function array, element-wise applicable to elements of a real matrix $Q = [q_{ij}] \in \mathbb{R}^{m \times n}$, i.e., $\mathcal{F}(Q) = [f(q_{ij})]$, $i = 1, \dots, m$, $j = 1, \dots, n$, such that $f()$ is an odd and monotonically increasing function. In the case when $\mathcal{F}$ is based on the linear function $f(x) = x$, the ZNN design (1) becomes $\dot{E}(t) = -kE(t)$, and it is called linear ZNN model.

The analytic solution to the ordinary differential equation $\dot{E}(t) = -kE(t)$ defining the linear ZNN evolution is equal to $E(t) = E(0)\mathrm{e}^{-kt}$. Thus $E(t)$ in the linear ZNN model exponentially converges with the convergence rate $k > 0$ to the zero equilibrium point $E(t) = 0$, as $t \to +\infty$. In addition, if an odd and monotonically increasing function $f()$ is used inside the entry-wise activation $\mathcal{F}(\cdot)$, then $E(t)$ again globally converges to $E(t) = 0$, starting from an arbitrary initial state $E(0)$. Typical AF accelerates the exponential convergence. Moreover, some activation functions can initiate a finite convergence time.

Machine learning and ZNN techniques often rely on differentiable functions, but this constraint is not always satisfied in practice. In fact, functions that are not differentiable at some points are very common. The goal of our study is to solve problems arising in the ZNN dynamical evolution in situations where $E(t)$ is non-differentiable in one or more singular points.

The hierarchy of sections is scheduled as follows. The research problem statement and research questions are described in Sect. 2. Section 3 explores possible solutions to the stated problem in the ZNN development and applications. Several models that involve singular (non-differentiable and discontinuous cases) situations are presented in Sect. 4, while some concluding observations are given in the concluding Sect. 5.

2   Problem Statement and Research Questions

The time-derivative $\dot{E}(t)$ in (1) assumes element-wise time derivative. Namely, the time derivative $\dot{E}(t)$ in the matrix case $E(t) = [e_{ij}(t)]$ assumes $\dot{E}(t) = [\dot{e}_{ij}(t)]$, while in the scalar case $E(t) = [e(t)]$ it follows $\dot{E}(t) = [\dot{e}(t)]$. We have singled out several possible situations in which a problem with the application of ZNN dynamics (1) occurs.

Problem 1. The problem for $E(t) = [e_{ij}(t)]$ occurs for the case where $\dot{e}_{ij}(t_0)$ does not exist, for any element $e_{ij}(t)$ and a time instant t_0 within the chosen time interval $[0, T]$.

Problem 2. The problem for $E(t) = [e_i(t)]$ will happen in the case when $\dot{e}_i(t_0)$ does not exist, for an arbitrary element $e_i(t)$ and an arbitrary time instant $t_0 \in [0, T]$.

Problem 3. The problem for $E(t) = [e(t)]$ occurs for the case where $\dot{e}(t_0)$ does not exist for any $t_0 \in [0, T]$.

Problem 4. Another problem appears in implementing the ZNN design for solving time-varying nonlinear equations (TVNE). The error function for solving TVNEs of the general form

$$\wp(x(t), t) = 0, \quad x(t) \in \mathbb{R}. \tag{2}$$

is defined in [8, 25] by $e(t) = \wp(x(t), t)$. The ZNN dynamics for solving TVNE is the following dynamical system

$$\dot{e}(t) = \dot{\wp}(x(t), t) = -k\mathcal{F}\left(\wp(x(t), t)\right) \iff \frac{\partial \wp}{\partial x}\dot{x}(t) = -k\mathcal{F}\left(\wp(x(t), t)\right) - \frac{\partial \wp}{\partial t}. \tag{3}$$

The mathematical properties of the ZF defined by $\wp^2(x(t), t)$ prompt its use in approximating the solution of (3). The subsequent model was proposed in [21] as a development of the ZNN design on the error function $e(t) = \wp(x(t), t)$:

$$\dot{x}(t) = -\frac{k}{2}\frac{\partial}{\partial x}\wp^2(x(t), t) = -k\wp(x(t), t)\frac{\partial \wp}{\partial t}. \tag{4}$$

The Eq. (4) represents the Gradient Neural Dynamics (GND) for solving TVNE. Continuing such research, the authors of [9] proposed the Improved Zeroing Neural Network (IZNN) model for solving TVNE

$$\frac{\partial \wp}{\partial x}\dot{x}(t) = -k\left(\alpha_1\wp(x(t), t) + \alpha_2\left(\wp(x(t), t)\right)^{\frac{1}{\omega}}\right) - \frac{\partial \wp}{\partial t}, \tag{5}$$

where $\alpha_1, \alpha_2 > 0$ and $\omega > 1$. The dynamical flow (5) possesses a finite time convergence.

The difficulty in the application of DNE dynamics (3), (4) and (5) occurs when at least one of the derivatives $\frac{\partial \wp}{\partial x}$, $\dot{x}(t)$ and $\frac{\partial \wp}{\partial t}$ is not defined. In this case, the gradient of $\wp(x(t), t)$ is not defined.

The non-differentiability of the function $f(t)$ at the argument t_0 occurs in several cases.

Situation S1. The function $f(t)$ is discontinuous at t_0, for example $t/|t|$;
Situation S2. The graph of $f(t)$ has a corner point, for example $|t|$;
Situation S3. The function $f(t)$ is unbounded and goes to infinity. An example is $1/t$;
Situation S4. The function $f'(t)$ is not defined at t_0, for example $\sqrt{t}$ at $t = 0$;
Situation S5. The function $f(t)$ can be defined and finite at t_0, but $f'(t)$ is infinite, which means that $f(t)$ has a vertical tangent at t_0. An example is $t^{1/3}$ at $t = 0$.

We will use the term *singular points* of ZNN in order to indicate points at which non-differentiability occurs in the ZNN design. One approach to solving the singularity is to use *subderivatives*. Another solution is based on using conditional shift to escape differentiation is singular points or division by zero (DBZ).

3   Possible Solutions

Division by zero (DBZ) is an attempt to calculate a quotient of two expressions in time-variant or invariant form whose denominator is zero. A possible solution in situations S1–S4 is based on the modified ZNN dynamics developed in [5] for solving the DBZ problem. These situations occur when the function $E(t)$ is unbounded and goes to infinity. The paper [23] overcomes the singularity in the DBZ using the approach based on GNNs. The DBZ control singularity phenomenon arising in the nonlinear control was studied in [26] using a suitable Zhang dynamical system. The strategy used in [5] is based on the *Matlab Fcn* block that implements the conditional shifting function $t + (t == s)\lambda$ in the singular point s, where $\lambda = 10^{-6}$ is the *Matlab* constant. The output of the equality operator $a == b$ in *Matlab* is 1 if values of operands a and b are of identical value; otherwise, the output is 0. As a result, the first possible solution to overcome the DBZ problem at any singular point s is based on the conditional shifting of the Zhagian $E(t)$ in the form $\tilde{E}(t) = E(t + (t == s)\lambda)$. More precisely, the following conditional shifting is used to avoid the singularity in the point s:

$$t + (t == s)\lambda = \begin{cases} t, & t \neq s, \\ s + \lambda, & t = s. \end{cases}$$

If the step λ in the shift $\tilde{E}(t) = E(t + (t == s)\lambda)$ is too small and can not solve the singularity, another step is to increase the value of λ.

The second approach is based on **subderivation and subdifferential**. The approach based on subderivative is applicable in finding zeros of ZFs $E(t)$ that involve convex but non differentiable entries.

In the one-dimensional case, a subderivative of a convex function $f : \beth \to \mathbb{R}$, $\beth \subseteq \mathbb{R}$, at a point x_0 in the open interval $\beth$ is a real number ς such that

$$f(x) \geq f(x_0) + \varsigma(x - x_0)$$

for all $x \in \beth$. According to the converse of the mean value theorem, the set of subderivatives at x_0 for f is a nonempty closed interval $[l, u]$, such that l and u are defined by the one-sided limits

$$l = \lim_{x \to x_0^-} \frac{f(x) - f(x_0)}{x - x_0}, \quad u = \lim_{x \to x_0^+} \frac{f(x) - f(x_0)}{x - x_0}.$$

The set $[l, u]$ containing all subderivatives is designated as the subdifferential of f at x_0 and denoted by $\partial f(x_0)$. If f is convex and its subdifferential at x_0 contains exactly one subderivative, then f is differentiable at x_0. A subdifferential of f at x_0 is the set of subderivatives

$$\partial f(x_0) := \{\varsigma \in \mathbb{R} |\ f(x) \geq f(x_0) + \varsigma(x - x_0), \ \forall x \in \beth\}.$$

Let us assume the existence of an element $e_{ij}(t) \in E(t)$ such that $\dot{e}_{ij}(t_0)$ does not exist at a time $t_0 \in [0, T]$. Our idea is to replace $\dot{e}_{ij}(t_0)$ by the subdifferential $\partial(e_{ij}(t))(t_0)$. In the convex case, instead of $\dot{e}_{ij}(t)$ it is reasonable to use

$$\dot{e}_{ij}(t) = \text{Sel}\left(\partial(e_{ij}(t))(t_0)\right) = \begin{cases} \dot{e}_{ij}(t), & t \neq t_0, \\ \text{Sel}\left(\partial(e_{ij}(t))(t_0)\right), & t = t_0, \end{cases} \tag{6}$$

where $\text{Sel}\left(\partial(e_{ij}(t))(t_0)\right)$ denotes an arbitrary element from the set $\partial(e_{ij}(t))(t_0)$. The ZNN (1) changes accordingly into

$$\acute{E}(t) = -k\mathcal{F}(E(t)), \tag{7}$$

where $\acute{E}(t) = [\acute{e}_{ij}(t)]$ if $E(t) = [e_{ij}(t)]$. The dynamical system (7) will be termed as S-ZNN dynamics. In the scalar case $E(t) = e(t)$, the modified S-ZNN dynamics becomes $\acute{E}(t) = [\acute{e}(t)]$ if.

Non-differentiable cases in the optimization problem (3) can be solved by two strategies. The first strategy is based on replacing $e(t) = \wp(x(t), t)$ by

$$\tilde{e}(t) = \wp\left(x(t + (t == s)\lambda),\ t + (t == 0)\lambda\right),$$

where s is an expected singular point. The second solution proposes the use of the subdifferential $\partial\wp(x(t), t)$ of $\wp(x(t), t)$, i.e., subderivatives of $\frac{\partial \wp}{\partial x}$, $\dot{x}(t)$ or $\frac{\partial \wp}{\partial t}$.

Subgradient methods are verified tools for optimizing a non-differentiable convex function $f(x)$, $x \in \mathbb{R}^n$. These methods iteratively update the approximation vector taking iterative steps in the negative subgradient direction using a positive step size. In most cases, the subgradient method uses a pre-specified step length rather than the exact or approximate line search typical for gradient-descent methods. The fundamentals of nonsmooth convex optimization methods were established and developed in [16,17,20]. Different directions have been taken for the construction of non-smooth optimization methods. Some of these methods are based on smooth approximations for nonsmooth functions [1,3,4,6,7,12,14,15]. The principle based on the approximation of nonsmooth functions initiated a number of methods intended for solving various convex optimization problems [4,7,14]. Some practical approaches in the field of non-smooth optimization arose as a result of the progress of subgradient methods with a space expansion [2,10,13,17–19]. In [11] a relaxation subgradient method (RSM) was proposed based on the rank-two correction matrices, which are analogous to the updates used in quasi-Newton (QN) optimization methods.

We assume that the proposed S-ZNN dynamics will be able to overcome difficulties in singular cases of the ZNN design.

4   Models with Singular Cases

The Simulink implementation of the linear ZNN flow $\dot{e} = -k\,e(t)$ for zeroing $e(t) = a(t)x(t) - 1$, $a(t), x(t) \in \mathbb{R}$, is presented in Fig. 1. In this case, $\dot{e}(t) = \dot{a}(t)x(t) + a(t)\dot{x}(t)$ and the implicit ZNN model $\dot{e} = -k\,e(t)$ is given by

$$a(t)\dot{x}(t) = -\dot{a}(t)x(t) - k[a(t)x(t) - 1],$$

which gives

$$\dot{x}(t) = [1 - a(t)]\dot{x}(t) - \dot{a}(t)x(t) - k[a(t)x(t) - 1].$$

The presented Simulimk is suitable for calculating $x(t) = a(t)^{-1}$. Singular cases appear in situations $a(t) = 0$.

Example 1. Our goal in this example is to calculate $x(t) = |t - 1|^{-1}$ using $a(t) = |t - 1|$ and the error function $e(t) = |t - 1|x(t) - 1$. If we just put this example using the pure ZNN model (1) we will obtain an error in the Simulink

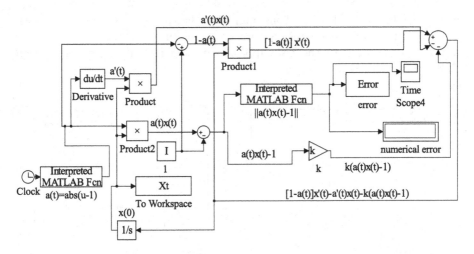

Fig. 1. Simulink of the ZNN model for zeroing $e(t) = 1 - a(t)x(t)$.

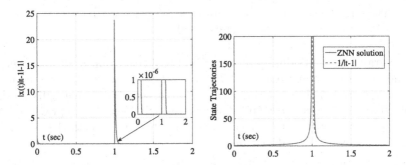

Fig. 2. The error function behaviour with $k = 100$ (left) and the trajectory of $x(t)$ (right) with $k = 100$ in Example 1. (Color figure online)

implementation. On the other hand, since the objective function $x(t) = |t-1|^{-1}$ is nonconvex, the S-ZNN model (7) is not applicable. So, it is necessary to apply shifting $\tilde{e}(t) = e(t + (t == 1)\lambda)$ to avoid the DBZ problem in the time $t = 1$. For $k = 100$ the error norm $|\tilde{e}(t)|$ and the trajectory of $x(t)$ are presented in Fig. 2.

Graph in Fig. 2 (left) indicates a sudden increase of $|\tilde{e}(t)|$ near the point $t = 1$. Graphs in Fig. 2 (right) indicate a small difference near the point $t = 1$ between the solid blue graph which represents the trajectory generated by the ZNN model (1) for $\tilde{e}$ and the dashed red graph which represents the theoretical solution.

Example 2. Choosing $a(t) = |t-1| - t$ and $e(t) = a(t)x(t) - 1$, our goal is to estimate the value of $x(t) = (|t-1|-t)^{-1}$. The term $|t-1|$ is non-differentiable in $t = 1$. Moreover, $a(0.5) = 0$, so $x(t)$ includes singularity. We apply the Simulink implementation presented in Fig. 1 of the linear ZNN flow under the zero initial condition and $k = 100$ to study the influence of the singularity $t = 0.5$ and non-differentiability in $t = 1$ on the behaviour of the linear ZNN formula $\dot{e}(t) - k\,e(t)$. Figure 3 (left) represents the residual error $|e(t)| = |(|t - 1| - t)\,x(t) - 1|$. Graphs in Fig. 3 (right) illustrate the ZNN approximation of $x(t)$ and the exact solution $x^*(t) = 1/(|t - 1| - t)$.

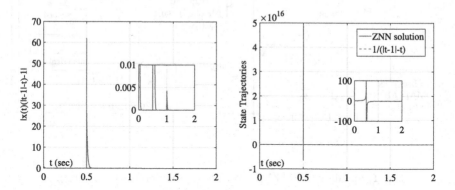

Fig. 3. The error function behaviour with $k = 100$ (left) and the trajectory of $x(t)$ with $k = 100$ in Example 2. (Color figure online)

The graph of the residual error in figure Fig. 3 (left) vanishes toward zero at all time points different from $t = 0.5$ and $t = 1$, while a significant fluctuation at $t = 0.5$ and a small fluctuation at $t = 1$ are observable. The graphs in Fig. 3 (right) show the perfect correspondence of the solid blue line representing the ZNN approximation and the dashed red line which represents the theoretical solution in all points deviating from $t = 0.5$.

Example 3. Choosing $a(t) = |t - 1|^{-1}$ in the scalar ZF $e(t) = 1 - a(t)x(t)$, $a(t), x(t) \in \mathbb{R}$, the linear ZNN dynamics can be used to estimate the time-variant function $|t - 1|$ which is non-differentiable at $t = 1$. Figure 4 (left) represents the behaviour of $|x(t)|t - 1|^{-1} - 1|$ for $k = 100$. The graph in Fig. 4 (left) approaches zero, except at time $t = 1$ where the error increases rapidly. State trajectories initiated by the linear ZNN formula $\dot{e}(t) = -k\,e(t)$ and the exact trajectory $x^*(t) = |t - 1|$ for $k = 100$ are shown in Fig. 4 (right). The solid blue line shows the ZNN approximation of $|t-1|$ and the red dashed line is the trajectory of $x^*(t)$. The graphs in Fig. 4 (right) confirm that the blue solid trajectory generated by the linear ZNN evolution (1) coincides with the red dashed line of the theoretical solution $|t - 1|$. The only exception is the immediate vicinity of the point $t = 1$ where a significant difference between the ZNN and exact trajectories occurs. These observations show that the proposed linear ZNN model (1) can be used to calculate $|t-1|$ in all points except $t = 1$. The value $t = 1$ is the appropriate point to replace $\dot{e}(1)$ by an appropriate element from the subderivative $\partial(e(t))(1)$. Such replacement leads to the S-ZNN design (7).

Example 4. Our intention is to apply linear ZNN dynamics for the computation of $A^{-1}(t)$ in the special case where some elements of the matrix $A(t)$

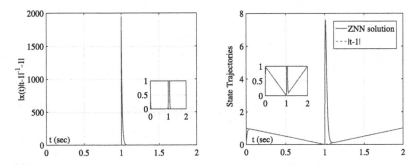

Fig. 4. The error function behaviour with $k = 100$ (left) and The trajectory of $x(t)$ with $k = 100$ (right) in Example 3. (Color figure online)

are expressions causing singularity at certain time instants. For this reason, the following matrix is considered

$$A(t) = \begin{bmatrix} \frac{1}{|t-1|} & 1 \\ 1 & 0 \end{bmatrix},$$

with singularity in $t = 1$ and theoretical inverse

$$A^{-1}(t) = \begin{bmatrix} 0 & 1 \\ 1 & -\frac{1}{|t-1|} \end{bmatrix}.$$

The parameter k is chosen as $k = 1000$ and the initial condition is $X(0) = 0$.

At the point $t = 1$, the matrix $A(t)$ contains an infinite expression $1/0$, causing the error status at $t = 1$ in the Simulink presented in Fig. 1. To overcome the problem of singularity in the function $1/|t - 1|$ for $t = 1$ we consider larger values *Relative tolerance* $=10^{-3}$, *Absolute tolerance* $= 10^{-3}$ in the configuration parameters in the Simulink model for matrix inversion and the conditional displacement

$$t = t + (t == 1)10^{-2} = \begin{cases} t + 10^{-2}, & \text{if } t = 1 \\ t, & \text{else.} \end{cases} \tag{8}$$

The shifting (8) in $t = 1$ enables overcoming the problem with the singularity in this time instant and apply the linear ZNN formula for the computation of matrix inverse problem and in this case where matrix $A(t)$ includes non-differential expression.

These specific modifications in the ZNN design generate the residual errors as in Fig. 5.

From Fig. 5 (left) it can be seen that the linear ZNN formula can be used to compute the inverse of $A(t)$, since $||A(t)X(t) - I||_F$ vanishes toward 0 as the parameter k increases and in the special case that one element of the matrix is non-differentiable at $t = 1$. Figure 5 (right) shows that elementwise state trajectories of the theoretical solution $A^{-1}(t)$ (red dashed lines) are in correspond

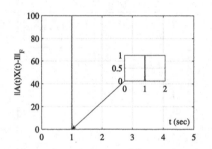 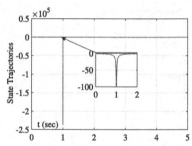

Fig. 5. The error function behaviour with $k = 1000$ (left) and trajectories of entries in $X(t)$ with 1000 (right) in Example 4. (Color figure online)

excellently to the blue lines generated by the linear ZNN formula and this substantiates the fact that we can use the ZNN design for the case where at least one element of a time-varying matrix is non-differentiable at some time points.

Example 5. It is useful to study the behaviour of the linear ZNN for solving the TVNE (2) of one variable. The ZNN dynamics (1) is based on the error function $e(t) = \wp(x(t), t)$. From (3) we derive the dynamics

$$\dot{x}(t) = \left(1 - \frac{\partial \wp}{\partial x}\right) \dot{x}(t) - k\wp(x(t), t) - \frac{\partial \wp}{\partial t}. \tag{9}$$

The TVNE $\wp(x(t), t) \equiv (x(t) - |t|)(x(t) - 1/|t - 2|) = 0$ is considered with the theoretical solutions $x^*(t) = |t|$ and $x^*(t) = 1/|t-2|$. It is noticed that the second solution contains a singularity at the time instant $t = 2$. The error function is $e(t) = \wp(x(t), t)$ while Fig. 6 represents the Simulink model of (9).

The ZNN design with configuration parameters *Relative tolerance=Absolute tolerance*=10^{-6} leads to the error at the point $t = 2$. We used large values *Relative tolerance=Absolute tolerance*=10^{-4} in the configuration parameters. The generated residual errors of $||\wp(x(t), t)||_F$ are shown in Fig. 7 for the gain parameters $k = 1000$ and initial conditions $x(0) = -10$, $x(0) = 10$.

Graphs included in Fig. 7 strengthen the fact that the ZNN formula with greater tolerances can be used for the solution of TVNE problems even in the case when one solution of the TVNE includes singularity. Furthermore, it is observable that for the choice of $x(0) = -10$ and $x(0) = 10$ the ZNN formula converges to the solution $x^*(t) = 1/|t - 2|$.

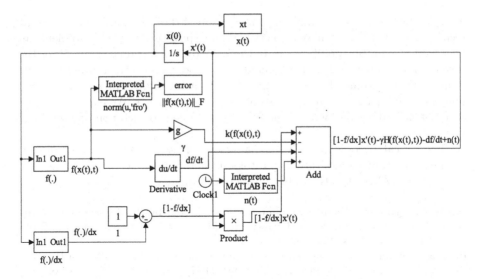

Fig. 6. Simulink of the ZNN formula (9).

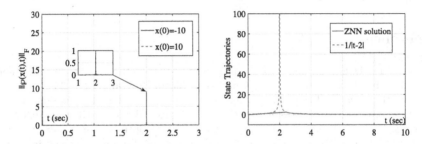

Fig. 7. The behaviour of $||\varphi(x(t),t)||_F$ (left) and the trajectory of $x(t)$ (right) with $k = 1000$ under the initial condition $x(0) = -10$ in Example 5.

5   Concluding Remarks

The ZNN dynamical flow requires the use of the element-wise time-derivative of the error function $\dot{E}(t)$ at each time t. The difficulty arises in the case where a time derivative does not exist, for each element of $E(t)$ and any time t_0 from a time interval $[0, T]$. In this study, we propose an extension of the ZNN formula for the case where the time-derivative of the Zhang function does not exist at some points. The non-differentiability occurs frequently and in various forms.

The aim of our study is to solve the problems that arise in the ZNN dynamic evolutions in such situations. We propose possible solutions to the problem when the indefinite error function $E(t)$ (which can be a matrix, a vector, or a scalar) contains non-diferentiable entries. This means that the time derivative $\dot{e}_{ij}(t)$ of some elements $e_{ij}(t) \in E(t)$ does not exist at some points within the chosen time interval $[0, T]$. The first approach is based on increasing relative and resolute error tolerances in the Simulink implementation of the ZNN model and the shift technique to avoid division by zero (DBZ). More precisely, the DBZ problem can be solved by suitable modifications of the ZNN dynamics proposed in [5] for the case when the function $E(t)$ is unbounded and goes to infinity. Another possible solution is the extended model S-ZNN based on the subderivation (6) at each point $t_0 \in [0, T]$ where $\dot{e}_{ij}(t_0)$ does not exist, for some i, j. We expect that the scope of ZNN dynamics will be successfully extended to such non-differentiable cases.

Theoretical analysis of S-ZNN dynamics requires a detailed analysis of convergence and convergence rate. Noise-tolerant and finite-time or time-predefined convergent S-ZNN models can be considered in further research.

Acknowledgments. Predrag Stanimirović is supported by the Science Fund of the Republic of Serbia, (No. 7750185, Quantitative Automata Models: Fundamental Problems and Applications - QUAM).

Predrag Stanimirović is supported by the Ministry of Science, Technological Development and Innovation, Registration number: 451-03-47/2023-01/ 200124. This work was supported by the Ministry of Science and Higher Education of the Russian Federation (Grant No. 075-15-2022-1121).

References

1. Boob, D., Deng, Q., Lan, G.: Stochastic first-order methods for convex and non-convex functional constrained optimization. Math. Program. **127** (2022)
2. Cao, H., Song, Y., Khan, K.: Convergence of subtangent-based relaxations of nonlinear programs. Processes **7**(4), 221 (2019). https://doi.org/10.3390/pr7040221
3. Fang, C., Li, C.J., Lin, Z., Zhang, T.: SPIDER: Near-optimal non-convex optimization via stochastic path-integrated differential estimator. In: Advances in Neural Information Processing Systems, pp. 687–697 (2018)
4. Gasnikov, A.V., Nesterov, Y.E.: Universal method for stochastic composite optimization (2016). https://arxiv.org/ftp/arxiv/papers/1604/1604.05275.pdf. Accessed 28 Jan 2021
5. Gerontitis, D., Behera, R., Sahoo, J.K., Stanimirović, P.S.: Improved finite-time zeroing neural network for time-varying division. Stud. Appl. Math. **146**(2), 526–549 (2021)
6. Ghadimi, S., Lan, G.: Accelerated gradient methods for nonconvex nonlinear and stochastic programming. Math. Program. **156**(1–2), 59–99 (2016)
7. Golshtein, E.G., Nemirovskii, A.S., Nesterov, Y.E.: Level method, its generalizations and applications. Econ. Math. Methods **31**(3), 164–180 (1995)
8. Jin, J.: A robust zeroing neural network for solving dynamic nonlinear equations and its application to kinematic control of mobile manipulator. Complex Intell. Syst. **7**, 87–99 (2021). https://doi.org/10.1007/s40747-020-00178-9
9. Jin, J., Zhao, L., Yu, F., Xi, Z.: Improved zeroing neural networks for finite time solving nonlinear equations. Neural Comput. Appl. **32**, 4151–4160 (2019)
10. Krutikov, V.N., Samoilenko, N.S., Meshechkin, V.V.: On the properties of the method of minimization for convex functions with relaxation on the distance to extremum. Autom. Remote. Control. **80**(1), 102–111 (2019)

11. Krutikov, V.N., Stanimirović, P.S., Indenko, O.N., Tovbis, E.M., Kazakovtsev, L.A.: Optimization of subgradient method based on Rank-two correction of metric matrices. J. Appl. Ind. Math. **16**(3), 427–439 (2022)
12. Lan, G.: First-order and Stochastic Optimization Methods for Machine Learning. Springer, Berlin (2020). https://doi.org/10.1007/978-3-030-39568-1
13. Nemirovskii, A.S., Yudin, D.B.: Complexity of problems and efficiency of optimization methods. Nauka, Moscow (1979)
14. Nesterov, Y.E.: Universal gradient methods for convex optimization problems. Math. Program. Ser. A **152**, 381–404 (2015)
15. Ouyang, H., Gray, A.: Stochastic smoothing for nonsmooth minimizations: accelerating SGD by exploiting structure. In: Proceedings of the 29th International Conference on Machine Learning (ICML), Edinburgh, Scotland, vol. 1, pp. 33–40 (2012)
16. Polyak, B.T.: A common method for extreme tasks solution. Rep. USSR Acad. Sci. **174**, 33–36 (1967)
17. Polyak, B.T.: Introduction to Optimization. Nauka, Moscow (1983)
18. Shor, N.Z.: Minimization Methods for Non-differentiable Functions and their Applications. Kiev (1979)
19. Polyak, B.T.: Optimisation of non-smooth composed functions. J. Comput. Math. Phys. **9**(3), 507–521 (1969)
20. Shor, N.Z.: Application of the gradient descent method for solving network transportation problems. In: Materials of the Seminar of Theoretical and Applied Issues of Cybernetics and Operational Research, no. 1, pp. 9–17 (1962)
21. Wang, J.: A recurrent neural network for real-time matrix inversion. Appl. Math. Comput. **55**, 89–100 (1993)
22. Zhang, Y., Guo, D.: Zhang Functions and Various Models. Springer, Heidelberg (2015). https://doi.org/10.1007/978-3-662-47334-4
23. Zhang, Y., Gong, H., Li, J., Huang, H., Yin, Z.: Symbolic solutions to division by zero problem via gradient neurodynamics. In: Liu, D., Xie, S., Li, Y., Zhao, D., El-Alfy, E.S. (eds.) ICONIP 2017. LNCS, vol. 10636, pp. 745–750. Springer, Cham (2017). https://doi.org/10.1007/978-3-319-70090-8_75
24. Zhang, Y., Yi, C.: Zhang Neural Networks and Neural-Dynamic Method. Nova Science Publishers Inc., New York (2011)
25. Zhang, Y., Yi, C., Guo, D.: Comparison on Zhang neural dynamics and gradient-based neural dynamics for online solution of nonlinear time-varying equation. Neural Comput. Appl. **20**(1), 1–7 (2011)
26. Zhang, Y., Zhang, Y., Chen, D., Xiao, Z., Yan, X.: Division by zero, pseudo-division by zero, Zhang dynamics method and Zhang-gradient method about control singularity conquering. Int. J. Syst. Sci. **48**, 1–12 (2017)

Mathematical Programming

An Algorithm for Decentralized Multi-agent Feasibility Problems

Olga Pinyagina[(✉)][iD]

Department of Data Mining and Programming Technologies,
Institute of Computational Mathematics and Information Technologies,
Kazan Federal University, Kazan, Russia
Olga.Piniaguina@kpfu.ru

Abstract. We consider the feasibility problem in a multi-agent decentralized form, where each agent has the personal information of the feasible subset, which is unknown to other agents. The common feasible set is composed of the agents' feasible subsets. For solving this problem, we reformulate it in the form of a variational inequality and propose an algorithm based on the projection method. Preliminary test calculations confirm the efficiency of the proposed approach.

Keywords: Feasibility problem · Decentralized system · Projection method · Variational inequality

1 Introduction

Now, decentralized systems are actual in different domains of modern human activities. In this connection, certain aspects become important. On the one hand, the volume of data can be so great that it is not efficient, or even it is not possible to realize the calculation at the unique center, at the main server. On the other hand, the aspect of security requires keeping the information in the private domain and forbids to share it with other users. Therefore, decentralized problems arise and have to be investigated for effective use.

Recently in [1], a decentralized penalty method was proposed for solving general convex constrained multi-agent optimization problems, where each agent has the personal information of the feasible set and goal function, which is unknown to other agents. The agents share their current states only. An important aspect of this method is the choice of communication structure between users and the construction of the corresponding penalty function. An interesting feature of this method is its convergence even in the case when the feasible set of the initial problem is empty. In [1], the application of this approach to the pure feasibility problem is also considered; this problem is reformulated as an optimization problem, which can be solved by the gradient projection method. In the present paper, we extend the idea of this approach concerning the feasibility problem. We reformulate the feasibility problem in the form of a variational inequality, which can be solved by the projection method. The numerical experiments show the efficiency of the proposed approach.

M. Khachay et al. (Eds.): MOTOR 2023, CCIS 1881, pp. 19–28, 2023.
https://doi.org/10.1007/978-3-031-43257-6_2

The paper is organized as follows. In Sect. 2, we recall the main scheme of the penalty method for decentralized problems from [1] applied to the feasibility problem. In Sect. 3, we propose a new reformulation of the feasibility problem in the form of a variational inequality and describe the projection method for solving it. Section 4 describes preliminary numeric experiments. The conclusion section briefly resumes the paper results.

2   The Penalty Approach for Multi-agent Decentralized System

There are different approaches to define and solve multi-agent optimization problems. The first decomposition methods were not really decentralized, the purpose of these methods was to reduce information flows, but they kept the coordination center (see [2]). The two main classes of modern decentralized methods are incremental methods (see [3–5] and references therein) and primal-dual decomposition methods (see [6,7] and references therein). An exhaustive overview of modern decentralized methods can be found in [8] and references therein.

Recently, in [1] a decentralized penalty method was proposed for general convex constrained multi-agent optimization problems. Let us recall the main idea of this approach applied to the feasibility problem.

The general feasibility problem has the form:

$$\text{Find } z \in D,$$

where D is a convex closed set defined in R^n.

In the decentralized problem definition, m agents have convex closed subsets X_i, for $i = 1, \ldots, m$, which are also defined in the space R^n and

$$D = \bigcap_{i=1}^{m} X_i.$$

Each agent has the personal information of the feasible subset, which is unknown to other agents. The agents share their states only and keep their feasible sets in secret. To consider the states of all users, the initial problem is reformulated in the Euclidean space with the dimension $N = nm$ as follows:

$$\text{Find } x \in X, \tag{1}$$

where now the feasible set has the form

$$X = \prod_{i=1}^{m} X_i. \tag{2}$$

Here $x^T = (x_1^T, \ldots, x_m^T)$, $x_i^T = (x_{i1}, \ldots, x_{in})$, for $i = 1, \ldots, m$.

To obtain the common solution, the agents have to exchange the current information. The structure of a communication network can have different forms;

the necessary requirement is that the network must be connected and each agent sends and receives information to and from some other agents.

In the present paper, we will define the network communication structure as an oriented graph $W = (V, A)$, where V is the set of m nodes-users (which are denoted by numbers $i = 1, \ldots, m$ for brevity) and A is the set of directed links, where each link (i, j) means that user i sends its state x_i to user j.

The most complex case is the full graph, where each pair of agents communicates (Fig. 1):

$$W' = (V, A'), \quad A' = \{(i, j) : i \neq j, \ i, j = 1, \ldots, m\}.$$

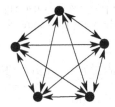

Fig. 1. The maximal communication network structure for network of 5 agents.

In [1], the network structure

$$W'' = (V, A'), \quad A' = \{(i, j), (j, i) : j = i \bmod m + 1, \ i = 1, \ldots, m\}.$$

is used (Fig. 2). It is a two-directed cycle.

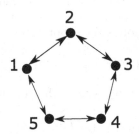

Fig. 2. A communication network structures for network of 5 users in the form of a two-directed cycle.

For the latter network, the penalty function is defined as follows in [1]

$$p(x) = (2\tau)^{-1}\|Px\|^2, = (2\tau)^{-1}\sum_{i=1}^{m}\|P_i x\|^2, \tag{3}$$

where

$$P = \begin{pmatrix} I & -I & \Theta & \ldots & \Theta & \Theta \\ \Theta & I & -I & \ldots & \Theta & \Theta \\ \ldots & \ldots & \ldots & \ldots & \ldots & \ldots \\ -I & \Theta & \Theta & \ldots & \Theta & I \end{pmatrix} = \begin{pmatrix} P_1 \\ P_2 \\ \ldots \\ P_m \end{pmatrix},$$

here I is the $n \times n$ unit matrix, Θ is the $n \times n$ zero matrix, P_i is the corresponding $n \times nm$ submatrix of P, $i = 1, \ldots, m$, and $\tau > 0$ is a fixed scaling parameter.

Then the initial feasibility problem (1)–(2) is reformulated as an optimization problem

$$\min_{x \in X} \longrightarrow p(x).$$

For this problem, in [1] one proposes to use the gradient projection method, which consists in solving the following problem with fixed $\alpha > 0$ and some point $x^k \in X$:

$$\min_{z \in X} \longrightarrow \{\langle p'(x^k), z \rangle + (2\alpha)^{-1}\|z - x^k\|^2\}.$$

Then the iterative process has the form

$$x^{k+1} = \pi_X[x^k - \alpha p'(x^k)], \quad k = 0, 1, 2, \ldots \tag{4}$$

where π_X denotes the operation of projection onto the set X.

The gradient for the penalty function (3) $p'(x)$ is defined as follows:

$$p'_i(x) = \begin{cases} \tau^{-1}(2x_1 - x_2 - x_m) & \text{if } i = 1; \\ \tau^{-1}(2x_i - x_{i+1} - x_{i-1}) & \text{if } i = 2, \ldots, m-1; \\ \tau^{-1}(2x_m - x_1 - x_{m-1}) & \text{if } i = m. \end{cases}$$

Here $p'(x) = (p'_1(x), \ldots, p'_m(x))^T$.

Therefore, each agent can calculate its own gradient based on its personal state and information about states of two neighbors, with which it shares data. Then each agent separately solves the subproblem

$$\min_{z_i \in X_i} \longrightarrow \{\langle p'_i(x^k), z_i \rangle + (2\alpha)^{-1}\|z_i - x^k_i\|^2\}.$$

In other words, the iterative process for each agent has the form

$$x^{k+1}_i = \pi_{X_i}[x^k_i - \alpha p'_i(x^k)], \quad k = 0, 1, 2, \ldots \quad i = 1, \ldots, m.$$

An interesting feature of this method is its convergence even the feasible set of the initial problem D is empty.

So, this approach involves some penalty function, whose gradient describes the communication network and is used in the gradient projection method. But there exist such forms of communication structure, which cannot be presented as a gradient of some function. We consider this case in the following section and propose a general formulation of the feasibility problem in the form of a variational inequality.

3 An Algorithm for Decentralized Multi-agent Feasible Problem

Note that the simplest graph of communication is a one-directed cycle (Fig. 2), for example:

$$W = (V, A), \quad A = \{(i, j) : j = i \bmod m + 1, \ i = 1, \ldots, m\}. \tag{5}$$

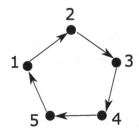

Fig. 3. The simplest communication network structure for 5 users.

Because in our multi-user system, any user solves its problem privately and corresponds only its current state to the outer world, the penalty was usually imposed on the distance between its state and the states of its neighbors, which receive this state. Case (5) corresponds to the minimal flows of information, where each user receives information from one user and sends information to another one. But this situation cannot be described by some penalty function and this problem cannot be reduced to some optimization problem. Therefore, we try to reformulate problem (1)–(2) in the form of a variational inequality.

Let a continuous mapping $G : R^n \longrightarrow R^n$ be given.

Then we can formulate the following variational inequality: Find $x^* \in X$ such that

$$\langle G(x^*), x - x^* \rangle \geq 0, \quad \forall x \in X. \tag{6}$$

For network (5), this mapping G has the form

$$G(x) = \begin{pmatrix} x_1 - x_2 \\ x_2 - x_3 \\ \cdots \\ x_m - x_1 \end{pmatrix}. \tag{7}$$

Evidently, if $D \neq \emptyset$, then variational inequality (6)–(7) has a solution x^*, where $x_i^* = x_j^*$, $i, j = 1, \ldots, m$. Let us prove the inverse assertion.

Proposition 1. *If $D \neq \emptyset$, then for any solution x^* of problem (6)–(7), we have $x_i^* = x_j^*$, $i, j = 1, \ldots, m$.*

Proof. Assume the opposite. Let there exist $i \in \{1, \ldots, m\}$ such that

$$x_i^* \neq x_{i \bmod m+1}^*.$$

Since $D \neq \emptyset$, we can take any $z \in D$ and obtain

$$\langle x_1^* - x_2^*, z - x_1^* \rangle + \langle x_2^* - x_3^*, z - x_2^* \rangle + \cdots + \langle x_m^* - x_1^*, z - x_m^* \rangle \geq 0. \quad (8)$$

On the one hand, after simple transformations we have

$$\|x_1^*\|^2 - \langle x_1^*, x_2^* \rangle + \|x_2^*\|^2 - \langle x_2^*, x_3^* \rangle + \cdots + \|x_m^*\|^2 - \langle x_m^*, x_1^* \rangle \leq 0.$$

On the other hand, we add the expression $-x_j^* + x_j^*$, where $j = (i+1) \bmod m$, to the right-hand side of each ith scalar product in (8). After simple transformations, we obtain

$$\|x_2^*\|^2 - \langle x_1^*, x_2^* \rangle + \|x_3^*\|^2 - \langle x_2^*, x_3^* \rangle + \ldots$$
$$+ \|x_m^*\|^2 - \langle x_{m-1}^*, x_m^* \rangle + \|x_1^*\|^2 - \langle x_m^*, x_1^* \rangle$$
$$\geq \|x_1^* - x_2^*\|^2 + \|x_2^* - x_3^*\|^2 + \cdots + \|x_m^* - x_1^*\|^2 > 0.$$

We get a contradiction, then we obtain that any solution of problem (6)–(7) gives a solution to problem (1)–(2), as desired.

The mapping G in (7) can be presented as

$$G(x) = Px,$$

where, as above,

$$P = \begin{pmatrix} I & -I & \Theta & \ldots & \Theta & \Theta \\ \Theta & I & -I & \ldots & \Theta & \Theta \\ \ldots & \ldots & \ldots & \ldots & \ldots \\ -I & \Theta & \Theta & \ldots & \Theta & I \end{pmatrix},$$

I is the $n \times n$ unit matrix, Θ is the $n \times n$ zero matrix.

For solving variational inequality (6)–(7), the projection method can be used. The iterative process of the projection method has the form for some point $x^k \in X$ and some fixed $\alpha > 0$

$$x^{k+1} = \pi_X [x^k - \alpha G(x^k)], \quad k = 0, 1, 2, \ldots \quad (9)$$

This problem is split into m separate subproblems for m users. At stage k, having the current point x^k, each user i constructs its direction by the rule:

$$G_i(x^k) = \begin{cases} x_i^k - x_{i+1}^k & \text{if } i = 1, \ldots, m-1; \\ x_m^k - x_1^k & \text{if } i = m \end{cases}$$

and solve the problem

$$x_i^{k+1} = \pi_{X_i} [x_i^k - \alpha G_i(x_i^k)].$$

Here

$$G(x) = \begin{pmatrix} G_1(x) \\ G_2(x) \\ \cdots \\ G_m(x) \end{pmatrix}.$$

Now let us prove the convergence of the proposed algorithm. Firstly, we consider the mapping $T : R^N \longrightarrow R^N$

$$T(x) = \begin{pmatrix} x_2 \\ x_3 \\ \cdots \\ x_m \\ x_1 \end{pmatrix}. \tag{10}$$

The mapping G and T are connected as follows

$$G(x) = x - T(x), \tag{11}$$

for all x. In the case $D \neq \emptyset$, variational inequality (6)–(7) is equivalent to the fixed point problem: Find $x^* \in X$ such that

$$x^* = T(x^*).$$

Recall certain definitions concerning the mapping properties.

Definition 1. *A mapping $B : R^N \longrightarrow R^N$ is said to be non-stretching if for all $x, y \in R^N$ it holds that*

$$\|B(x) - B(y)\| \leq \|x - y\|.$$

Definition 2. *A mapping $B : R^N \longrightarrow R^N$ is said to be inversely strongly monotone with a constant γ on the set $X \in R^N$ (or shortly ISM-mapping), if*

$$\langle B(x) - B(y), x - y \rangle \geq \gamma \|B(x) - B(y)\|^2, \quad \forall x, y \in X.$$

Let us show that mapping (10) is non-stretching. In fact,

$$\|T(x) - T(y)\|^2 = \|x_2 - y_2\|^2 + \|x_3 - y_3\|^2 + \cdots + \|x_m - y_m\|^2 + \|x_1 - y_1\|^2 = \|x - y\|^2.$$

The following proposition establishes the dependence between properties of mappings G and T, which are connected by correlation (11).

Proposition 2 *([9], Proposition 1.1). Let $T : R^N \longrightarrow R^N$ be a non-stretching mapping. Then the mapping G from (11) is an ISM mapping with constant $\gamma = 0.5$.*

Therefore, we are ready to prove the convergence of the projection method (9).

Theorem 1. *Let a sequence $\{x^k\}$ be generated by algorithm (9) and there exist a solution to problem (6)–(7). Then the sequence $\{x^k\}$ converges to a solution x^* to problem (6)–(7).*

Proof. Firstly, note that the mapping G in problem (6)–(7) is an ISM mapping with constant $\gamma = 0.5$ due to Proposition 2. Then due to Theorem 11.2 from [9], the sequence $\{x^k\}$ converges to a solution to problem (6)–(7), as desired.

4   Computational Experiments

In this section, we describe preliminary numeric experiments for the considered algorithm. In all the following series of computational tests, we compared algorithms (4) (based on the communication structure of Fig. 2) and (9) (based on the communication structure of Fig. 3). In all experiments, the parameters of methods $\tau = 1$, $\alpha = 0.4$. We used the stop criterion

$$\|x^{k+1} - x^k\| \le \varepsilon,$$

where $\varepsilon = 0.001$.

We considered the feasibility problem, where the set D has the form

$$Ax \le b, \quad x \in R^n,$$

Here A is an $m \times n$ matrix, b is an m-dimensional vector. Or, in other words,

$$X_i = \{x \in R^n : \langle a_i, x \rangle \le b_i\}, \quad i = 1, \ldots, m,$$

where a_i is the ith row of the matrix A, b_i is the ith component of the vector b.

The coefficients a_{ij} $i = 1, \ldots, m$, $j = 1, \ldots, n$ of the matrix A were uniformly distributed random values from the segment $[-20, 20]$. The coefficients b_j of the vector b were uniformly distributed random values from the segment $[0, 20]$. We considered the problems for $m = 10, 15, 20, 25, 30, 35, 40$ users and the space dimension $n = 15, 20, 25, 30, 35, 40, 45, 50$.

The computational results are presented in Table 1, which has the following structure. The first two columns contain the dimensions of problems, m is the number of agents, n is the dimension of agents space. Each row presents the aggregate results for 100 problem instances: mean values (mean val.) and standard deviations (st.dev.) of iterations numbers and calculation time.

The program was written in Visual C# with double precision, tested on an Intel i3-4170 CPU at 3.7 GHz, 4 Gb, running under Windows 7.

In all cases, the best results were obtained for the simplest structure of the communication network and algorithm (9). Note also that the first network structure (Fig. 2) requires twice as much information flow as compared to the second structure (Fig. 3).

Table 1. Computational results.

m	n	Algorithm (4)				Algorithm (9)			
		iterations		time (ms)		iterations		time (ms)	
		mean val.	st.dev.	mean val.	st.dev.	mean val.	st.dev.	mean val.	st.dev.
10	15	281	155	3,160	1,765	192	69	2,010	0,975
10	20	252	90	3,750	1,558	180	20	2,450	0,623
10	25	221	69	3,690	1,302	179	18	2,710	0,516
15	20	633	270	12,761	5,613	383	44	7,150	0,973
15	25	538	160	13,470	5,642	379	16	8,810	3,161
15	30	521	178	16,190	8,045	385	17	11,120	3,401
20	25	1 146	443	39,740	16,193	656	48	21,250	4,524
20	30	1 027	438	40,740	17,816	666	71	24,740	5,863
20	35	953	260	42,360	12,710	665	9	28,060	4,012
25	30	1 616	537	78,750	27,615	991	15	45,080	5,205
25	35	1 591	531	89,310	29,222	1 006	22	52,150	4,244
25	40	1 465	379	93,410	24,663	1 016	12	59,050	3,576
30	35	2 482	738	168,650	50,988	1 406	20	88,400	8,345
30	40	2 120	648	162,800	49,622	1 427	18	102,570	7,440
30	45	2 015	525	173,290	47,699	1 444	15	114,960	7,141
35	40	3 201	847	289,631	84,979	1 894	30	157,470	10,443
35	45	2 888	739	292,030	78,395	1 918	23	178,920	11,977
35	50	2 797	734	317,621	87,685	1 934	24	201,510	14,340

5 Conclusions

In the present paper, we considered the feasibility problem in multi-agent decentralized form, where each agent has the personal information of the feasible subset, which is unknown to other agents. The common feasible set is composed of the agents feasible subsets. For solving this problem, we reformulated it in the form of a variational inequality and proposed an algorithm based on the projection method. Preliminary test calculations are also given. The proposed approach has shown to be efficient and is promising for practical use.

References

1. Konnov, I.: Decentralized multi-agent optimization based on a penalty method. Optimization **71**(15), 4529–4553 (2022). https://doi.org/10.1080/02331934.2021.1950151
2. Bertsekas, D.P., Tsitsiklis, J.N.: Parallel and Distributed Computation: Numerical Methods. Prentice-Hall, London (1989)

3. Lobel, I., Ozdaglar, A., Feijer, D.: Distributed multi-agent optimization with state-dependent communication. Math. Program. **129**, 255–284 (2011)
4. Duchi, J., Agarwal, A., Wainwright, M.: Dual averaging for distributed optimization: convergence analysis and network scaling. IEEE Trans. Autom. Control **57**, 592–606 (2012)
5. Nedić, A., Olshevsky, A.: Distributed optimization over time-varying directed graphs. IEEE Trans. Autom. Control **60**, 601–615 (2015)
6. Boyd, S., Parikh, N., Chu, E., Peleato, B., Eckstein, J.: Distributed optimization and statistical learning via the alternating direction method of multipliers. Found. Trends Mach. Learn. **3**, 1–122 (2011)
7. Lan, G., Lee, S., Zhou, Y.: Communication-efficient algorithms for decentralized and stochastic optimization. Math. Program. **180**, 237–284 (2011)
8. Gorbunov, E., Rogozin, A., Beznosikov, A., Dvinskikh, D., Gasnikov, A.: Recent theoretical advances in decentralized distributed convex optimization. In: Nikeghbali, A., Pardalos, P.M., Raigorodskii, A.M., Rassias, M.T. (eds.) High-Dimensional Optimization and Probability. Springer Optimization and Its Applications, vol. 191, pp. 253–325. Springer, Cham (2022). https://doi.org/10.1007/978-3-031-00832-0_8
9. Konnov, I.V.: Nonlinear Optimization and Variational Inequalities. Kazan University Press, Kazan (2013). [in Russian]

Online Optimization Problems with Functional Constraints Under Relative Lipschitz Continuity and Relative Strong Convexity Conditions

Oleg Savchuk[1,2] , Fedor Stonyakin[1,2(✉)] , Mohammad Alkousa[1,3] ,
Rida Zabirova[1], Alexander Titov[1] , and Alexander Gasnikov[1,3,4,5]

[1] Moscow Institute of Physics and Technology, 9 Institutsky lane, Dolgoprudny
141701, Russia
fedyor@mail.ru, {mohammad.alkousa,zabirova.rr,gasnikov.av}@phystech.edu
[2] V. I. Vernadsky Crimean Federal University, 4 Academician Vernadsky Avenue,
Simferopol 295007, Republic of Crimea, Russia
[3] HSE University, Moscow, 20 Myasnitskaya street, Moscow 101000, Russia
[4] Institute for Information Transmission Problems RAS, 11 Pokrovsky boulevard,
109028 Moscow, Russia
[5] Caucasus Mathematical Center, Adyghe State University, 208 Pervomaiskaya
street, Maykop, Republic of Adygea 385000, Russia

Abstract. In this work, we consider the problem of strongly convex
online optimization with convex inequality constraints. A scheme with
switching over productive and non-productive steps is proposed for these
problems. The convergence rate of the proposed scheme is proven for the
class of relatively Lipschitz-continuous and strongly convex minimization problems. Moreover, we study the extensions of the Mirror Descent
algorithms that eliminate the need for a priori knowledge of the lower
bound on the (relative) strong convexity parameters of the observed functions. Some numerical experiments were conducted to demonstrate the
effectiveness of one of the proposed algorithms with a comparison with
another adaptive algorithm for convex online optimization problems.

Keywords: Online Optimization · Strongly Convex Programming
Problem · Relatively Lipschitz-Continuous Function · Relatively
Strongly Convex Function · Mirror Descent · Regularization

Introduction

The development of numerical methods for solving non-smooth online optimization problems presents a great interest nowadays due to the appearance of many

The research was supported by Russian Science Foundation (project No. 21-71- 30005),
https://rscf.ru/en/project/21-71-30005/..

applied problems with the corresponding statement [3,5–7,11]. Online optimization plays a key role in solving machine learning, finance, networks, and other
problems. As some examples of such problems, we can mention multi-armed
bandits, job-shop scheduling and ski rental problems, search games, etc. One of
the most popular methods of solving online optimization problems is the Mirror
Descent method [13]. Indeed, the usual Lipschitz condition with respect to the
Euclidean norm may not be very convenient, and the using of other norms leads
to the need to consider the Mirror Descent method

$$x_{k+1} := \arg\min_{x \in Q}\{\gamma\langle \nabla f_k(x_k), x - x_k\rangle + V(x, x_k)\},$$

for some convenient $\gamma > 0$ and $V(x, y)$ is a Bregman divergence (see (2)), instead
of the usual subgradient method

$$x_{k+1} := \arg\min_{x \in Q}\{\gamma\langle \nabla f_k(x_k), x - x_k\rangle + \frac{1}{2}\|x - x_k\|_2^2\}.$$

Let us note, that Mirror Descent can be also applied for solving online optimization problems in a stochastic setting [1,4], which allows using an arbitrary,
not necessarily 1–strongly convex, distance-generating function (see (2)).

Remind, that the online optimization problem represents the problem of min-
imizing the sum (or the arithmetic mean) of T convex functionals $f_t : Q \longrightarrow \mathbb{R}$
$(t = 1, \ldots, T)$ given on some compact convex set $Q \subset \mathbb{R}^n$

$$\min \sum_{t=1}^{T} f_t(x), \quad s.t. \quad g(x) \leq 0, \quad x \in Q, \tag{1}$$

where $g : Q \longrightarrow \mathbb{R}$ is a convex functional constraint. The key feature of the
problem statement consists in the possibility of calculating the (sub)gradient
$\nabla f_t(x)$ of each functional f_t only once.

Online optimization problems with convex constraints-inequalities were considered, for example, in [8]. As an example of an applied problem, it is natural
to note the sparse online binary classification problem [8].

Recently, in [15] there were proposed some modifications of the Mirror
Descent method for solving online optimization problems in the case, if all the
convex functions $f_t(x)$ and convex functional constraint $g(x)$ satisfy Lipschitz
condition, i.e. there exists a constant $M > 0$, such that

$$|g(x) - g(y)| \leq M\|x - y\|,$$

$$|f_t(x) - f_t(y)| \leq M\|x - y\|, \quad \forall t = 1, \ldots, T.$$

In the case of non-negativity of regret

$$Regret_T := \sum_{t=1}^{T} f_t(x_t) - \min_{x \in Q} \sum_{t=1}^{T} f_t(x),$$

where x_t are the obtained points from the work of the proposed methods for the problem (1), these methods are optimal for the considered class of problems accordingly to [6], the number of non-productive steps during their work is $O(T)$. In the case of negative regret, the number of non-productive steps for the proposed methods is $O(T^2)$ [15].

Later, in [16] the smoothness class for the applicability of such approaches has been extended by reducing the requirement of Lipschitz continuity of functions to the recently proposed concept of relative Lipschitz continuity [9,12]. Therefore a few years ago, the optimization field introduced classes of relatively smooth [2], relatively Lipschitz-continuous, and relatively strongly convex optimization problems [9,10]. These concepts have expanded the class of problems to which optimal complexity estimates of gradient-type methods in high-dimensional spaces can be applied.

Let $h : Q \longrightarrow \mathbb{R}$ be a distance-generating function (or prox-function) that is continuously differentiable and convex. For all $x, y \in Q$ we consider the corresponding Bregman divergence

$$V(y, x) = h(y) - h(x) - \langle \nabla h(x), y - x \rangle. \tag{2}$$

Definition 1. *Let us call a convex function* $f : Q \longrightarrow \mathbb{R}$ *to be an M-relatively Lipschitz-continuous for some* $M > 0$, *if the following inequality holds*

$$\langle \nabla f(x), y - x \rangle + M \sqrt{2V(y, x)} \geq 0, \quad \forall x, y \in Q,$$

where $V(y, x)$ *is a corresponding Bregman divergence.*

This concept has been widely used in many applied problems and has also enabled the proposal of subgradient methods for both non-differentiable and non-Lipschitz Support Vector Machine (SVM) and for problems of the intersection of n ellipsoids while maintaining optimal convergence rate estimates for the class of simply Lipschitz-continuous functions. It is worth noting that the proposed methods also allowed the use of an imprecisely defined function (more exactly, a function that admits a representation in a model form), nevertheless, the method was also optimal.

In this paper, we improve the existing estimates of the convergence rate by considering a class of strongly convex functions and generalize the obtained problem statement to the case of problems with functional constraints.

Definition 2. *Let* $\mu > 0$. *A function* f *over a convex set* Q *is called* μ-*strongly convex with respect to a convex function* h *if*

$$f(x) \geq f(y) + \langle \nabla f(y), x - y \rangle + \mu V(x, y) \quad \forall x, y \in Q,$$

where $V(x, y)$ *is a Bregman divergence corresponding to* h.

In this paper we investigate an alternative approach for strongly convex problems, which guarantees exactly the estimate for the regret $O(\log T)$, where T is the number of terms f_i. In [8] the worst estimates were obtained. Moreover, in our work, we obtained more flexible estimates for online optimization problems with constraints that allow us to use the difference in the parameters of the strong convexity of each of the functions f_i.

It is also worth noting that the approach proposed in this article based on schemes with switching over productive and non-productive steps may allow avoiding an additional projection operation on an admissible set (if it is described by a system of inequalities) during iterations.

More precisely, we present a novel theorem that provides a tighter bound on regret, in terms of the number of productive steps taken by the algorithm. Specifically, the theorem proves that if the algorithm completes exactly T productive steps and has a non-negative regret, then the number of non-productive steps satisfies $T_J \leq CT$, where C is a constant. This result significantly improves existing convergence rate estimates for the Mirror Descent method with functional constraints. In addition, we obtain the complexity of the bound in terms of T and some other problem parameters. This corollary allows us to determine the number of productive steps needed to achieve the desired accuracy of regret in practice.

We also consider some modifications of the Mirror Descent method for solving non-smooth online optimization problems [5]. Specifically, the paper introduces two algorithms for solving strongly convex minimization problems with and without regularization. The first algorithm, called General-Norm Online Gradient Descent: Relatively Strongly Convex and Relatively Lipschitz-Continuous Case, is based on a convex function h and updates the solution iteratively using predictions and observations of the objective function f_t. The second algorithm, called Adaptive General-Norm Online Gradient Descent with Regularization, extends the first algorithm by introducing an adaptive regularization term that depends on a function d that is both relatively Lipschitz-continuous and relatively strongly convex.

For each algorithm, we provide the theoretical justification of bounds on the regret. These theorems guarantee upper bounds on the regret for each algorithm and can be used to analyze the performance of the algorithms. Overall, the paper presents a comprehensive framework for solving non-smooth online optimization problems with functional constraints, and the results have practical implications for a broad range of applications.

The paper consists of an introduction and 4 main sections. In Sect. 1 we consider the basic statement of the constrained online optimization problem and propose a modification of the Mirror Descent method for minimizing the arithmetic mean of relatively strongly convex and relatively Lipschitz-continuous functionals, supposing that functional constraint satisfies the same conditions. We also provide a theoretical justification for the convergence rate of the proposed method. Section 2 is devoted to some modifications of the algorithms, proposed in [5] for the corresponding class of problems with regularization. In Sect. 3 we combine the above-mentioned ideas and propose algorithms with switching over productive and non-productive steps both with and without iterative regularization during the work of algorithms. In Sect. 4 we present some numerical experiments which demonstrate the effectiveness of one of the proposed algorithms and a comparison with another adaptive algorithm for the considered optimization problems.

To sum it up, the contributions of the paper can be stated as follows:

- We proposed an optimal method (Algorithm 1) for solving a constrained online optimization problem with relatively strongly convex and relatively Lipschitz-continuous objective functionals and functional constraints. For the case of non-negative regret, the number of non-productive steps is bounded by $O(T)$.
- We proposed two algorithms (Algorithm 2 and Algorithm 3) for solving strongly convex minimization problems with and without regularization based on iteratively updating steps by using some auxiliary functions. Similar to [5], we present extensions of Mirror Descent (Algorithm 4 and Algorithm 5) that exclude the need for a priori knowledge of the lower bound on the (relatively) strong convexity parameters of the observed functions.
- We provided the results of numerical experiments demonstrating the advantages of using the proposed methods.

1   Mirror Descent for Relatively Strongly Convex and Relatively Lipschitz-Continuous Online Optimization Problems with Inequality Constraints

In this section, we present a scheme for solving the problem (1), when the f_t and g are relatively strongly convex and relatively Lipschitz-continuous. The ensure points of the proposed scheme are selected among the points x_t for which $g(x_t) \leq \varepsilon$, therefore, we will call step t productive if $g(x_t) \leq \varepsilon$ and if the reverse inequality $g(x_t) > \varepsilon$ holds then step t will be called non-productive. Let I and J denote the set of indexes of productive and non-productive steps, respectively. Let $T := |I|, T_J := |J|$, and x^* be a solution of (1), i.e. $x^* = \arg\min_{x \in Q} \sum_{t=1}^{T} f_t(x)$, and $g(x^*) \leq 0$. Throughout this article, $V(x, y)$ is the Bregman divergence corresponding to the convex function h (see (2)). Let us consider a subgradient method with switching over productive and non-productive steps. As a result of this method, we get a sequence $\{x_k\}_{k \in I}$ on productive steps, which can be considered as a solution to the problem (1) with accuracy δ (see (3)).

Theorem 1. *Suppose that, for each t, f_t is an M_f-relatively Lipschitz-continuous and μ-strongly convex function with respect to the prox-function h. Let $g(x)$ be M_g-relatively Lipschitz-continuous and μ-strongly convex function with respect to h. Suppose that Algorithm 1 for*

$$\varepsilon = \frac{M^2}{\mu} \frac{1 + \ln T}{T},$$

Algorithm 1. Constrained Online Optimization: Mirror Descent for Relatively Lipschitz-Continuous and Relatively Strongly Convex Problems (MDRL-RS).

Require: $\varepsilon > 0, \mu > 0, M > 0, T, x_1 \in Q$.

1: $i := 1, t := 1$;
2: **repeat**
3:     Observe f_t.
4:     **if** $g(x_t) \leq \varepsilon$ **then**
5:         $\eta_t = \frac{1}{t\mu}$;
6:         $x_{t+1} := \arg\min_{x \in Q}\{\eta_t\langle \nabla f_t(x_t), x\rangle + V(x, x_t)\}$;   "productive step"
7:         $i := i + 1$;
8:         $t := t + 1$;
9:     **else**
10:        $\eta_t = \frac{1}{t\mu}$;
11:        $x_{t+1} := \arg\min_{x \in Q}\{\eta_t\langle \nabla g(x_t), x\rangle + V(x, x_t)\}$;   "non-productive step"
12:        $t := t + 1$;
13:    **end if**
14: **until** $i = T + 1$.
15: Guaranteed accuracy:

$$\delta := \frac{M^2}{\mu T}\left(1 + \ln(T + T_J)\right) - \varepsilon\frac{T_J}{T}. \tag{3}$$

Ensure: $x_k, k \in I$.

---

where $M = \max\{M_f, M_g\}$, works exactly T productive steps and $Regret_T \geq 0$. Then there exists a constant $C \in (2; 3)$ such that the number of non-productive steps satisfies $T_J \leq CT$, moreover, the following inequality holds:

$$Regret_T := \sum_{t=1}^{T} f_t(x_t) - \min_{x \in Q}\sum_{t=1}^{T} f_t(x) \leq \frac{M^2}{\mu}\left(1 + \ln\left((C + 1)T\right)\right) = O(T\varepsilon),$$

where $g(x_t) \leq \varepsilon \quad \forall t = 1, \ldots, T$.

Proof. The proof is given in [14].

Remark 1. Let us show that our algorithm will necessarily make at least one productive steps. Indeed, suppose, that the number of productive steps equals zero, then

$$\varepsilon T_J \leq \sum_{t=1}^{T_J}\left(g(x_t) - g(x^*)\right) \leq \frac{M^2}{\mu}\left(1 + \ln T_J\right).$$

It is obvious, that for a sufficiently large T_J, the above inequality does not hold. Thus, for a sufficiently large number of non-productive steps, there will be at least one productive step.

Let us find out how many non-productive steps need to be taken to achieve the inequality:

$$\varepsilon T_J = \frac{T_J M^2}{\mu}\frac{1 + \ln T}{T} > \frac{M^2}{\mu}(1 + \ln T_J),$$

Thus,

$$\frac{1 + \ln T}{T} > \frac{1 + \ln T_J}{T_J}.$$

Then $T_J \leq CT$, where C is a constant, which proves that the number of non-productive steps is bounded until at least one productive step is made.

2 Online Mirror Descent with Regularization

In this section, we propose some modifications of the algorithms proposed in [5] for relatively strongly convex and relatively Lipschitz-continuous online optimization problems and provide theoretical estimates of the quality of the solution. Let us consider the following strongly convex minimization problem

$$\min_{x \in Q} \sum_{t=1}^{T} f_t(x), \qquad (4)$$

where $Q \subset \mathbb{R}^n$ is some compact convex set, $f_t : Q \longrightarrow \mathbb{R}$ are relatively strongly convex and relatively Lipschitz-continuous functions. Define $\mu_{1:t} := \sum_{s=1}^{t} \mu_s$, where μ_s is the parameter of relative strong convexity of the function f_s. Let $\mu_{1:0} = 0$. Remind that $V(x, y)$ is the Bregman divergence corresponding to the convex function h (see (2)).

Algorithm 2. General-Norm Online Gradient Descent: Relatively Strongly Convex and Relatively Lipschitz-Continuous Case (OGDRS-RL).

1: Input: convex function h.
2: Initialize x_1 arbitrarily.
3: **for** $t = 1, \ldots, T$ **do**
4:     Observe f_t.
5:     Compute $\eta_{t+1} = \frac{1}{\mu_{1:t}}$ and let y_{t+1} be such that $\nabla h(y_{t+1}) = \nabla h(x_t) - \eta_{t+1} \nabla f_t(x_t)$.
6:     Let $x_{t+1} = \arg\min_{x \in Q} V(x, y_{t+1})$ be the projection of y_{t+1} onto Q.
7: **end for**
8: Output: $x_k, k = 1, \ldots, T$.

Theorem 2. *Suppose that, for each t, f_t is an M_t-relatively Lipschitz-continuous and μ_t-strongly convex function with respect to prox-function h. Applying the Algorithm 2, we have*

$$Regret_T \leq \sum_{t=1}^{T} \frac{M_t^2}{\mu_{1:t}}.$$

Proof. The proof is given in [14].

Let's now consider an analogue of Algorithm 2 for relatively strongly convex and relatively Lipschitz-continuous problems with iterative regularization. Define $\lambda_{1:t} := \sum_{s=1}^{t} \lambda_s, \lambda_{1:0} = 0$. The proposed algorithm is listed as Algorithm 3.

Algorithm 3. Adaptive General-Norm Online Gradient Descent with Regularization (AOGD-R).

1: Input: convex function h.
2: Initialize x_1 arbitrarily.
3: **for** $t = 1, \ldots, T$ **do**
4:    Observe f_t.
5:    Compute $\lambda_t = \frac{1}{2}\left(\sqrt{(\mu_{1:t} + \lambda_{1:t-1})^2 + 8M_t^2/(A^2 + 2M_d^2)} - (\mu_{1:t} + \lambda_{1:t-1})\right)$.
6:    Compute $\eta_{t+1} = \frac{1}{\mu_{1:t} + \lambda_{1:t}}$ and let y_{t+1} be such that

$$\nabla h(y_{t+1}) = \nabla h(x_t) - \eta_{t+1}\left(\nabla f_t(x_t) + \lambda_t \nabla d(x_t)\right).$$

7:    Let $x_{t+1} = \arg\min_{x \in Q} V(x, y_{t+1})$ be the projection of y_{t+1} onto Q.
8: **end for**
9: Output: $x_k, k = 1, \ldots, T$.

For Algorithm 3, we have the following result.

Theorem 3. *Suppose that, for each t, f_t is M_t-relatively Lipschitz-continuous and μ_t-relatively strongly convex function with respect to the prox-function h. Let $d : Q \longrightarrow \mathbb{R}$ be M_d-relatively Lipschitz-continuous and 1-strongly convex function with respect to h. Suppose that $d(x) \geq 0$, $\forall x \in Q$ and $A^2 = \sup_{x \in Q} d(x)$. Applying Algorithm 3, the following inequalities hold*

$$Regret_T \leq \lambda_{1:T}A^2 + \sum_{t=1}^{T} \frac{(M_t + \lambda_t M_d)^2}{\mu_{1:t} + \lambda_{1:t}},$$

and

$$Regret_T \leq 2 \inf_{\lambda_1^*, \ldots, \lambda_T^*}\left((A^2 + 2M_d^2)\lambda_{1:T}^* + \sum_{t=1}^{T} \frac{(M_t + \lambda_t^* M_d)^2}{\mu_{1:t} + \lambda_{1:t}^*}\right).$$

Proof. The proof is given in [14].

3   The Case of Online Optimization Problems with Inequality Constraints

In this section, we consider a scheme with switching over productive and non-productive steps both with and without iterative regularization for a relatively strongly convex and relatively Lipschitz-continuous constrained online optimization problem. Similarly to Sect. 2, we define $\mu_{1:t} := \sum_{s=1}^{t} \mu_s$, where μ_s is the parameter of relative strong convexity of the function f_s and let $\mu_{1:0} = 0$. If t is the number of non-productive steps, then $\mu_t = \mu_g$, where μ_g is the parameter of relative strong convexity of the function g. As a result of the proposed algorithms, we get a sequence $\{x_k\}_{k \in I}$ on productive steps, which can be considered as a solution of the problem (1) with accuracy δ (see (5) and (6).

Algorithm 4. Mirror Descent for Constrained Online Optimization Problems with Relatively Lipschitz-Continuous and Relatively Strongly Convex Functions (MDCOORL-RS).

Require: $\varepsilon > 0, M > 0, T, x_1 \in Q$.

1: $i := 1, t := 1$;
2: **repeat**
3:     Observe f_t.
4:     **if** $g(x_t) \leq \varepsilon$ **then**
5:         $\eta_t = \frac{1}{\mu_{1:t}}$;
6:         $x_{t+1} := \arg\min_{x \in Q}\{\eta_t \langle \nabla f_t(x_t), x \rangle + V(x, x_t)\}$;   "productive step"
7:         $i := i + 1$;
8:         $t := t + 1$;
9:     **else**
10:        $\eta_t = \frac{1}{\mu_{1:t}}$;
11:        $x_{t+1} := \arg\min_{x \in Q}\{\eta_t \langle \nabla g(x_t), x \rangle + V(x, x_t)\}$;   "non-productive step"
12:        $t := t + 1$;
13:    **end if**
14: **until** $i = T + 1$.
15: Guaranteed accuracy:

$$\delta := \frac{1}{T}\left(\sum_{t=1}^{T+T_J} \frac{M^2}{\mu_{1:t}} - \varepsilon T_J\right). \tag{5}$$

Ensure: $x_k, k \in I$.

Theorem 4. *Suppose that, for each t, f_t is an M_t-relatively Lipschitz-continuous and μ_t-strongly convex function with respect to the convex function h. Let $g(x)$ be M_g-relatively Lipschitz-continuous and μ_g-strongly convex function with respect to h. If Algorithm 4 works exactly T productive steps and $Regret_T \geq 0$, then the following inequality holds:*

$$Regret_T \leq \sum_{t=1}^{T+T_J} \frac{M^2}{\mu_{1:t}} - \varepsilon T_J,$$

where $M = \max\{M_t, M_g\}$ and $g(x_t) \leq \varepsilon$, $\forall t = 1, \ldots, T$.

Proof. The proof is given in [14].

Corollary 1. *Assume that all conditions of Theorem 4 hold and suppose* $\mu_t \geq \mu > 0$ *for all* $1 \leq t \leq T + T_J$. *If*

$$\varepsilon = \frac{M^2}{\mu}\frac{1 + \ln T}{T},$$

then the bound on the regret of Algorithm 4 is $O(\ln T)$.

Proof.

$$0 \leq Regret_T \leq \sum_{t=1}^{T+T_J}\frac{M^2}{\mu_{1:t}} - \varepsilon T_J \leq \sum_{t=1}^{T+T_J}\frac{M^2}{\mu t} - \varepsilon T_J \leq \frac{M^2}{\mu}\Big(\ln(T+T_J)+1\Big) - \varepsilon T_J,$$

hence $\varepsilon T_J \leq \dfrac{M^2}{\mu}\Big(1 + \ln(T + T_J)\Big)$. Let $\varepsilon = \dfrac{M^2}{\mu}\dfrac{1 + \ln T}{T}$. Then we have

$$\frac{1 + \ln T}{T}T_J \leq 1 + \ln(T + T_J),$$

and

$$\frac{T_J}{T} \leq \frac{1 + \ln(T + T_J)}{1 + \ln T} = \frac{1 + \ln T + \ln(1 + \frac{T_J}{T})}{1 + \ln T} \leq 1 + \ln(1 + \frac{T_J}{T}).$$

Since the linear function grows faster than the logarithmic one, it is obvious, that with a sufficiently large T_J, the above inequality does not hold, and then $\dfrac{T_J}{T}$ is bounded. Thus we proved that there exists such a constant $C > 0$, that $T_J \leq CT$. So, we have

$$Regret_T \leq \frac{M^2}{\mu}\Big(1 + \ln\big((C + 1)T\big)\Big) = O(\ln T) = O(T\varepsilon).$$

Let's consider an analogue of Algorithm 4 for relatively strongly convex and relatively Lipschitz-continuous problems with iterative regularization. Similarly to Sect. 2, we define $\lambda_{1:t} := \sum_{s=1}^{t}\lambda_s, \lambda_{1:0} = 0$.

Theorem 5. *Suppose that, for each* t, f_t *is an* M_t-*relatively Lipschitz-continuous and* μ_t-*relatively strongly convex function with respect to the prox-function* h. *Let* $g(x)$ *be* M_g-*relatively Lipschitz-continuous and* μ_g-*relatively strongly convex function with respect to* h. *Let* $d : Q \longrightarrow \mathbb{R}$ *be* M_d-*relatively Lipschitz-continuous and* 1-*relatively strongly convex function with respect to* h. *Suppose also that* $d(x) \geq 0$, $\forall x \in Q$. *If Algorithm 5 works exactly* T *productive steps and* $Regret_T \geq 0$, *then the following inequalities hold:*

$$Regret_T \leq \lambda_{1:T+T_J}A^2 + \sum_{t=1}^{T+T_J}\frac{(M + \lambda_t M_d)^2}{\mu_{1:t} + \lambda_{1:t}} - \varepsilon T_J,$$

Algorithm 5. Constrained Online Optimization: Mirror Descent for Relatively Strongly Convex and Relatively Lipschitz-Continuous Problems with Regularization (MDCOORL-RS-R).

Require: $\varepsilon > 0, T, x_1 \in Q$.
1: $i := 1, t := 1$;
2: **repeat**
3:     Observe f_t.
4:     **if** $g(x_t) \le \varepsilon$ **then**
5:         $\lambda_t = \frac{1}{2}\left(\sqrt{(\mu_{1:t} + \lambda_{1:t-1})^2 + 8M^2/(A^2 + 2M_d^2)} - (\mu_{1:t} + \lambda_{1:t-1})\right)$;
6:         $\eta_t = \frac{1}{\mu_{1:t} + \lambda_{1:t}}$;
7:         $x_{t+1} := \arg\min_{x \in Q}\{\eta_t\langle \nabla f_t(x_t) + \lambda_t \nabla d(x_t), x\rangle + V(x, x_t)\}$;   "productive step"

8:         $i := i + 1$;
9:         $t := t + 1$;
10:    **else**
11:        $\lambda_t = \frac{1}{2}\left(\sqrt{(\mu_{1:t} + \lambda_{1:t-1})^2 + 8M^2/(A^2 + 2M_d^2)} - (\mu_{1:t} + \lambda_{1:t-1})\right)$;
12:        $\eta_t = \frac{1}{\mu_{1:t} + \lambda_{1:t}}$;
13:        $x_{t+1} := \arg\min_{x \in Q}\{\eta_t\langle \nabla g(x_t) + \lambda_t \nabla d(x_t), x\rangle + V(x, x_t)\}$;   "non-productive step"
14:        $t := t + 1$;
15:    **end if**
16: **until** $i = T + 1$.
17: Guaranteed accuracy:

$$\delta := \frac{1}{T}\left(\lambda_{1:T+T_J}A^2 + \sum_{t=1}^{T+T_J}\frac{(M + \lambda_t M_d)^2}{\mu_{1:t} + \lambda_{1:t}} - \varepsilon T_J\right). \tag{6}$$

Ensure: $x_k, k \in I$.

---

and

$$Regret_T \le 2 \inf_{\lambda_1^*, \ldots, \lambda_{T+T_J}^*}\left((A^2 + 2M_d^2)\lambda_{1:T+T_J}^* + \sum_{t=1}^{T+T_J}\frac{(M + \lambda_t^* M_d)^2}{\mu_{1:t} + \lambda_{1:t}^*}\right) - \varepsilon T_J.$$

where $A^2 = \sup_{x \in Q} d(x)$, $M = \max\{M_t, M_g\}$ *and* $g(x_t) \le \varepsilon$, $\forall t = 1, \ldots, T$.

Proof. The proof is given in [14].

We can formulate the following statement for concrete values of μ_t. Partially, we can achieve intermediate rates for regret between T and $\log T$.

Corollary 2. *Assume that all conditions of Theorem 5 hold and* $\mu_t = t^{-\alpha}$ *for all* $1 \le t \le T + T_J$.

1. *If* $\alpha = 0, \lambda_t = 0 \ \forall 1 \le t \le T + T_J$, *and* $\varepsilon = M^2\frac{1 + \ln T}{T}$, *then the bound on the regret of Algorithm 5 is* $O(\ln T)$.

2. *If $\alpha > 1/2, \lambda_1 = \sqrt{T+T_J}, \lambda_t = 0$ for $1 < t \leq T+T_J$, and*

$$\varepsilon = \frac{A^2 + 2(M_d^2 + M^2)}{\sqrt{T}},$$

then the bound on the regret of Algorithm 5 is $O(\sqrt{T})$.

3. *If $0 < \alpha \leq 1/2, \lambda_1 = (T+T_J)^\alpha, \lambda_t = 0 \quad \forall 1 \leq t \leq T+T_J$ and*

$$\varepsilon = \left(A^2 + 2M_d^2 + \frac{4M^2}{\alpha} \right) T^{\alpha-1},$$

then the bound on the regret of Algorithm 5 is $O(T^\alpha)$.

Proof. The proof is given in [14].

4   Numerical Experiments

In this section, to demonstrate the performance of the proposed Algorithm MDCOORL-RS, we conduct some numerical experiments for the considered problem (1) and make a comparison with an adaptive Algorithm 2, proposed in [15]. All experiments were implemented in Python 3.4, on a computer fitted with Intel(R) Core(TM) i7-8550U CPU @ 1.80GHz, 1992 Mhz, 4 Core(s), 8 Logical Processor(s). RAM of the computer is 8 GB.

Let us consider the following function

$$f(x) = \frac{1}{T} \sum_{i=1}^{T} \left(|\langle a_i, x \rangle - b_i| + \frac{\mu_i}{2} \|x\|_2^2 \right), \tag{7}$$

where $a_i \in \mathbb{R}^n, b_i \in \mathbb{R}, \mu_i > 0$. Functional constraints are defined as follows

$$g(x) = \max_{1 \leq i \leq m} \left\{ \langle \alpha_i, x \rangle - \beta_i + \frac{\widehat{\mu}_i}{2} \|x\|_2^2 \right\}, \tag{8}$$

where $\alpha_i \in \mathbb{R}^n, \beta_i \in \mathbb{R}, \widehat{\mu}_i > 0$.

Function f is the arithmetic mean of the functions $f_i(x) = |\langle a_i, x \rangle - b_i| + \frac{\mu_i}{2} \|x\|_2^2$, $i = 1, \ldots, T$. Each of these functions is M_i-Lipschitz-continuous and μ_i-strongly convex. Also, function g is M_g-Lipschitz-continuous and μ_g-strongly convex. Coefficients $a_i, \alpha_i \in \mathbb{R}^n$ and constants $b_i, \beta_i \in \mathbb{R}$ in (7) and (8) are randomly generated from the uniform distribution over $[0, 1)$. Also, the strong convexity parameters μ_i and $\widehat{\mu}_i$ are randomly chosen in the interval $(0, 1)$.

We choose a standard Euclidean proximal setup as a prox-function, starting point $x_0 = \left(\frac{1}{\sqrt{n}}, \ldots, \frac{1}{\sqrt{n}} \right) \in \mathbb{R}^n$ and Q is the unit ball in $\mathbb{R}^n$.

We run Algorithm MDCOORL-RS and adaptive Algorithm 2 from [15] with $n = 1000$ and $m = 10$ and different values of T with $\varepsilon = 1/\sqrt{T}$. The results of the work of these algorithms are represented in Fig. 1, below. These results demonstrate the number of non-productive steps, the running time is given in seconds, the guaranteed accuracy δ of the approximated solution (sequence $\{x_t\}_{t \in I}$ on productive steps), and the values $\frac{1}{T} \sum_{i=1}^{T} f_i(x_i)$, where x_i is productive, as a function of T. The dotted curve represents the results of the proposed Algorithm MDCOORL-RS, whereas the dashed curve represents the results of the adaptive Algorithm 2 in [15].

From the conducted experiments, we can see that the adaptive Algorithm 2 in [15], works faster than Algorithm MDCOORL-RS, with a smaller amount of non-productive steps. But when increasing the number of functionals f_i in (7), the guaranteed accuracy δ and values of the objective function at productive steps, produced by Algorithm MDCOORL-RS is better.

Note that from Fig. 1, we can see that increasing of T (the number of functionals f_i) leads to an increase of δ (the accuracy of the solution). In other words, increasing the number of functionals f_i in the objective function (7), which in fact is increasing information about the objective function or actually enlarging data about the problem, leads to increasing the accuracy of the solution.

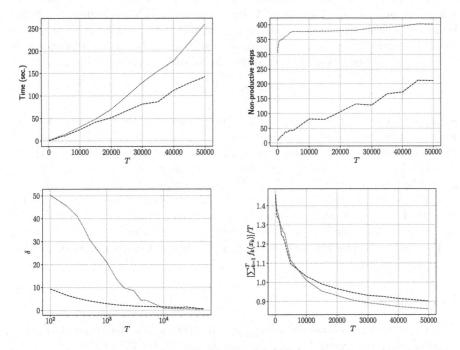

Fig. 1. The results of Algorithm MDCOORL-RS (dots) and adaptive Algorithm 2 in [15] (dashed) for the objective function (7) with constraints (8).

Conclusions

In this paper, we considered relatively strongly convex and relatively Lipschitz-continuous constrained online optimization problems. We proposed some methods with switching over productive and non-productive steps and provided corresponding estimates of the quality of the solution. We also presented analogues of the methods proposed earlier in [5], for solving relatively strongly convex and relatively Lipschitz-continuous online optimization problems with and without regularization. Furthermore, for the problems with functional constraints, we have proposed a scheme with switching over productive and non-productive steps with adaptive regularization. We also proved that if the algorithm runs exactly T productive steps and has a non-negative regret, then the number of non-productive steps satisfies $T_J \leq CT$, where C is a constant. In particular, for the proposed methods, we obtained some bounds on the algorithm's regret in terms of the number of productive steps made by the algorithm under specific assumptions about the parameters of relative strong convexity and some other parameters of the problem.

The key idea of the considered methods is that at each step of the algorithm for each selected f_t, we determine the corresponding parameter of the relative strong convexity μ_t. Thus, it is possible to take into account the parameter of relative strong convexity of each of the functions f_t. This is highly significant because the functions are selected during the method's working process, and it would be a mistake to assume that some strong convexity can be set initially. It is important to note, that if we consider the following functional constraint

$$g(x) = \max_{1 \leq i \leq m} \{g_i(x)\},$$

where each g_i is μ_i-relatively strongly convex function, then in the process of working of the algorithm at this particular non-productive step t, it makes sense to consider the first of the constraints $g_i(x)$ for which the condition $g_i(x_t) \leq \varepsilon$ is violated and the corresponding parameter μ_i, i.e. $\mu_t = \mu_i$. We do not initially know which constraint will be violated in the process of working of the method, and it is logical to take into account its relative strong convexity parameter instead of the global relative strong convexity one, which may turn out to be much larger. We have analyzed the results of the given numerical experiments and compared the effectiveness of one of the proposed algorithms with Algorithm 2 proposed in [15].

References

1. Alkousa, M.S.: On some stochastic mirror descent methods for constrained online optimization problems. Comput. Res. Model. **11**(2), 205–217 (2019)
2. Bauschke, H.H., Bolte, J., Teboulle, M.: A descent lemma beyond Lipschitz gradient continuity: first-order methods revisited and applications. Math. Oper. Res. **42**(2), 330–348 (2017)
3. Bubeck, S., Cesa-Bianchi, N.: Regret analysis of stochastic and nonstochastic multi-armed bandit problems. Found. Trends Mach. Learn. **5**(1), 1–122 (2012)
4. Gasnikov, A.V., Lagunovskaya, A.A., Usmanova, I.N., Fedorenko, F.A., Krymova, E.A.: Stochastic online optimization. Single-point and multi-point non-linear multi-armed bandits. Convex and strongly-convex case. Autom. Remote Control **78**(2), 224–234 (2017)

5. Hazan, E., Rakhlin, A., Bartlett, P.: Adaptive online gradient descent. In: Advances in Neural Information Processing Systems, vol. 20 (2007)
6. Hazan, E., Kale, S.: Beyond the regret minimization barrier: optimal algorithms for stochastic strongly-convex optimization. JMLR **15**, 2489–2512 (2014)
7. Hazan, E.: Introduction to online convex optimization. Found. Trends Optim. **2**(3–4), 157–325 (2015)
8. Jenatton, R., Huang, J., Archambeau, C.: Adaptive algorithms for online convex optimization with long-term constraints. In: Proceedings of the 33rd International Conference on Machine Learning, PMLR, vol. 48, pp. 402–411 (2016)
9. Lu, H.: Relative continuity for non-lipschitz nonsmooth convex optimization using stochastic (or deterministic) mirror descent. Inf. J. Optim. **1**(4), 288–303 (2019)
10. Lu, H., Freund, R., Nesterov, Yu.: Relatively smooth convex optimization by first-order methods and applications. SIOPT **28**(1), 333–354 (2018)
11. Lugosi, G., Cesa-Bianchi, N.: Prediction, Learning and Games. Cambridge University Press, New York (2006)
12. Nesterov, Y.: Relative smoothness: new paradigm in convex optimization. Conference report, EUSIPCO-2019, A Coruna, Spain, 4 September 2019 (2019). http://eusipco2019.org/wp-content/uploads/2019/10/Relative-Smoothness-New-Paradigm-in-Convex.pdf
13. Orabona, F., Crammer, K., Cesa-Bianchi, N.: A generalized online mirror descent with applications to classification and regression. Mach. Learn. **99**, 411–435 (2015)
14. Savchuk, O., Stonyakin, F., Alkousa, M., Zabirova, R., Titov, A., Gasnikov, A.: Online optimization problems with functional constraints under relative Lipschitz continuity and relative strong convexity conditions. arXiv preprint (2023). https://arxiv.org/pdf/2303.02746.pdf
15. Titov, A.A., Stonyakin, F.S., Gasnikov, A.V., Alkousa, M.S.: Mirror descent and constrained online optimization problems. In: Evtushenko, Y., Jaćimović, M., Khachay, M., Kochetov, Y., Malkova, V., Posypkin, M. (eds.) OPTIMA 2018. CCIS, vol. 974, pp. 64–78. Springer, Cham (2019). https://doi.org/10.1007/978-3-030-10934-9_5
16. Titov, A.A., Stonyakin, F.S., Alkousa, M.S., Ablaev, S.S., Gasnikov, A.V.: Analogues of switching subgradient schemes for relatively lipschitz-continuous convex programming problems. In: Kochetov, Y., Bykadorov, I., Gruzdeva, T. (eds.) MOTOR 2020. CCIS, vol. 1275, pp. 133–149. Springer, Cham (2020). https://doi.org/10.1007/978-3-030-58657-7_13

A Cutting Method with Successive Use of Constraint Functions in Constructing Approximating Sets

Rashid Yarullin[✉] and Igor Zabotin

Kazan (Volga region) Federal University, Kazan, Russia
yarullinrs@gmail.com

Abstract. We propose a cutting method with an approximation of the constraint region for solving a conditional minimization problem. The developed method is characterized by the fact that when constructing approximating sets, there is a consistent use of constraint functions. This approach of considering constraints is implemented in such a way that only one constraint is used at the initial stage, and the number of involved constraints is increased as iteration points reach the admissible set. As a result, the proposed method uses a less amount of computational operations for constructing approximating sets, which is favorably distinguished from the known cutting methods. We discuss various implementations of the sequential use of constraint functions. The convergence of the developed cutting method is proved, and an estimation of the solution accuracy is obtained for the proposed method.

Keywords: nondifferentiable optimization · high-dimensional optimization problems · convex function · subgradient · cutting method · cutting plane · approximating set · polyhedral approximation · consistent use of constraints · estimation of the solution accuracy

1 Introduction

Nowadays, when solving optimization problems, it is necessary to take into account a large number of variables and constraint functions that define an admissible set. In this regard, there are encounter problems associated with the numerical implementation of many optimization methods in practice, since when moving from the current iteration point to the next, the applied minimization method may "freeze" during the recalculation of intermediate data. To overcome such computational problems, the development of new optimization algorithms is currently being carried out either taking into account the specific structures of

This paper has been supported by the Kazan Federal University Strategic Academic Leadership Program ("PRIORITY-2030").

solved problems, or using special methods to reduce the number of restrictions involved to construct the next approximation.

Many high-dimensional optimization problems have a sparse structure. Due to this feature, it is possible to ignore a significant part of the components of the objective function and constraint functions in computational processes. In particular, a subgradient method is proposed for unconditionally minimizing a piecewise linear function in the paper [1]. Since the minimized function is given by sparse vectors, the used subgradients are dominated by zero elements, and when constructing the next approximation, a small number of components of the current iteration point are recalculated. Another way is proposed for solving high-dimensional optimization problems with a specific structure in the article [2]. In this work, a modification of the Frank-Wolfe conditional gradient method was developed to minimize quadratic function defined by a doubly sparse matrix. This matrix is characterized by the predominance of zero elements not only in rows but also in columns. Because of this feature of the matrix, the gradient of the minimized function is not computed at each step, but is recalculated based on the gradient obtained at the previous iteration. Note that in practice, it is necessary to organize the correct work with the memory of the computer for implementing methods from [1,2], since the complexity of recalculating (sub)gradients and approximations depends on the used structure of the data storage.

In certain situations, when the structure of the solved optimization problem is unknown or the minimized function and the constraint functions are given in a general form, fundamentally different approaches become in demand for using the objective function and constraints in constructing approximations. For example, various modifications of the mirror descent method are proposed for solving convex optimization problems with several functional constraints in the article [3], where the approximations are found as follows. If the current iteration point is close enough to the admissible set, then the next approximation is calculated based on the (sub)gradient of the objective function, and this step is considered productive. Otherwise, any constraint is chosen from the set of ε-unviolated constraints in the neighborhood of the current approximation to calculate the next iteration point, and this step is unproductive. Note that the feature of calculating subgradients allows to use only one constraint on unproductive steps, therefore the algorithms from [3] favorably distinguish from some variants of the mirror descent method (see, for example, [4,5]). However, the operation of determining a productive step is very laborious, since all the constraint functions are used.

One more approach of taking into account constraints is proposed for conditional minimization problems in the paper [6]. The constraints are aggregated according to this approach as follows. Active constraints are determined in the neighborhood of the current iteration point at each step, and one "surrogate" constraint is formed as a linear combination of these active constraints. Then, in a certain way, the next approximation is constructed by minimizing the objective function on the auxiliary set specified by this "surrogate" constraint. In this regard, the process of finding iteration points is simplified, but it is necessary to use all the constraints for determining the active ones, and such an operation will require significant computing resources.

In this paper, we propose a cutting method with approximation of the constraint region by polyhedral sets to minimize a convex nondifferentiable function. The developed method is characterized by the fact that the constraint functions are introduced sequentially for constructing approximating sets. This approach of using constraints is because of the following circumstance. The point is that it is required to determine a certain subset of constraints for constructing the next approximating set by separating the current approximation from the admissible set with cutting planes in many cutting methods with approximation of the feasible region (see, e.g., [7–13]), and this constraint selection is carried out by enumeration of all constraint functions, i.e. there are some encounter computational difficulties appeared in practice for taking into account the constraints as in [3,6]. Therefore, only one constraint is used to implement the cutoff at the initial step of the proposed cutting method, and the number of involved constraints increases as iteration points reach the admissible set, and in view of this fact the amount of computational operations is decreased that occur in the process of constructing approximating sets.

2   The Problem Settings

Let $f(x)$, $g_j(x)$, $j \in J = \{1, \ldots, m\}$, $m \geq 1$, be convex functions defined in an n-dimensional Euclidian space R_n, $G_j = \{x \in R_n : g_j(x) \leq 0\}$,

$$G' = \bigcap_{j \in J} G_j,$$

$G'' \subset R_n$ be a convex closed set,

$$G = G' \bigcap G''.$$

We immediately note that the interior of the set G may be empty. We solve the problem

$$\min\{f(x) : x \in G\}. \tag{1}$$

Suppose $f^* = \min\{f(x) : x \in G\}$, $X^* = \{x \in G : f(x) = f^*\}$, $x^* \in X^*$, $g(x) = \max_{j \in J} g_j(x)$, $\partial f(x)$, $\partial g_j(x)$, $j \in J$, are subdifferentials of the functions $f(x)$, $g_j(x)$, $j \in J$, at the point $x \in R_n$, respectively, $K = \{0, 1, \ldots\}$, $|\tilde{J}|$ is a cardinality of the set $\tilde{J}$, $L(y) = \{x \in R_n : f(x) \leq f(y)\}$, $I(x, \tilde{J}) = \{j \in \tilde{J} : x \notin G_j\}$, where $\tilde{J} \subset J$.

3   The Minimization Method

The proposed method constructs a sequence of points $\{x_k\}$, $k \in K$, as follows.

Step 0. Choose a convex closed set $M_0 \subseteq R_n$ such that $x^* \in M_0$. Assign $k = 0$,

$$J_0^1 = \{1\}, \quad J_0^2 = J \setminus \{1\}.$$

Step 1. Find a point

$$x_k \in M_k \bigcap G'' \bigcap L(x^*). \tag{2}$$

Step 2. Form the subsets J_k', $J_k'' \subset J$ as follows. Assign

$$J_k' = I(x_k, J_k^1).$$

If $I(x_k, J_k^2) = \emptyset$, then assign $J_k'' = \emptyset$. Otherwise, choose J_k'' as any non-empty subset of the set $I(x_k, J_k^2)$.

Step 3. Assign $J_k = J_k' \bigcup J_k''$. If

$$J_k = \emptyset, \tag{3}$$

then the iterative process is stopped, and x_k is a solution of the problem (1).

Step 4. Choose finite sets $A_k^j \subset \partial g_j(x_k)$ for all $j \in J_k$.

Step 5. Assign

$$M_{k+1} = M_k \bigcap T_k \tag{4}$$

where

$$T_k = \bigcap_{j \in J_k} \{x \in R_n : g_j(x_k) + \langle a, x - x_k \rangle \leq 0 \ \forall a \in A_k^j\}. \tag{5}$$

Step 6. Assign

$$J_{k+1}^1 = J_k^1 \cup J_k'', \tag{6}$$

$$J_{k+1}^2 = J_k^2 \setminus J_k'', \tag{7}$$

$k := k + 1$, and go Step 3.

Let us make some remarks concerning the proposed method.

Cutting-plane methods with approximation of the constraint region (see, e.g., [7–12]) usually form some subset of constraint indices to construct the next approximating set at each step. The process of forming such a subset of indices can be very laborious, since these methods involve all the constraint functions that define the admissible set. Therefore, there is a requirement to consistently take into account the constraints involved in the construction of approximating sets, and this approach of using constraints is implemented in the proposed method as follows.

The transition from the current approximating set M_k to M_{k+1} occurs by separating the point x_k from the set G' by cutting planes according to (4), (5). The set J is divided into two parts in order not to use all the constraints that determine G'. The first part J_k^1 includes the numbers of constraints allowed to participate in the construction of cutting planes at the k-th iteration, and the second part J_k^2 contains unused numbers of constraints, and in a certain way, as will be discussed below, some subset of indices will be moved from J_k^2 to the first part J_{k+1}^1 in order to sequentially take into account all the functions $g_j(x)$, $j \in J$.

Definition 1. *Denote by $\pi(I(x_k, \tilde{J}))$ the procedure which is implemented as a process of finding all numbers $i \in \tilde{J} \subseteq J$ such that $x_k \notin G_i$.*

The sets J'_k, J''_k are results of executing the procedure $\pi(I(x_k, J^1_k))$, $\pi(I(x_k, J^2_k))$, respectively, for each $k \in K$ in the proposed method. The sets of indices J'_k, J''_k determine the list of active constraints that are involved in separating the point x_k from the set G' to construct the approximating set M_{k+1}. Note that J'_k stores the numbers of involved constraints, and J''_k stores the numbers of constraints from the part of J^2_k that are unused at previous iterations, and in the case of $I(x_k, J^2_k) \neq \emptyset$ the set J''_k can include any indices from J^2_k depending on the complexity of the implementation of the procedure $\pi(I(x_k, J^2_k))$ according to Step 3 of the proposed method.

Remark 1. Suppose that $I(x_k, J^2_k) \neq \emptyset$ is fulfilled for some $k \in K$. It is clear that the execution of the procedure $\pi(I(x_k, J^2_k))$ can be laborious under the large value of $|J^2_k|$. In this connection, at Step 3 the J''_k can include not all the numbers of constraints obtained as a result of executing $\pi(I(x_k, J^2_k))$, but only a part of these numbers. In particular, it is possible to find any constraint with the number j_k that satisfies the condition $x_k \notin G_{j_k}$, and the procedure $\pi(I(x_k, J^2_k))$ can be stopped after obtaining this number j_k. In addition, J''_k can contain a small group of constraint numbers, for which the values of constraint functions and subgradients can be easily calculated at the point x_k. Due to this feature of forming the set J''_k, the consistent consideration of the constraint functions is realized for constructing the approximating sets, and as a result the computer resources are used to a lesser extent.

It is admissable to place all elements obtained by running $\pi(I(x_k, J^2_k))$ into J''_k. Therefore, all functions $g_j(x)$, $j \in J$ are used to form the set of indices of active constraints J_k. In this case, the complexity implementation of separating the point x_k from G' increases, so it is recommended to use this method of forming the set J''_k for a small value of m. Note that if the set J_k is formed in the method by using all constraints at each step $k \in K$, then the developed method is conceptually close to [10].

Lemma 1. *For any $k \in K$ the equality*

$$J^1_k \cup J^2_k = J \tag{8}$$

is fulfilled.

Proof. According to construction we have

$$J^1_k, J'_k, J^2_k, J''_k \subset J \quad \forall k \in K. \tag{9}$$

Suppose $k = 0$. Then, according to the property of the set subtraction operation (see, e.g., [14, c. 20]) and taking into account Step 3 of the proposed method, it follows that

$$J^1_0 \cup J^2_0 = \{1\} \cup (J \setminus \{1\}) = \{1\} \cup J = J.$$

Now suppose that the assertion of the lemma holds for some $k = l \geq 0$. Let us prove the assertion of the lemma for $k = l + 1$. According to the induction hypothesis we have $J_l^1 \cup J_l^2 = J$. Hence and from (6), (7), (9), taking into account the property of the set difference operation, we have

$$J_{l+1}^1 \cup J_{l+1}^2 = J_l^1 \cup (J_l'' \cup (J_l^2 \setminus J_l'')) = J_l^1 \cup J_l'' \cup J_l^2 = J.$$

The lemma is proved.

4   The Convergence Proof

Theorem 1. *Suppose the point x_k is constructed for some $k \in K$ by the proposed method, and the equality (3) is fulfilled. Then x_k is a solution of the problem (1).*

Proof. According to the condition of the theorem we have

$$J_k = J_k' \cup J_k'' = \emptyset.$$

The equality $J_k = \emptyset$ is true if and only if $J_k' = \emptyset$ and $J_k'' = \emptyset$. In view of $J_k' = \emptyset$ we have

$$x_k \in G_j \quad \forall j \in J_k^1,$$

and from the construction of the set J_k'' it follows that $I(x_k, J_k^2) = \emptyset$, i.e.

$$x_k \in G_j \quad \forall j \in J_k^2.$$

Consequently, taking into account Lemma 1, we have

$$x_k \in G_j \quad \forall j \in J.$$

Therefore, $x_k \in G'$. In addition, according to Step 3 of the method, the inclusion $x_k \in G''$ and the inequality $f(x_k) \leq f^*$ are fulfilled. Thus,

$$x_k \in G, \quad f(x_k) \leq f^*,$$

i.e. $f(x_k) = f^*$. The theorem is proved.

Lemma 2. *For all $k \in K$ the inclusion*

$$x^* \in M_k. \tag{10}$$

holds true.

Proof. For $k = 0$, the inclusion (10) is satisfied by the condition of choosing the set M_0 in Step 3 of the method. Let us assume that (10) is fulfilled for some $k = p \geq 0$. Let us obtain the validity of the inclusion (10) for $k = p + 1$. Then the lemma will be proved.

In view of the inductive assumption we have $x^* \in M_p$. Then in view of (4) to prove the inclusion $x^* \in M_{p+1}$ it is required to show that

$$x^* \in T_p. \tag{11}$$

Note that $J_p \neq \emptyset$, otherwise the iterative process is stopped and we have $x_p = x^*$. For all $a \in A_p^j$, $j \in J_p$, due to the convexity of the functions $g_j(x)$ the inequalities

$$g_j(x^*) - g_j(x_p) \geq \langle a, x^* - x_p \rangle$$

hold. But $g_j(x^*) \leq 0$ for all $j \in J$. Consequently, $g_j(x_p) + \langle a, x^* - x_p \rangle \leq 0$ for all $a \in A_p^j$, $j \in J_p$, i.e. the inclusion $x^* \in T_p$ is fulfilled. The lemma is proved.

Lemma 3. *Suppose the sequence $\{x_k\}$, $k \in K$, is constructed by the proposed method. Then for any numbers k', $k'' \in K$ such that*

$$k'' > k'$$

we have the inclusion

$$x_{k''} \in T_{k'}. \tag{12}$$

Proof. In view of Step 3 we have $x_k \in M_k$, $\forall k \in K$. Therefore, $x_{k''} \in M_{k''}$. In addition, according to (4) we get

$$M_{k''} \subset M_{k''-1} \subset \cdots \subset M_{k'+1} = M_{k'} \bigcap T_{k'}.$$

Hence it follows that the inclusion (12) holds true. The lemma is proved.

We will further assume that the sequence $\{x_k\}$, $k \in K$, is bounded. Note that the boundedness of the sequence $\{x_k\}$, $k \in K$, can be achieved by choosing M_0 as a bounded set.

Remark 2. Since the number of constraints $g_j(x)$, $j \in J$ is finite, and according to Step 3 of the proposed method, at least one constraint is transferred from the index set J_k^2 to J_{k+1}^1 at each iteration $k \in K$ under $I(x_k, J_k^2) \neq \emptyset$, then we get

$$J_k^1 = J, \quad J_k^2 = \emptyset$$

for iterations $k \geq m$, consequently, the procedure $\pi(I(x_k, J_k^1))$ involves all the functions $g_j(x)$, $j \in J$ to form the numbers of constraints needed to separate the point x_k from the set M_k.

Lemma 4. *Suppose the sequence $\{x_k\}$, $k \in K$, is constructed by the proposed method. Then any limit point of the sequence $\{x_k\}$, $k \in K$, belongs to the set G.*

Proof. By construction, for each point x_k the inclusion $x_k \in G''$ is satisfied. Therefore, to prove the assertion, it suffices to show that any limit point will belong to the set G'.

Assume that the assertion of the lemma is false. Then there exists a convergent subsequence $\{x_k\}$, $k \in K_1 \subset K$ of the sequence $\{x_k\}$, $k \in K$ such that its limit point $\bar{x}$ satisfies the inequality $g(\bar{x}) > 0$. Assign $g(\bar{x}) = \gamma$. Since the function $g(x)$ is continuous, there is a neighborhood ω of the point $\bar{x}$ such that

$$g(x) \geq \frac{\gamma}{2}$$

for all $x \in \omega$. Fix a number $\hat{k} \geq m$, $\hat{k} \in K_1$ such that for all $k \in K_1$, $k \geq \hat{k}$ the inclusion $x_k \in \omega$ is fulfilled. Then we have

$$g(x_k) = \max_{j \in J} g_j(x_k) \geq \frac{\gamma}{2} \quad \forall k \in K_1, \quad k \geq \hat{k}. \tag{13}$$

In view of (13) and the finiteness of the set J, there exists $r \in J$ such that for an infinite number of numbers $k \in K_1$, $k \geq \hat{k}$, the inequality

$$g_r(x_k) \geq \frac{\gamma}{2} \tag{14}$$

is fulfilled.

Assign

$$K_2 = \{k \in K_1 : k \geq \hat{k}, \ g_r(x_k) \geq \frac{\gamma}{2}\}.$$

Choose numbers $k', k'' \in K_2$ such that $k'' > k'$. Then according to Lemma 3 the inclusion (12) is defined. Hence it follows that the inequality

$$g_r(x_{k'}) + \langle a, x_{k''} - x_{k'} \rangle \leq 0 \tag{15}$$

is fulfilled for all $a \in A_{k'}^r$. Since $g_r(x_{k'}) > 0$ and $r \in J_{k'}$, then from inequalities (14), (15) under $k = k'$ it follows that

$$\langle a, x_{k''} - x_{k'} \rangle \leq -\frac{\gamma}{2} \tag{16}$$

for all $a \in A_{k'}^r$.

Further, since the sequence $\{x_k\}$, $k \in K$ is bounded, there exists (see, e.g., [15, p. 121]) number $\theta > 0$ such that for all $a \in \partial g_r(x_k)$, $k \in K_2$, we have the inequality

$$\|a\| \leq \theta. \tag{17}$$

Then from (16), (17) it follows that

$$\theta \|x_{k''} - x_{k'}\| \geq \frac{\gamma}{2}. \tag{18}$$

Now choose for each $k \in K_2$ an index $p_k \in K_2$ such that $p_k \geq k + 1$. According to (18)

$$\theta \|x_{p_k} - x_k\| \geq \gamma/2.$$

The last inequality is contradictory because $x_k \to \bar{x}$ and $x_{p_k} \to \bar{x}$ under $k \to \infty$, $k \in K_2$. The lemma is proved.

Theorem 2. *Suppose the sequence $\{x_k\}$, $k \in K$ is constructed by the proposed method. Then any limit point of this sequence belongs to the set X^*.*

Proof. Let $\{x_k\}$, $k \in K_1 \subset K$ be any convergent subsequence of the sequence $\{x_k\}$, $k \in K$, and $\bar{x}$ be its limit point. Then according to Lemma 4 the inclusion $\bar{x} \in G$ holds true, therefore,

$$f(\bar{x}) \geq f^*.$$

On the other side, for each $k \in K$ we have (2), i.e.

$$f(x_k) \leq f^*, \quad k \in K_1.$$

Passing in this inequality to the limit in $k \in K_1$, we obtain $f(\bar{x}) \leq f^*$. Thus, $f(\bar{x}) = f^*$, and the assertion of the theorem is proved.

Theorem 3. *Let $g_{\bar{j}}(x)$, $\bar{j} \in J$, be a strongly convex function with strong convexity constant $\mu_{\bar{j}} > 0$, $\bar{x} \in G$, and for some $k \in K$ the inclusion $\bar{j} \in J_k$ holds, the points x_k, $y_k = \bar{x} + \bar{\alpha}(x_k - \bar{x})$ are constructed, where $\bar{\alpha} \in (0,1)$ is a fixed number. If the inequality*

$$g_{\bar{j}}(x_k) - g_{\bar{j}}(y_k) \leq \xi, \tag{19}$$

is fulfilled for some $\xi > 0$, then we get the estimation

$$f^* - f(x_k) \leq \|\bar{s}\| \sqrt{\frac{2\xi}{\mu_{\bar{j}} \bar{\alpha}(1 - \bar{\alpha})}}, \tag{20}$$

where $\bar{s} \in \partial f(\bar{x})$.

Proof. Since $\bar{x} \in G$, then

$$g_{\bar{j}}(\bar{x}) \leq 0, \tag{21}$$

$$f^* - f(x_k) \leq f(\bar{x}) - f(x_k) \leq \|\bar{s}\| \|\bar{x} - x_k\|. \tag{22}$$

In addition, in view of $\bar{j} \in J_k$ we have $x_k \notin G'$, consequently,

$$g_{\bar{j}}(x_k) > 0. \tag{23}$$

Further, by the definition of a strongly convex function, we have

$$\frac{\mu_{\bar{j}}}{2} \bar{\alpha}(1 - \bar{\alpha}) \|x_k - \bar{x}\|^2 \leq \bar{\alpha} g_{\bar{j}}(x_k) + (1 - \bar{\alpha}) g_{\bar{j}}(\bar{x}) - g_{\bar{j}}(y_k).$$

Hence and from (19), (21), (23) taking into account $\bar{\alpha} \in (0,1)$ we get

$$\|x_k - \bar{x}\| \leq \sqrt{\frac{2\xi}{\mu_{\bar{j}} \bar{\alpha}(1 - \bar{\alpha})}}.$$

Finally, the validity of the estimation (20) follows from the last inequality and the relation (22). The theorem is proved.

References

1. Nesterov, Y.: Subgradient methods for huge-scale optimization problems. Math. Program. **146**, 275–297 (2014)
2. Anikin, A.S., Gasnikov, A.V., Gornov, A.Y., Kamzolov, D.I., Maksimov, Y.V., Nesterov, Y.E.: Effective numerical methods for huge-scale linear systems with double-sparsity and applications to PageRank. Proc. Moscow Inst. Phys. Technolo. **7**(4), 74–94 (2015) [in Russian]
3. Stonyakin, F.S., Alkousa, M.S., Stepanov, A.N., Barinov, M.A.: Adaptive mirror descent algorithms in convex programming problems with Lipschitz constraints. Trudy Instituta Matematiki I Mekhaniki URO RAN **24**(2), 266–279 (2018)
4. Beck, A., Ben-Tal, A., Guttmann-Beck, N., Tetruashvili, L.: The coMirror algorithm for solving nonsmooth constrained convex problems. Oper. Res. Lett. **38**(6), 493–498 (2010)
5. Nemirovsky, A., Yudin, D.: Problem Complexity and Method Efficiency in Optimization. Wiley & Sons, New York (1983)
6. Ermoliev, Y.M., Kryazhimskii, A.V., Ruszczyński, A.: Constraint aggregation principle in convex optimization. Math. Program. **76**(3), 353–372 (1997)
7. Bulatov, V.P.: Embedding Methods in Optimization Problems. Nauka, Novosibirsk (1977). [in Russian]
8. Levitin, E.S., Polyak, B.T.: Constrained minimization methods. USSR Comput. Math. Math. Phys. **6**(5), 1–50 (1966)
9. Topkis, D.M.: Cutting plane methods without nested constraint sets. Oper. Res. **18**, 404–413 (1970)
10. Zabotin, I.Y., Yarullin, R.S.: A cutting-plane method without inclusions of approximating sets for conditional minimization. LJM **36**(2), 132–138 (2015)
11. Shulgina, O.N., Yarullin, R.S., Zabotin, I.Y.: A cutting method with approximation of a constraint region and an epigraph for solving conditional minimization problems. LJM **39**(6), 847–854 (2018)
12. Zabotin, I., Shulgina, O., Yarullin, R.: Procedures for updating immersion sets in the cutting method and estimating the solution accuracy. CCIS **1661**, 218–229 (2022)
13. Khamisov, O.V.: Development optimization methods in investigations of VP Bulatov. Bull. Irkutsk State Univ. Ser. Math. **4**(2), 6–15 (2011) [in Russian]
14. Kuratowski, K., Mostowski, A.: Set theory. Mir, Moscow (1970). [in Russian]
15. Polyak, B.T.: Introduction to Optimization. Nauka, Moscow (1983). [in Russian]

Implementing One Variant of the Successive Concessions Method for the Multi-objective Optimization Problem

Igor Zabotin⬤, Oksana Shulgina, and Rashid Yarullin⁽✉⁾⬤

Kazan (Volga region), Federal University, Kazan, Russia
yarullinrs@gmail.com

Abstract. We propose one variant of the successive concessions method for solving a multi-objective optimization problem, which differs from the mentioned famous method by the approach of determining concessions. The concessions are set in the proposed method in such a way that the solutions found for the problems of the current and previous stages differ from each other not by the value of the objective functions, but by some distance no more than a given value. We describe the implementation of the method in cases when the problem of each stage is a convex programming problem. This implementation uses the algorithm proposed and proved in this paper, which belongs to the class of cutting methods with approximating the feasible set by polyhedral sets.

Keywords: Multi-objective optimization · Successive concessions method · Convex programming · Approximating set · Polyhedral approximation · Cutting plane · Approximations sequence · Convergence · Generalized-support vectors

1 Introduction

There are a significant number of schemes and methods based on various principles (for example, [1–6]) for solving multi-objective optimization problems. These principles are essentially based, in particular, on whether all the particular criteria of the problem are equivalent to each other or whether there is a ranking criterion according to their significance.

The problem to be solved in this note belongs to the second of these types. One method for solving multi-criteria problems, in which particular optimization criteria differ in importance, is the so-called concessions method (for example, [4,6]). Formally, we will describe this method later, and the idea behind it is as follows.

This paper has been supported by the Kazan Federal University Strategic Academic Leadership Program ("PRIORITY-2030").

At the first stage, we find the optimal value of the objective function, related to the main particular criterion, under the given constraints of the original problem. At each next stage, under the same basic restrictions, the optimal value of the objective function of the next most important criterion is found with the additional condition that it differs from the optimal values already found at the previous stages by no more than given positive values. The problem solution of the last stage, related to the least significant criterion, is taken as the solution to the original multi-criteria problem.

Note that the solution of the last stage may differ significantly in distance from the problems solutions of the previous stages and, in particular, from the problem solution of the first, i.e. main stage. This circumstance does not allow the method to be applied when a solution to a multi-criteria problem is considered as a point of the admissible set, which in a certain sense is close to the sets of solutions to particular problems. In this regard, we propose to change the approach of specifying concessions in the method so that it is possible to guarantee the given deviations between the solutions of adjacent stages precisely by the distance, and not by the values of the objective functions in them.

After setting the problem, we will represent a general scheme of the method with the modified approach of specifying concessions, and then we will propose an implementation of the scheme for the case when the problem of each stage is a convex programming problem. The implementation is based on the developed here method for solving the convex programming problem, which belongs to the class of cutting methods. This method approximates the constraint region by polyhedral sets to construct points, and it is characterized by the fact that the constructed main iteration points belong to the admissible set.

2 Problem Setting

We solve the multi-objective optimization problem with m particular criteria, $m \geq 2$. Let these criteria be given by continuous functions

$$f_1(x), \ldots, f_m(x)$$

defined in n-dimensional Euclidian space R_n, and the feasible set $D \subset R_n$ is bounded and closed.

We immediately agree that the criteria are unequal with each other. Namely, the first criterion defined by the function $f_1(x)$ is main. The other criteria determined by the functions $f_2(x), \ldots, f_m(x)$, respectively, are numbered in descending order of their importance, if they are not equivalent to each other.

Without loss of generality, suppose for each $j = 1, \ldots, m$ the particular problem of the related j-th optimization criterion is defined by the following form:

$$\min\{f_j(x) : x \in D\}. \tag{1}$$

Set

$$f_j^* = \min\{f_j(x) : x \in D\}, \quad X_j^* = \{x \in D : f_j(x) = f_j^*\}$$

for each $j = 1, \ldots, m$.

Lets consider the point $x^* \in D$ as some solution of the initial problem such that the distance from this point to the solution set X_1^* of the main problem among (1) does not exceed the pre-fixed number ε. The formalized rule of choosing the point x^* and approaches of considering another criteria used to find this point will be proposed later while describing the variant of the mentioned famous method. We immediately note that another point is taken as a solution in the famous method.

3   The Variant of the Successive Concessions Method

Firstly, we describe and analyze the classic successive concessions method. This method consists of m stages. Denote by k the stage numbers from $1, \ldots, m$. The method is as follows. Choose the numbers $\varepsilon_j > 0$, $j = 1, \ldots, m - 1$.

At the first stage (i.e. under $k = 1$) the point $x_1 \in D$ is found as a solution of the problem

$$\min\{f_1(x) : x \in D\},$$

and set

$$\tilde{f}_1 = f_1(x_1).$$

Then the solution $x_k \in D$ is found for each $k = 2, \ldots, m$ successively by solving the problem

$$\min\{f_k(x) : x \in D, f_j(x) \leq \tilde{f}_j + \varepsilon_j, j = 1, \ldots, k - 1\},$$

and set

$$\tilde{f}_k = f_k(x_k).$$

The point x_m is considered as a solution $\tilde{x}$ of the initial multi-objective problem which is obtained at the last stage.

Note the method does not garante that the number $\tilde{f}_k$, $1 \leq k \leq m - 1$, will be nearer to the optimal value f_k^* of problem (1) than the value $\tilde{f}_{k+1}$ of the less important criterion to its optimal value f_{k+1}^*.

Further, for all $j = 1, \ldots, m - 1$ the value $f_j(x_m)$ differs from $\tilde{f}_j$ no more than the value of ε_j. In particular, we have

$$f_1(\tilde{x}) - \tilde{f}_1 \leq \varepsilon_1$$

for the main criterion of the problem, i.e. taking into account the equality $\tilde{f}_1 = f_1^*$, the difference between $f_1(\tilde{x})$ and f_1^* is insignificant under the small value of ε_1. However, it is easy to give an example, when the distance from the point $\tilde{x}$ to the solution set X_1^* of the problem $\min\{f_1(x) : x \in D\}$ of the main stage is significant under the small value ε_1. Such a significant difference may turn out to be unacceptable due to the above definition of the solution to the original optimization problem.

Thus, the method cannot guarantee a predetermined deviation of the point $\tilde{x}$ from the set X_1^*, since the specified distance between the solutions x_k, x_{k+1},

$1 \leq k \leq m-1$, is not provided. Therefore, the variant of the method is described below, in which the fixed deviation between the points x_k, x_{k+1} is guaranteed instead of taking into account the deviation of the values of the functions, and not the values of the functions at these points.

Firstly, we describe the principal schema of such a variant. This schema consists of the m stages, and denote by k the numbers of these stages.

Choose the numbers $\varepsilon_k > 0$, $k = 1, \ldots, m - 1$. Set

$$D_1 = D$$

for $k = 1$, i.e. at the first stage, and find the point $x_k \in D_1$ as a solution to the problem

$$\min\{f_1(x) : x \in D_1\}.$$

Then construct the set

$$D_k = D_{k-1} \cap U_{k-1}$$

for $k = 2, \ldots, m$ successively, where

$$U_{k-1} = \{x \in \mathrm{R}_n : \|x - x_{k-1}\| \leq \varepsilon_{k-1}\},$$

and obtain the point $x_k \in D_k$ as a solution to the problem

$$\min\{f_k(x) : x \in D_k\}. \tag{2}$$

The point $x^* = x_m$ is considered to be a solution to the initial multi-objective problem.

Note that the inequalities

$$\|x_k - x_j\| \leq \varepsilon_j \quad \forall j = 1, \ldots, k - 1$$

hold for each $k = 2, \ldots, m$. Thus, taking into account $x_m = x^*$ we have

$$\|x^* - x_k\| \leq \varepsilon_k \quad \forall k = 1, \ldots, m - 1,$$

and, in particular, the solutions of the main stage and the initial problem satisfy

$$\|x^* - x_1\| \leq \varepsilon_1.$$

The sets U_k, $k = 1, \ldots, m - 1$, are constructed in R_n as balls with centers at the points x_k in the proposed variant. It is clear that the sets U_k can be constructed by another ways. Suppose that all particular criteria of the initial problem are defined by linear and convex quadratic functions, and D is a convex polyhedral set. Then, it is convenient to choose the sets U_k as convex polyhedra, for example, n-dimensional cubes with centers at the points $x_k = (\xi_1^k, \ldots, \xi_n^k)$, i.e.

$$U_k = \{x = (\xi_1, \ldots, \xi_n) \in \mathrm{R}_n : \xi_i^k - \varepsilon_k \leq \xi_i \leq \xi_i^k + \varepsilon_k, \ i = 1, \ldots, n\}.$$

In this case, the problems of finding the points x_k will be linear or, respectively, quadratic programming problems at each stage.

In this partial case, there is no problems to implement the described schema in practice. But in the general case, it is almost impossible to find the points x_k as an exact solution of problem (2). Lets describe below and prove the algorithm which implements the general schema proposed above with taking into account approximate solution of problem (2).

4 Implementing the Method and Its Discussion

Further, it is assumed that the functions $f_j(x)$, $j = 1, \ldots, m$, are convex in R_n. The set $D \subset R_n$ is convex, bounded, closed, and has a nonempty interior int D.

If, in addition, the functions $f_j(x)$ are, for example, linear, then from the practical viewpoint it is convenient for approximately solving each problem of (2) to apply so-called cutting methods which approximate the set D by some polyhedral sets for constructing iteration points (e.g., [7–15, 17]). Their convenience lies in the fact that each iteration point is found by solving the linear programming problem. However, the sequence of iteration points is constructed in these methods not belonging to the region of constraints. Therefore, the solution x_k of problem (2) is chosen as some approximation not belonging to the set D_k after applying the scheme of such a method at the k-th stage. Then the set

$$D_{k+1} = D_k \cap U_k$$

can be empty for some small $\varepsilon_k > 0$, and the problem of the $(k + 1)$-th stage will not have a solution.

In connection with this remark, we propose the cutting method which is used to implement the described scheme of the successive concessions method, where all iteration points are constructed belonging to the admissible set.

Construct a convex bounded closed set $M \subset R_n$ such that $D \subset M$. Define numbers

$$\varepsilon_k > 0, \quad k = 1, \ldots, m - 1, \quad \delta_k > 0, \quad k = 1, \ldots, m, \quad q > 1.$$

Put $D_1 = D$. Choose a number $v_1 \in$ int D_1. Assign $k = 1$.

Step 1. Put $i = 0$, $M_k^0 = M$, $w_k^{-1} = v_k$.
Step 2. Find a point

$$y_k^i \in M_k^i \cap E_k, \tag{3}$$

where $E_k = \{x \in R_n : f_k(x) \le f_k'\}$, $f_k' = \min\{f_k(x) : x \in D_k\}$. If $y_k^i \in D_k$, then put $\tilde{x}_k = y_k^i$, and go to Step 7.
Step 3. Compute

$$z_k^i = \lambda_k^i v_k + (1 - \lambda_k^i) y_k^i, \quad u_k^i = y_k^i + q_k^i(z_k^i - y_k^i),$$

where $\lambda_k^i \in (0, 1)$ and $q_k^i \in [1, q]$ are chosen in accordance with

$$z_k^i \notin \text{int } D_k, \quad u_k^i \in \text{int } D_k.$$

Step 4. The point w_k^i is fixed as a record point from u_k^i and w_k^{i-1} for the function $f_k(x)$, i.e.

$$f_k(w_k^i) = \min\{f_k(u_k^i), f_k(w_k^{i-1})\}. \tag{4}$$

Step 5. If

$$f_k(w_k^i) - f_k(y_k^i) \leq \delta_k, \tag{5}$$

then put $\tilde{x}_k = w_k^i$, and go to Step 7. Otherwise, perform Step 6.

Step 6. Choose a finite set A_k^i of generalized-support vectors to the set D_k at the point z_k^i, and assign

$$M_k^{i+1} = M_k^i \cap \{x \in \mathrm{R_n} : \langle a, x - z_k^i \rangle \leq 0 \ \forall a \in A_k^i\}. \tag{6}$$

Increase the value of i by one, and go to Step 2.

Step 7. If $k = m$, then put

$$x^* = \tilde{x}_k,$$

and the solution process of the initial multi-objective problem is finished. Otherwise, put

$$D_{k+1} = D_k \cap \{x \in \mathrm{R_n} : \|x - \tilde{x}_k\| \leq \varepsilon_k\}, \tag{7}$$

choose a point $v_{k+1} \in \operatorname{int} D_{k+1}$, increase the value of k by one, and go to Step 1.

We make some remarks concerning the proposed cutting method assuming the fact that the number k is fixed, $1 \leq k \leq m$.

Remark 1. It is easy to prove the inclusion $D_k \subset M_k^i$ for all $i \geq 0$ based on determine condition of the set M and construction approach (6) of the sets M_k^i, $i \geq 1$. Therefore,

$$M_k^i \cap E_k \neq \emptyset,$$

and it is admissible to choose the points y_k^i according to (3). In particular, the point y_k^i can be found in accordance with the condition

$$f_k(y_k^i) = \min\{f_k(x) : x \in M_k^i\}. \tag{8}$$

If $f_k(x)$ is linear or convex quadratic, and M is a polyhedron, then (8) is a linear or quadratic programming problem, respectively, for all $i \geq 0$.

Remark 2. In view of (3) the inclusion $y_k^i \in E_k$ is fulfilled, $i \geq 0$. Therefore, if $y_k^i \in D_k$ for some i, then y_k^i is a problem solution of the k-th stage, and there is a transition from Step 2 to Step 7 in order to solve the problem of the next stage.

Remark 3. If we put $q_k^1 = 1$ at Step 3 of the method, then z_k^i will be an intersection point of the segment $[v_k, y_k^i]$ with the boundary of the set D_k. The condition $q_k^i > 1$ allows to find such an intersection point approximately.

Remark 4. Since $\tilde{x}_k \in D_k$ and int $D \neq \emptyset$, then from (7) it follows that int $D_{k+1} \neq \emptyset$ under any $\varepsilon_k > 0$, and it is admissible to choose the point v_{k+1} at Step 7 of the method. Moreover, $u_k^i \in$ int D_k, $i \geq 0$. Thus, $w_k^i \in$ int D_k, $i \geq 0$, and $\tilde{x}_k \in$ int D_k according to Step 5. Then in view of (7) $\tilde{x}_k$ will be an interior point of the set D_{k+1}, and it is possible to put

$$v_{k+1} = \tilde{x}_k$$

at Step 7 of the method.

Remark 5. Since $f_k(y_k^i) \leq f_k'$ taking into account (3), and the inclusion $w_k^i \in D_k$ is fulfilled according to Step 4 of the method, then the estimation

$$f_k(y_k^i) \leq f_k' \leq f_k(w_k^i), \tag{9}$$

$i \geq 0$, holds true for the value f_k'. While fulfilling inequality (5) under some i, it indicates that w_k^i is a δ_k-solution of problem (2), and it is possible to move for solving the problem of $k + 1$-th stage through Step 7. If (5) is fulfilled under $k = m$, then the δ_m-solution of the last stage is obtained, and the obtained point $\tilde{x}_m$ is accepted as a solution x^* of the initial problem. Note that we will prove an existence of the number $i \geq 0$ below such that inequality (5) will be fulfilled under any fixed $\delta_k > 0$.

Remark 6. When solving problem (2) of the k-th stage, we have the relaxed optimization process in view of choosing the point w_k^i, $i = 0, 1, \ldots$, from condition (4). The proposed algorithm differs from the mentioned above famous cutting methods by this feature. Also note that we can skip the definition of the point w_k^{-1} at Step 1 of the method, and it is admissible immediately to assign

$$w_k^0 = u_k^0$$

under $i = 0$ for each $k = 1, \ldots, m$ at Step 4 considering $w_k^{-1} = u_k^0$ in (4).

5   The Convergence Proof

Further, lets prove the fact that for each $k = 1, 2, \ldots, m$ there is a number $i = i_k$ such that inequality (5) holds, and in this way we will obtain the approximation solution $\tilde{x}_k = w_k^{i_k}$ of problem (2). In fact, we will prove below the convergence of the method described at Steps 1–4, 5 for solving problem (2).

Suppose $I = \{0, 1, \ldots\}$. Assume that Steps 2–4, 6 are executed successively infinite many times independence of condition (5) under the fixed k ($1 \leq k \leq m$). In this way the sequences $\{y_k^i\}$, $\{z_k^i\}$, $\{u_k^i\}$, $\{w_k^i\}$, $i \in I$, will be constructed. Note that these sequences are bounded in view of the boundedness of the sets D, M. Let's represent the following auxiliary assertions concerning them.

Lemma 1. *Let $\{y_k^i\}$, $i \in I' \subset I$, be a convergence subsequence of the sequence $\{y_k^i\}$, $i \in I$. Then*

$$\lim_{i \in I'} \|z_k^i - y_k^i\| = 0.$$

Proof. Fix numbers l, $p_l \in I'$ such that $l < p_l$. In view of (3), (6) we have the inclusions

$$y_k^{p_l} \in M_k^{p_l} \subset M_k^l.$$

Since any vector of the set A_k^l is generalized-support at the point z_k^l not only to the set D_k but taking into account (6) also to the set $M_k^{p_l}$, then

$$\langle a, y_k^{p_l} - z_k^l \rangle \le 0 \quad \forall a \in A_k^l.$$

But from Step 3 of the method it follows that

$$z_k^i = y_k^i + \lambda_k^i (v_k - y_k^i), \quad i \in I, \tag{10}$$

thus,

$$\langle a, y_k^l - y_k^{p_l} \rangle \ge \lambda_k^l \langle a, y_k^l - v_k \rangle \quad \forall a \in A_k^l.$$

Further, according to Lemma 1 [16] there is a number $\Delta > 0$ such that the inequality $\langle a, y_k^i - v_k \rangle \ge \Delta$ holds for all $i \in I'$, $i \in A_k^i$. Consequently,

$$\langle a, y_k^l - y_k^{p_l} \rangle \ge \Delta \lambda_k^l \quad \forall a \in A_k^l.$$

Hence, we get

$$\| y_k^{p_l} - y_k^l \| \ge \Delta \lambda_k^l.$$

From the last inequality obtained for any l, $p_l \in I'$ such that $l < p_l$ and the convergence of the sequence $\{y_k^i\}$, $i \in I'$, it follows that the limit ratio $\lambda_k^i \to 0$ holds under $i \to \infty$, $i \in I'$. Then in accordance with equality (10) and the boundedness of the sequence $\{\| v_k - y_k^i \|\}$, $i \in I'$, we prove the assertion of the lemma.

Lemma 2. *The equality*

$$\lim_{i \to \infty} f_k(w_k^i) = f_k'$$

holds for the sequence $\{w_k^i\}$, $i \in I$.

Proof. Suppose the subset $I' \subset I$ is constructed in accordance with fact that $\{w_k^i\}$, $\{y_k^i\}$, $i \in I'$, are convergence subsequences. Denote by $\bar{w}_k$, $\bar{y}_k$ their limit points, respectively. Since $w_k^i \in D_k$, $i \in I'$, but the set D_k is bounded, then $\bar{w}_k \in D_k$, and, consequently,

$$f_k(\bar{w}_k) \ge f_k'. \tag{11}$$

Since $f_k(w_k^i) \le f_k(u_k^i)$ according to (4), then taking into account Lemma 1 and the boundedness of the sequence $\{q_k^i\}$, $i \in I'$, we have the inequality

$$f_k(\bar{w}_k) \le f_k(\bar{y}_k).$$

But from $f_k(y_k^i) \le f_k'$, $i \in I'$, it follows that $f_k(\bar{y}_k) \le f_k'$, and, therefore,

$$f_k(\bar{w}_k) \le f_k'.$$

Hence and from (11) we obtain

$$f_k(\bar{w}_k) = f'_k. \tag{12}$$

Further, in view of (4) the sequence $\{f_k(w_k^i)\}$, $i \in I$, is decreases monotonically, and it convergences because of its boundedness. Then taking into account continuity of the function $f_k(x)$ and equality (12) we get

$$\lim_{i \to \infty} f_k(w_k^i) = \lim_{i \in I'} f_k(w_k^i) = f_k(\bar{w}_k) = f'_k.$$

The lemma is proved.

Lemma 3. *The equality*

$$f_k(\bar{y}_k) = f'_k \tag{13}$$

holds for any limit point $\bar{y}_k$ of the sequence $\{y_k^i\}$, $i \in I$, and if (8) is fulfilled for all $i \in I$, then the whole sequence $\{f_k(y_k^i)\}$, $i \in I$, convergences to f'_k.

Proof. Let $\bar{y}_k$ be a limit point of the convergence subsequence $\{y_k^i\}$, $i \in I' \subset I$. Since

$$u_k^i = y_k^i + q_k^i(z_k^i - y_k^i), \quad i \in I',$$

then from Lemma 1 and the boundness of the sequence $\{q_k^i\}$, $i \in I'$ it follows that

$$\|u_k^i - y_k^i\| \to 0, \quad i \to \infty, \quad i \in I'. \tag{14}$$

Note that any limit point of the sequence $\{u_k^i\}$, $i \in I'$, belongs to D_k according to the closedness of the set D_k and the inclusions $u_k^i \in D_k$, $i \in I'$. Then in view of (14) $\bar{y}_k \in D_k$, and, thus,

$$f_k(\bar{y}_k) \geq f'_k.$$

On the other hand, $f_k(\bar{y}_k) \leq f'_k$, because $f_k(y_k^i) \leq f'_k$, $i \in I'$, according to (3). Consequently, equality (13) holds true.

Further, suppose the sequence $\{y_k^i\}$, $i \in I$, is constructed in accordance with condition (8). In view of (6) we have $M_k^{i+1} \subset M_k^i$, $i \in I$, and, thus,

$$f_k(y_k^{i+1}) \geq f_k(y_k^i), \quad i \in I.$$

Hence, taking into account the boundness of $\{y_k^i\}$, $i \in I$, the sequence $\{f_k(y_k^i)\}$, $i \in I$, converges. Then according to the first proved assertion of the lemma $\{f_k(y_k^i)\}$, $i \in I$, is a minimized sequence. The second assertion of the lemma is also proved.

Theorem 1. *There is a number $i = i_k \in I$ for each $k = 1, \ldots, m$ such that the point*

$$\tilde{x}_k = w_k^{i_k} \tag{15}$$

is fixed as a δ_k-solution of problem (2) of the k-th stage at Step 5 of the method.

Proof. Fix the number k. From inequalities (9) and Lemmas 2, 3 it follows that there is some $i = i_k$ such that inequality (5) is fulfilled for the points y_k^i, w_k^i, consequently, we get equality (15). The theorem is proved.

Suppose, besides the mentioned above conditions, the functions $f_k(x)$, $k = 1, \ldots, m - 1$, also satisfy the Lipschitz condition with the constants L_k, respectively, in the set D. Since

$$\tilde{x}_k \in D_k, \quad k = 1, \ldots, m,$$

and the sets D_k are constructed according to (7), then the inequalities

$$\|\tilde{x}_k - \tilde{x}_j\| \le \varepsilon_j \quad j = 1, \ldots, k - 1$$

hold under each $k = 2, \ldots, m$ for the points $\tilde{x}_1, \ldots, \tilde{x}_m$. Then taking into account $\tilde{x}_m = x^*$ we have

$$\|x^* - \tilde{x}_k\| \le \varepsilon_k$$

for $k = 1, \ldots, m - 1$. Hence we obtain the estimations

$$|f_k(x^*) - f_k(\tilde{x}_k)| \le L_k \varepsilon_k, \quad k = 1, \ldots, m - 1.$$

Thus, if the constants L_k are known, then it is possible by choosing the numbers ε_k to ensure the deviation of the values of $f_k(\tilde{x}_k)$ and $f_k(x^*)$ from each other not exceeding the value of $\Delta_k = L_k \varepsilon_k$ as in the classical method of successive concessions.

References

1. Ehrgott, M.: Multicriteria Optimization, 2nd edn. Springer, Heidelberg (2005). https://doi.org/10.1007/3-540-27659-9
2. Ralph, L., Keeny, Y.: Decisions with Multiple Objectives: Preferences and Value Tradeoffs. Radio and communications, Moscow (1981). [in Russian]
3. Lotov, A.V., Pospelova, I.I.: Lecture notes on the theory and methods of the multi-objective optimization. Moscow (2005). [in Russian]
4. Podinovsky, V.V., Gavrilov, V.M.: Optimization by Consistently Applied Criteria. Sov. radio, Moscow (1975). [in Russian]
5. Podinovsky, V.V.: Ideas and Methods of Criteria Importance Theory in Multicriteria Decision-Making Problems. Nauka, Moscow (2019). [in Russian]
6. Sobol, I.M.: Selection of Optimal Parameters in Problems with Many Criteria. Drofa, Moscow (2006). [in Russian]
7. Bulatov, V.P.: Embedding Methods in Optimization Problems. Nauka, Novosibirsk (1977). [in Russian]
8. Bulatov, V.P.; Belykh, T.I.; Yas'kova, Eh.N.: Efficient methods for solving convex programming problems that apply the embedding of the admissible set into simplices. Diskretn. Anal. Issled. Oper. **15** (3), 3–10 (2008). [in Russian]
9. Demyanov, V.F., Vasiliev, L.V.: Nondifferentiable Optimization. Nauka, Moscow (1981). [in Russian]
10. Levitin, E.S., Polyak, B.T.: Constrained minimization methods. USSR Comput. Math. Math. Phys. **6**(5), 1–50 (1966)
11. Polyak, B.T.: Introduction to Optimization. Nauka, Moscow (1983). [in Russian]
12. Nesterov, Yu.: Introductory Lectures on Convex Optimization. Kluwer Academic Publishers, Boston (2004)

13. Topkis, D.M.: A note on cutting-plane methods without nested constraint sets. Oper. Res. **18**, 1216–1220 (1970)
14. Zabotin, I. Ya., Yarullin, R. S.: One approach to constructing cutting algorithms with dropping of cutting planes. Russian Math. (Iz. VUZ). **57**(3), 60–64 (2013)
15. Zabotin, I.Ya., Yarullin, R.S.: Cutting Methods Without Nested Approximating Sets for Mathematical Programming Problems. Kazan University Publisher, Kazan (2019). [in Russian]
16. Zabotin, I.Ya.: On the several algorithms of immersion-severances for the problem of mathematical programming. Bull. Irkutsk State Univ. Ser. "Mathematics" **4**(2), 91–101 (2011). [in Russian]
17. Zabotin, I.Ya., Shulgina, O.N., Yarullin, R.S.: A cutting method with approximation of a constraint region and an epigraph for solving conditional minimization problems. LJM **39**(6), 847–854 (2018)

Stochastic Optimization

UCB Strategy for Gaussian and Bernoulli Multi-armed Bandits

M. A. Ershov⬤ and A. S. Voroshilov$^{(\boxtimes)}$⬤

Yaroslav-the-Wise Novgorod State University,
Veliky Novgorod 173003, Russian Federation
{s244525,s244528}@std.novsu.ru

Abstract. We have considered a modification of the UCB strategy for a multi-armed bandit having a Gaussian or Bernoulli distribution of one-step incomes. This strategy involves choosing an action that corresponds to the current highest value of the upper confidence bound (UCB) of the interval estimates of mathematical expectations of one-step income. The control goal is a minimax strategy, that means in minimizing maximum regrets. We computed the maximum regrets using Monte-Carlo simulations. We also performed a regression analysis of the function of the dependence of the maximum of the regret function on the strategy parameter.

Keywords: multi-armed bandit · UCB Strategy · minimax approach · regression analysis

1 Introduction

We consider a J-armed bandit problem [1–3], i.e., a slot machine with $J \geq 2$ arms, hereinafter referred to as actions. When choosing an action, the player wins a one-step income $\xi(n)$, and in total player can choose actions N times, where N is the control horizon. The player's goal is to maximize the mathematical expectation of total income. To do this, during the game, it is necessary to determine the action corresponding to the highest income and ensure its preferential choice. Each choice of one of the actions brings random income, the distribution of which depends only on the currently selected action. If the best action was known, then it should be applied all the time, but the mathematical expectations of income $m_1, \ldots, m_J$ are a priori unknown to the player.

The problem is also known as the problem of expedient behavior and adaptive control in a random environment [4,5] and is accompanied by the solution of the "information or control" dilemma, which consists in the fact that to maximize the total expected income, the player would like to apply only the action that corresponds to the highest expected one-step income. However, to determine this action, it is necessary to compare it with the rest, the use of which leads to a decrease in total income.

Supported by Russian Science Foundation, project number 23-21-00447, https://rscf. ru/en/project/23-21-00447/.

To solve the problem, we use the UCB strategy. Note that UCB implies the choice of an action based on the upper confidence bound of the interval of the mathematical expectation of total income. This strategy is quite simple to apply and has been considered, for example, in [6–9]. In these works, it is established that in the case of fixed income distributions of a multi-armed bandit, the regret function has an asymptotic order of growth $\ln N$. In this article, we consider another approach, in which the main task of the strategy is to select the optimal value of the interval estimation additive so that the maximum loss (regrets) is minimal. This approach is called minimax, since it is quite cautious in making decisions and considers all possible risks; it was considered, for example, in [2, 10]. In practice, it was found that with conventional modeling, the optimal value of the interval estimation additive is determined with a relatively small error. In this article, for a more accurate search and reduction of the error, we propose to use a regression analysis of the maximum regrets depending on the UCB additive.

The structure of the article is presented below. Section 2 describes the classical formulation of the problem of a multi-armed bandit having a Gaussian distribution and the UCB strategy. Section 3 describes a modification of the problem when the machine has a Bernoulli distribution. This modification is designed for large control horizons and allows you to significantly reduce the calculation time of the best action due to the batch version of the UCB strategy. Note that the batch version was also considered in [11–13] and the order of minimax risk is equal to $\sqrt{N}$ or close to $\sqrt{N}$. Section 4 describes the calculation of the regret function for multi-armed bandits. In the Bernoulli distribution, the mathematical expectation is a probability, which, of course, takes a value from 0 to 1. As a result, there is a limitation for the set of definitions of the regret function. Section 5 presents a general view of the graph of the regret function, as well as a description and an example of the result of regression analysis. Section 6 presents numerical simulation results. We have considered cases when $J = \{2, 3, 4, 6, 8\}$. The standard normal distribution with parameters $m = 0, D = 1$ and horizon $N = 1000$ was chosen for the Gaussian distribution. For the Bernoulli distribution, the mathematical expectation p took values from the set $p = \{1/4, 1/2, 3/4\}$, horizon $N = 5000$. Note that the calculations took less time for the Bernoulli distribution than for the Gaussian distribution. Section 7 presents the conclusion.

2 Gaussian Multi-armed Bandit

Let's consider a Gaussian multi-armed bandit with $J \geq 2$ actions. Formally, it is a controlled random process ξ_n, $n = 1, 2, \ldots, N$, where N is a control horizon. Random variable ξ_n depends only on currently chosen action y_n and is normally distributed with a density

$$f(x|m_i) = \frac{\exp\left(-\frac{(x-m_i)^2}{2D_i}\right)}{\sqrt{2\pi D_i}},$$

where $y_n = i$, $i = 1, 2, \ldots, J$. Variances $D_1, \ldots, D_J$ are assumed to be known and their values are ordered in descending order ($D_1 \geq D_2 \geq \cdots \geq D_J$). In what follows, this requirement can be canceled, since the algorithm is not sensitive to a significant change in variance, so they can be evaluated at the initial stage of control. Mathematical expectations $m_1, \ldots, m_J$ are assumed to be unknown and are not ordered.

Let n decisions were made and the action i was chosen n_i times, denote by $X_i(n)$ the current total income for use. Then $X_i(n)/n_i$ is a point estimate of the mathematical expectation m_i. Since the goal is to maximize the mathematical expectation of total income, it may seem obvious in the future to choose the action corresponding to the highest current value of the point estimate $X_i(n)/n_i$. However, at the initial stage, the best action may receive a lower estimate [6], therefore, to choose the best action, instead of point estimates of mathematical expectations, we will use the upper bounds of their interval estimates

$$U_i(n) = \frac{X_i(n)}{n_i} + a\sqrt{\frac{D \ln n}{n_i}},$$

where $a > 0$ is strategy parameter, $D = \max(D_1, \ldots, D_J)$, n_i is the number of choices of action i, $n = n_1 + n_2 + \cdots + n_J$. This strategy is called UCB (upper confidence bound) strategy.

3   Bernoulli Multi-armed Bandit

Let's consider a Bernoulli multi-armed bandit described by the distribution

$$\begin{cases} P(\xi_n = 1 | y_n = i) = p_i, \\ P(\xi_n = 0 | y_n = i) = q_i, \end{cases}$$

where $p_i + q_i = 1$, $y_n = i$, $i = 1, \ldots, J$, $n = 1, 2, \ldots, N$ with variance $D = p_i q_i$ (see., e.g., [3,14,15]). As an application of the multi-armed bandit problem, we consider binary data processing in the article. The values of the process 1 and 0 correspond to successfully and unsuccessfully processed data number n.

If there is too much data, then processing them one by one will take a lot of time, in which case batch data processing should be used [11,16]. Batch processing is a convenient control strategy because it allows the player to change the action less often and the time of full processing will be determined by the number of batches. Let's consider control strategy that apply actions M times in a row. Let there be K batches of dimension M each. In this case, the control horizon will be defined as $N = MK$. Then the upper confidence bounds will take the form

$$U_i(k) = \frac{X_i(k)}{k_i} + a\sqrt{\frac{DM \ln k}{k_i}}$$

where k_i is the number of processed batches when selecting action i, accordingly $k = 1, \ldots, K$ is the total number of processed batches.

4   Calculation of Regrets

The main objective of the strategy is to minimize the maximum regrets by finding the optimal strategy parameter. This strategy prescribes to apply all the actions once at the start of the control and then at each instant of time $n+1$ to choose the action corresponding to the maximum of the values $U_i(n)$.

Upon completion of control, the total income will be equal to $\sum_{i=1}^{J} X_i(n)$. If the mathematical expectations $m_1, \ldots, m_J$ were known, then the optimal strategy is always to use the action corresponding to the maximum of $m_1, \ldots, m_J$ and the total income would be equal to $N \max(m_1, \ldots, m_J)$. Then regrets can be defined as the normalized difference between the maximum income and the mathematical expectation of total income $\mathbf{E}(\sum_{i=1}^{J} X_i(N))$

$$l_N = \frac{N \max(m_1, \ldots, m_J) - \mathbf{E}(\sum_{i=1}^{J} X_i(N))}{\sqrt{DN}}.$$

Let's limit ourselves to such a set of parameters as was suggested in [11,16]. In this case we consider such collections of mathematical expectations, which contain one greater and other equal smaller ones, e.g., $m_1 = m + d\sqrt{D/N}$, $m_k = m - d\sqrt{D/N}$, $k = 2, \ldots, J$, $d \geq 0$. This set describes "close" distributions, on which the regret attains its maximum values. In the case of a Gaussian distribution, when modeling without limitation of generality, we assume that $m = 0$, $D = 1$.

For the Bernoulli distribution, the regrets will take the form

$$l_N = \frac{N \max(p_1, \ldots, p_J) - \mathbf{E}(\sum_{i=1}^{J} X_i(N))}{\sqrt{DN}},$$

where $p_1 = p + d\sqrt{D/N}$, $p_k = p - d\sqrt{D/N}$, $k = 2, \ldots, J$, $d \geq 0$, $D = 0.25$ is the maximum variance of one-step income, which is achieved at $p = 0.5$.

Since $0 \leq p_i \leq 1$ the value of d must be less than some positive value.

$$\begin{cases} p + d\sqrt{D/N} \leq 1, \\ p - d\sqrt{D/N} \geq 0. \end{cases} \quad \rightarrow \quad \begin{cases} d \leq (1-p)\sqrt{N/D}, \\ d \leq p\sqrt{N/D}. \end{cases}$$

In this case $1 - p = q$. Therefore $0 \leq d \leq \min(p, q)\sqrt{N/D}$.

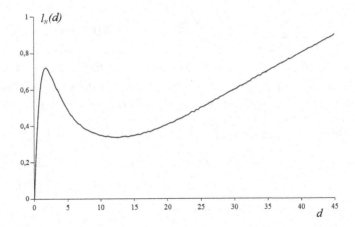

Fig. 1. Regrets for Gaussian two-armed bandit.

5 Search for Optimal Values

In Fig. 1, we present regrets for a two-armed bandit with a Gaussian distribution and with strategy parameter $a = 1$ and horizon $N = 100$.

According to the graph, the value of the function tends to infinity and there is no global maximum, but we will prove it analytically. Consider the example of a two-armed bandit. The one-step income from the choice of action i is defined as $m_i + \varepsilon_i$, where ε_i is a random variable with a Gaussian distribution. The values of $U_i(n)$ at the initial stage are equal

$$\begin{cases} U_1(2) = \frac{m_1 + \varepsilon_1}{1} + a\sqrt{\frac{1 \cdot \ln 2}{1}} = d\sqrt{D/N} + \varepsilon_1 + a\sqrt{\ln 2}, \\ U_2(2) = \frac{m_2 + \varepsilon_2}{1} + a\sqrt{\frac{1 \cdot \ln 2}{1}} = -d\sqrt{D/N} + \varepsilon_2 + a\sqrt{\ln 2}. \end{cases}$$

$$\begin{cases} \lim\limits_{d \to +\infty} U_1(2) = \lim\limits_{d \to +\infty} \left(d\sqrt{D/N} + \varepsilon_1 + a\sqrt{\ln 2} \right), \\ \lim\limits_{d \to +\infty} U_2(2) = \lim\limits_{d \to +\infty} \left(-d\sqrt{D/N} + \varepsilon_2 + a\sqrt{\ln 2} \right). \end{cases}$$

$$\begin{cases} \lim\limits_{d \to +\infty} U_1(2) = \sqrt{D/N} \lim\limits_{d \to +\infty} d + \varepsilon_1 + a\sqrt{\ln 2} = +\infty, \\ \lim\limits_{d \to +\infty} U_2(2) = \sqrt{D/N} \lim\limits_{d \to +\infty} (-d) + \varepsilon_2 + a\sqrt{\ln 2} = -\infty. \end{cases}$$

Therefore, as $d \to +\infty$, the best action is determined at the initial stage and the worst action is applied only once.

The maximum possible income corresponds to the first action and is equal to $Nd\sqrt{D/N}$. Then the total income for a two-armed bandit can be represented as

$$X_1 + X_2 = \sum_{i=1}^{n_1}(m_1 + \varepsilon_i) + \sum_{j=1}^{n_2}(m_2 + \varepsilon_j) = n_1 m_1 + n_2 m_2 + E$$

$$= n_1 d\sqrt{D/N} - n_2 d\sqrt{D/N} + E = (n_1 - n_2)d\sqrt{D/N} + E,$$

where E is the sum of N quantities with a Gaussian distribution. Then E_i is the sum of n_i quantities with a Gaussian distribution. For a two-armed bandit, at the end of the simulation $n_1 + n_2 = N$.

Then as $d \to +\infty$ total income will be equal to

$$X_1 + X_2 = (n_1 - n_2)d\sqrt{(D/N)} + E = (N - 1 - 1)d\sqrt{(D/N)}$$
$$+ E = (N - 2)d\sqrt{(D/N)} + E.$$

Then as $d \to +\infty$ regrets will be equal to

$$l_N = \frac{Nd\sqrt{D/N} - (N - 2)d\sqrt{D/N} - E}{\sqrt{DN}} = \frac{2d\sqrt{D/N} - E}{\sqrt{DN}}.$$

$$\lim_{d \to +\infty} l_N = \lim_{d \to +\infty} \left(\frac{2d\sqrt{D/N} - E}{\sqrt{DN}} \right) = \frac{\lim_{d \to +\infty} \left(2d\sqrt{D/N} - E \right)}{\lim_{d \to +\infty} \left(\sqrt{DN} \right)}$$

$$= \frac{2\sqrt{D/N} \lim_{d \to +\infty} (d) - E}{\sqrt{DN}} = +\infty.$$

It's the same for a multi-armed bandit. Therefore, if we consider the regret at the initial stage, there is no global maximum. Let's limit ourselves to searching for a local maximum on the set Q

$$Q = \{d : 0 \le d \le C < +\infty\},$$

where $C \ne \arg\max_{d \in Q}(l_N)$. This means that the function reaches a local maximum inside the segment $[0, C]$.

Let's introduce the function of dependence of the local maximum of the regrets on the value a and denote it as $\varphi(a)$

$$\varphi(a) = \max_{d \in Q} l_N(d, a).$$

Then the optimal value of a is

$$\tilde{a} = \arg\min\{\varphi(a)\}.$$

In practice, we have computed that in the vicinity of the minimum point, the function $\varphi(a)$ is approximated well by a polynomial of the second degree $\tilde{\varphi}(a) = b_2 a^2 + b_1 a + b_0$, where $b_2 > 0$. Thus, the optimal value of a can be found as the minimum $\tilde{\varphi}(a)$:

$$\tilde{\varphi}'(a) = 2b_2 a + b_1 = 0,$$

$$\tilde{a} \approx -\frac{b_1}{2b_2}.$$

For each case, we selected 30 values of a in increments of 0.01 in the vicinity of the minimum point. In Table 1 regression analysis results for a two-armed bandit with a Gaussian distribution and horizon $N = 1000$ are presented.

Table 1. Regression analysis of the function $\tilde{\varphi}(a)$

Parameter	Value
Function $\tilde{\varphi}(a)$	$1.44a^2 - 2.95a + 2.26$
Optimal $\tilde{a}$	1.02
Coefficient of determination R^2	0.997
Correlation coefficient r	0.998
Approximation error	0.12%
F-test	8458.987

The correlation coefficient is almost equal to 1, which indicates a strong relationship between a and $\tilde{\varphi}(a)$. R^2 is also almost equal to 1, therefore, the regression equation should be considered statistically significant. The approximation error is less than 10%, which indicates a well-chosen model of the equation. Since there is only 1 independent variable in the regression equation, according to the F-table of Critical Values for Significance Level $\alpha = 0.05$, $F_{crit} = 4.2$. $F \gg F_{crit}$, therefore, null hypothesis H_0 about the statistical insignificance of the regression equation is rejected. Similar results were obtained for the remaining cases.

In Fig. 2, the solid line corresponds to the function $\varphi(a)$ and the dashed line corresponds to the function $\tilde{\varphi}(a)$.

6   Numerical Results

We computed the regrets using Monte-Carlo simulations. In all cases, 400000 averaging were performed. The deviation from the mathematical expectation d varied from 0 to 15 in increments of 0.3. Let $\tilde{d} = \arg\max(l_N)$, $\tilde{a} = \arg\min_{d \in Q} \tilde{\varphi}(a)$.

In Fig. 3, the numerical results for multi-armed bandit with the Gaussian distribution in $N = 1000$ are presented. The values of the parameters and the maximum normalized regrets corresponding to lines 1, 2, 3, 4, 5 are presented in Table 2.

In Fig. 4, we present the results of calculations for the Bernoulli distribution with batch processing of data with the number of batches $K = 50$ over $M = 100$ ($N = 5000$) data and mathematical expectation $p = 0.5$. Then $d \leq 70.71$. The values of the parameters of lines 1, 2, 3, 4, 5 are presented in Table 3. Normalized regrets for the Gaussian and Bernoulli bandits are almost identical. We can see that the maximum regrets for a bandit with a Gaussian distribution turned out to be greater. However, with large deviations from the mathematical expectation d, regrets for a bandit with a Bernoulli distribution increase significantly faster.

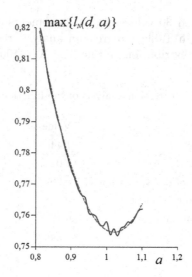

Fig. 2. Graph of the maxima of the regrets function.

Table 2. Regrets of a multi-armed bandit with a Gaussian distribution

Line	1	2	3	4	5
J	2	3	4	6	8
$\tilde{d}$	1.8	2.1	2.7	3.0	3.3
$\max\limits_{d \in Q} l_N$	0.75	1.27	1.67	2.30	2.79
$\tilde{a}$	1.02	0.90	0.79	0.71	0.64

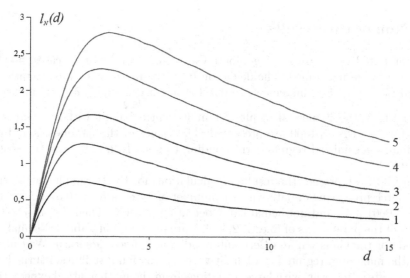

Fig. 3. Regrets of a multi-armed bandit with a Gaussian distribution.

This is since at the initial stage, the worst action is applied M times. In Fig. 5 and 6, and in Tables 4 and 5, we present the results of calculations for the Bernoulli distribution at $p = 0.25$ and $p = 0.75$ ($K = 50$, $M = 100$, $N = 5000$), which have equal deviations. Then $d \leq 35.36$. These values provide lower maximum normalized regrets than at $p = 0.5$, since the variance is less than the largest value $D = 0.25$.

Table 3. Regrets of a multi-armed bandit with Bernoulli distribution and $p = 0.5$.

Line	1	2	3	4	5
J	2	3	4	6	8
$\tilde{d}$	1.8	2.1	2.7	3.0	3.3
$\max\limits_{d \in Q} l_N$	0.70	1.18	1.56	2.18	2.72
$\tilde{a}$		0.91	0.70	0.57	0.40	0.29

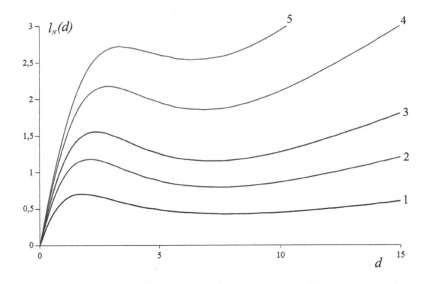

Fig. 4. Regrets of a multi-armed bandit with Bernoulli distribution and $p = 0.5$.

Table 4. Regrets of a multi-armed bandit with Bernoulli distribution and $p = 0.25$.

Line	1	2	3	4	5
J	2	3	4	6	8
$\tilde{d}$	1.5	1.8	2.1	2.4	3.0
$\max\limits_{d \in Q} l_N$	0.61	1.02	1.34	1.88	2.35
$\tilde{a}$	0.78	0.61	0.50	0.35	0.27

Table 5. Regrets of a multi-armed bandit with Bernoulli distribution and $p = 0.75$.

Line	1	2	3	4	5
J	2	3	4	6	8
$\tilde{d}$	1.5	1.8	2.1	2.4	2.7
$\max\limits_{d \in Q} l_N$	0.62	1.02	1.34	1.88	2.35
$\tilde{a}$	0.78	0.59	0.47	0.32	0.24

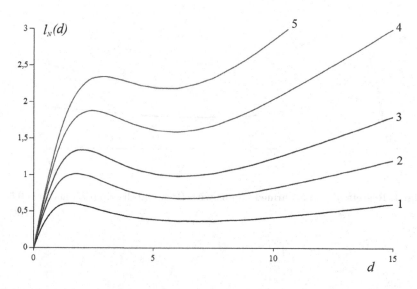

Fig. 5. Regrets of a multi-armed bandit with Bernoulli distribution and $p = 0.25$.

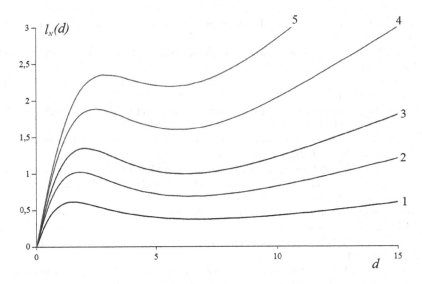

Fig. 6. Regrets of a multi-armed bandit with Bernoulli distribution and $p = 0.75$.

7   Conclusion

We considered a modification of the UCB strategy for multi-armed bandits having a Gaussian and Bernoulli distribution. Batches processing of data with the highest variance (achieved at $p = 0.5$), the maximum regrets of a bandit with a Bernoulli distribution are obtained less than for a bandit with a normal distribution. But with large deviations of d, batch processing leads to significantly greater regrets than processing data one by one. For the cases $p = 0.25$ and $p = 0.75$, the maximum normalized regrets are close in values, since in this case the variances are equal.

References

1. Presman, E., Sonin, I.: Sequential Control with Incomplete Information. Academic Press, New York (1990)
2. Bather, J.: The minimax risk for the two-armed bandit problem. In: Herkenrath, U., Kalin, D., Vogel, W. (eds.) Mathematical Learning Models Theory and Algorithms. LNCS, vol. 20, pp. 1–11. Springer, New York (1983). https://doi.org/10.1007/978-1-4612-5612-0_1
3. Berry, D.A., Fristedt, B.: Bandit Problems: Sequential Allocation of Experiments. Chapman and Hall, New York (1985)
4. Nazin, A., Poznyak, A.: Adaptivnyi vybor variantov: rekurrentnye algoritmy (Adaptive Choice between Alternatives: Recurrent Algorithms). Nauka, Moscow (1986). (in russian)
5. Sragovich, V.: Mathematical Theory of Adaptive Control. World Scientific, Singapore (2006)

6. Lai, T.: Adaptive treatment allocation and the multi-armed bandit problem. Ann. Stat. **15**(3), 1091–1114 (1987)
7. Auer, P.: Using confidence bounds for exploitation-exploration trade-offs. J. Mach. Learn. Res. **3**, 397–422 (2002)
8. Auer, P., Cesa-Bianchi, N., Fischer, P.: Finite-time analysis of the multi-armed bandit problem. Mach. Learn. **47**(2–3), 235–256 (2002)
9. Lugosi, G., Cesa-Bianchi, N.: Prediction, Learning and Games. Cambridge University Press, Cambridge (2006)
10. Vogel, W.: An asymptotic minimax theorem for the two-armed bandit problem. Ann. Math. Stat. **31**(2), 444–451 (1960)
11. Kolnogorov, A.V.: Robust parallel control in a random environment and data processing optimization. Autom. Remote. Control. **75**(12), 2124–2134 (2014). https://doi.org/10.1134/S0005117914120042
12. Kolnogorov, A.: Gaussian two-armed bandit: limit description. Prob. Inf. Transm. **56**(3), 278–301 (2020)
13. Perchet, V., Rigollet, P., Chassang, S., Snowberg, E.: Batched bandit problems. Ann. Stat. **44**(2), 660–681 (2016)
14. Varshavsky, V.: Kollektivnoe povedenie avtomatov (Collective Behavior of Automata). Nauka, Moscow (1973). (in russian)
15. Tsetlin, M.: Automation Theory and Modeling of Biological Systems. Academic Press, New York (1973)
16. Garbar, S., Kolnogorov, A.: Customization of J. Bather UCB strategy for a Gaussian multi-armed bandit. Math. Game Theory Appl. **14**(2), 3–30 (2022)

Estimation of Both Unknown Parameters in Gaussian Multi-armed Bandit for Batch Processing Scenario

Sergey Garbar$^{(\boxtimes)}$ (ID)

Yaroslav-the-Wise Novgorod State University, Velikiy Novgorod 173003, Russia
Sergey.Garbar@novsu.ru

Abstract. We consider a Gaussian multi-armed bandit problem with both reward means and variances unknown. A Gaussian multi-armed bandit is considered because in case of batch processing the cumulative rewards for the batches are distributed close to normally. A batch version of the UCB strategy is proposed. Strategy's description that is invariant in regards to the horizon size is obtained. We consider different approaches to the task of estimating unknown variances of rewards and study their effect on the normalized regret. A set of Monte-Carlo simulations is performed to study the batch strategy and illustrate the results for the two-armed bandit.

Keywords: multi-armed bandit · two-armed bandit · Gaussian multi-armed bandit · UCB · batch processing · Monte Carlo simulation

1 Introduction

The multi-armed bandit (MAB) problem [1] is considered. J-armed bandit is imagined as a slot machine with J arms. A gambler (decision-making agent) is able to choose one of the arms and collect the associated reward. The goal of the gambler is to maximize the cumulative reward during the course of control. This problem is also known as the problem of adaptive control in a random environment [2] and the problem of expedient behavior [3]. It is a reinforcement learning problem, so it is also studied in machine learning [4,5]. MABs have also been used to model such problems as managing research projects in a large organization like a science foundation or a pharmaceutical company [6,7].

Formally the multi-armed bandit is a controlled random process $\xi(n)$, $n = 1, 2, \ldots, N$. Number of steps N is called the control horizon size. Value $\xi(n)$ at step n only depends on the chosen arm. We assume that there is some prior assumptions regarding the class of distribution of rewards of arms, and it can be described by some parameter θ.

Strategy σ applied by the gambler should solve the exploration-exploitation trade-off: he or she should send some of the available opportunities to exploit

Supported by Russian Science Foundation, project number 23-21-00447, https://rscf.ru/en/project/23-21-00447/.

the best (according to the currently obtained information) arm to explore the arms that at the moment are evaluated to be less ludicrous, so the information about the θ can be clarified.

We are considering the batch processing scenario, when the arm is chosen not for a single use, but for a batch of M uses. Further we assume $N = MK$ for the simplicity of reasoning and notation. K is therefore the number of processed batches. The upside of batch processing is the lack of need to decide which arm to use after each step. Also, if arms can be used in parallel, the possibility to choose an arm for the batch gives an advantage as total processing time depends on the number of batches rather than on number of single arm uses. Batch processing can be used in internet advertising [8] and other data processing scenarios. One can intuitively expect to observe much higher regret for the batch processing, as the arm choice is not altered as often as in sequential strategies, but as shown in [9–15], this is not always the case.

If M is sufficiently large, the cumulative reward for the batch will be distributed close to normally, according to the central limit theorem. That justifies further study of Gaussian multi-armed bandits.

The variance of the batch reward depends on the variance of a single reward. Further we denote variances for arms' rewards $D_l = d_l D$, where $D = \max_{l=1...J} D_l$, and therefore $d_l \leq 1, l = 1, \ldots, J$.

We are considering the case of "close" distribution of rewards, for which mean rewards are described as

$$\left\{ m_l = m + c_l \sqrt{D/N}; m \in \mathbf{R}, |c_l| \leq C < \infty, l = 1, 2, \ldots, J \right\} \qquad (1)$$

where c_l is a measure of "closeness" of reward means. The definitive feature of "close" distributions is the difference between expected values of order $N^{-1/2}$. Maximal normalized regrets are observed on that domain and have the order $N^{1/2}$ (see [16]). For "distant" distributions the normalized regrets have smaller values.

The probability density function for cumulative reward of processed batch for l-th arm is

$$f_l(x) = (2\pi M D_l)^{-1/2} \exp\left(-(x - M m_l)^2 / (2M D_l)\right), \ l = 1, 2, \ldots, J. \qquad (2)$$

Therefore the Gaussian MAB is described by the vector parameter $\theta = (m_1, \ldots, m_J, D_1, \ldots, D_J)$.

We define the regret as expected difference between maximally obtainable rewards and the rewards that were gained. If the gambler had full information about the bandit, the best strategy would be to choose the arm with grater associated reward on each step. Therefore the regret can be calculated as

$$L_N(\sigma, \theta) = E_{\sigma,\theta} \left(\sum_{n=1}^{N} \left(\max_{l=1...J} m_l - \xi(n) \right) \right) \qquad (3)$$

By $E_{\sigma,\theta}(\cdot)$ we understand the expected value calculated in respect to measure measure generated by strategy σ and parameter θ.

Normalized regret (to the horizon size and variance D) is considered as it is is useful for comparison of various strategies if they are applied on different horizon sizes:

$$\hat{L}_N (\sigma, \theta) = \frac{1}{\sqrt{DN}} L_N (\sigma, \theta). \tag{4}$$

The goal of the gambler therefore may be stated as minimizing the normalized regret (4).

In [13, 14] an invariant description independent of the control horizon size is given for the batch version of strategy [17] and associated normalized regret (for case of batch processing for MABs). In these works the case of known rewards variances is studied.

Contributions of this paper include invariant description of the batch version of a variant of UCB strategy for the case of unknown variances of rewards and set of Monte Carlo simulations justified by the obtained description which allows to estimate the optimal strategy parameters values and compare different approaches to estimate the variances.

In the following sections we introduce and study an invariant description of control strategy [17] for the case of unknown variances of rewards. Section 2 contains description of the UCB strategy and its batch version. Section 3 builds an invariant description of the batch version of the strategy on the unit horizon and is summarized as Theorem 1. Section 4 presents Monte Carlo simulation results which allow to find optimal parameter values and to compare different ways to estimate the variances.

2   UCB Strategy

It is obvious that the gambler should choose the arm with higher associated mean reward for the purpose of minimizing the regret. Suppose that after step n the l-th arm was in total used n_l times ($l = 1, \ldots, J$). Let $X_l(n)$ denote cumulative reward for the corresponding arm. Then $X_l(n)/n_l$ estimates the mean reward m_l. However, simply using the arm with highest associated value of $X_l(n)/n_l$ can result in significant losses. That can happen in case if initial estimate $X_l(n)/n_l$, corresponding to the largest m_l, by chance takes a lower value. Consequently this action will be never applied, which can entail significant losses. Therefore, the strategy should allow each arm to be played infinitely many times on the infinite control horizon for its mean reward to be estimated correctly.

UCB strategies propose to choose the arm with highest associated upper confidence bound (hence the name) of interval estimation for the mean reward. In [17] the following version of UCB strategy is described:

$$U_l (n) = \frac{X_l (n)}{n_l} + \frac{\sqrt{2D_l \log (n/n_l)}}{\sqrt{n_l}}, \tag{5}$$

for $l = 1, 2, \ldots, J$, $n = 1, 2, \ldots, N$.

Confidence bound term $\sqrt{2D_l \log (n/n_l)/n_l}$ grows slowly as the gambler plays more rounds (i.e. as n increases), ensuring that no arm will be excluded from selection. Term $\sqrt{D_l/n_l}$ characterizes the width of the confidence interval for the mean reward estimate, and $\sqrt{\log (n/n_l)}$ ensures the growth of the confidence interval width as the step number increases.

That is the way how UCB strategy negotiates the exploration-exploitation trade-off: it strives to greedily choose the arm with highest point estimation of mean reward, but also there is an exploration term which makes it possible to play an arm that was not explored thoroughly enough. The second term is used for exploration and grows with time.

Strategy prescribes to use each of the arms once for the initial estimation of bounds.

2.1 Batch Version of the Strategy

Batch version of the strategy (5) would find upper confidence bounds for cumulative batch rewards, therefore the bounds should be found as

$$U_l(k) = \frac{X_l(k)}{k_l} + \frac{a_l\left(\hat{d}_1, \ldots, \hat{d}_J\right)\sqrt{M\hat{D}_l \, \log(k/k_l)}}{\sqrt{k_l}}, \tag{6}$$

where k is the number of processed batches, k_l is the number of batches for which the l-th arm was selected, $k = 1, 2, \ldots, K$. $\hat{D}_l$ is the estimation of variance of a single reward for l-th arm. Expression

$$\hat{d}_l = \hat{D}_l / \max_{l=1\ldots J} \hat{D}_l \tag{7}$$

describes variance estimation in relation to the maximum variance estimate.

Function $a_l\left(\hat{d}_1, \ldots, \hat{d}_J\right)$ substitutes constant $\sqrt{2}$ as the original strategy was applied to Bernoulli MABs and depends on variances of arm rewards. Transitioning to Gaussian MABs requires to determine its optimal values. Note that we take into account not the estimations of variances D_l themselves, but of the ratios d_l that determine the variances in relation to the maximal one. Also note that $a_l(d_1, \ldots, d_i, \ldots, d_j, \ldots, d_J) = a_l(d_1, \ldots, d_j, \ldots, d_i, \ldots, d_J)$ for $i \neq l$, $j \neq l$, that is changing numbering for the arms does not affect the parameter values.

In [18] it was shown that if we split the processing steps into a fairly small amount of batches (e.g. $N = 50$), the normalized regret will not grow significantly compared to the situation when number of steps tends to infinity. Therefore the steps can be split in relatively low number of batches. We should also take into the consideration that increasing the batch size will also increase the variance of its cumulative reward and while the single arm rewards could be considered "close", batch rewards can be fairly "distant", which can yield significant losses.

3   Invariant Description

The indicator $I_l(k)$ is function which equals 1 if arm l was chosen for k-th batch and 0 otherwise:

$$I_l(k) = \begin{cases} 1, \text{if } U_l(k) = \max_{l=1..J} U_l(k), \\ 0, \text{otherwise.} \end{cases} \tag{8}$$

Only one of indicators can equal 1 for the k-th batch. Reward for k-th processed batch depends on the selected arm and can be expressed as

$$\xi(k) = \sum_{l=1}^{J} I_l(k)(Mm_l + \sqrt{MD_l}\eta_{l,k}), \tag{9}$$

where $\eta_{l,k}$ is the standard normal random variable, and $m_l M + \sqrt{MD}\eta_{l,k} \sim N(Mm_l, \sqrt{MD})$ is the random reward for the k-th batch and l-th arm.

Cumulative rewards associated with each of the arms after processing k batches are

$$X_l(k) = k_l Mm_l + \sum_{i=1}^{k} I_l(i)\sqrt{MD_l}\eta_{l,i}. \tag{10}$$

Note that for $i = 1..k$, indicator $I_l(k)$ equals 1 exactly k_l times, which means that $\sum_{i=1}^{k} I_l(i)\sqrt{MD}\eta_{l,k}$ is a normally distributed random variable with variance $k_l MD$.

Therefore

$$X_l(k) = k_l M\left(m + c_l\sqrt{D/MK}\right) + \sqrt{k_l MD}\eta, \tag{11}$$

where η is a standard normal random variable.

Confidence bounds (6) are therefore equal to

$$U_l(k) = M\left(m + c_l\sqrt{D_l/MK}\right) + \frac{\sqrt{MD_l}\eta}{\sqrt{k_l}} + \frac{a_l\left(\hat{d}_1, \ldots, \hat{d}_J\right)\sqrt{\hat{D}_l \log(k/k_l)}}{\sqrt{k_l}}. \tag{12}$$

Next apply the transformations that do not change the arrangement of bounds: $u_l(k) = \sqrt{K/(MD)}(U_l(K) - Mm)$ and $t_l = k_l/K, t = k/K$ and get

$$u_l(k) = c_l + \frac{\eta}{\sqrt{t_l}} + \frac{a_l\left(\hat{d}_1, \ldots, \hat{d}_J\right)\sqrt{\hat{d}_l \log \frac{t}{t_l}}}{\sqrt{t_l}}. \tag{13}$$

The obtained formulas describe the batch version of UCB strategy on the control horizon of a unit size: variables t, t_l change their value in interval $[0, 1]$, $l = 1..J$. This description depends only on the number of processed batches K.

3.1   Estimating the Variance

Further we consider two ways to estimate the variance.

First, we can use estimation of variance of batches' rewards to calculate the variance of the single reward. We divide the estimate of batch variance by batch size M for that purpose. Note that in this case to make the initial estimation it is required to use each arm twice.

$$\hat{D}_l = \frac{k_l}{M(k_l - 1)} \left(\frac{\sum_{i=1}^{k} I_l(i)(Mm_l + \sqrt{MD_l}\eta_{l,i})^2}{k_l} \right.$$
$$\left. - \left(\frac{\sum_{i=1}^{k} I_l(i)(Mm_l + \sqrt{MD_l}\eta_{l,i})}{k_l} \right)^2 \right).$$

Taking into account that sum of squares of k_l standard normally distributed random variables is a random variable χ_{k_l} with a chi-squared distribution with k_l degrees of freedom, the estimate for variance is

$$\hat{D}_l = \frac{k_l}{M(k_l - 1)} \left(M^2 m_l^2 + \frac{MD_l\chi_{k_l}}{k_l} + \frac{\sum_{i=1}^{k} 2I_l(i)Mm_l\sqrt{MD_l}\eta_{l,i})}{k_l} \right.$$
$$\left. -M^2 m_l^2 - \frac{MD_l\chi_1}{k_l} - \frac{\sum_{i=1}^{k} 2I_l(i)Mm_l\sqrt{MD_l}\eta_{l,i})}{k_l} \right) = D_l \frac{\chi_{(k_l-1)}}{k_l - 1} \quad (14)$$

$$= d_l D \frac{\chi_{(k_l-1)}}{k_l - 1}.$$

Using this estimate, the upper confidence bound in invariant form (13) is

$$u_l(k) = c_l + \frac{\eta}{\sqrt{t_l}} + \frac{a_l\left(\hat{d}_1, \ldots, \hat{d}_J\right) \sqrt{d_l \frac{\chi_{(t_l K-1)}}{t_l K-1} \log \frac{t}{t_l}}}{\sqrt{t_l}}. \quad (15)$$

This expression depends only on the number of processed batches K, not on the number of single steps N or batch size M. Note that $\hat{a}_l$ also only depends on the number of processed batches according to (7) and (14), $l = 1, \ldots, J$.

The second approach aims to get more accurate estimation of the rewards' variances by taking into account the fact that single rewards in batches can be split into batches of smaller sizes. If the batch size is large enough, we can split it into M_2 parts of size M_1 such that sum of M_1 single rewards is distributed close to normally. In practice one can estimate the variances using the values of single rewards no matter its distribution, but the following description will not hold, and regret may differ slightly.

Note that the first approach is the special case of the second with $M_2 = 1$ and $M_1 = M$.

We continue to evaluate the variances of these parts during the full time of control.

$$\hat{D} = \frac{k_l M_2}{M_1(k_l M_2 - 1)} \left(\frac{\sum_{i=1}^{kM_2} I_l(i)(M_1 m_l + \sqrt{M_1 D_l}\eta(k))^2}{k_l M_2} \right.$$
$$\left. - \left(\frac{\sum_{i=1}^{kM_2} I_l(i)(M_1 m_l + \sqrt{M_1 D_l}\eta(k))}{k_l M_2} \right)^2 \right) = d_l D \frac{\chi_{k_l M_2 - 1}}{k_l M_2 - 1}. \quad (16)$$

Using this estimate, the upper confidence bound in invariant form (13) is

$$
u_l(k) = c_l + \frac{\eta}{\sqrt{t_l}} + \frac{a_l\left(\hat{d}_1, \ldots, \hat{d}_J\right)\sqrt{d_l\frac{X(t_l K M_2 - 1)}{t_l K M_2 - 1}\log\frac{t}{t_l}}}{\sqrt{t_l}}. \tag{17}
$$

For these two approaches the estimations of variances depend only on the number of processed batches K. If the batch is big enough, we can use the second approach for more accurate estimation.

To implement the strategy algorithmically one can estimate variances by storing the information not only about the sum of rewards, but also about sum of squares of rewards for batches of size M_1.

3.2   Invariant Description for Normalized Regret

As we can describe the batch processing using only the number of processed batches K, the following description of the normalized regret is invariant of the horizon size N.

$$
\begin{aligned}
\hat{L}_N(\sigma, \theta) &= \frac{1}{\sqrt{DN}}E_{\sigma,\theta}\left(\sum_{k=1}^{K}\left(M\max_{l=1..J}m_l - \xi(k)\right)\right) \\
&= \frac{1}{\sqrt{DN}}E_{\sigma,\theta}\left(\sum_{k=1}^{K}\left(M\max_{l=1..J}m_l - \sum_{l=1}^{J}Mm_l I_l(k)\right)\right) \\
&= \frac{1}{\sqrt{DN}}E_{\sigma,\theta}\left(\sum_{l=1}^{J}M\left(\sqrt{\frac{D}{N}}\max_{i=1..J}c_J - c_l\right)\left(\sum_{n=1}^{K}I_l(k)\right)\right) \\
&= \sum_{l=1}^{J}\left(\max_{i=1..J}c_J - c_l\right)E_{\sigma,\theta}(t_l).
\end{aligned} \tag{18}
$$

The reasoning in this section can be summarized as a theorem.

Theorem 1. *For multi-armed bandit with J arms, mean rewards m_l with "close" distribution (1) and unknown variances D_l ($l = 1, 2, \ldots, J$), the usage of the batch version of UCB strategy (6) with batch size $M = M_1 M_2$ with variance estimated as (16) has an invariant description on the unit control horizon defined by bounds (17). For normalized regret (4) expression (18) holds.*

The obtained description of the batch strategy depends only on the number of batches, i.e. is invariant of the horizon size. If the batch size is large enough, each individual batch can be split into smaller parts for more accurate estimation of the rewards variances. Stating the theorem allows to use Monte Carlo simulations to determine the optimal strategy parameters values and to study the normalized regret for different approaches of variance estimation. Therefore the following section is justified by the Theorem 1.

4   Simulation Results

For all the illustrations on figures we use a two-armed bandit and assume that the second arm is the best with $m_1 = 0$ and $m_2 > 0$, i.e. $c_1 = 0$ and $c_2 > 0$.

To assure the ability to notice the differences in normalized regrets for different ways of estimating the variances of rewards, first we need to determine the optimal parameter of the strategy $a(d_1, d_2)$.

Figure 1 illustrates the approach used to determine the optimal parameter values. Figure shows plots for normalized regrets for parameter c_2, which defines the difference between mean rewards, and different values of $a(d_1, d_2)$ and $d_1 = d_2 = 1$. The results of simulations show that for "close" distribution of rewards for $a(1, 1) \in (0.875, 0.925)$ the maximal normalized regret is no greater than 0.697. When the variance is known, the invariant description of the strategy is independent of the number of processed batches K. Therefore, if it is sufficiently large, any number can be used to find the optimal parameter values. Case of $K = 100$ is presented of that figure. Data for the figure are averaged over 1000000 simulations.

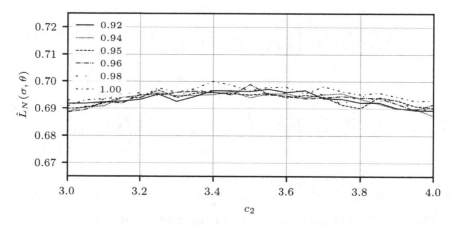

Fig. 1. Normalized regret $\hat{L}_N(\sigma, \theta)$ vs. c_2 for different values of strategy parameter $a(D_1, D_2)$ for $d_1 = d_2 = 1$

We compared different values of $a(d_1, d_2)$ and present the results as Table 1. First we assumed without loss of generality that $d_1 = 1 \geq d_2$. Otherwise note that $a(d_1, d_2) = a(d_2, d_1)$. The optimal values for the first arm's strategy parameter (with greater estimated variance) are denoted a_1. For the second arm notation a_2 is used. Though we did not only search the equal values of a_1, a_2, the lowest regret was observed in that case. On the other hand we cannot guarantee that some set of unequal values won't be a better choice, but the advantage of choosing a different set of parameters is not expected to be significant. Also note that as the strategy is not very sensitive to the variance estimations, it is also not very sensitive to the values of a_1 and a_2. Table also contains maximum observed normalized regret $\max \hat{L}_N(\sigma, \theta)$, which is stated only for the domain of "close" distribution of rewards. For the greater distances between mean rewards higher magnitude of normalized regret is possible. We find the values where the local minima for normalized regret are located and denote them c_+ and c_- for the positive and the negative. Figure 2 shows normalized regret for different values of d_2. Data the figure are averaged over 200000 simulations.

Table 1. Optimal values for $a(d_1, d_2)$ for two-armed bandit.

d_1	d_2	a_1	a_2	$\max \hat{L}_N(\sigma, \theta)$	c_-	c_+
1.0	1.000	0.93	0.93	0.696	−3.5	3.5
1.0	0.750	0.92	0.92	0.652	−3.0	3.3
1.0	0.500	0.89	0.89	0.598	−2.8	3.2
1.0	0.250	0.88	0.88	0.537	−2.7	3.1
1.0	0.125	0.86	0.86	0.497	−2.3	2.7

For demonstration purposes the case when single rewards have normal distribution with variances $D_1 = D_2 = 1$ (and so $d_1 = d_2 = 1$) is considered on the following plots on Figs. 3 and 4. Therefore we can estimate the variances of single arm uses as having the same invariant description as obtained above, i.e. $M_1 = 1, M_2 = M$. We did not interpolate parameter values from Table 1, but used the closest to the current variance estimates ratio. We consider the case of equal variances as the highest normalized regret is observed in this situation, as shown in [13] and Table 1. We choose $a = 0.93$ for further simulations.

Figures 3 and 4 show normalized regret for different methods of estimating the variance in batch processing. Batch sizes of $M = 16$ (Fig. 3) and $M = 4$ (Fig. 4) are selected. In each case $K = 50$ batches were processed. Data the figures are averaged over 1000000 simulations.

Line 1 shows the case when variances are estimated for batches during the control time. Which means variances for rewards are estimated based on k_1 and k_2 data points correspondingly. On initial stages of control estimations are the least accurate, then they gets better during the course of control. To estimate the variance at least two measures are required, therefore no matter how much worse one arm is, it will be used for at least two batches to estimate its variance. Therefore normalized regret is higher for greater values of c_2, as less profitable arm is used $2M$ times. Higher regret in that case is therefore the more noticeable the larger batch size is. Maximum normalized regret is 0.704 for $M = 4$ and 0.705 for $M = 16$. We observe little regret difference for different batch sizes as the number of processed batches is the same and variance is estimated based on that data.

All regret values in text are presented in Table 2 for convenience.

In [19] it is shown that the normalized regret does not depend very strongly on the precision of the variance estimation, which means that in some cases it

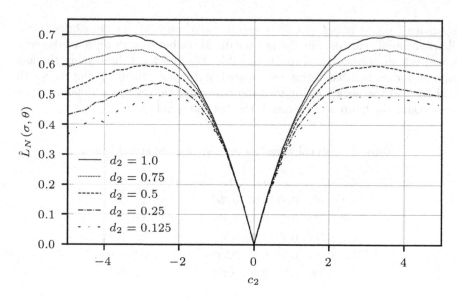

Fig. 2. Normalized regret $\hat{L}_N(\sigma, \theta)$ vs. c_2 for different values of d_2 with parameter a values from Table 1

may be enough to only estimate the variance on initial stages of control. Line 2 shows the case when variance is estimated for single rewards, and only for the first batch for each arm. Variances are therefore estimated based on M points each. That is the reason expected normalized regret is significantly higher for $M = 4$ (0.747) than for $M = 16$ (0.707), as the variance is estimated less accurately in this case. This approach is obviously very sensitive to the batch size. Nevertheless its use can be justified when the batch size is big enough as the variance estimation during the full time of control requires significant volume of computations.

Line 3 shows the case when the variance is estimated during the full time of control for single rewards. At the end of control variances are estimated based on $k_1 M$ and $k_2 M$ points. For $M = 4$ normalized regret is 0.701, for $M = 16$ it is 0.696.

Line 4 shows regret in case when the exact value of variance is used instead of estimation. Maximum regret is 0.691 for $M = 4$ and 0.690 for $M = 16$.

Overall it is preferred to estimate reward variances based on batches of smaller sizes when possible. If the batch size is big enough, it is possible to only estimate variances on for first few batches.

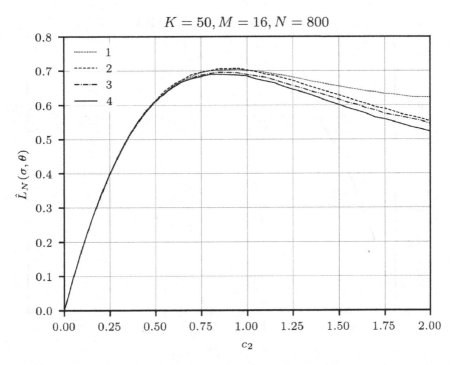

Fig. 3. Normalized regret $\hat{L}_N(\sigma, \theta)$ vs. c_2 for $M = 16$ and different approaches to estimate the variance: 1 - estimating variances for whole batches, 2 - estimating variances for single rewards on initial stage of control, 3 - estimating variances for single rewards during the full time of control, 4 - using the exact values

Table 2. Maximum normalized regrets for different approaches to estimate variances.

Used approach	$M = 4$	$M = 16$
Based on batch rewards	0.704	0.705
Based on single rewards of the first batch	0.747	0.707
Based on single rewards for the full time of control	0.701	0.696
Known variance	0.691	0.690

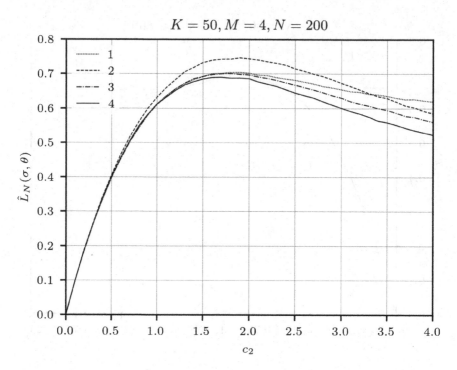

Fig. 4. Normalized regret $\hat{L}_N(\sigma, \theta)$ vs. c_2 for $M = 4$ and different approaches to estimate the variance: 1 - estimating variances for whole batches, 2 - estimating variances for single rewards on initial stage of control, 3 - estimating variances for single rewards during the full time of control, 4 - using the exact values

5   Conclusions

Batch version of UCB strategy [17] for the case of unknown reward variances is described and studied. Two approaches for estimation of variance are proposed. An invariant descriptions on a unit control horizon (dependent only on number of processed batches) for the strategy and normalized regret are given.

A set of Monte Carlo simulations was performed for the case of two-armed bandit. Optimal strategy parameters values for various sets of arms' rewards variances were determined. Normalized regrets were studied for the worst case of equal variances. A comparison of different ways to estimate the variances of arms' rewards is presented for different batch sizes.

References

1. Lattimore, T., Szepesvari, C.: Bandit Algorithms. Cambridge University Press, Cambridge (2020)
2. Sragovich, V.: Mathematical Theory of Adaptive Control. World Scientific, Singapore (2006)

3. Tsetlin, M.: Automaton Theory and Modeling of Biological Systems. Academic Press, New York (1973)
4. Auer, P.: Using confidence bounds for exploitation-exploration trade-offs. J. Mach. Learn. Res. **3**, 397–422 (2002)
5. Lugosi, G., Cesa-Bianchi, N.: Prediction, Learning and Game. University Press, New York (2006)
6. Berry, D., Fristedt, B.: Bandit Problems: Sequential Allocation of Experiments. Chapman and Hall, London (1985)
7. Gittins, J.: Multi-armed bandit allocation indices. In: Wiley-Interscience Series in Systems and Optimization. John Wiley & Sons, Ltd., Chichester (1989)
8. Zhang, D., Lu, J.: Batch-mode computational advertising based on modern portfolio theory. In: Azzopardi, L., et al. (eds.) ICTIR 2009. LNCS, vol. 5766, pp. 380–383. Springer, Heidelberg (2009). https://doi.org/10.1007/978-3-642-04417-5_44
9. Kolnogorov, A.V.: Parallel design of robust control in the stochastic environment (the two-armed bandit problem). Autom. Remote Control **73**, 689–701 (2012)
10. Perchet, V., Rigollet, P., Chassang, S., Snowberg, E.: Batched bandit problems. Ann. Stat. **44**(2), 660–681 (2016)
11. Kolnogorov, A.: Gaussian two-armed bandit and optimization of batch data processing. Prob. Inf. Trans. **54**, 84–100 (2018)
12. Gao, Z., Han, Y., Ren, Z., Zhou, Z.: Batched multi-armed bandits problem. In: NeurIPS (2019)
13. Garbar, S.: Invariant description of UCB strategy for multi-armed bandits for batch processing scenario. In: 2020 24th International Conference on Circuits, Systems, Communications and Computers (CSCC), pp. 75–78 (2020)
14. Garbar, S.: Invariant description for batch version of UCB strategy for multi-armed bandit. J. Phys.: Conf. Ser. **1658**, 012015 (2020)
15. Kolnogorov, A.V., Nazin, A.V., Shiyan, D.N.: Two-armed bandit problem and batch version of the mirror descent algorithm. Autom. Remote Control **83**, 1288–1307 (2022)
16. Vogel, W.: An asymptotic minimax theorem for the two-armed bandit problem. Ann. Math. Statist. **31**, 444–451 (1960)
17. Lai, T.L.: Adaptive treatment allocation and the multi-armed bandit problem. Ann. Stat. **25**, 1091–1114 (1987)
18. Garbar, S.: Stochastic differential equations for limiting description of UCB rule for Gaussian multi-armed bandits. arXiv:2112.06423 (2021)
19. Garbar, S.: Dependency of regret on accuracy of variance estimation for different versions of UCB strategy for Gaussian multi-armed bandits. J. Phys.: Conf. Ser. **2052**, 012013 (2021)

Zero-Order Stochastic Conditional Gradient Sliding Method for Non-smooth Convex Optimization

Aleksandr Lobanov[1,2,3]($\boxtimes$) (iD), Anton Anikin[4] (iD), Alexander Gasnikov[1,5,6] (iD), Alexander Gornov[4] (iD), and Sergey Chukanov[7]

[1] Moscow Institute of Physics and Technology, Dolgoprudny, Russia
anikin@icc.ru
[2] ISP RAS Research Center for Trusted Artificial Intelligence, Moscow, Russia
[3] Moscow Aviation Institute, Moscow, Russia
lobanov.av@mipt.ru
[4] Matrosov Institute for System Dynamics and Control Theory, Irkutsk, Russia
gasnikov.av@mipt.ru, gornov@icc.ru
[5] Institute for Information Transmission Problems RAS, Moscow, Russia
[6] Caucasus Mathematical Center, Adyghe State University, Maikop, Russia
[7] Federal Research Center "Computer Science and Control" RAS, Moscow, Russia

Abstract. The conditional gradient idea proposed by Marguerite Frank and Philip Wolfe in 1956 was so well received by the community that new algorithms (also called Frank–Wolfe type algorithms) are still being actively created. In this paper, we study a non-smooth stochastic convex optimization problem with constraints. Using a smoothing technique and based on an accelerated batched first-order Stochastic Conditional Gradient Sliding method, we propose a novel gradient-free Frank–Wolfe type algorithm called Zero-Order Stochastic Conditional Gradient Sliding (ZO-SCGS). This algorithm is robust not only for the class of non-smooth problems, but surprisingly also for the class of smooth black box problems, outperforming the SOTA algorithms in the smooth case in term oracle calls. In practical experiments we confirm our theoretical results.

Keywords: Frank–Wolfe type algorithms · Non-smooth convex optimization · Gradient-free method

1 Introduction

The history of the conditional gradient method begins with the Frank–Wolfe algorithm proposed in 1956 [16]. Marguerite Frank and Philip Wolfe proposed an alternative to the gradient descent method for solving a class of quadratic constrained optimization problems that uses linear optimization on a convex

The work was supported by the Ministry of Science and Higher Education of the Russian Federation (Goszadaniye) 075-00337-20-03, project No. 0714-2020-0005.

compact set, avoiding projection. A little later in 1966, Evgenii Levitin and Boris Polyak in [34] investigated the Frank–Wolfe method (named Conditional Gradient), obtaining the rate of convergence and showed that this rate is optimal for the class of smooth convex problems and for all algorithms that use linear minimization oracle. Since then, the conditional gradient algorithm has gained much interest in the community, because in some cases it is computationally cheaper to solve the linear minimization problem over the feasible set (thereby guaranteeing a presence over the feasible set) than to perform a projection over the feasible set. Currently, the conditional gradient method is actively used in solving practical problems of network routing [21,35,38], matrix completion [17,20], as well as in problems of machine learning [30,40], federated learning [12], online optimization [9,18,28], standard optimization [19,24,39] and huge-scale optimization [3,7,11].

However, as far as we know, there are no gradient-free algorithms (based on the conditional gradient method) to solve the black box problem in the non-smooth case. Where the black box problem means that only the zero-order oracle [42] is available to us, i.e. we have access to the value of the objective function, not its gradient. This class of problems is a particular case of the practical problems above, when the gradient calculation procedure is too expensive [1,43] or not available at all [8,14]. Already in November 2022, a survey appeared [23], which provides various techniques for creating optimal gradient-free algorithms (based generally on accelerated batched first-order methods) to solve the black-box problem. The optimal for a gradient-free algorithm is usually understood by three criteria: iteration complexity, oracle complexity, and maximum level of adversary noise. Thus, by choosing the accelerated batched conditional gradient method and using the smoothing technique from the survey, it is possible to develop a gradient-free algorithm to solve black-box problem in non-smooth case.

In this paper, we focus on black-box problems in the non-smooth case, namely, non-smooth convex stochastic optimization problems. To solve this problem, we use a smoothing scheme approach with l_2 randomization. Based on the accelerated batched conditional gradient method, also known as the Stochastic Conditional Gradient Sliding Method from [33], we create an algorithm and derive optimal estimates: iteration complexity, oracle complexity, and maximum adversary noise level. As far as we know, this is the first gradient-free algorithm for solving a non-smooth convex optimization problem. We show in theory that Zero-Order Stochastic Conditional Gradient Sliding Method outperforms the oracle complexity of gradient-free algorithms (which are state of the art algorithms) in a smooth setting, which is a surprising fact. In practical experiments we confirm our theoretical results.

1.1 Our Contributions

Our contributions can be summarized as:

- We present the first gradient-free algorithm based on the conditional gradient method "Zero-Order Stochastic Conditional Gradient Sliding Method" (ZO-

SCGS) for solving a non-smooth convex stochastic optimization problem with constraints.

- Our theoretical results show that the algorithm is robust for black-box problems not only in the non-smooth case, but also for the smooth setting case. That is, our algorithm outperforms state of the art algorithms on oracle calls. In particular, the SOTA algorithm Zero-Order Conditional Gradient Method (ZSCG) from [5] has an estimation of oracle complexity $\sim \varepsilon^{-3}$, while our algorithm has an estimation of oracle complexity $\sim \varepsilon^{-2}$.
- We empirically test our theoretical results by comparing the Zero-Order Stochastic Conditional Gradient Sliding Method (ZO-SCGS) with the Zero-Order Conditional Gradient Method (ZSCG) on a model case in a smooth setting. We explain the reason for the advantage of the proposed algorithm.

1.2   Paper Organization

This paper has the following structure. In Sect. 2 we provide related works. In Sect. 3 we consider the formulation of the problem. We present the novel gradient-free algorithm in Sect. 4. In Sect. 5 we discuss the theoretical results obtained. We verify our results with a model experiment in Sect. 6. While Sect. 7 concludes the paper. We provide a detailed proof of the Theorem 1 in the supplementary materials (Appendix A)[4].

2   Related Works

Conditional Gradient Methods. There are many works [5,10,13,24,29,33, 37,39,49,50] in the field of conditional gradient methods research. The latest research results in this area are presented in a recent survey on conditional gradient methods [6]. For instance, the Stochastic Frank–Wolf algorithm from [29], which is a generalization of the Frank–Wolf algorithm to stochastic optimization by replacing the gradient in the update with its stochastic approximation, requires $\sim \varepsilon^{-3}$ calls of stochastic gradients and performing $\sim \varepsilon^{-1}$ linear optimization. Also, for instance, the Stochastic Away Frank–Wolfe algorithm from [24], which is derived from combining the Away-Step Stochastic Frank–Wolfe algorithm [27] and the Pairwise Stochastic Frank–Wolfe algorithm [32], requires $\sim \varepsilon^{-4} \log^6 \left(\varepsilon^{-1} \right)$ calls of stochastic gradients and performing $\sim \log \left(\varepsilon^{-1} \right)$ linear optimization. In another work [39], the Momentum Stochastic Frank–Wolf algorithm, which is obtained from the Stochastic Frank–Wolfe algorithm by replacing the gradient estimator with the momentum estimator, requires $\sim \varepsilon^{-3}$ calls of stochastic gradients and linear optimization and in [33] the Stochastic Conditional Gradient Sliding algorithm was proposed, which is an accelerated batched method and requires $\sim \varepsilon^{-2}$ calls of stochastic gradients and performing $\sim \varepsilon^{-1}$ linear optimization. The above algorithms solve the problem of convex stochastic optimization and are first-order methods, but the Zeroth-Order

---

[4] The full version of this article, which includes the Appendix A can be found at the following link: https://arxiv.org/abs/2303.02778.

Stochastic Conditional Gradient Method from [5], which solves the black box problem in the smooth case, requires $\sim \varepsilon^{-3}$ calls of stochastic gradients and performing $\sim \varepsilon^{-1}$ linear optimization. In this paper, we choose the accelerated batched first order method: Stochastic Conditional Gradient Sliding algorithm from [33] as the basis for creating a novel gradient-free algorithm, since it has the best number of stochastic gradient calls presented. We will compare the efficiency of our algorithm to the Zeroth-Order Stochastic Conditional Gradient Method from [5], which is one of the SOTA algorithms.

Gradient-Free Methods. The research field of gradient-free algorithms can be traced back to at least 1952 [31]. Recent works [2,4,14,15,22,36,47] are heavily focused on creating optimal gradient-free algorithms based on three criteria: iteration complexity, oracle complexity, and maximum level of adversary noise. For black-box problems, a gradient approximation is usually used instead of an exact gradient in first-order algorithms. For instance, work [47] investigated gradient approximation via coordinate-wise randomization, and work [15] investigated gradient approximation via random search randomization. Also, for instance, in [4] the gradient approximation via a "kernel-based" approximation is studied, the feature of which is to take into account the advantages for the case of increased smoothness. Some works use smoothing schemes via l_1 or l_2 randomization. For instance, paper [2] studied l_1 randomization as an alternative to the exact gradient for solving smooth optimization problems. Another paper [22] explained the advantages of solving non-smooth problems using a smoothing scheme with l_2 randomization. And in [36] the smoothing scheme through l_1 randomization for non-smooth optimization problems is investigated and it is shown that in practice there are no significant advantages of l_1 randomization over l_2 randomization. In this paper, we use a smoothing scheme with l_2 randomization to create a gradient-free algorithm for solving a non-smooth convex stochastic optimization problem.

3 Setup

We study a non-smooth convex stochastic optimization problem with constraints

$$f^* := \min_{x \in Q} [f(x) := \mathbb{E}_\xi [f(x,\xi)]] \tag{1}$$

where $Q \subseteq \mathbb{R}^d$ is a convex compact set and $f : Q \to \mathbb{R}$ is a convex function. This problem is also known as the black box problem, where a zero-order (gradient-free) oracle returns a function value $f(x,\xi)$ at the requested point x, possibly with some adversarial noise $\delta(x)$. We now formally introduce the definition of a gradient-free oracle.

Definition 1 (Gradient-free oracle). *Let gradient-free oracle returns a noise value of $f(x,\xi)$, i.e. for all $x \in Q$*

$$f_\delta(x,\xi) := f(x,\xi) + \delta(x).$$

Next, we consider the assumptions we use in our theoretical results.

3.1   Assumptions

We assume that the function is Lipschitz continuous and is convex on set Q_γ.

Assumption 1 (Lipschitz continuity of the function). *Function $f(x,\xi)$ is Lipschitz continuous with constant M, i.e. for all $x,y \in Q$:*

$$|f(y,\xi) - f(x,\xi)| \leq M(\xi)\|y - x\|_p.$$

Moreover, there exists a positive constant M such that $\mathbb{E}\left[M^2(\xi)\right] \leq M^2$.

Assumption 2 (Convexity on the set Q_γ). *Let $\gamma > 0$ a small number to be defined later and $Q_\gamma := Q + B_2^d(\gamma)$, then the function f is convex on the set Q_γ.*

We also assume that adversarial noise is bounded.

Assumption 3 (Boundedness of noise). *For all $x \in Q$, it holds $|\delta(x)| \leq \Delta$.*

Our Assumption 1 of a Lipschitz continuity of the function is similar as in [22] and generalizes to a stochastic setting. For the special case when $p = 2$ we use the notation M_2 for the Lipschitz constant (see e.g. [14]). Assumption 2 is quite standard in the literature (see e.g. [41,48]). We used l_2-ball here since we use l_2 randomization in this paper. In more general the formulation of the assumption depends on the choice of gradient approximation (see e.g. [36]). So much prior work in the context of stochastic optimization often assumed the boundedness of stochastic or deterministic noise (such as e.g. [1,44,46]). In Assumption 3, we consider bounded deterministic noise.

3.2   Notation

We use $\langle x, y \rangle := \sum_{i=1}^d x_i y_i$ to denote standard inner product of $x, y \in \mathbb{R}^d$, where x_i and y_i are the i-th component of x and y respectively. We denote l_p-norms (for $p \geq 1$) in $\mathbb{R}^d$ as $\|x\|_p := \left(\sum_{i=1}^d |x_i|^p\right)^{1/p}$. Particularly for l_2-norm in $\mathbb{R}^d$ it follows $\|x\|_2 := \sqrt{\langle x, x \rangle}$. We denote l_p-ball as $B_p^d(r) := \left\{x \in \mathbb{R}^d : \|x\|_p \leq r\right\}$ and l_p-sphere as $S_p^d(r) := \left\{x \in \mathbb{R}^d : \|x\|_p = r\right\}$. Operator $\mathbb{E}[\cdot]$ denotes full mathematical expectation. We notation $\tilde{O}(\cdot)$ to hide logarithmic factors. To define the diameter of the set Q we introduce $D := \max_{x,y \in Q} \|x - y\|_p$.

4   Main Result

In this section, we present a novel algorithm (see Algorithm 1) that is optimal in terms of iterative complexity, the number of gradient-free oracle calls, and the maximum value of adversarial noise. This algorithm is based on an accelerated first-order Stochastic Conditional Gradient Sliding (SCGD) method from [33]. This section is structured as follows: in Subsect. 4.1 we introduce the basic idea of the smoothing scheme, in Subsects. 4.2 and 4.3 we consider the main elements of the smoothing scheme via l_2 randomization, and in Subsect. 4.4 we present the new gradient-free method (see Algorithm 1 for more details).

We start with the main idea of solving problem (1) via the smoothing scheme.

4.1 Smoothing Scheme Intuition

The main idea of solving problem (1) via the smoothing scheme is to replace the problem. That is, instead of solving the non-smooth problem we will solve its smoothed problem:

$$\min_{x \in Q} f_\gamma(x), \tag{2}$$

where f_γ a smooth approximation of the non-smooth function f, which we define below. Thus, to solve the smooth problem (2) it is sufficient to choose the accelerated batched algorithm $\mathbf{A}(L_{f_\gamma}, \sigma^2)$. Next, we introduce the assumptions of smoothness of the function f_γ and bounded variance of the gradient $\nabla f_\gamma(x, \psi)$.

Assumption 4 (L_{f_γ}-smoothness). *Function $f_\gamma(x)$ is differentiable and there exists a constant $L_{f_\gamma} \geq 0$ such that for $x, y \in Q$:*

$$\|\nabla f_\gamma(y) - \nabla f_\gamma(x)\|_q \leq L_{f_\gamma} \|y - x\|_p.$$

Assumption 5 (Bounded variance and unbiased). *Gradient $\nabla f_\gamma(x, \xi)$ has bounded variance such that for $x \in Q$:*

$$\mathbb{E}_\psi \left[\|\nabla_x f_\gamma(x, \psi) - \nabla f_\gamma(x)\|_q^2 \right] \leq \sigma^2, \quad \mathbb{E}_\psi \left[\nabla f_\gamma(x, \psi) \right] = \nabla f_\gamma(x).$$

Assumptions 4 and 5 are quite common in the literature (see e.g. [26,33,44]). Here q is such that $1/p + 1/q = 1$. And a random variable ψ we define below.

The connection between Problems (1) and (2) is as follows: to solve a non-smooth problem with ε-accuracy, it is necessary to solve a smooth problem with $(\varepsilon/2)$-accuracy, where ε-suboptimality is the accuracy of the solution in terms of expectation (see Appendix A for the proof of this statement). So, to solve Problem (1) (under the Assumption 4 and 5) with Algorithm $\mathbf{A}(L_{f_\gamma}, \sigma^2)$, we need to know the gradient of the smoothed function $\nabla f_\gamma(x, \psi)$, L_{f_γ}-smoothness constant, and the variance estimate σ^2.

In the following subsections we will define these elements.

4.2 Smooth Approximation

Since problem (1) is non-smooth, we introduce a smooth approximation of the non-smooth function f as follows:

$$f_\gamma(x) := \mathbb{E}_{\tilde{e}} \left[f(x + \gamma \tilde{e}) \right], \tag{3}$$

where $\gamma > 0$ is smoothing parameter, $\tilde{e}$ is random vector uniformly distributed on $B_2^d(\gamma)$. Here $f_\gamma(x) := \mathbb{E} \left[f(x, \xi) \right]$. The following lemma provides the connection between non-smooth function f and smoothed function f_γ.

Lemma 1. *Let Assumptions 1, 2 it holds, then for all $x \in Q$ we have*

$$f(x) \leq f_\gamma(x) \leq f(x) + \gamma M_2.$$

Proof. For the first inequality we use the convexity of the function $f(x)$

$$f_\gamma(x) = \mathbb{E}_{\tilde{e}}\left[f(x + \gamma\tilde{e})\right] \geq \mathbb{E}_{\tilde{e}}\left[f(x) + \langle \nabla f(x), \gamma\tilde{e}\rangle\right] = \mathbb{E}_{\tilde{e}}\left[f(x)\right] = f(x).$$

For the second inequality we have

$$|f_\gamma(x) - f(x)| = |\mathbb{E}_{\tilde{e}}\left[f(x + \gamma\tilde{e})\right] - f(x)| \leq \mathbb{E}_{\tilde{e}}\left[|f(x + \gamma\tilde{e}) - f(x)|\right]$$
$$\leq \gamma M_2 \mathbb{E}_{\tilde{e}}\left[\|\tilde{e}\|_2\right] \leq \gamma M_2,$$

using the fact that f is M_2-Lipschitz function.

$\square$

The next lemmas provide properties of the smoothed function f_γ.

Lemma 2. *Let Assumptions 1, 2 it holds, then for $f_\gamma(x)$ from (3) we have*

$$|f_\gamma(y) - f_\gamma(x)| \leq M\|y - x\|_p, \quad \forall x, y \in Q.$$

Proof. Using M-Lipschitz continuity of function f we obtain

$$|f_\gamma(y) - f_\gamma(x)| \leq \mathbb{E}_{\tilde{e}}\left[|f(y + \gamma\tilde{e}) - f(x + \gamma\tilde{e})|\right] \leq M\|y - x\|_p.$$

$\square$

Lemma 3 (Theorem 1, [22]). *Let Assumptions 1, 2 it holds, then $f_\gamma(x)$ has $L_{f_\gamma} = \frac{\sqrt{d}M}{\gamma}$-Lipschitz gradient*

$$\|\nabla f_\gamma(y) - \nabla f_\gamma(x)\|_q \leq L_{f_\gamma}\|y - x\|_p, \quad \forall x, y \in Q.$$

4.3 Gradient via l_2 Randomization

The gradient of $f_\gamma(x, \xi)$ can be estimated by the following approximation:

$$\nabla f_\gamma(x, \xi, e) = \frac{d}{2\gamma}\left(f_\delta(x + \gamma e, \xi) - f_\delta(x - \gamma e, \xi)\right)e, \tag{4}$$

where $f_\delta(x, \xi)$ is gradient-free oracle from Definition 1, e is a random vector uniformly distributed on $S_2^d(\gamma)$. The following lemma provides properties of the gradient $\nabla f_\gamma(x, \xi, e)$.

Lemma 4 (Lemma 2,[36]). *Gradient $\nabla f_\gamma(x, \xi, e)$ has bounded variance (second moment) for all $x \in Q$*

$$\mathbb{E}_{\xi, e}\left[\|\nabla f_\gamma(x, \xi, e)\|_q^2\right] \leq \kappa(p, d)\left(dM_2^2 + \frac{d^2\Delta^2}{\sqrt{2}\gamma^2}\right),$$

where $1/p + 1/q = 1$ and

$$\kappa(p, d) = \sqrt{2}\min\{q, \ln d\} d^{1 - \frac{2}{p}}.$$

Remark 1. Using the fact that the second moment is the upper estimate of the variance for the unbiased gradient and assuming that Δ is sufficiently small we obtain the following estimate of the variance from Lemma 4:

$$\sigma^2 \leq 2\sqrt{2}\min\{q, \ln d\} d^{2 - \frac{2}{p}} M_2^2.$$

4.4   Zero-Order Stochastic Conditional Gradient Sliding Method

Now we present gradient-free algorithm (see Algorithm 1) to solve problem (1). We chose Stochastic Conditional Gradient Sliding Method as accelerated batched Algorithm $\mathbf{A}(L_{f_\gamma}, \sigma^2)$. Substituting the approximation of the gradient via l_2 randomization $\nabla f_\gamma(x, \xi, e)$ ($\nabla f_\gamma(x, \psi)$ from Subsect. 4.1, where $\psi = (\xi, e)$ is not only the random value ξ, but also the randomization on the l_2-sphere e, which was introduced in Subsect. 4.3) instead of the exact gradient, we obtain a new ZO-SCGD Algorithm 1 to solve the non-smooth problem (1).

Algorithm 1 Zero-Order Stochastic Conditional Gradient Sliding (ZO-SCGS)

Input: Start point $x_0 \in Q$, maximum number of iterations $N \in \mathbb{Z}_+$.
    Let stepsize $\zeta_k \in [0, 1]$, learning rate $\eta_k > 0$, accuracies β_k, batch size $B_k \in \mathbb{Z}_+$,
    smoothing parameter $\gamma > 0$.
Initialization: Generate independently vectors $e_1, e_2, ...$ uniformly distributed on unit
    l_2-sphere, and set $y_0 \leftarrow x_0$
1: **for** $k = 1, ..., N$ **do**
2:     $z_k \leftarrow (1 - \zeta_k)x_{k-1} + \zeta_k y_{k-1}$
3:     Sample $\{e_1, ..., e_{B_k}\}$ and $\{\xi_1, ..., \xi_{B_k}\}$ independently
4:     $g_k \leftarrow \frac{1}{B_k} \sum_{i=1}^{B_k} \left[\frac{d}{2\gamma} (f(z_k + \gamma e_i, \xi_i) - f(z_k - \gamma e_i, \xi_i)) e_i \right]$
5:     $y_k \leftarrow \text{CG}(g_k, y_{k-1}, \eta_k, \beta_k)$ $\triangleright$ See CG in Algorithm 2
6:     $x_k \leftarrow (1 - \zeta_k)x_{k-1} + \zeta_k y_k$
7: **end for**
Output: x_N.

Algorithm 1 has such parameters as number of iterations N, batch size B, stepsize ζ, learning rate η, accuracies β. The recommendations for selecting these parameters can be found in Theorem 1. To prove theorem we also need to know the values of the following parameters: constant of Lipschitz gradient $L_{f_\gamma} = \frac{2\sqrt{d}MM_2}{\varepsilon}$ (by substituting $\gamma = \varepsilon/(2M_2)$ in Lemma 3), where constant of Lipschitz continuity M defined in Lemma 2 under Assumption 1, and estimate of the variance $\sigma^2 \leq 2\sqrt{2} \min\{q, \ln d\} d^{2-\frac{2}{p}} M_2^2$ (from Remark 1).

Next theorem provides estimates of the convergence rate of Algorithm 1.

Theorem 1. *Let ε be desired accuracy to solve problem (1) and γ be chosen as $\gamma = \varepsilon/(2M_2)$. Let function $f(x, \xi)$ satisfy the Assumptions 1, 2 and 3. Then Zero-Order Stochastic Conditional Gradient Sliding algorithm (see Algorithm 1) with $\zeta_k = 3/(k+3)$, $\eta_k = 8\sqrt{d}MM_2/(\varepsilon(k+3))$, $\beta_k = 2\sqrt{d}MM_2D^2/(\varepsilon(k+1)(k+2))$, and $B_k = \left\lceil \min\{q, \ln d\} d^{1-\frac{2}{p}}(k+3)^3 \varepsilon^2/(MD)^2 \right\rceil$ achieves $\mathbb{E}[f(x_k)] - f^* \leq \varepsilon$ after*

$$N = \mathcal{O}\left(\frac{d^{1/4}\sqrt{MM_2}D}{\varepsilon} \right), \quad T = \mathcal{O}\left(\frac{\min\{q, \ln d\} d^{2-\frac{2}{p}} M_2^2 D^2}{\varepsilon^2} \right)$$

number of iterations and gradient-free oracle calls respectively.

See Appendix A for detailed proof.

The results of Theorem 1 show that Zero-Order Stochastic Conditional Gradient Sliding algorithm converges with ε-accuracy in $N \sim d^{1/4}\varepsilon^{-1}$ iterations. The number of solutions to linear optimization problems, also known as the linear minimization oracle (LMO), is $\mathcal{O}\left(\sqrt{d}\varepsilon^{-2}\right)$. Batch size $B_k \in \mathbb{Z}_+$ must be chosen integer, so in Theorem 1 $\lceil \cdot \rceil$ denotes the whole part of the next integer number. The number of oracle calls T requiring the Algorithm 1 to solve a non-smooth problem (1) with ε-accuracy is $T \sim \min\{q, \ln d\}d^{2-\frac{2}{p}}\varepsilon^{-2}$.

Remark 2 (Smooth setting). In Theorem 1, we presented the convergence results of Algorithm 1 in the non-smooth setting, since in this paper we focus on solving non-smooth convex stochastic optimization problems. However, the algorithm proposed in this paper is robust to the smooth setting as well. To obtain similar estimates of the algorithm for smooth setting, it is sufficient not to change constant of Lipschitz gradient (i.e., it is not necessary to substitute the value obtained in Lemma 3). Therefore, Algorithm 1 with parameters $\zeta_k = 3/(k+3)$, $\eta_k = 4L/(k+3)$, $\beta_k = LD^2/((k+1)(k+2))$, and $B_k = \left\lceil \min\{q, \ln d\}d^{2-\frac{2}{p}}M_2^2(k+3)^3/(LD)^2 \right\rceil$ achieves $\mathbb{E}[f(x_k)] - f^* \leq \varepsilon$ after $N \sim \varepsilon^{-1/2}$ iterations, performs $\sim \varepsilon^{-1}$ linear optimization and requires $T \sim \min\{q, \ln d\}d^{2-\frac{2}{p}}\varepsilon^{-2}$ gradient-free oracle calls.

Remark 3. In Subsect. 4.4, we focus on obtaining optimal estimates of iterative N and oracle T complexities, so in proving the Theorem 1 we considered the case $\Delta = 0$. However, an optimal estimate of the maximum adversarial noise can be obtained by performing a similar convergence analysis of the Stochastic Conditional Gradient Sliding Method for the biased stochastic oracle (see example analysis in [25]). For brevity, we omit this analysis, stating that the estimate of maximum adversarial noise is $\Delta \lesssim \varepsilon^2 d^{-1/2}$ for gradient-free algorithms created by applying smoothing scheme via l_2 randomization (see e.g. [14,22]).

5   Discussion

As far as we know, Zero-Order Stochastic Conditional Gradient Sliding (ZO-SCGS) is the first gradient-free conditional gradient-type algorithm that solves a non-smooth convex stochastic optimization problem (1). This algorithm, as Theorem 1 shows, is robust for solving non-smooth black-box problems. But most interestingly, this algorithm is also robust for smooth black box problems, because it is superior in terms of the number of oracle calls to the state of the art algorithms. For instance, the Zeroth-Order Stochastic Conditional Gradient Method (ZSCG) from [5], which is a SOTA algorithm, has the following oracle complexity of $T \sim d\,\varepsilon^{-3}$ in any setting, while Algorithm 1 has oracle complexity of $T \sim d\,\varepsilon^{-2}$ in the Euclidean setting $p = 2$ $(q = 2)$ and $T \sim \ln(d)\,\varepsilon^{-2}$ in the simplex setting $p = 1$ $(q = \infty)$. One reason for the advantage of our algorithm may be that the ZSCG method uses Direct Finite Difference (FFD), while the

ZO-SCGS method (see Algorithm 1) uses Central Finite Difference (CFD). It is worth noting that [45] explains why it is worth estimating the gradient via central finite difference. Another possible reason may be the choice of Gaussian smoothing instead of smoothing via l_2 randomization, because in practical examples it often happens that the algorithm whose gradient is approximated over l_2 randomization works better than the algorithm whose gradient is approximated over Gaussian smoothing. Last but not least, a possible reason is that the Zeroth-Order Stochastic Conditional Gradient Method (ZSCG) used the unaccelerated first-order Stochastic Frank–Wolfe (SFW) method of [28] as its base. Since the Stochastic Conditional Gradient Method already has an estimate on the number of calls to the stochastic gradient as $\sim \varepsilon^{-3}$. It is hard to expect an improvement in estimate of oracle complexity when creating a gradient-free method based on it. Therefore, in this paper we created an optimal gradient-free method based on an accelerated batched first-order algorithm. So far we have observed theoretical advantages of Algorithm 1 (robust for solving non-smooth black box problems) in terms of oracle complexity over SOTA algorithms, which are robust for solving smooth black box problems. Therefore, in Sect. 6 we will verify our theoretical results with a model example of a convex stochastic optimization problem in a smooth setting.

6   Experiments

In this section we focus on verifying our theoretical results obtained in Sect. 4 via experiments[1]. In particular, we numerically compare the Zero-Order Stochastic Conditional Gradient Sliding Method (ZO-SCGS) proposed in this paper (see Algorithm 1) with the Zeroth-Order Stochastic Conditional Gradient Method (ZSCG) from [5]. We consider a standard model example of a black box problem in a smooth setting, which has the following form:

$$\min_{x \in Q} f(x) := \frac{1}{2}\langle x, Ax \rangle - \langle b, x \rangle,$$

where $Q = \{x \in \mathbb{R}^d : \|x\|_1 = 1, x \geq 0\}$ is a simplex set, $A \in \mathbb{R}^{d \times d}$ is a random positively determined matrix, $b \in \mathbb{R}^d$ is a vector such that $b = Ax_*$, and x_* is a solution to the problem $x_* = \arg\min_{x \in Q} f(x)$. In all tests, the dimensionality of the problem is $d = 100$, we fix the maximum number of calls to the gradient-free oracle $T_{max} = 10^7$, and the parameters of the algorithms are taken according to theoretical recommendations: for instance, parameters for Algorithm 1, see Remark 2, and parameters for Zeroth-Order Stochastic Conditional Gradient Method, see [5]. In Fig. 1 we compare the ZO-SCGS method with the ZSCG method. In particular, Fig. 1a shows the dependence of the optimal error $(f(x_k) - f^*)$ on the number of calls of the gradient-free oracle T. And Fig. 1b examines the dependence of the optimal error $(f(x_k) - f^*)$ on the number of iterations N. We observe that Algorithm 1 significantly outperforms the ZSCG method in the

---

[1] Code repository link: https://github.com/htower/zo-scgs.

number of oracle calls. Also, when the maximal value of the gradient-free oracle call is fixed, we see that the Algorithm 1 is first inferior to the ZSCG method in the number of iterations.

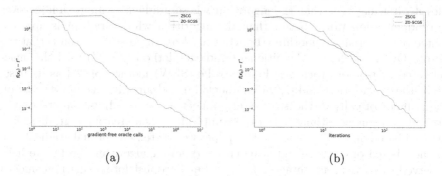

(a)   (b)

Fig. 1. Comparison of convergence result of Algorithm 1 with ZSCG method [5].

According to theoretical estimates for the ZO-SCGS and ZSCG methods, the batch size should be taken at a large size, which is a disadvantage of these algorithms. In Fig. 2 we compare Algorithm 1 with ZSCG methods using the fixed batch-size $B_k = 100$. Figure 2a shows the dependence of the optimal error $(f(x_k) - f^*)$ on the number of calls of the gradient-free oracle T. And Fig. 2b examines the dependence of the optimal error $(f(x_k) - f^*)$ on the number of iterations N. We see that for a fixed (small) batch size, both algorithms have convergence, which is a positive result for practical experiments to use. We also see that ZO-SCGS and ZSCG methods require the same number of calls to the gradient-free oracle, since we have fixed the batch size in contrast to Fig. 1. We can also observe that Algorithm 1 significantly outperforms the method both in the number of to gradient-free oracle calls and in iterations.

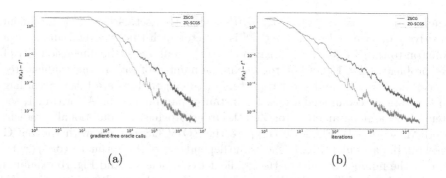

(a)   (b)

Fig. 2. Comparison of convergence result of algorithms with fixed batch size.

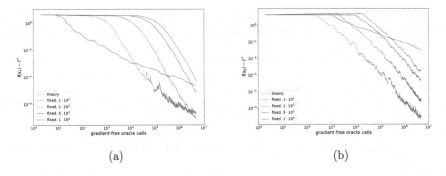

(a)   (b)

Fig. 3. Effect of the batch size parameter B_k on convergence results.

Figure 3 shows the effect of the batch size parameter B_k on convergence. Where 'theory' means that the batch size corresponds to theoretical estimates, 'fixed b' means that the batch size corresponds to the value of b. Figure 3a explores the dependence of Zero-Order Stochastic Conditional Gradient Sliding (ZO-SCGS) on batch size B_k, and Fig. 3b explores the dependence of the Zeroth-Order Stochastic Conditional Gradient Method (ZSCG) on batch size B_k. We see that theoretical estimates of the batch size slow down the convergence rate of both methods. And we can also observe a tendency that the smaller the batch size, the faster the algorithms converge. However, it is worth observing the golden mean, because with a very small batch size the positive convergence effect will not be observed, as well as with a very large batch size.

7 Conclusion

We presented, as far as we know, the first gradient-free algorithm of the conditional gradient type, which is robust for solving non-smooth convex stochastic optimization problems (black-box problems in a non-smooth setting). Using a smoothing scheme with l_2 randomization and basing on an accelerated batched first-order algorithm, we showed that Zero-Order Stochastic Conditional Gradient Sliding (ZO-SCGS) is the optimal algorithm for three criteria: total number of iterations, oracle complexity, and maximum adversarial noise. Our theoretical results show that Algorithm 1 is a robust method not only for non-smooth black box problems, but also for black box problems with a smooth setting. We verified our theoretical results on a practical experiment in a smooth setup by comparing our algorithm with the state of the art algorithm. We have shown that using a fixed (small enough) batch size achieves better accuracy than with batch size derived from theoretical estimates.

References

1. Akhavan, A., Pontil, M., Tsybakov, A.: Exploiting higher order smoothness in derivative-free optimization and continuous bandits. Adv. Neural. Inf. Process. Syst. **33**, 9017–9027 (2020)

2. Akhavan, A., Chzhen, E., Pontil, M., Tsybakov, A.B.: A gradient estimator via L1-randomization for online zero-order optimization with two point feedback. arXiv preprint arXiv:2205.13910 (2022)
3. Anikin, A., et al.: Efficient numerical methods to solve sparse linear equations with application to pagerank. Optim. Methods Softw. **37**(3), 907–935 (2022). https://doi.org/10.1080/10556788.2020.1858297
4. Bach, F., Perchet, V.: Highly-smooth zero-th order online optimization. In: Conference on Learning Theory, pp. 257–283. PMLR (2016)
5. Balasubramanian, K., Ghadimi, S.: Zeroth-order nonconvex stochastic optimization: handling constraints, high dimensionality, and saddle points. Found. Comput. Math., 1–42 (2022)
6. Braun, G., et al.: Conditional gradient methods. arXiv preprint arXiv:2211.14103 (2022)
7. Bubeck, S.: Convex optimization: algorithms and complexity. Found. Trends Mach. Learn. **8**(3–4), 231–357 (2015). https://doi.org/10.1561/9781601988614
8. Bubeck, S., Jiang, Q., Lee, Y. T., Li, Y., Sidford, A.; Complexity of highly parallel non-smooth convex optimization. In: Advances in Neural Information Processing Systems, vol. 32 (2019)
9. Chen, L., Zhang, M., Karbasi, A.: Projection-free bandit convex optimization. In: The 22nd International Conference on Artificial Intelligence and Statistics, pp. 2047–2056. PMLR (2019)
10. Combettes, C. W., Spiegel, C., Pokutta, S.: Projection-free adaptive gradients for large-scale optimization. arXiv preprint arXiv:2009.14114 (2020)
11. Cox, B., Juditsky, A., Nemirovski, A.: Decomposition techniques for bilinear saddle point problems and variational inequalities with affine monotone operators. J. Optim. Theory Appl. **172**, 402–435 (2017). https://doi.org/10.1007/s10957-016-0949-3
12. Dadras, A., Prakhya, K., Yurtsever, A.: Federated frank-wolfe algorithm. In: In Workshop on Federated Learning Recent Advances and New Challenges (in Conjunction with NeurIPS) (2022)
13. Duchi, J., Hazan, E., Singer, Y.: Adaptive subgradient methods for online learning and stochastic optimization. J. Mach. Learn. Res. **12**(7) (2011)
14. Dvinskikh, D., Tominin, V., Tominin, I., Gasnikov, A.: Noisy zeroth-order optimization for non-smooth saddle point problems. In: Pardalos, P., Khachay, M., Mazalov, V. (eds.) Mathematical Optimization Theory and Operations Research. Lecture Notes in Computer Science, vol. 13367, pp. 18–33. Springer, Cham (2022). https://doi.org/10.1007/978-3-031-09607-5_2
15. Dvurechensky, P., Gorbunov, E., Gasnikov, A.: An accelerated directional derivative method for smooth stochastic convex optimization. Eur. J. Oper. Res. **290**(2), 601–621 (2021). https://doi.org/10.1016/j.ejor.2020.08.027
16. Frank, M., Wolfe, P.: An algorithm for quadratic programming. Naval Res. Logistics Q. **3**(1–2), 95–110 (1956). https://doi.org/10.1002/nav.3800030109
17. Freund, R.M., Grigas, P., Mazumder, R.: An extended Frank-Wolfe method with "in-face" directions, and its application to low-rank matrix completion. SIAM J. Optim. **27**(1), 319–346 (2017). https://doi.org/10.1137/15m104726x
18. Garber, D., Kretzu, B.: Improved regret bounds for projection-free bandit convex optimization. In: International Conference on Artificial Intelligence and Statistics, pp. 2196–2206. PMLR (2020)
19. Garber, D., Wolf, N.: Frank-Wolfe with a nearest extreme point oracle. In: Conference on Learning Theory, pp. 2103–2132. PMLR (2021)

20. Garber, D.: Linear convergence of Frank-Wolfe for rank-one matrix recovery without strong convexity. Math. Program. **199**, 1–35 (2022). https://doi.org/10.1007/s10107-022-01821-8

21. Gasnikov, A., Gasnikova, E.: Traffic assignment models. Numerical aspects. arXiv preprint arXiv:2003.12160 (2020)

22. Gasnikov, A., et al.: The power of first-order smooth optimization for black-box non-smooth problems. arXiv preprint arXiv:2201.12289 (2022)

23. Gasnikov, A., Dvinskikh, D., Dvurechensky, P., Gorbunov, E., Beznosikov, A., Lobanov, A.: Randomized gradient-free methods in convex optimization. arXiv preprint arXiv:2211.13566 (2022)

24. Goldfarb, D., Iyengar, G., Zhou, C.: Linear convergence of stochastic frank Wolfe variants. In: Artificial Intelligence and Statistics, pp. 1066–1074. PMLR (2017)

25. Gorbunov, E., Dvinskikh, D., Gasnikov, A.: Optimal decentralized distributed algorithms for stochastic convex optimization. arXiv preprint arXiv:1911.07363 (2019)

26. Gorbunov, E., Danilova, M., Gasnikov, A.: Stochastic optimization with heavy-tailed noise via accelerated gradient clipping. Adv. Neural. Inf. Process. Syst. **33**, 15042–15053 (2020)

27. Guélat, J., Marcotte, P.: Some comments on Wolfe's 'away step'. Math. Program. **35**(1), 110–119 (1986). https://doi.org/10.1007/bf01589445

28. Hazan, E.: Introduction to online convex optimization. Found. Trends® Optim. **2**(3–4), 157–325 (2016). https://doi.org/10.1561/2400000013

29. Hazan, E., Luo, H.: Variance-reduced and projection-free stochastic optimization. In: International Conference on Machine Learning, pp. 1263–1271. PMLR (2016)

30. Jaggi, M.: Sparse convex optimization methods for machine learning (No. ETH Zürich, THESIS LIB) (2011)

31. Kiefer, J., Wolfowitz, J.: Stochastic estimation of the maximum of a regression function. Ann. Math. Statist., 462–466 (1952). https://doi.org/10.1214/aoms/1177729392

32. Lacoste-Julien, S., Jaggi, M.: On the global linear convergence of Frank-Wolfe optimization variants. In: Advances in Neural Information Processing Systems, vol. 28 (2015)

33. Lan, G., Zhou, Y.: Conditional gradient sliding for convex optimization. SIAM J. Optim. **26**(2), 1379–1409 (2016). https://doi.org/10.1137/140992382

34. Levitin, E.S., Polyak, B.T.: Constrained minimization methods. USSR Comput. Math. Math. Phys. **6**(5), 1–50 (1966)

35. LeBlanc, L.J., Helgason, R.V., Boyce, D.E.: Improved efficiency of the Frank-Wolfe algorithm for convex network programs. Transp. Sci. **19**(4), 445–462 (1985). https://doi.org/10.1287/trsc.19.4.445

36. Lobanov, A., Alashqar, B., Dvinskikh, D., Gasnikov, A.: Gradient-Free Federated Learning Methods with l_1 and l_2-Randomization for Non-Smooth Convex Stochastic Optimization Problems. arXiv preprint arXiv:2211.10783 (2022)

37. McMahan, H.B., Streeter, M.: Adaptive bound optimization for online convex optimization. arXiv preprint arXiv:1002.4908 (2010)

38. Mitradjieva, M., Lindberg, P.O.: The stiff is moving-Conjugate direction Frank-Wolfe methods with applications to traffic assignment. Transp. Sci. **47**(2), 280–293 (2013). https://doi.org/10.1287/trsc.1120.0409

39. Mokhtari, A., Hassani, H., Karbasi, A.: Stochastic conditional gradient methods: From convex minimization to submodular maximization. J. Mach. Learn. Res. **21**(1), 4232–4280 (2020)

40. Négiar, G., et al.: Stochastic Frank-Wolfe for constrained finite-sum minimization. In: International Conference on Machine Learning, pp. 7253–7262. PMLR (2020)

41. Risteski, A., Li, Y.: Algorithms and matching lower bounds for approximately-convex optimization. In: Advances in Neural Information Processing Systems, vol. 29 (2016)
42. Rosenbrock, H.: An automatic method for finding the greatest or least value of a function. Comput. J. **3**(3), 175–184 (1960). https://doi.org/10.1093/comjnl/3.3.175
43. Saha, A., Tewari, A.: Improved regret guarantees for online smooth convex optimization with bandit feedback. In: Proceedings of the Fourteenth International Conference on Artificial Intelligence and Statistics, pp. 636–642. JMLR Workshop and Conference Proceedings (2011)
44. Stich, S.U., Karimireddy, S.P.: The error-feedback framework: better rates for sgd with delayed gradients and compressed updates. J. Mach. Learn. Res. **21**(1), 9613–9648 (2020)
45. Scheinberg, K.: Finite difference gradient approximation: to randomize or not? INFORMS J. Comput. **34**(5), 2384–2388 (2022). https://doi.org/10.1287/ijoc.2022.1218
46. Vasin, A., Gasnikov, A., Spokoiny, V.: Stopping rules for accelerated gradient methods with additive noise in gradient, vol. 2812, no. 2812. Weierstraß-Institut für Angewandte Analysis und Stochastik, Berlin (2021)
47. Vaswani, S., Bach, F., Schmidt, M.: Fast and faster convergence of sgd for overparameterized models and an accelerated perceptron. In :The 22nd International Conference on Artificial Intelligence and Statistics, pp. 1195–1204. PMLR (2019)
48. Yousefian, F., Nedić, A., Shanbhag, U.V.: On stochastic gradient and subgradient methods with adaptive steplength sequences. Automatica **48**(1), 56–67 (2012). https://doi.org/10.1016/j.automatica.2011.09.043
49. Yurtsever, A., Sra, S., Cevher, V.: Conditional gradient methods via stochastic path-integrated differential estimator. In: International Conference on Machine Learning, pp. 7282–7291. PMLR (2019)
50. Zhang, M., Shen, Z., Mokhtari, A., Hassani, H., Karbasi, A.: One sample stochastic frank-Wolfe. In: International Conference on Artificial Intelligence and Statistics, pp. 4012–4023. PMLR (2020)

Discrete and Combinatorial Optimization

Tabu Search Metaheuristic
for the Penalty Minimization Personnel
Task Scheduling Problem

Ivan Davydov[1]([✉])(ID), Igor Vasilyev[2,3](ID), and Anton V. Ushakov[2](ID)

[1] Sobolev Institute of Mathematics of SB RAS, 4 Acad. Koptyug av.,
630090 Novosibirsk, Russia
iadavydov@math.nsc.ru
[2] Matrosov Institute for System Dynamics and Control Theory of SB RAS,
134 Lermontov str., 664033 Irkutsk, Russia
{vil,aushakov}@icc.ru
[3] Novosibirsk Research Center, Huawei Russian Research Institute,
630090 Novosibirsk, Russia

Abstract. Personnel scheduling is an active research field motivated by not only economic considerations but also the understanding of importance of improving working conditions and fairness in assigning employees to tasks. A large number of daily tasks and an expanding staff require effective automation of the task allocation process. In the problem under consideration, given sets of tasks and staff, it is required to assign the certain number of employees to each task, taking into account their skills. The goal is to minimize penalties induced by conflicting assignments as well as by uneven workload of the staff. To solve the problem, a two-phase heuristic consisting of a greedy heuristic followed by a randomize tabu search has been developed. Computational experiments shows that the proposed approach allows us to find optimal or near-optimal solutions on instances corresponding to real-life problems.

Keywords: crew scheduling · personnel task scheduling · tabu search · MIP formulation · greedy algorithm · rostering

1 Introduction

Personnel scheduling problems have been attracting a growing attention from both research community and industry in the last decades [3,8,11,12,35]. The active research in this field is motivated by not only economic considerations but also the understanding of importance of improving working conditions and fairness in assigning employees to tasks. This problem is common for many companies regardless of the branch of economy. Personnel costs may make up a

The study of the first author was carried out within the framework of the state contract of the Sobolev Institute of Mathematics (project FWNF-2022-0019). The research of the third author was funded by the Ministry of Education and Science of the Russian Federation No. 121041300065-9.

M. Khachay et al. (Eds.): MOTOR 2023, CCIS 1881, pp. 109–121, 2023.
https://doi.org/10.1007/978-3-031-43257-6_9

significant part of operational expenses, hence it is of utter importance to schedule the workforce in a most efficient way in order to reduce the corresponding costs. At the same time, if work schedules take into account only costs, then employees may often be unsatisfied. This in turn may provoke the turnover, increase work-related fatigue [37], substantially decrease productivity, and even cause staff absenteeism.

Though personnel scheduling problems may include different objectives and constraints, they all suppose that there is a set of staff members or employees, who must be assigned to a set of tasks or shifts subject to specific industrial constraints. In particular, unique schedule requirements specific to many industries and business enterprises demand appropriate mathematical models and algorithms. The scheduling process is quite complex and involves many sub-tasks like modelling demand in workforce over the planning horizon, planning days off, constructing shifts (what tasks must be executed and by what number of employees in order to satisfy demand), task assignment, etc. [12].

Due to a large number of variations and goals of personnel scheduling, there is a huge amount of research in this direction. As the problem is highly application-oriented, most of the related research papers address specific problems arisen in the corresponding industry. For example, a huge body of the literature deals with personnel scheduling in the transportation industry, e.g. railways [20] and airlines [19,22,36], where it is often referred to as crew scheduling. Another traditional, vital application is to roster shifts for nurses and physicians [6,10]. Actually, the nurse rostering problem is one of the traditional widely addressed personnel scheduling problems dated back to 1960s.

Personal scheduling is usually carried out in three principal stages: (i) staffing, (ii) shift scheduling, and (iii) rostering [4,32]. They may be considered separately (one by one) or combined. The first one is to determine the number of personnel of the required skills sufficient to meet the requirements during a given time interval. The second stage aims at identifying shifts or tasks and determining the required skills and the number of staff members of those skills needed to meet the workforce demand. The last stage is to schedule all the employees over the time horizon, i.e. assign them to the shifts subject to specific workplace rules (for example, the number of days off).

In this paper, we consider a personnel rostering problem that is supposed to be solved on a daily basis. Each day, a new set of tasks is allocated. They must be executed by a set of employees or staff members. The set of employees is divided into several subsets. Employees in the same subset have the same skills, i.e. they are able to execute the same tasks. A task is assumed to be completed if the required number of staff members is assigned to it. Some limited amount of tasks can be failed or left uncompleted, i.e. there are no employees assigned to these tasks. This number is given and is a parameter of the problem. One employee can be assigned to several tasks. However, some assignments may induce penalties, e.g. if tasks have the same or close starting time, or they are far away from each other (spaced in time). or some of the tasks require violating day off requirements, etc. The problem assumes additional constraints on the workload of the employee subsets, i.e. the workload (the number of tasks assigned) of all employees in the same group should be as close as possible. Difference in workload is also penalized.

The aim of the problem is to assign staff to all necessary tasks minimizing the total penalty value, i.e. minimize the penalty induced by assigning staff members to conflicting pairs of tasks and by unbalanced workload of the employee subsets. One can see that the problem relates to the third stage of the personal scheduling process, and the outputs of the two previous stages are supposed to be known. As far as we know, exactly the same personnel rostering problem has not been considered in the literature.

Real-life instances of the problem can involve more than 250 employees and more than 1000 tasks (which leads to huge amounts of variables in the model) are hard to solve by a general MIP solver. Such instances appear to be hard to solve by a general MIP solvers. To obtain high-quality feasible solutions, the problem requires developing a problem specific solution approach.

As far as we know, the aforementioned problem can be viewed as a variation and extension of several problems proposed in the literature. First of all, the problem is closely related to the fixed job scheduling problem, where there is a set of tasks with given starting and ending times that must be executed nonpreemptively. Given a set of identical processors that can perform one task per time, one has to assign tasks to processors in a feasible manner in order to minimize the number of processors. A very similar problem, called the heterogeneous workforce scheduling problem or the numbers of workers minimization problem, was addressed in [34]. Here, an additional requirement is that each task can be assigned not to all processors but only to some pre-specified subset. Apart from minimizing the number of employed processors, the authors propose to use several other objective functions, e.g. cost minimization or making the best use of a given workforce. In [14], the authors extend the conventional fixed job scheduling problem by introducing several specific constraints related to the bus transportation industry. Another closely related problem was studied in [1,27], where each job is assigned a weight. As opposed to the previous problem, the goal is to minimize the total value of the jobs left uncompleted.

Recently, the problem from [34] was reintroduced in [26] as the shift minimization personnel task scheduling problem (SMPTSP). Indeed, the problem objective is to minimize the number of personnel assigned to complete a given set of tasks with starting time and ending time provided. Each staff member is supposed to have skills to perform only some subset of tasks. The authors claim that such a problem is a variant of a more general personnel task scheduling problem [25], which can incorporate various objectives. The authors develop a Lagrangian heuristic and Wedelin's type algorithm to find quality solutions for large scale instances. In [33] the authors proposed a two-stage heuristic for SMPTSP composed of three types of constructive heuristics followed by a local branching heuristic. Other solution approaches to SMPTSP include a three-stage metaheuristic [30], a relax-and-fix heuristic [5], 3-phase algorithm based on the iterated greedy technique [29], a decomposition-based greedy algorithm [21], a constraint programming technique [13], etc.

Note that the aforementioned problems and the problem studied in this paper can also be viewed as a interval scheduling problem. A thorough survey of such problems is given in [24].

A more general variant of SMPTSP is studied in [28] where the goal is to design shifts and assign staff members in order to achieve the equity objective function, i.e. minimize the difference between obtained and targeted workload of employees.

There are also some general solution frameworks suitable for solving a wide range of different employee scheduling problems [23].

An example of employee workload balancing aspect is addressed in [9] for airline crew rostering. The goal is to minimize the total deviation of working time obtained from the standard working time.

For the introduced personnel scheduling problem, we propose a two-phase approach consisting of a greedy constructive heuristic followed by a randomized Tabu Search approach (TS). Note that TS is considered as one of the most popular metaheuristics widespread in diverse applications. The idea of TS was presented and formalized by Glover [16–18]. Comprehensive surveys about the approach details and its numerous applications can be found in [15,31].

The paper is structured as follows. The problem statement and its MIP formulation is described in Sect. 2. The details of the solution approach are given in Sect. 3. The effectiveness and efficiency of the proposed approach are illustrated in a series of computational experiments on real life instances in Sect. 4.

2   Problem Statement and MIP Formulation

In this section, we provide a mathematical formulation of the proposed personnel rostering problem. Recall that there is a set of tasks I and a set of staff members S (divided into groups G) who have skills to execute only some subset of tasks. Each task has the starting time and ending time and requires a certain number of staff members to be completed. A pre-specified number of tasks can be failed. A staff member can be assigned to multiple tasks, although some assignments lead to penalty. Penalty value becomes higher if tasks starting times are close to each other or the time space between tasks is large enough. In both cases, an employee wouldn't be able to complete both task in time and will cause a delay. Note that we suppose that the penalties are already computed using an interval graph that defines conflicting tasks. Fairness aspect also plays an important role in big teams. It is preferable if all the staff members within a group have the same workload.

First of all, let us introduce the following additional notations:

1. G is a set of groups of staff,
2. a_i denote the number of staff members required to complete task i, $a_i \in \mathbb{Z}_{\geq 1}$,

3. S_i is a set of staff from S that can be assigned to task $i \in I$ (have skills to execute the task), $S_i \subset S$.
4. w_{ij} is a penalty for taking both tasks i and j. Penalty values are computed based on time space between tasks and their execution time. $w_{ij} \in \mathbb{Z}_{\geq 0}$, $i, j \in I$.
5. K is the number of tasks that can be failed.
6. C is a penalty for unbalanced load in the group.

Let us introduce the binary variables:

$$x_{is} = \begin{cases} 1, & \text{if staff member } s \text{ is assigned to task } i, \\ 0, & \text{otherwise.} \end{cases}$$

$$z_{sij} = \begin{cases} 1, & \text{if staff member } s \text{ is assigned to conflicted pair of tasks } i \text{ and } j, \\ 0, & \text{otherwise.} \end{cases}$$

$$d_i = \begin{cases} 1, & \text{if task } i \text{ is failed,} \\ 0, & \text{otherwise.} \end{cases}$$

We also introduce the variable y_{gmin} equals minimal load of the staff member in group g and variable y_{gmax} equals maximal load of the staff member in group g.

With these notations, the problem is

$$\min \left(\sum_{i,j \in I} \sum_{s \in S} w_{ij} z_{sij} + C \sum_{g \in G} (y_{gmax} - y_{gmin}) \right) \tag{1}$$

$$\sum_{s \in S_i} x_{is} = (1 - d_i) a_i, \qquad\qquad i \in I, \tag{2}$$

$$\sum_{i \in I} d_i \leq K, \tag{3}$$

$$z_{sij} \geq x_{is} + x_{js} - 1, \qquad\qquad i, j \in I, s \in S_i \cap S_j, \tag{4}$$

$$y_{gmax} \geq \sum_{i \in I} \sum_{s \in g} x_{is}, \qquad\qquad g \in G, \tag{5}$$

$$y_{gmin} \leq \sum_{i \in I} \sum_{s \in g} x_{is}, \qquad\qquad g \in G, \tag{6}$$

$$z_{sij}, x_{is}, d_i \in \{0, 1\}, \qquad\qquad i, j \in I, s \in S, \tag{7}$$

$$y_{gmax}, y_{gmin} \in \mathbb{Z}_+, \qquad\qquad g \in G. \tag{8}$$

Goal function (1) calculates the total penalty induced by the staff assignment plus the total penalty for unequal workload within each group. Set of constraints (2) forces to assign the required number of staff members to each task, if the task is not failed. Constraint (3) allows to fail at most K tasks. Inequalities (4) bind variables x_{is} and z_{sij}, i.e. they force $z_{sij} = 1$ if both assignments x_{is} and x_{js} are made. Finally, inequalities (5) and (6) define the values of y_{gmin} and y_{gmax} variables.

The considered problem has the same basic structure as many rostering problems (e.g. [2]) and appears to be not tractable by MIP solvers due to high number of variables and specific landscape of the goal function. An instance of the problem becomes even more challenging when the total amount of tasks is moderate, thus allowing allocating the staff members without conflicts. The reason is probably in the presence of numerous allocations with the same goal function value that happens due to the high symmetry level of the model. We note that although the statement of the problem implies integer values of all variables, one can omit these constraints for z_{sij}, y_{gmax} and y_{gmin} variables for $i, j \in I, s \in S, g \in G$. We also note that the dimension of the problem can be sufficiently reduced on instances, where the w_{ij} matrix is sparse, as there is no need to introduce z_{sij} variable and the corresponding constraint (4) if the value of w_{ij} is zero. However, even after these simplifications, the problem still remains difficult for MIP solvers.

3   Solution Approach

In this section, we will describe a heuristic approach to solve the problem. The first phase of the approach is devoted to the construction of the initial solution. To this end, we propose a greedy constructive heuristic which performs as follows. It begins the construction of a solution from scratch. In each step, it picks a task from the task set in a predefined order and seeks for a staff member who is suited most for the task. To this end, for each candidate, it calculates the penalty value that will be induced by the assignment of the candidate to this task. If there are several candidates with the same minimal penalty value, the task is assigned to the less loaded candidate, i.e. the one whose number of already assigned tasks is the smallest.

Algorithm 1. Constructive greedy heuristic

1: **for** $i \in I$ **do**
2:     **for** $s \in S_i$ **do**
3:         calculate penalty P_{is} for x_{is}
4:         **if** $P_{is} \leq bestval$ **then**
5:             $bestcand = \emptyset$, $bestcand \leftarrow s$, $bestval = P_{is}$
6:         **end if**
7:         **if** $P_{is} = bestval$ **then**
8:             $bestcand \leftarrow s$
9:         **end if**
10:     **end for**
11:     assign s from $bestcand$ with smallest load to task i
12: **end for**

The procedure can be easily randomized to provide a variety of outputs by disturbing the sequence of the tasks in the outer loop. This allows us to run the procedure several times with different input task sequences and choose the best solution found as an output.

Note that at this step a simplified version of the goal function is used, i.e. we do not consider the second therm directly. The procedure is focused on minimization of the penalty value induced by task assignments, while the workload balancing is carried out only indirectly.

If the value of $K > 0$, i.e. a number of tasks can be omitted is non-zero, the presented approach is followed by a similar post-processing procedure: during at most K steps, we choose the task with the largest penalty value and exclude it from the solution. The procedure terminates when either K tasks are excluded from the solution or the total penalty value is zeroed. Although there is no straightforward motivation in the model to complete as many tasks as possible, such a solution will always be preferable.

3.1   Tabu Search

In order to improve the solution S_0 found by the greedy algorithm, it is fed as a starting point for the Tabu search approach. We develop a simple and fast variant of tabu search that employs a randomized neighborhood search and a simple tabu list structure. A general outline of our procedure is presented in Algorithm 2. Here $Z(S)$ is the goal function value of the solution S.

Algorithm 2. Randomized Tabu Search

1: Initialization: $S^* \leftarrow S_0$; $S \leftarrow S_0$; $tabuList \leftarrow S_0$.
2: **while** Stopping Condition does not hold **do**
3:     Generate a randomized neighborhood $N_s \leftarrow getNeighbors(S)$.
4:     **for** $(S' \in N_s)$ **do**
5:         **if** $(S' \notin tabuList) \wedge (Z(S') < Z(S))$ **then**
6:             $S \leftarrow S'$
7:         **end if**
8:     **end for**
9:     **if** $Z(S) < Z(S^*)$ **then**
10:         $S^* \leftarrow S$
11:     **end if**
12:     Update $tabuList$
13: **end while**
14: **return** S^*.

We use two types of neighborhoods. The first one, *Swap* neighborhood, consists of all solutions which are obtained from the incumbent by reassigning one task of one staff member to another staff member. All solutions in such neighborhood have the same number of completed tasks. The second one, *Switch* neighborhood, is constructed from the incumbent solution by "switching off" one (i.e. we decide to skip this task and do not assign any employee to it) of the tasks and "switching on" another one. This neighborhood is used only if $K > 0$. We note that both neighbourhoods do not change the number of completed tasks. This number is defined and fixed during the run of a greedy heuristic. In order to reduce the calculation time, we exploit the idea of randomized neighbourhood. During the search, we examine not the whole set of neighbouring solutions, but only a small random subset, i.e. each solution from the *Swap* or *Switch* neighbourhoods is examined only with a given probability $p_n > 0$. It is known that such a trick allows one to sufficiently reduce the calculation time without any significant reduction in solution quality [7].

As was noted above, we use a simplified version of the tabu list to reduce the computational efforts on checking whether a solution is contained in the list or not. Thus, tabu list contains not the entire solutions, but only last moves, i.e. the task that has been reassigned to another staff or "switched off" becomes "tabooed" for a number of consecutive iterations. We also note that in our tabu search procedure, we use a modified version of the objective function to overcome the high symmetry of the model. To this end, we introduce another term to the objective function that evaluates not only the difference in load between the members of a group but also the amount of members whose load is different from the biggest load and smallest load within the group. I.e. for each group with imperfect workload balance we calculate $s_{gdiff} = \min n_g - n_{gmin}, n_g - n_{gmax}$, where n_g is the size of the group g, while n_{gmin} and n_{gmax} define the number of staff members in the group with minimal and maximal load correspondingly. These values are then added to the goal function during the evaluation of the neighbouring solution. Such modification allows us to drive the search process towards more balanced solutions, as it tends to reduce the number of staff members with extremal workload values.

4   Computational Experiments

In this section, we present the results of computational experiments, performed on the data samples related to a real-life personnel scheduling.

In the first series of experiments, we consider the data set, which consists of 7 instances of the problem: **test0, test2, test3, test4, test5, test6, test9**. These instances are related to a planning of a huge international airport. The dimensions of the instances are presented in Table 1. Bigger instances (test6, test9) are related to a full 24 h schedule of the airport, while smaller ones represent some parts of the day. Total load here refers to the sum of all a_i values, i.e. the total number of required assignments to be made.

Table 1. Instance size

Param	test0	test2	test3	test4	test5	test6	test9
Num of staff	18	18	18	249	249	249	64
Num of groups	2	2	2	6	6	6	4
Num of tasks	10	10	30	151	591	1385	1465
Total load	15	15	30	250	975	2215	2327

4.1 MIP Solvers

Using the formulation proposed above, we tried to solve the instances with two well-known MIP solvers: free-distributed SCIP solver and commercial IBM ILOG CPLEX. The results are as follows. In 1 h of running time both solvers were able to solve to optimality instances **test0**, **test2**, which are of small size (10 tasks, 18 staff members). Instance **test3** appeared to be infeasible: 11 out of 30 tasks have no performers listed. If we exclude these tasks, the problem can also be solved optimally by both solvers in less than 1 sec.

Remaining 4 instances appeared to be more complex for MIP solvers. Indeed, in 1 h of running time optimal solution was found only for **test5** instance and only by the CPLEX solver. The scope of the obtained results is presented in the Table 2. Instances $test9\_0$, $test9\_10\_4$ and $test9\_10\_5$ refer here to the same instance $test9$ but with a different penalty for imbalanced workload - 0, 10^4 and 10^5 respectively. In all other instances, this value equals 10^4. Due to the extremely overloaded schedule in $test9$, in this instance different penalty values provide a sufficient difference in solutions obtained. Scope of these results are presented in Table 2. Here $nvars$ and $ncons$ provides the number of variables and constraints in the corresponding MIP model. Column $non-zeroes$ gives the information of the density of the constraint matrix, the number of non-zero elements. The upper bound value, obtained by SCIP solver, is given in S_ub. The lower and upper bounds, provided by CPLEX solver are contained in CP_lb and CP_ub correspondingly.

Table 2. Results obtained with the MIP solvers

Name	nvars	ncons	non-zeroes	S_ub	CP_ub	CP_lb
test4	174479	274883	898079	26668	3516	0
test5	174479	274883	898079	26668	10000 (opt)	10000 (223s)
test6	2236247	3785121	1893253	2770472	141600	0
test9_0	833601	1481403	4603812	5695152	5141160	0
test9_10_4	833601	1481403	4603812	5695152	1732004	0
test9_10_5	833601	1481403	4603812	5695152	48350040	0

Although we observe that CPLEX is able to provide much better upper bounds than SCIP, the lower bounds found by both solvers in 1 h remain zeroes, thus making the $GAP \geq 100\%$.

4.2 Tabu Search

In this subsection, we provide the results obtained with Tabu search approach. During the initial testing, we set the following values of the parameters. We set $p_n = 0.05$, thus we exploit only 5% of the neighborhood. Length of the tabu list is set to 7 iterations. The termination criterion is the running time in seconds. In particular, we set it to 10 s.

The approach is able to find the optimal solutions for instances **test0, test2, test3*** even if we reduce the calculation time to 1 s. Results, obtained for bigger instances are presented in Table 3. Columns *ntasks, nstaff, load* represent the size of the instance, i.e. the number of tasks, the number of staff members available, and the total amount of workload. The latter is the sum of task requirements. *CFvalue* represents the total conflict penalty induced by assignments in the resulting solution. *Imb* shows the total imbalance in working groups, and *ImbP* is the penalty for one unit of imbalance in the goal function. Column *total* is the sum of *CFvalue* and *ImbP*, which is the goal function value. Column *totalGR* is added to the table for comparison. It shows the goal function value of the solution, obtained with the greedy approach. Finally, the MIP column contains the best upper bound for the problem found within an hour by a MIP Solver. We note that, although the presented approach is randomized, and thus does not guarantee the same output on different runs with the same input data, during this experiment, we observed identical results during 20 runs on each instance. The calculation time provided is also seems to be enough as the TS approach tends to find the best known solution in less than 2 s of computational time. As it can be observed from the table, for instance *test4* our approach was unable to find the optimal solution. The best solution found so far contains no conflicts, but one group of workers has an imbalanced load, while the solution, found by CPLEX solver is perfectly balanced, but contains a number of conflicting assignments. Test instance 5 is solved optimally by both MIP solver and our approach. On all other instances, the proposed approach significantly outperforms the commercial CPLEX solver. In our opinion, it happens due to the significant growth of the dimension of the instance and thus the number of variables and constraints in the model. On these instances, even an initial solution, obtained with greedy approach, appears to be better, although it still can be significantly improved by local search.

Table 3. TS results, 10 sec run time

Name	ntasks	nstaff	load	CFvalue	Disb	DisbP	totalGR	total	MIP
test4	249	151	250	0	1	10000	40000	10000	3516
test5	249	591	975	0	1	10000	30000	10000	10000
test6	249	1385	2215	43500	3	10000	130580	73500	141600
test9	64	1465	2327	1198920	27	0	1495860	1198920	5141160
test9	64	1465	2327	1222560	9	10000	1685860	1312560	1732004
test9	64	1465	2327	1264320	8	100000	3395860	1364320	48350040

5   Conclusion

In this paper, we consider a new personnel rostering problem aimed at minimizing the total penalty value induced by assignments of employees to specific pairs of tasks and by imbalanced workload in the groups of employees. We propose an integer programming formulation of this problem. As real-life problems instances turn out to be not tractable by general MIP solvers, we developed a fast two-phase solution approach based on a randomized tabu search heuristic.

Our future research may be focused on extending the proposed problem formulation by incorporating new objectives or considering joint shift scheduling and personnel rostering.

References

1. Arkin, E.M., Silverberg, E.B.: Scheduling jobs with fixed start and end times. Discret. Appl. Math. **18**(1), 1–8 (1987). https://doi.org/10.1016/0166-218X(87)90037-0

2. Azadeh, A., Farahani, M.H., Eivazy, H., Nazari-Shirkouhi, S., Asadipour, G.: A hybrid meta-heuristic algorithm for optimization of crew scheduling. Appl. Soft Comput. **13**(1), 158–164 (2013). https://doi.org/10.1016/j.asoc.2012.08.012, https://www.sciencedirect.com/science/article/pii/S1568494612003596

3. Brucker, P., Qu, R., Burke, E.: Personnel scheduling: models and complexity. Eur. J. Oper. Res. **210**(3), 467–473 (2011). https://doi.org/10.1016/j.ejor.2010.11.017

4. Burke, E.K., Causmaecker, P.D., Berghe, G.V., Landeghem, H.V.: The state of the art of nurse rostering. J. Sched. **7**, 441–499 (2004). https://doi.org/10.1023/B:JOSH.0000046076.75950.0b

5. Chandrasekharan, R.C., Smet, P., Wauters, T.: An automatic constructive matheuristic for the shift minimization personnel task scheduling problem. J. Heuristics **27**, 205–227 (2021). https://doi.org/10.1007/s10732-020-09439-9

6. Cheang, B., Li, H., Lim, A., Rodrigues, B.: Nurse rostering problems - a bibliographic survey. Eur. J. Oper. Res. **151**(3), 447–460 (2003). https://doi.org/10.1016/S0377-2217(03)00021-3

7. Davydov, I., Kochetov, Y., Dempe, S.: Local search approach for the competitive facility location problem in mobile networks. Int. J. Artif. Intell. **16**(1), 130–143 (2018)

8. De Bruecker, P., Van den Bergh, J., Beliën, J., Demeulemeester, E.: Workforce planning incorporating skills: state of the art. Eur. J. Oper. Res. **243**(1), 1–16 (2015). https://doi.org/10.1016/j.ejor.2014.10.038

9. Doi, T., Nishi, T., Voss, S.: Two-level decomposition based matheuristic for airline crew rostering problems with fair working time. Eur. J. Oper. Res. **267** (2017). https://doi.org/10.1016/j.ejor.2017.11.046

10. Erhard, M., Schoenfelder, J., Fügener, A., Brunner, J.O.: State of the art in physician scheduling. Eur. J. Oper. Res. **265**(1), 1–18 (2018). https://doi.org/10.1016/j.ejor.2017.06.037

11. Ernst, A.T., Jiang, H., Krishnamoorthy, M., Owens, B., Sier, D.: An annotated bibliography of personnel scheduling and rostering. Ann. Oper. Res. **127**, 21–144 (2004). https://doi.org/10.1023/B:ANOR.0000019087.46656.e2

12. Ernst, A.T., Jiang, H., Krishnamoorthy, M., Sier, D.: Staff scheduling and rostering: a review of applications, methods and models. Eur. J. Oper. Res. **153**(1), 3–27 (2004). https://doi.org/10.1016/S0377-2217(03)00095-X, timetabling and Rostering

13. Fages, J.G., Lapègue, T.: Filtering atmostnvalue with difference constraints: application to the shift minimisation personnel task scheduling problem. Artif. Intell. **212**, 116–133 (2014). https://doi.org/10.1016/j.artint.2014.04.001

14. Fischetti, M., Martello, S., Toth, P.: Approximation algorithms for fixed job schedule problems. Oper. Res. **40**(1-supplement-1), S96–S108 (1992). https://doi.org/10.1287/opre.40.1.S96

15. Gendreau, M., Potvin, J.Y.: Tabu search. In: Burke, E.K., Kendall, G. (eds.) Search Methodologies: Introductory Tutorials in Optimization and Decision Support Techniques, pp. 165–186. Springer, US, Boston, MA (2005). https://doi.org/10.1007/0-387-28356-0_6

16. Glover, F.: Future paths for integer programming and links to artificial intelligence. Comput. Oper. Res. **13**(5), 533–549 (1986). https://doi.org/10.1016/0305-0548(86)90048-1

17. Glover, F.: Tabu search-part ii. ORSA J. Comput. **2**(1), 4–32 (1990). https://doi.org/10.1287/ijoc.2.1.4

18. Glover, F.: Tabu search-part i. ORSA J. Comput. **1**(3), 190–206 (1989). https://doi.org/10.1287/ijoc.1.3.190

19. Gopalakrishnan, B., Johnson, E.: Airline crew scheduling: state-of-the-art. Ann. Oper. Res. **140**(1), 305–337 (2005). https://doi.org/10.1007/s10479-005-3975-3

20. Heil, J., Hoffmann, K., Buscher, U.: Railway crew scheduling: models, methods and applications. Eur. J. Oper. Res. **283**(2), 405–425 (2020). https://doi.org/10.1016/j.ejor.2019.06.016

21. Hojati, M.: A greedy heuristic for shift minimization personnel task scheduling problem. Comput. Oper. Res. **100**, 66–76 (2018). https://doi.org/10.1016/j.cor.2018.07.010

22. Kasirzadeh, A., Saddoune, M., Soumis, F.: Airline crew scheduling: models, algorithms, and data sets. EURO J. Transp. Logist. **6**(2), 111–137 (2017). https://doi.org/10.1007/s13676-015-0080-x

23. Kletzander, L., Musliu, N.: Solving the general employee scheduling problem. Comput. Oper. Res. **113**, 104794 (2020). https://doi.org/10.1016/j.cor.2019.104794

24. Kolen, A.W.J., Lenstra, J.K., Papadimitriou, C.H., Spieksma, F.C.R.: Interval scheduling: a survey. Nav. Res. Logist. Q. **54**(5), 530–543 (2007). https://doi.org/10.1002/nav.20231

25. Krishnamoorthy, M., Ernst, A.T.: The personnel task scheduling problem. In: Yang, X., Teo, K.L., Caccetta, L. (eds.) Optimization Methods and Applications, Applied Optimization, vol. 52, pp. 343–368. Springer, Boston (2001). https://doi.org/10.1007/978-1-4757-3333-4_20

26. Krishnamoorthy, M., Ernst, A.T., Baatar, D.: Algorithms for large scale shift minimisation personnel task scheduling problems. Eur. J. Oper. Res. **219**(1), 34–48 (2012). https://doi.org/10.1016/j.ejor.2011.11.034

27. Kroon, L.G., Salomon, M., Wassenhove, L.N.V.: Exact and approximation algorithms for the tactical fixed interval scheduling problem. Oper. Res. **45**(4), 624–638 (1997). https://doi.org/10.1287/opre.45.4.624

28. Lapègue, T., Bellenguez-Morineau, O., Prot, D.: A constraint-based approach for the shift design personnel task scheduling problem with equity. Comput. Oper. Res. **40**(10), 2450–2465 (2013). https://doi.org/10.1016/j.cor.2013.04.005

29. Lin, S.W., Ying, K.C.: Minimizing shifts for personnel task scheduling problems: a three-phase algorithm. Eur. J. Oper. Res. **237**(1), 323–334 (2014). https://doi.org/10.1016/j.ejor.2014.01.035

30. Nurmi, K., as, N.K.: A successful three-phase metaheuristic for the shift minimiza-tion personal task scheduling problem. Adv. Oper. Res. **2021**, 8876990 (2021). https://doi.org/10.1155/2021/8876990
31. Prajapati, V.K., Jain, M., Chouhan, L.: Tabu search algorithm (TSA): a compre-hensive survey. In: Proceedings of the 3rd International Conference on Emerging Technologies in Computer Engineering: Machine Learning and Internet of Things, pp. 1–8, February 2020. https://doi.org/10.1109/ICETCE48199.2020.9091743
32. Örmeci, E.L., Salman, F.S., Yücel, E.: Staff rostering in call centers provid-ing employee transportation. Omega **43**, 41–53 (2014). https://doi.org/10.1016/j.omega.2013.06.003
33. Smet, P., Wauters, T., Mihaylov, M., Vanden Berghe, G.: The shift minimisation personnel task scheduling problem: a new hybrid approach and computational insights. Omega **46**, 64–73 (2014). https://doi.org/10.1016/j.omega.2014.02.003
34. Valls, V., Pérez, A., Quintanilla, S.: A graph colouring model for assigning a het-erogeneous workforce to a given schedule. Eur. J. Oper. Res. **90**(2), 285–302 (1996). https://doi.org/10.1016/0377-2217(95)00355-X
35. Van den Bergh, J., Beliën, J., De Bruecker, P., Demeulemeester, E., De Boeck, L.: Personnel scheduling: a literature review. Eur. J. Oper. Res. **226**(3), 367–385 (2013). https://doi.org/10.1016/j.ejor.2012.11.029
36. Wen, X., Sun, X., Sun, Y., Yue, X.: Airline crew scheduling: models and algorithms. Transp. Res. E **149**, 102304 (2021). https://doi.org/10.1016/j.tre.2021.102304
37. Xu, S., Hall, N.G.: Fatigue, personnel scheduling and operations: review and research opportunities. Eur. J. Oper. Res. **295**(3), 807–822 (2021). https://doi.org/10.1016/j.ejor.2021.03.036

An $O(n \log n)$-Time Algorithm for Linearly Ordered Packing of 2-Bar Charts into $OPT + 1$ Bins

Adil Erzin[1]([✉])(iD), Alexander Kononov[1](iD), Stepan Nazarenko[2],
and Konstantin Sharankhaev[2]

[1] Sobolev Institute of Mathematics, SB RAS, Novosibirsk 630090, Russia
adilerzin@math.nsc.ru
[2] Novosibirsk State University, Novosibirsk 630090, Russia

Abstract. Given a sequence of bins and n bar charts consisting of two bars each (2-BCs). Every bar has a positive height not exceeding 1. Each bin can contain any subset of bars of total height at most 1. It is required to pack all 2-BCs into the minimal number of bins so that the bars of each 2-BC do not change their order and occupy adjacent bins. Previously, a special case of the problem was considered where the first bars of any two 2-BCs cannot be placed into the same bin. For this case an $O(n^2)$-time algorithm that constructs a packing of length at most $OPT + 1$, where OPT is the optimum, was presented. In this paper, we propose a new, less time-consuming algorithm that also constructs a packing of length at most $OPT + 1$ for the same case of the problem with time complexity equals to $O(n \log n)$.

Keywords: Bar charts · Packing · Approximation

1 Introduction

The following problem of investment portfolio optimization in the oil and gas field was studied in [7]. All projects are characterized by the annual oil or gas production which can be represented using a bar chart (BC). The height of each bar corresponds to the volume of production during the current year. The total production volume of all projects must not exceed a given value for each year, which is due, for example, to the throughput of the pipe. The problem is to determine the start year of each project in such a way as to finish them in the shortest possible time.

This problem was called a Bar Charts Packing Problem (BCPP) and formulated as follows. Given a set of n bar charts, each bar has a height at most 1 and unit length. All BCs are required to pack in a unit-height strip of minimum length. If we split the strip into equal unit-width bins of height 1, then the packing length is the number of bins containing at least one bar. When packing BCs,

The study was carried out within the framework of the state contract of the Sobolev Institute of Mathematics (project FWNF-2022-0019).

crossing bars are naturally prohibited. Each BC's bar can move vertically, but they are inseparable horizontally and cannot change their order.

The BCPP was first formulated in [8] and its various subproblems were studied in [9–13]. In this paper, we focus on the special case of the Two-Bar Charts Packing problem (2-BCPP), where each BC consists of two bars. This problem is a generalization of the Bin Packing Problem (BPP) [20] and a relaxation of the Two-Dimensional Vector Packing Problem (2-DVPP) [23].

1.1   Related Results

In the BPP, items with given sizes must be packed in the minimal number of unit-capacity bins. BPP is a particular case of 2-BCPP when each 2-BC consists of equal bars. Assuming $P \neq NP$, BPP cannot be approximated within the ratio less than $3/2$ [15,27]. As shown in [26], this ratio is achieved by First Fit Decreasing algorithm in which items are sorted in non-increasing order of their sizes, and the current item is placed in the first suitable bin. Note that the proof of non-approximability uses very special instances when OPT is a small number, such as 2 or 3, even though the number of items unbounded. Therefore, to estimate the accuracy of approximation algorithms for the bin packing problem, an asymptotic approximation is used. For example, in the worst case, the First Fit Decreasing algorithm uses at most $11/9 OPT + 6/9$ bins to pack all items [5]. Moreover, Fernandez de la Vega and Lueker [14] proposed an asymptotical polynomial-time approximation scheme for the BPP. They showed that for any fixed $\varepsilon > 0$, there exists a polynomial-time algorithm with asymptotic worst-case ratio not exceeding $1 + \varepsilon$. Also, BPP admits an AFPTAS [22], and an additive approximation algorithm which packs any instance I in at most $OPT(I) + O(\log(OPT(I)))$ bins [19].

In the 2-DVPP, which is a generalization of the BPP, there are two attributes for items and bins. The problem is to minimize the number of used bins when all items are packed, considering both attributes of the bin's capacity limits. In 2003, an $O(n \log n)$-time algorithm for the 2-DVPP with absolute performance guarantee 2 was proposed [23]. A detailed survey of approximation algorithms for the 2-DVPP can be found in [4]. The best known algorithm builds a $(3/2+\varepsilon)$-approximate solution, for any $\varepsilon > 0$ [2]. On the other hand, the 2-BCPP is a relaxation of the 2-DVPP and any ρ-approximation algorithm for the 2-DVPP is a 2ρ-approximation algorithm for the 2-BCPP [12]. In 1997, Woeginger showed that there is no asymptotic polynomial time approximation scheme for the 2-DVPP unless $P = NP$ [28].

The Two-Bar Charts Packing problem is also a special case of the Resource-Constrained Project Scheduling Problem (RCPSP) with one renewable resource [3]. Each project is consisting of two unit-duration jobs, which must be executed without delay consuming a limited non-accumulative resource. Each project can be represented as a 2-BC in which the value of the consumed resource corresponds to the height of the bars. It is required to find the starting time for each project in order to finish all projects during the minimum time and no more than given amount of resource consumes during each time slot. The RCPSP problem is

NP-hard and polynomial algorithms with worst-case performance are not known for it. Generally, for such a problem heuristic algorithms are used followed by a posteriori analysis [16–18, 24].

As far as we know, there are only few papers devoted to the 2-BCPP problem. Since this problem is a generalization of the bin packing problem, which is strongly NP-hard, then 2-BCPP is strongly NP-hard as well. In [8], a linear time algorithm for the 2-BCPP was proposed. It builds a packing of length at most $2OPT + 1$, where OPT is the minimum packing length. Later, it was shown that the additive constant could be removed and the estimate could be reduced to $2OPT$.

The following subproblem of the 2-BCPP was studied in [9–12]. Given a set S of n 2-BCs. Let a_i be the height of the first bar and b_i be the height of the second bar of the 2-BC $i \in S$. A bar is called *big* if it is higher than $1/2$. A 2-BC is called *big* if at least one of its bars is big, i.e., $\max\{a_i, b_i\} > 1/2$. Let us denote the 2-BCPP problem with big 2-BCs as 2-BCPP$^>$. As shown in [12] the 2-BCPP$^>$ is strongly NP-hard. In the same paper, the authors presented an $O(n^2)$-time $(4/3\, OPT + 2/3)$-approximation algorithm for the 2-BCPP$^>$.

The 3-BCPP problem of packing bar charts with three bars in each BC was considered in [13], and a linear time algorithm, which constructs a packing of length at most $3OPT + 2$, was proposed. Later, the additive constant was reduced to 1. If at least two bars of each 3-BC are big and a small bar, if it exists, is not a second, it was shown how to find a 9/8-approximate solution with time complexity $O(n^{2.5})$. Moreover, if at least one bar of each 3-BC is big it was proven that 3-BCPP remains strongly NP-hard.

In [12], the particular case of the 2-BCPP$^>$, in which an additional restriction was imposed on the set of feasible solutions, was considered. In what follows, we will need the following definitions.

Definition 1. Packing *is a function* $p : S \to [1, 2n - 1]$, *which associates with each 2-BC* $i \in S$ *the bin number* $p(i)$ *where its first bar falls. The packing is feasible if the sum of the bar's heights that fall into each bin does not exceed 1, i.e., for each bin* e, $1 \leq e \leq 2n$, *the inequality*

$$\sum_{i \in S : p(i) = e} a_i + \sum_{i \in S : p(i) + 1 = e} b_i \leq 1$$

holds.

Definition 2. *The* packing length $L(p)$ *is the number of bins in which at least one bar falls.*

As a result of packing p, the first bar of 2-BC i falls into the bin $p(i)$ and the second bar falls into the bin $p(i) + 1$. We will consider only feasible packings; therefore, the word "feasible" will be omitted further.

Definition 3. *A packing* p *is* linearly ordered *if for any two 2-BCs* $x, y \in S$ *we have* $p(x) \neq p(y)$.

We note that with a linearly ordered packing, each bin contains no more than two bars. In particular, if the first (second) bar of each 2-BCs is big, then this condition holds. For this case Erzin et al. [12] proposed an $O(n^2)$-time algorithm that finds a solution with packing length of at most $OPT + 1$. In this paper, we present a new algorithm for the same case that constructs a solution with identical error with complexity equals $O(n \log n)$. Our algorithm based on construction a max-cardinality matching in a special bipartite graph with further alteration of the resulting cycles (if any) into one path. In [8], the algorithm A, which constructs a packing of arbitrary 2-BCs into at most $2OPT + 1$ bins, was presented. It consists of three stages. The new algorithm can be used at the second stage of the algorithm A instead of greedy procedure while packing the big 2-BCs. We are planning to compare these two approaches in the future.

Let us denote the problem of finding a min-length linearly ordered packing of 2-BCs as 2-BCPP[1].

The rest of the paper is organized as follows. Section 2 provides a statement of the problem as a Boolean Linear Programming (BLP). In Sect. 3, we describe the constructing of the special bipartite graph. Section 4 is devoted to finding the max-cardinality matching in the constructed graph. Section 5 describes the algorithm that forms one path from the cycles, if they exist, and the last section concludes the paper.

2 Formulation of the Problem

The BLP formulation for 2-BCPP was first presented in [8]. We present it here for the reader's convenience. To do this, we introduce the variables:

$$x_{ij} = \begin{cases} 1, \text{ if the first bar of the } i\text{th 2-BC falls into the bin } j; \\ 0, \text{ else.} \end{cases}$$

$$y_j = \begin{cases} 1, \text{ if the bin } j \text{ contains at least one bar;} \\ 0, \text{ else.} \end{cases}$$

Then 2-BCPP in the form of BLP is as follows.

$$\sum_j y_j \rightarrow \min_{x_{ij}, y_j \in \{0,1\}}; \tag{1}$$

$$\sum_j x_{ij} = 1, \ i \in S; \tag{2}$$

$$\sum_i a_i x_{ij} + \sum_k b_k x_{kj-1} \leq y_j, \ \forall j. \tag{3}$$

In this formulation, criterion (1) is the minimization of the packing length. Constraints (2) require each 2-BC to be packed into a strip once. Constraints (3) ensure that the sum of the bar's heights in each bin does not exceed 1.

3   Bipartite Graph Constructing

In this section, we show how to construct a special bipartite graph for a given set of 2-BCs, where the maximal matching can be found in linear time.

Definition 4. *Two 2-BCs i and j form a 1-union if as a result of packing p either $p(i) = p(j) + 1$ or $p(j) = p(i) + 1$. 2-BC i forms a 1-union on the left (on the right) with 2-BC j if the first bar of i is located on the left (right) of the first bar of j. We denote such union as $i \leftarrow j$ ($i \rightarrow j$).*

In the linear ordering packing each 2-BC can participate in the left and right 1-unions no more than once. We sort 2-BCs in non-increasing order of the height of the first bar and number them with integers $1, \ldots, n$. Let us construct a weighted complete directed *1-union graph* $G = (V, E)$ as follows. The vertices in V are the images of 2-BCs, and an arc $(i, j) \in E$ has weight 1 if and only if 1-union $i \rightarrow j$ is possible (Fig. 1a). The weights of other arcs are equal to 0. For convenience, we will further identify the vertices and 2-BCs, i.e., vertex i forms a 1-union with vertex j corresponds to 2-BCs i and j form a 1-union. Then the 2-BCPP[1] is reduced to the maximum asymmetric traveling salesman problem with arcs weights 0 or 1 (MaxATSP(0,1)) [11]. Indeed, a max-weight Hamiltonian cycle in G defines a min-length packing for the 2-BCPP[1], since each 1-union decreases the packing length by 1.

Let us define a *correct* bipartite graph $G_1 = (S_1, S_2, E_1)$ as follows. For definiteness, we call set S_1 the *left* and S_2 the *right* parts of the bipartite graph. Each vertex in S_1 corresponds to 2-BC which can form a 1-union on the left, and each vertex in S_2 corresponds to 2-BC which can form a 1-union on the right. Note that for each 2-BC it is possible to be included into both sets. The vertices in S_2 are ordered from top to bottom in non-increasing order of the height of the first bar. The edge $(i, j) \in E_1$ if $i \neq j$ and $i \leftarrow j$ (Fig. 2a). We assign to each vertex in S_1 a label equals to the smallest number of the adjacent vertex in S_2. The vertices of the left part are ordered from top to bottom in non-decreasing order of the assigned labels.

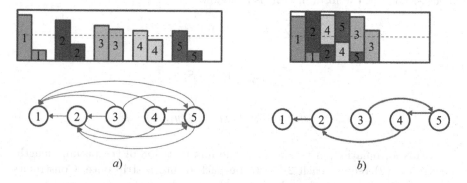

a)   b)

Fig. 1. *a*) The set of 2-BCs and corresponding 1-union graph; *b*) Optimal packing and max-weight path.

Fig. 2. a) Bipartite graph for 1-union graph in Fig. 1; b) Maximum matching (red bold lines).

Maximum matching in the correct bipartite graph defines a subset of arcs. If this subset does not form any cycle in the 1-union graph, then it defines an optimal packing (Fig. 1b), since cardinality of the maximum matching in G_1 is an upper bound on the number of 1-unions (Fig. 2b). Another case, when cycles exist, will be considered in Sect. 5.

3.1   Algorithm *Construct*

The algorithm *Construct* builds a correct bipartite graph with $O(n \log n)$ time complexity. At the first stage, elements of S are numbered with integers $1, \ldots, n$ in non-increasing order of the heights of the first bar and form the set S_2 placing vertices from top to bottom according to their numbers. Then, algorithm sets $S_1 = S_2$. After that, the elements of the left part S_1 are rearranged from top to bottom in non-decreasing order of the second bar's height keeping the vertex numbering. The second stage consists of a sequence of steps and allows to find edges of the graph $(i, j) \in E_1$, $i \neq j$. Starting with the highest vertex i in the left part, algorithm scans in numerical order the vertices of the right part until it finds the first node $j \neq i$, which can form a 1-union $i \leftarrow j$. Algorithm assigns a label $h(i) = j$ to the vertex i. As a result, the "highest" incident edge (i, j) for the top left vertex i is found. Note that the highest edge defines all other incident edges, i.e., if $(i, j) \in E_1$, then $(i, k) \in E_1$ for any $k > j$, $k \neq i$, since $a_k \leq a_j$. For the next vertex in the left part algorithm finds a highest vertex in the right part, which can form 1-union, starting from one node higher from the last found. Algorithm stops when next vertex in the left part cannot form a 1-union on the left with the last vertex in the right part or when all elements in S_1 have assigned labels. Then, all isolated vertices in S_1 and S_2 are deleted. Any 2-BC which do not participate in any 1-union evidently occupies two bins. For the constructed graph it is still possible to be incorrect. In this case, algorithm rearranges the elements of the left part from top to bottom in non-decreasing order of the assigned labels. In Fig. 3, we show the example of incorrect graph before last rearrangement. The nodes 1 and 3 are lower than vertex 2, although they are adjacent to the node 2, which is higher than the vertex 3 in the right part. After

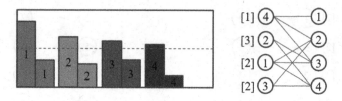

Fig. 3. Example of incorrect graph.

placing the vertices of the left part from top to bottom in non-decreasing order of labels (4,1,3,2), a correct graph will be obtained.

Lemma 1. *Algorithm* **Construct** *constructs correct bipartite graph in* $O(n \log n)$ *time.*

Proof. The graph is correct by construction. Ordering of sets S_1 and S_2 can be implemented in $O(n \log n)$ time. Algorithm finds the edges of the bipartite graph in linear time. The vertices of the left part are reordered, if necessary, at most n times. The lemma is proved.

4   Maximal Matching in the Correct Bipartite Graph

As mentioned in the previous section, cardinality of a maximum matching in G_1 is an upper bound on the number of 1-unions. Let M be the maximal matching. We introduce an $O(n)$-time algorithm **Max_Matching** to find M as follows. Given the list S_1, each vertex i of which is assigned a label $h(i)$, and the list S_2. Set $M = \emptyset$. Algorithm performs a typical procedure, which consists of the following. It considers the last (lowest) element i in S_1. Since the label $h(i)$ is assigned to it, the vertex i is adjacent to each vertex $k \geq h(i), k \neq i$ in the list S_2 including the last (lowest) vertex $j \neq i$. The edge (i, j) is added to the matching, i.e., $M = M \cup \{(i, j)\}$. Both vertices i and j are deleted from the S_1 and S_2, respectively, together with their incident edges. The procedure is repeated for the updated lists S_1 and S_2 until $S_1 \neq \emptyset$ or $S_2 \neq \emptyset$. It is possible for the vertices to become isolated while removing other vertices together with their incident edges. Algorithm deletes isolated vertices from the S_1 and S_2 and repeats the procedure for the next last element in the S_1.

Lemma 2. *Algorithm* **Max_Matching** *constructs a max-cardinality matching in the correct bipartite graph with $O(n)$ time complexity.*

Proof. Recall that if $(i, j) \in E_1$, then $(i, k) \in E_1$ for any $k \geq j$, $k \neq i$. Given correct bipartite graph, let i is the lowest vertex in the left part and (i, j) is its lowest incident edge. We will show that there exists a max-cardinality matching that contains the edge (i, j). Let M be an arbitrary maximal matching. If $(i, j) \notin M$, then two following cases are possible:

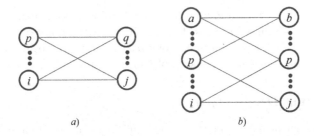

Fig. 4. Illustration for the proof of Lemma 2. *a)* Replacing (p, j) and (i, q) (red edges) with (p, q) and (i, j) (blue edges); *b)* Replacing (a, b), (p, j), and (i, p) (red edges) with (a, p), (p, b), and (i, j) (blue edges).

1. Vertex i is not incident to any edge in M;
2. Vertex i is incident to some edge in M but not to the edge (i, j).

In the first case, vertex j is adjacent to another vertex of left part, say p. If (p, j) replace with (i, j), the cardinality of M does not change.

In the second case, let $(i, q) \in M$ and $(p, j) \in M$. We get one of the following cases:

a) $p \neq q$. Then, if (p, j) and (i, q) replace with (p, q) and (i, j), the cardinality of M does not change (Fig. 4a).
b) $p = q$. The vertex i is lower than p in the left part of the graph. Therefore, p is adjacent to some vertex in the right part which is higher than p, say b. Otherwise, p is located lower than i in the left part. If vertex b is not incident to any edge in the matching, replacing (p, j) with (p, b) does not change the cardinality of M. Else, if b is incident in the matching to some edge (a, b), replacing (a, b), (p, j), and (i, p) with (a, p), (p, b), and (i, j) does not change the cardinality of M (Fig. 4b).

Thus, there is always max-cardinality matching with the lowest edge (i, j) incident to the lowest vertex $i \in S_1$. Algorithm includes the edge (i, j) in M and deletes the vertices i and j with all incident edges from the graph. Then, it considers next lowest vertex in the left part and repeats the same procedure until $S_1 \neq \emptyset$ or $S_2 \neq \emptyset$. Max-cardinality matching in the correct bipartite graph is constructed after one browsing of the vertices of the left part. So, the running time of the algorithm *Max_Matching* equals $O(n)$. The lemma is proved.

5   Constructing a Solution to the 2-BCPP[1]

In this section we propose algorithm *Delete_Cycles*, which constructs an approximate solution to the 2-BCPP[1] from the max-cardinality matching M. Recall that we identify the vertices and 2-BCs. Each edge in the M defines a corresponding 1-union. Since vertices can belong to both parts of bipartite graph, matching edges can form not only paths, but also cycles in a 1-union graph. If the edges of M do not form any cycle in the 1-union graph, then this matching determines an optimal solution to the 2-BCPP[1]. If there are cycles in 1-union graph formed by M, algorithm makes one path from them, losing no more than one arc. As a result, we get a packing of length at most $OPT + 1$.

5.1 Algorithm *Delete_Cycles*

Note, that algorithm **Construct** numbers all 2-BCs in non-increasing order of the height of the first bar. Algorithm **Delete_Cycles** constructs the following digraph $G_2 = (V_2, E_2)$. The elements of the set V_2 are the images of 2-BCs included in the M. The vertices are placed from left to right according to their numbers. The arc $(i, j) \in E_2$ if $(j, i) \in M$. Let us set a cycle counter $m = 0$. Algorithm builds a sequence of arcs starting from the rightmost vertex until it finds the end-node of the path or returns to the rightmost vertex forming a cycle. In the first case, algorithm packs 2-BCs which are the prototypes of vertices forming the path to the first empty bin according to the 1-unions in the path. Let us call the vertices in this path *visited*. In the second case, algorithm increases cycle counter by 1 and assigns to this cycle the counter value. The nodes in the cycle are visited too. This procedure is repeated for the rightmost unvisited vertex if it exists. Thus, after one browsing of the vertices, the quantity m of cycles in G_2 is known. Algorithm **Delete_Cycles** rebuilds existing cycles into one path decreasing the amount of arcs in the max-cardinality matching by one. Let us consider the possible cases.

1. $m = 1$. In this case, algorithm deletes arbitrary arc in the cycle forming a path. The number of arcs is reduced by one.
2. $m = 2$. Let (i, j) be the incoming arc to the vertex j with the smallest number (leftmost) in the first cycle, and (k, l) be the incoming arc to the vertex l with the smallest number (leftmost) in the second cycle. One of the arcs (i, j) or (k, l) is outgoing from the vertex with greater number. Without loss of generality, we assume that $i > k$. Algorithm deletes both arcs (i, j) and (k, l) and adds an arc (i, l) decreasing the amount of arcs by one. In Fig. 5a, we show the example with two cycles. Each subset of vertices $\{1, 3, 4, 9\}$ and $\{2, 5, 6, 7, 8\}$ forms a cycle. In the example, algorithm deletes the arcs $(3,1)$ and $(6,2)$ to the leftmost vertices in each cycle (Fig. 5b). Let us show that adding the arc (i, l) is always correct. Three cases are possible:
   (a) $l < j < k < i$ (Fig. 6a);
   (b) $l < k < j < i$ (Fig. 6b);
   (c) $j < l < k < i$ (Fig. 6c).
   Since all vertices are sorted in non-increasing order of the height of the first bar, $i > k$, and there is a 1-union $k \rightarrow l$, then a 1-union $i \rightarrow l$ is possible too. One arc is removed in each cycle, namely (i, j) (red one) and (k, l) (blue one), resulting in two paths. The algorithm combines these two paths into one by adding arc (i, l) (green one) (Fig. 6).
3. $m > 2$. Algorithm forms one path from the m-th and $(m - 1)$-th cycles in the same way as it was done for the case $m = 2$. For each next cycle

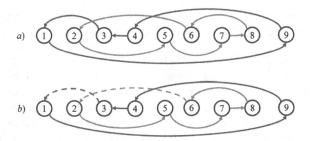

Fig. 5. $a)$ Example of two cycles in the graph; $b)$ Arcs $(3,1)$ and $(6,2)$ are deleted.

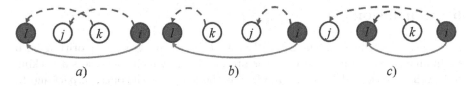

Fig. 6. Illustration to the algorithm **Delete_Cycles**. l is the leftmost vertex in the first cycle, vertex i is adjacent to the leftmost vertex in the second cycle. $a)$ case (a); $b)$ case (b); $c)$ case (c). (Color figure online)

$c = (m - 2), \ldots, 1$ algorithm performs the following procedure. The outgoing from the rightmost vertex r arc is removed. A new arc from vertex r to the end-node of the path built from the $(c + 1)$-th cycle is added. It is possible, since the vertex with the highest number in the c-th cycle is located to the right of the end-vertex of the built path. Thus, there is no decreasing of the number of arcs when algorithm rebuilds the cycles $c = (m - 2), \ldots, 1$.

Lemma 3. *Algorithm **Delete_Cycles** rebuilds existing cycles into one path decreasing the amount of arcs of the max-cardinality matching exactly by one in $O(n)$ time.*

Proof. As a result of one scan of $m < n$ cycles, they were rebuilt in one path with decreasing of the amount of arcs exactly by one. The lemma is proved.

5.2 $OPT + 1$ Approximation

Algorithm **APX** constructs an approximate packing.

Algorithm APX

Construct a bipartite graph by algorithm **Construct**
Find a max-cardinality matching by algorithm **Max_Matching**
Rebuild all cycles in one path by algorithm **Delete_Cycles**

Theorem 1. *Algorithm APX builds a packing of length at most $OPT + 1$ in $O(n \log n)$ time, where OPT is the optimum of the 2-BCPP[1].*

Proof. Lemmas 1, 2 and 3 imply that the running time of algorithm APX is $O(n \log n)$. Cardinality of a maximum matching in the correct bipartite graph is an upper bound on the number of 1-unions. If the found maximal matching forms any cycle in 1-union graph, then algorithm **Delete_Cycles** decreases the amount of arcs in the matching by one, resulting in the loss of only one possible 1-union. Hence, the length of the built packing is at most $OPT + 1$. The theorem is proved.

6 Conclusion

Previously, for the problem 2-BCPP[1], an $O(n^2)$-time greedy algorithm with preliminary lexicographic sorting was proposed [12], which constructs a packing of length at most $OPT + 1$, where OPT is the length of the optimal packing. In this paper, we propose a new, less time-consuming algorithm that also constructs a packing of length at most $OPT + 1$ for the same problem. The running time of our new algorithm is $O(n \log n)$. Reducing the time complexity of linear ordering packing is important, since in several packing algorithms it is used as a procedure at some step.

References

1. Baker, B.S.: A new proof for the first-fit decreasing bin-packing algorithm. J. Algorithms **6**, 49–70 (1985)
2. Bansal, N., Eliás, M., Khan, A.: Improved approximation for vector bin packing. In: SODA, pp. 1561–1579 (2016)
3. Brucker, P., Knust, S.: Complex Scheduling. Springer-Verlag, Berlin, Heidelberg (2006). https://doi.org/10.1007/3-540-29546-1
4. Christensen, H.I., Khanb, A., Pokutta, S., Tetali, P.: Approximation and online algorithms for multidimensional bin packing: a survey. Comput. Sci. Rev. **24**, 63–79 (2017)
5. Dósa, G.: The tight bound of first fit decreasing bin-packing algorithm is $FFD(I) \leq 11/9OPT(I) + 6/9$. In: Chen, B., Paterson, M., Zhang, G. (eds.) ESCAPE 2007. LNCS, vol. 4614, pp. 1–11. Springer, Heidelberg (2007). https://doi.org/10.1007/978-3-540-74450-4_1
6. Delorme, M., Iori, M., Martello, S.: Bin packing and cutting stock problems: mathematical models and exact algorithms. Eur. J. Oper. Res. **225**(1), 1–20 (2016)
7. Erzin, A., Plotnikov, R., Korobkin, A., Melidi, G., Nazarenko, S.: Optimal investment in the development of oil and gas field. In: Kochetov, Y., Bykadorov, I., Gruzdeva, T. (eds.) MOTOR 2020. CCIS, vol. 1275, pp. 336–349. Springer, Cham (2020). https://doi.org/10.1007/978-3-030-58657-7_27
8. Erzin, A., Melidi, G., Nazarenko, S., Plotnikov, R.: Two-bar charts packing problem. Optim. Lett. **15**(6), 1955–1971 (2021)
9. Erzin, A., Melidi, G., Nazarenko, S., Plotnikov, R.: A 3/2-approximation for big two-bar charts packing. J. Comb. Optim. **42**, 71–84 (2021)

10. Erzin, A., Melidi, G., Nazarenko, S., Plotnikov, R.: A posteriori analysis of the algorithms for two-bar charts packing problem. Commun. Comput. Inf. Sci. **1514**, 201–216 (2021)
11. Erzin, A., Shenmaier, V.: An improved approximation for packing big two-bar charts. J. Math. Sci. **267**(4), 465–473 (2022)
12. Erzin, A., Kononov, A., Melidi, G., Nazarenko, S.: A $4/3OPT + 2/3$ approximation for big two-bar charts packing problem. J. Math. Sci. **269**(6), 813–823 (2023)
13. Erzin, A., Sharankhaev, K.: Three-bar charts packing problem. Commun. Comput. Inf. Sci. **1739**, 61–75 (2023)
14. Fernandez de la Vega, W., Lueker, G.S.: Bin packing can be solved within $1 + \epsilon$ in linear time. Combinatorica **1**, 349–355 (1981)
15. Garey, M.R., Johnson, D.S.: Computers and intractability. A Guide to the Theory of NP-Completeness, Freeman, San Francisco (1979)
16. Goncharov, E.: A greedy heuristic approach for the resource-constrained project scheduling problem. Stud. Inf. Univ. **9**(3), 79–90 (2011)
17. Goncharov, E.: A stochastic greedy algorithm for the resource-constrained project scheduling problem. Discret. Anal. Oper. Res. **21**(3), 11–24 (2014)
18. Hartmann, S.: A self-adapting genetic algorithm for project scheduling under resource constraints. Nav. Res. Logist. **49**, 433–448 (2002)
19. Hoberg, R., Rothvos, T.: A logarithmic additive integrality gap for bin packing. In: Proceedings of the Twenty-Eighth Annual ACM-SIAM Symposium on Discrete Algorithms, pp. 2616–2625. SIAM (2017)
20. Johnson D.S.: Near-optimal bin packing algorithms. Massachusetts Institute of Technology. PhD thesis (1973)
21. Johnson, D.S., Garey, M.R.: A 71/60 theorem for bin packing. J. Complex. **1**(1), 65–106 (1985)
22. Karmarkar, N., Karp, R.M.: An efficient approximation scheme for the one dimensional bin-packing problem. In: 23rd Annual Symposium on Foundations of Computer Science, pp. 312–320. IEEE (1982)
23. Kellerer, H., Kotov, V.: An approximation algorithm with absolute worst-case performance ratio 2 for two-dimensional vector packing. Oper. Res. Lett. **31**, 35–41 (2003)
24. Kolisch, R., Hartmann, S.: Experimental investigation of heuristics for resource-constrained project scheduling: an update. Eur. J. Oper. Res. **174**, 23–37 (2006)
25. Li, R., Yue, M.: The proof of $FFD(L) \leq 11/9OPT(L) + 7/9$. Chin. Sci. Bull. **42**(15), 1262–1265 (1997)
26. Simchi-Levi, D.: New worst-case results for the bin-packing problem. Nav. Res. Logist. **41**(4), 579–585 (1994)
27. Vazirani, V.V.: Approximation Algorithms. Springer, Berlin, Heidelberg (2001). https://doi.org/10.1007/978-3-662-05269-3_4
28. Woeginger, G.J.: There is no asymptotic PTAS for two-dimensional vector packing. Inf. Process. Lett. **64**(6), 293–297 (1997)
29. Yue, M.: A simple proof of the inequality $FFD(L) \leq 11/9OPT(L) + 1, \forall L$, for the FFD bin-packing algorithm. Acta Math. Appl. Sin. **7**(4), 321–331 (1991)
30. Yue, M., Zhang, L.: A simple proof of the inequality $MFFD(L) \leq 71/60OPT(L) + 1, \forall L$, for the MFFD bin-packing algorithm. Acta Math. Appl. Sin. **11**(3), 318–330 (1995)

Approximation Algorithms for Graph Cluster Editing Problems with Cluster Size at Most 3 and 4

Victor Il'ev[1,2](✉) [iD] and Svetlana Il'eva[1] [iD]

[1] Dostoevsky Omsk State University, Omsk, Russia
`iljev@mail.ru`
[2] Sobolev Institute of Mathematics SB RAS, Omsk, Russia

Abstract. In clustering problems, one has to partition a given set of objects into pairwise disjoint subsets (clusters) taking into account only similarity of objects. In the graph cluster editing problem similarity relation on the set of objects is given by an undirected graph whose vertices are in one-to-one correspondence with objects and edges correspond to pairs of similar objects. The goal is to find a nearest to a given graph $G = (V, E)$ cluster graph, i.e., a graph on the vertex set V each connected component of which is a complete graph. The distance between graphs is understood as the Hamming distance between their incidence vectors.

We consider a variant of the cluster editing problem in which the size of each cluster is bounded from above by a positive integer s. In 2011, Il'ev and Navrotskaya proved that this problem is NP-hard for any fixed $s \geqslant 3$. In 2015, Puleo and Milenkovic proposed a 6-approximation algorithm for this problem. In 2016, Il'ev, Il'eva and Navrotskaya presented an approximation algorithm that is 3-approximation in case of $s = 3$ and 5-approximation in case of $s = 4$.

Now we propose simple greedy-type 2-approximation algorithms for these cases with tight performance guarantees.

Keywords: Graph · Cluster editing · Approximation algorithm · Performance guarantee

1 Introduction

In clustering problems, one has to partition a given set of objects into pairwise disjoint subsets (clusters) taking into account only similarity of objects. In graph clustering problems similarity relation on the set of objects is given by an undirected graph whose vertices are in one-to-one correspondence with objects and edges correspond to pairs of similar objects. A version of this problem is known as the Graph Approximation problem [1,4–6,12,14]. In this problem, the goal is to find a nearest to a given graph $G = (V, E)$ cluster graph, i.e., a graph on the vertex set V each connected component of which is a complete graph. The distance between graphs is understood as the Hamming distance between their incidence vectors.

© The Author(s), under exclusive license to Springer Nature Switzerland AG 2023
M. Khachay et al. (Eds.): MOTOR 2023, CCIS 1881, pp. 134–145, 2023.
https://doi.org/10.1007/978-3-031-43257-6_11

Later, the Graph Approximation problem was repeatedly and independently rediscovered and studied under various names (Correlation Clustering, Cluster Editing, etc. [2,3,11]). Recently, the unweighted version of the problem was named the Cluster Editing problem, whereas the Correlation Clustering problem refers to statements with arbitrary weights of edges [10,11,13].

In different traditional statements of the graph Cluster Editing (**CE**) problem the number of clusters may be given, bounded, or undefined. We focus our attention on a relatively new version of problem **CE** in which the size of every cluster is bounded from above by a positive integer s.

Introduce now some definitions and notation.

We consider only ordinary graphs, i.e., the graphs without loops and multiple edges. Denote by K_2, K_3 and K_4 complete 2-, 3- and 4-vertex graphs, respectively (edge, triangle and tetrahedron). Let Q_4 be a graph obtained from K_4 by removing precisely one edge (absent edge).

An ordinary graph $G = (V, E)$ is called a *cluster graph* if every connected component of G is a complete graph [11]. Let $\mathcal{M}(V)$ be the family of all cluster graphs on the set of vertices V.

If $G_1 = (V, E_1)$ and $G_2 = (V, E_2)$ are ordinary graphs both on the set of vertices V, then the *distance* $d(G_1, G_2)$ between them is defined as

$$d(G_1, G_2) = |E_1 \Delta E_2| = |E_1 \setminus E_2| + |E_2 \setminus E_1|,$$

i.e., $d(G_1, G_2)$ is the number of distinct edges in G_1 and G_2. Evidently, $d(G_1, G_2)$ equals the Hamming distance between the incidence vectors of the graphs G_1 and G_2.

In the 1960–1980s the following Graph Approximation problem was under study. It can be considered as one of formalizations of the graph clustering problem [1,4–6,12,14]:

Problem CE (Cluster Editing). Given a graph $G = (V, E)$, find a graph $M^* \in \mathcal{M}(V)$ such that

$$d(G, M^*) = \min_{M \in \mathcal{M}(V)} d(G, M).$$

The versions of Problem **CE**, where the number of clusters is equal to a given positive integer k (**CE$_k$**) or is bounded from above by k (**CE$_{\leqslant k}$**) were also studied, $2 \leqslant k \leqslant |V|$.

Problem **CE** is NP-hard, problems **CE$_k$** and **CE$_{\leqslant k}$** are NP-hard for any fixed $k \geqslant 2$. Main results on computational complexity and approximation algorithms with performance guarantees for these problems can be found in surveys [7,13].

In this paper, we consider the following statement of the problem. Let $\mathcal{M}^{\leqslant s}(V)$ be the family of all cluster graphs on V such that the size of each connected component is at most an integer s, $2 \leqslant s \leqslant |V|$.

Problem CE$^{\leqslant s}$. Given a graph $G = (V, E)$ and an integer s, $2 \leqslant s \leqslant |V|$, find $M^* \in \mathcal{M}^{\leqslant s}(V)$ such that

$$d(G, M^*) = \min_{M \in \mathcal{M}^{\leqslant s}(V)} d(G, M).$$

In [2], in the proof of NP-hardness of Problem **CE** without any constraints on the number and sizes of clusters, it was actually shown that Problem $\mathbf{CE}^{\leqslant 3}$ is NP-hard. In 2011, Il'ev and Navrotskaya [8] proved that Problem $\mathbf{CE}^{\leqslant s}$ is NP-hard for any fixed $s \geqslant 3$, whereas Problem $\mathbf{CE}^{\leqslant 2}$ is polynomially solvable. In 2015, Puleo and Milenkovic [10] proposed a 6-approximation algorithm for problem $\mathbf{CE}^{\leqslant s}$. In 2016, Il'ev, Il'eva and Navrotskaya [9] presented for this problem an approximation algorithm that is 3-approximation in case of $s = 3$ and 5-approximation in case of $s = 4$.

Apart from theoretical interest, cases of small cluster sizes are also interesting from a practical point of view (e.g., allocation of bulky objects into containers of bounded capacity).

In this paper, we propose simple approximation algorithms for cases $s = 3$ and $s = 4$ with better performance guarantees.

The paper is organized as follows. In Sect. 2, we propose a greedy-type 2-approximation algorithm for Problem $\mathbf{CE}^{\leqslant 3}$. In Sect. 3, a similar polynomial-time 2-approximation algorithm is presented for Problem $\mathbf{CE}^{\leqslant 4}$. Performance guarantees of both algorithms are tight. Conclusion summarizes the results of the work.

2   An Approximation Algorithm for Problem $\mathbf{CE}^{\leqslant 3}$

In [9], for Problem $\mathbf{CE}^{\leqslant s}$ ($s \geqslant 3$) a polynomial-time approximation algorithm with the performance guarantee $\lfloor \frac{(s-1)^2}{2} \rfloor + 1$ was proposed. For cases $s = 3$ and $s = 4$ it yields performance guarantees 3 and 5, respectively. In this section, we offer a polynomial-time 2-approximation algorithm for Problem $\mathbf{CE}^{\leqslant 3}$.

Algorithm A1
Input: An arbitrary graph $G = (V, E_G)$.
Output: A cluster graph $M = (V, E_M) \in \mathcal{M}^{\leqslant 3}(V)$ – an approximate solution to Problem $\mathbf{CE}^{\leqslant 3}$.

0   $G' \leftarrow G$, $M \leftarrow (V, \emptyset)$.
1   **While** there is a clique K_3 in G' **do**
    add a K_3 to M and remove it from G' with all incident edges.
2   **While** there is a clique K_2 in G' **do**
    add a K_2 to M and remove it from G' with all incident edges.
End.

Time complexity of Algorithm A1 is $\mathcal{O}(n^3)$, where $n = |V|$.

Remark 1. Without loss of generality suppose that the optimal graph M^* is a subgraph of G. Note that in Problem $\mathbf{CE}^{\leqslant 3}$ such M^* always exists [8].

Further, we consider only such optimal solutions.

Lemma 1. *Let G be an arbitrary graph, M^* be an optimal solution to Problem $\mathbf{CE}^{\leqslant 3}$ on G. Denote by*

– *E_1 the set of edges of the graph G that are not placed in M^*, but are included in M by Algorithm A1 (i.e., $E_1 = (E_G \cap E_M) \setminus E_{M^*}$);*

- E^* the set of edges of G that are not included in M, but are placed in M^* (i.e., $E^* = (E_G \cap E_{M^*}) \setminus E_M$). Then

$$|E^*| \leqslant 2|E_1|.$$

Proof. Consider the following mental procedure of labelling edges of the set E^* whose steps correspond to steps of Algorithm A1. The procedure assigns labels to all edges of the set E^* that are removed from the current graph G' at every step of Algorithm A1. Labelling are realized with using edges of the set E_1. Each edge $e \in E_1$ gives its labels of the form $e*$ to at most 2 adjacent to e unlabelled edges of the set E^*. At the beginning all edges of E^* are unlabelled.

Note that any clique K_3 found by Algorithm A1 at Step 1 can have 0, 1 or 3 common edges with cliques K_3 of the graph M^*. Hence only 3 or 2 edges of any clique $K_3 \subseteq M$ may belong to the set E_1. Note also that if the edge e of a clique K_2 found by Algorithm A1 at Step 2 belongs to E_1, then at most 2 edges of E^* are adjacent to e in the current graph G'. Therefore, only the following cases are possible.

a) Let K_3 be a clique found by Algorithm A1 at Step 1, where all 3 its edges a, b, c belong to E_1. Then in the graph G' at most 6 unlabelled edges of E^* are adjacent to these 3 edges. They get at most 6 labels of the form $a*, b*, c*$ (Fig. 1a).

b) Let K_3 be a clique found by Algorithm A1 at Step 1, where only 2 its edges a, b belong to E_1. Then in the graph G' at most 4 edges of E^* are adjacent to these 2 edges, and they get labels $a*, b*$ (Fig. 1b).

c) Let K_2 be a clique found by Algorithm A1 at Step 2, and its edge a belongs to E_1. Then in the graph G' at most 2 edges of E^* are adjacent to a, and they get labels $a*$ (Fig. 1c).

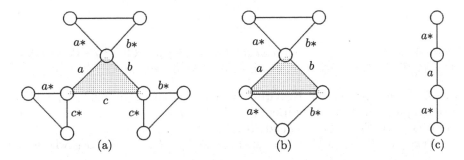

Fig. 1. Gray triangles are cliques found by Algorithm A1, white triangles are cliques of M^* (in M^* some cliques K_3 and K_2 can be replaced by K_2 and K_1, respectively). The double line represents the common edge of the graphs M and M^*. (Color figure online)

On completion of Algorithm A1 the graph G' becomes empty. As far as all edges of the set E^* that are removed from the current graph G' get labels, and each edge $e \in E_1$ gives its labels to at most 2 adjacent to e unlabelled edges of the set E^*, then $|E^*| \le 2|E_1|$.

Lemma 1 is proved.

Example 1. Example of working Algorithm A1 (Fig. 2).

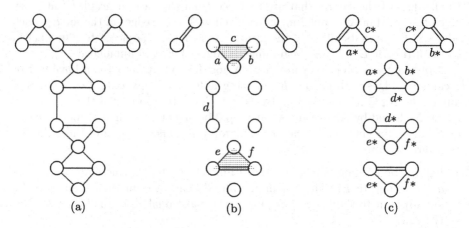

(a)   (b)   (c)

Fig. 2. Graphs G (a), M (b), and M^* (c). Gray triangles are K_3 found by Algorithm A1. Double lines represent common edges of the graphs M and M^*. (Color figure online)

Theorem 1. *Let $G = (V, E_G)$ be an arbitrary graph. Then*

$$\frac{d(G, M)}{d(G, M^*)} \le 2, \tag{1}$$

where $M^* = (V, E_{M^*})$ is an optimal solution to Problem $\mathbf{CE}^{\le 3}$ on the graph G, $M = (V, E_M)$ is the cluster graph constructed by Algorithm A1.

Proof. Taking into account Remark 1 we suppose that the optimal graph M^* is a subgraph of G. By the definition, $d(G, M^*) = |E_G \setminus E_{M^*}| + |E_{M^*} \setminus E_G|$. Since $M^* \subseteq G$, then $E_{M^*} \setminus E_G = \emptyset$, and $d(G, M^*) = |E_G \setminus E_{M^*}|$.

Write the difference $E_G \setminus E_{M^*}$ in the form $E_G \setminus E_{M^*} = E_0 \cup E_1$, where

- E_0 is the set of edges of the graph G that are not placed neither in M^*, nor in M;
- E_1 is the set of edges of the graph G that are not placed in M^*, but are included in M by Algorithm A1 (i.e., $E_1 = (E_G \cap E_M) \setminus E_{M^*}$).

Then

$$d(G, M^*) = |E_0| + |E_1|. \tag{2}$$

By constructing, the graph M is a subgraph of the graph G, therefore $d(G, M) = |E_G \setminus E_M|$. Write the difference $E_G \setminus E_M$ in the following form: $E_G \setminus E_M = E_0 \cup E^*$, where E^* is the set of edges of G not included in M, but placed in M^* (i.e., $E^* = (E_G \cap E_{M^*}) \setminus E_M$).

Then
$$d(G, M) = |E_0| + |E^*|. \tag{3}$$

By (2), (3), and Lemma 1, we obtain

$$\frac{d(G, M)}{d(G, M^*)} = \frac{|E_0| + |E^*|}{|E_0| + |E_1|} \leqslant \frac{|E^*|}{|E_1|} \leqslant \frac{2|E_1|}{|E_1|} = 2.$$

Theorem 1 is proved.

Remark 2. Example 1 shows that bound (1) is tight.

3   An Approximation Algorithm for Problem $CE^{\leqslant 4}$

In [9], a polynomial-time 5-approximation algorithm was proposed for Problem $CE^{\leqslant 4}$. In this section, we offer a polynomial-time 2-approximation algorithm for this problem.

Algorithm A2
Input: An arbitrary graph $G = (V, E_G)$.
Output: A cluster graph $M = (V, E_M) \in \mathcal{M}^{\leqslant 4}(V)$ – an approximate solution to Problem $CE^{\leqslant 4}$.

0    $G' \leftarrow G$, $M \leftarrow (V, \emptyset)$.
1   **While** there is a clique K_4 in G' **do**
     add a K_4 to M and remove it from G' with all incident edges.
2   **While** there is a clique K_3 in G' **do**
     add a K_3 to M and remove it from G' with all incident edges.
3   **While** there is a clique K_2 in G' **do**
     add a K_2 to M and remove it from G' with all incident edges.
End.

Time complexity of Algorithm A1 is $\mathcal{O}(n^4)$, where $n = |V|$.

Remark 3. Without loss of generality suppose that in Problem $CE^{\leqslant 4}$ any 4-vertex component of the optimal solution M^* either coincides with a clique K_4 of a given graph G, or is obtained from a subgraph Q_4 of G by adding the absent edge. Besides that, in Problem $CE^{\leqslant 4}$ such M^* always exists.

Proof. If a clique $K_4 \subseteq M^*$ would be obtained from some 4-vertex subgraph H of $G = (V, E_G)$ by adding at least 2 edges, then instead of their adding we would remove from H at most 2 edges and obtain another cluster graph $M' \in \mathcal{M}^{\leqslant 4}(V)$ such that $d(G, M') \leqslant d(G, M^*)$.

Further, we consider only such optimal solutions to Problem $\mathbf{CE}^{\leqslant 4}$.

Lemma 2. *Let G be an arbitrary graph, M^* be an optimal solution to Problem $\mathbf{CE}^{\leqslant 4}$ on G. Denote by*

- E_1 *the set of edges of the graph G that are not placed in M^*, but are included in M by Algorithm A1 (i.e., $E_1 = (E_G \cap E_M) \setminus E_{M^*}$);*
- $E_2 = E_{M^*} \setminus E_G$;
- E^* *the set of edges of G that are not included in M, but are placed in M^* (i.e., $E^* = (E_G \cap E_{M^*}) \setminus E_M$). Then*

$$|E^*| \leqslant 2(|E_1| + |E_2|).$$

Proof. Once again consider a mental procedure of labelling edges of the set E^* whose steps correspond to steps of Algorithm A2. The procedure assigns labels to all edges of the set E^* that are removed from the current graph G' at every step of Algorithm A2. Labelling are realized with using edges of the sets E_1 and E_2. Each edge $e \in E_1$ or E_2 gives its labels of the form $e*$ to at most 2 adjacent to e unlabelled edges of the set E^*. At the beginning all edges of E^* are unlabelled.

Reasoning as in Lemma 1, one can show that quantity of edges of the set E_1 is sufficient to labelling edges belonging to all cliques K_4, K_3 and K_2 of the graph M^* that are subgraphs of G (Fig. 1, 3). Only some edges of cliques $K_4 \subseteq M^*$ obtained from subgraphs $Q_4 \subset G$ can remain unlabelled with using edges of E_1 at the moment of removing from the current graph G'.

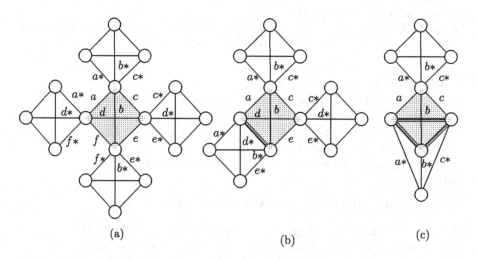

(a)   (b)   (c)

Fig. 3. Gray tetrahedrons are K_4 found by Algorithm A2, white tetrahedrons are cliques of M^* (in M^* some cliques K_4 can be replaced by K_3, K_2 and K_1). Double lines represent common edges of the graphs M and M^*. (Color figure online)

List all cases when quantity of edges of E_1 may be insufficient to labelling all edges of the set E^* that are removed from the current graph G' at some step of

Algorithm A2. In all these cases, edges of the set E^* remaining unlabelled with using edges of E_1 belong to some $Q_4 \subset G$ with absent edge $x \in E_2$.

Case 1. At Step 1 Algorithm A2 finds a cligue $K_4 \subset G'$ three edges a, b, c of which are incident to vertex 1 of Q_4. Then $a, b, c \in E_1$ and they give their labels $a*, b*, c*$ to edges 12, 13 and 14, respectively (it is possible that these are their second labels, and first labels $a*, b*, c*$ were given to some edges incident to other endpoints of edges a, b, c) (Fig. 4a).

After removing K_4 and all incident to it edges from G', the next graph G' will contain edges $23, 34 \in Q_4$. Note that at least one of these edges must belong to E^*.

Two cases are possible.

1) Only one of edges 23, 34 belongs to E^*. W.l.o.g. suppose that $23 \in E^*$.

It means that Algorithm A2 includes in M some clique containing edge 34. If this clique is K_4 or K_3, then it also contains some edge d incident to vertex 3. Note that $d \in E_1$ and it gives its label $d*$ to edge 23.

The only situation, where edge 23 at the moment of its removing from the current graph G' can remain without label getting from an edge of the set E_1 is the following.

Algorithm A2 finds at Step 3 the clique $K_2 =< 34 >$, and so edge 34 is removed from G' with all incident edges. But in this case $34 \notin E_1$, hence edge 23 remains unlabelled with using edges of E_1.

In order to avoid this situation the procedure assigns to edge 23 label $x*$, where $x = 24 \in E_2$ (Fig. 4b).

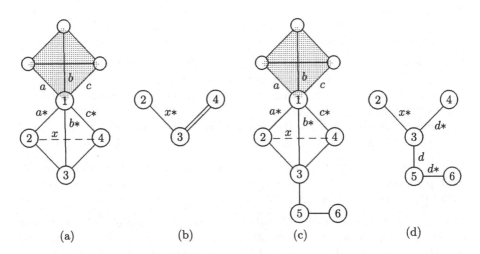

(a)   (b)   (c)   (d)

Fig. 4. Gray tetrahedrons are K_4 found by Algorithm A2. The dotted line represents the edge $x \in E_{M*} \setminus E_G = E_2$. The double line represents the common edge of the graphs M and M^*. (Color figure online)

2) Both edges 23, 34 belong to E^*.

It means that Algorithm A2 includes in M some edges incident to one, two or three vertices of $\{2, 3, 4\}$. These edges belong to E_1 and give theirs labels to both edges 23, 34, except the only situation, when Algorithm A2 finds at Step 3 some clique $K_2 =< 35 >$, where $5 \notin \{2, 4\}$, and besides in G' there is an edge $56 \in E^*$ (Fig. 4c). Then $d = 35 \in E_1$ and d gives its labels $d*$ to edge 56 and one of edges 23, 34, let say 34. In this situation edge 23 remains unlabelled with using edges of E_1.

In order to avoid such situation the procedure assigns to edge 23 label $x*$, where $x = 24 \in E_2$ (Fig. 4d).

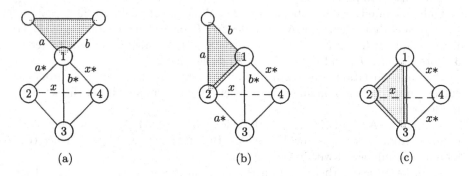

Fig. 5. Gray triangles are K_3 found by Algorithm A2. The dotted line represents the edge $x \in E_{M^*} \setminus E_G = E_2$. Double lines represent common edges of the graphs M and M^*. (Color figure online)

Case 2 differs from case 1 in that Algorithm A2 finds in G' at Step 2 a clique K_3 instead of K_4 (Fig. 5a). Here as in case 1 edges 12 and 13 get labels $a*$ and $b*$, but edge 14 can't be labelled with using edges of the set E_1. So the procedure assigns to 14 label $x*$, where $x = 24 \in E_2$.

And just like in case 1, in order to avoid unwanted situations in the next graph G' listed in subcases 1), 2) of case 1, edge 23 gets label $x*$, where $x = 24 \in E_2$.

Case 3 differs from previous one in that the clique K_3 found by Algorithm A2 at Step 2 has a common edge with Q_4, let say 12 (Fig. 5b). In this case as above edge 14 gets label $x*$, where $x = 24 \in E_2$.

Case 4. At Step 2 Algorithm A2 finds the clique K_3 with edges 12, 13, 23 and removes it from the graph G' with all incident edges. But before removing, the procedure assigns to edges 14 and 34 labels $x*$, where $x = 24 \in E_2$ (Fig. 5c).

On completion of Algorithm A2 the graph G' becomes empty. As far as all edges of the set E^* that are removed from the current graph G' get labels, and each edge $e \in E_1$ or E_2 gives its labels to at most 2 adjacent to e unlabelled edges of the set E^*, then $|E^*| \le 2(|E_1| + |E_2|)$.

Lemma 2 is proved.

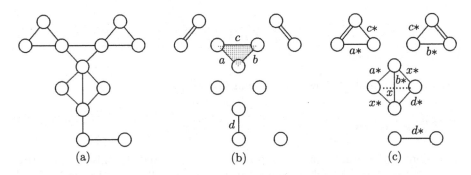

Fig. 6. Graphs G (a), M (b), and M^* (c). The gray triangle is K_3 found by Algorithm A2. The dotted line represents the edge $x \in E_{M^*} \setminus E_G = E_2$. Double lines represent common edges of the graphs M and M^*. (Color figure online)

Example 2. Example of working Algorithm A2 (Fig. 6).

Theorem 2. *Let $G = (V, E_G)$ be an arbitrary graph. Then*

$$\frac{d(G, M)}{d(G, M^*)} \leqslant 2, \qquad (4)$$

where $M^* = (V, E_{M^*})$ is an optimal solution to Problem $\mathbf{CE}^{\leqslant 4}$ on the graph G, $M = (V, E_M)$ is the cluster graph constructed by Algorithm A2.

Proof. By the definition, $d(G, M^*) = |E_G \setminus E_{M^*}| + |E_{M^*} \setminus E_G|$. Write the difference $E_G \setminus E_{M^*}$ in the form $E_G \setminus E_{M^*} = E_0 \cup E_1$, where

- E_0 is the set of edges of the graph G that are not placed neither in M^*, nor in M by Algorithm A2;
- E_1 is the set of edges of the graph G that are not placed in M^*, but are included in M by Algorithm A2.

  Denote $E_2 = E_{M^*} \setminus E_G$. Then

$$d(G, M^*) = |E_0| + |E_1| + |E_2|. \qquad (5)$$

By constructing, the graph M is a subgraph of the graph G, therefore $d(G, M) = |E_G \setminus E_M|$. Write the difference $E_G \setminus E_M$ in the following form: $E_G \setminus E_M = E_0 \cup E^*$, where E^* is the set of edges of G not included in M, but placed in M^*.

Then

$$d(G, M) = |E_0| + |E^*|. \qquad (6)$$

Therefore, by (5), (6), and Lemma 2, we obtain

$$\frac{d(G, M)}{d(G, M^*)} = \frac{|E_0| + |E^*|}{|E_0| + |E_1| + |E_2|} \leqslant \frac{|E^*|}{|E_1| + |E_2|}$$

$$\leqslant \frac{2(|E_1| + |E_2|)}{|E_1| + |E_2|} = 2.$$

Theorem 2 is proved.

Remark 4. Example 2 shows that bound (4) is tight.

4   Conclusion

A version of the graph clustering problem is considered. In this version sizes of all clusters don't exceed a given positive integer s. This problem is NP-hard for every fixed $s \geqslant 3$. New polynomial-time 2-approximation greedy-type algorithms with tight performance guarantees are proposed for the cases $s = 3$ and $s = 4$. Performance guarantees of these algorithms are better than ones of earlier presented approximation algorithms.

Acknowledgement. The research of the first author was funded in accordance with the state task of the IM SB RAS, project FWNF-2022-0020.

References

1. Ageev, A.A., Il'ev, V.P., Kononov, A.V., Talevnin, A.S.: Computational complexity of the graph approximation problem. Diskretn. Anal. Issled. Oper. Ser. 1. **13**(1), 3–11 (2006). (in Russian). English transl. in: J. Appl. Indust. Math. 1(1), 1–8 (2007)
2. Bansal, N., Blum, A., Chawla, S.: Correlation clustering. Mach. Learn. **56**, 89–113 (2004)
3. Ben-Dor, A., Shamir, R., Yakhimi, Z.: Clustering gene expression patterns. J. Comput. Biol. **6**(3–4), 281–297 (1999)
4. Fridman, G.Š.: A graph approximation problem. Upravlyaemye Sistemy. Izd. Inst. Mat. Novosib. **8**, 73–75 (1971). (in Russian)
5. Fridman, G.Š.: Investigation of a classifying problem on graphs. Methods Model. Data Process. (Nauka, Novosibirsk) 147–177 (1976). (in Russian)
6. Il'ev, V.P., Fridman, G.Š.: On the problem of approximation by graphs with a fixed number of components. Dokl. Akad. Nauk SSSR. **264**(3), 533–538 (1982). (in Russian). English transl. in: Sov. Math. Dokl. 25(3), 666–670 (1982)
7. Il'ev, V., Il'eva, S., Kononov, A.: Short survey on graph correlation clustering with minimization criteria. In: Kochetov, Y., Khachay, M., Beresnev, V., Nurminski, E., Pardalos, P. (eds.) DOOR 2016. LNCS, vol. 9869, pp. 25–36. Springer, Cham (2016). https://doi.org/10.1007/978-3-319-44914-2_3
8. Il'ev, V.P., Navrotskaya, A.A.: Computational complexity of the problem of approximation by graphs with connected components of bounded size. Prikl. Diskretn. Mat. **3**(13), 80–84 (2011). (in Russian)
9. Il'ev, V.P., Il'eva, S.D., Navrotskaya, A.A.: Graph clustering with a constraint on cluster sizes. Diskretn. Anal. Issled. Oper. **23**(3), 50–20 (2016). (in Russian). English transl. in: J. Appl. Indust. Math. 10(3), 341–348 (2016)
10. Puleo, G.J., Milenkovic, O.: Correlation clustering with constrained cluster sizes and extended weights bounds. SIAM J. Optim. **25**(3), 1857–1872 (2015)

11. Shamir, R., Sharan, R., Tsur, D.: Cluster graph modification problems. Discret. Appl. Math. **144**(1–2), 173–182 (2004)
12. Tomescu, I.: La reduction minimale d'un graphe à une reunion de cliques. Discret. Math. **10**(1–2), 173–179 (1974)
13. Wahid, D.F., Hassini, E.: A literature review on correlation clustering: cross-disciplinary taxonomy with bibliometric analysis. Oper. Res. Forum **3**(47), 1–42 (2020)
14. Zahn, C.T.: Approximating symmetric relations by equivalence relations. J. Soc. Ind. Appl. Math. **12**(4), 840–847 (1964)

On Cone Partitions for the Min-Cut and Max-Cut Problems with Non-negative Edges

Andrei V. Nikolaev[1]([⊠]) and Alexander V. Korostil[2]

[1] P.G. Demidov Yaroslavl State University, Yaroslavl, Russia
a.nikolaev@uniyar.ac.ru
[2] Tinkoff Bank, Moscow, Russia

Abstract. We consider the classical minimum and maximum cut problems: find a partition of vertices of a graph into two disjoint subsets that minimize or maximize the sum of the weights of edges with endpoints in different subsets. It is known that if the edge weights are non-negative, then the min-cut problem is polynomially solvable, while the max-cut problem is NP-hard.

We construct a partition of the positive orthant into convex cones corresponding to the characteristic cut vectors, similar to a normal fan of a cut polyhedron. A graph of a cone partition is a graph whose vertices are cones, and two cones are adjacent if and only if they have a common facet. We define adjacency criteria in the graphs of cone partitions for the min-cut and max-cut problems. Based on them, we show that for both problems the vertex degrees are exponential, and the graph diameter equals 2. These results contrast with the clique numbers of graphs of cone partitions, which are linear for the minimum cut problem and exponential for the maximum cut problem.

Keywords: Min-cut and max-cut problems · Cut polytope · Cone partition · 1-skeleton · Vertex adjacency · Graph diameter · Vertex degree · Clique number

1 Introduction

We consider two classical problems of finding a cut in an undirected graph.
MINIMUM AND MAXIMUM CUT PROBLEMS.
INSTANCE. Given an undirected graph $G = (V, E)$ with an edge weight function $w : E \to \mathbb{R}_{\geq 0}$.
QUESTION. Find a subset of vertices $S \subset V$ such that the sum of the weights of the edges from E with one endpoint in S and another in $V \backslash S$ is as small as possible (*minimum cut* or *min-cut*) or as large as possible (*maximum cut* or *max-cut*).

The research is supported by the P. G. Demidov Yaroslavl State University Project VIP-016.

Both problems have many practical applications. The minimum cut problem is most often associated with the max-flow min-cut theorem of Ford and Fulkerson: the maximum flow in the flow network is equal to the total weight of the edges in a minimum cut [18]. In particular, it is used in planning communication networks and determining their reliability [23,28]. In turn, the maximum cut problem arises in cluster analysis [12], the Ising model in statistical physics [3], the VLSI design [3,13], and the image segmentation [32].

In terms of computational complexity, the min-cut problem with non-negative edges is polynomially solvable, for example, by Dinic-Edmonds-Karp flow algorithm in $O(|V|^3|E|)$ time [16], or by Stoer-Wagner algorithm in $O(|V||E| + |V|^2 \log |V|)$ time [33]. On the other hand, the min-cut and max-cut problems with arbitrary edges, and the max-cut problem with non-negative edges are known to be NP-hard [20,22]. More background information on the min-cut and max-cut problems can be found in the Encyclopedia of Optimization [17] and the Handbook of Combinatorial Optimization [27].

In this paper, we approach to the cut problem from a polyhedral point of view by studying the properties of the graphs of cone partitions of a positive orthant with respect to the characteristic cut vectors. The results of the research are summarized in Table 1 and highlighted in bold.

Table 1. Pivot table of properties of the graphs of cone partitions for the cut problems in the complete graph K_n

	Arbitrary cut	Minimum non-negative cut	Maximum non-negative cut
Vertex adjacency	$O(1)$ [4]	$O(n)$	$O(n)$
Diameter	1 [4]	2	2
Vertex degree	$2^{n-1} - 1$ [4]	$2^{n-k} + 2^k - 4$	$2^{n-1} - 2^k - 2^{n-k} + 2 + n$
Clique number	2^{n-1} [4]	$2n - 3$ [9]	$\binom{n}{\frac{n}{2}-1}$ or $\binom{n}{\frac{n-1}{2}}$ [6]

2   Cut Polytope and Cone Partition

We consider a complete graph $K_n = (V, E)$ on n vertices. With each subset $S \subseteq V$ we associate the characteristic 0/1−vector $\mathbf{v}(S) \in \{0, 1\}^d$, where $d = C_n^2$ and

$$\mathbf{v}(S)_{i,j} = \begin{cases} 1, & \text{if } |S \cap \{i, j\}| = 1, \\ 0, & \text{otherwise.} \end{cases}$$

Thus, the coordinates of the characteristic vector (also known as the *cut vector*) indicate whether the corresponding edges are in the cut or not.

The *cut polytope* CUT(n) (see Barahona and Mahjoub [4]) is defined as the convex hull of all characteristic (cut) vectors:

$$\text{CUT}(n) = \text{conv} \{\mathbf{v}(S) : S \subseteq V\} \subset \mathbb{R}^d.$$

An example of constructing the cut polytope CUT(3) is shown in Fig. 1.

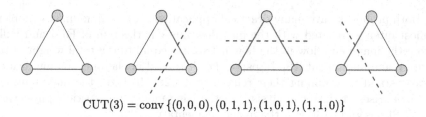

$$\text{CUT}(3) = \text{conv}\left\{(0,0,0),(0,1,1),(1,0,1),(1,1,0)\right\}$$

Fig. 1. An example of constructing a cut polytope for K_3

The cut polytope and its various relaxations often serve as the linear programming models for the cut problem. See, for example, the polynomial time algorithm by Barahona for the max-cut problem on graphs not contractible to K_5 [5].

We consider a dual construction of a cone partition (see Bondarenko [7,8]), similar to a *normal fan* (see, for example, Ziegler [34]). Let X be some set of points in $\mathbb{R}^d$ (for example, all cut vectors $\mathbf{v}(S)$ for $S \subseteq V$), and $\mathbf{x} \in X$. Denote by

$$K(\mathbf{x}) = \{\mathbf{c} \in \mathbb{R}^d : \langle \mathbf{c}, \mathbf{x} \rangle \geq \langle \mathbf{c}, \mathbf{y} \rangle, \ \forall \mathbf{y} \in X\},$$

where $\langle \mathbf{c}, \mathbf{x} \rangle = \mathbf{c}^T \mathbf{x}$ is the scalar product. Thus, $K(\mathbf{x})$ as a set of solutions to a system of linear homogeneous inequalities is a convex polyhedral cone that includes all points $\mathbf{c} \in \mathbb{R}^d$, for which the linear function $\mathbf{c}^T \mathbf{x}$ achieves its maximum on the set X at the point $\mathbf{x}$. The collection of all cones $K(\mathbf{x})$ is called the *cone partition of the space* $\mathbb{R}^d$ *with respect to the set* X (Fig. 2). The cone partition is analogous to the Voronoi diagram, exactly coinciding with it if the Euclidean norms of all points of the set X are equal.

The 1-*skeleton* of a polytope P is the graph whose vertex set is the vertex set of P and the edge set is the set of geometric edges or one-dimensional faces of P. The cone partition of space is directly related to the 1-skeleton of a polytope since two vertices $\mathbf{x}_1$ and $\mathbf{x}_2$ of the polytope conv(X) are adjacent if and only if the cones $K(\mathbf{x}_1)$ and $K(\mathbf{x}_2)$ have a common facet (see Bondarenko [7]):

$$\mathbf{x}_1 \text{ and } \mathbf{x}_2 \text{ adjacent } \Leftrightarrow \dim\left(K\left(\mathbf{x}_1\right) \cap K\left(\mathbf{x}_2\right)\right) = d - 1.$$

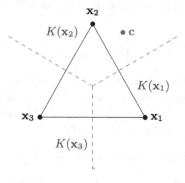

Fig. 2. An example of a 1-skeleton and cone partition

We call such two cones *adjacent* and consider *the graph of a cone partition* of the space $\mathbb{R}^d$ with respect to the set X. In the general case, it coincides with the 1-skeleton of a polytope.

The study of the 1-skeleton is of interest, since, on the one hand, the vertex adjacency can be directly applied to develop simplex-like combinatorial optimization algorithms that move from one feasible solution to another along the edges of the 1-skeleton. See, for example, the set partitioning algorithm by Balas and Padberg [1], Balinski's algorithm for the assignment problem [2], Ikura and Nemhauser's algorithm for the set packing [21], etc.

On the other hand, some characteristics of the 1-skeleton estimate the time complexity for different computation models and classes of algorithms. In particular, *the diameter* (the greatest distance between any pair of vertices) is a lower bound for the number of iterations of the simplex method and similar algorithms. Indeed, let the shortest path between a pair of vertices $\mathbf{u}$ and $\mathbf{v}$ of a polytope P consist of $d(P)$ edges. If a simplex-like algorithm chooses $\mathbf{u}$ as the initial solution, and the optimal solution is $\mathbf{v}$, then no matter how successfully the algorithm chooses the next adjacent vertex of the 1-skeleton, the number of iterations cannot be less than $d(P)$ (see, for example, Dantzig [15]).

Note that although the diameter of a graph can easily be found in polynomial time in the number of vertices, combinatorial polytopes tend to have exponentially many vertices. In general, it is NP-hard to determine the diameter of a 1-skeleton of a polytope specified by linear inequalities with integer data (see Frieze and Teng [19]).

Another important characteristic is the *clique number* of the 1-skeleton of the polytope P (the number of vertices in the largest clique), which serves as a lower bound on the worst-case complexity in the class of algorithms based on linear decision trees. See Bondarenko [8] for more details. Besides, for all known cases, it has been established that the clique number of 1-skeleton of a polytope is polynomial for polynomially solvable problems [10, 24, 25] and superpolynomial for intractable problems [7, 11, 26, 30].

Returning to the cut polytope, the properties of its 1-skeleton were studied by Barahona and Mahjoub in [4].

Theorem 1 (Barahona and Mahjoub [4]). *The 1-skeleton of the* $\mathrm{CUT}(n)$ *polytope is a complete graph.*

Thus, any two vertices of the cut polytope $\mathrm{CUT}(n)$ are adjacent, which makes the 1-skeleton not very useful in this case.

However, the polytope $\mathrm{CUT}(n)$ is associated with the cut problem in the graph with arbitrary edge weights and does not reflect the differences between max-cut and min-cut problems with non-negative edges. Since with arbitrary edges, both min-cut and max-cut problems are equivalent and NP-hard [20].

To take into account the specifics of the cut problem, the construction of a *cut polyhedron* is introduced (see Conforti et al. [14]), which is the dominant of a cut polytope:

$$\mathrm{dmt}(\mathrm{CUT}(n)) = \mathrm{CUT}(n) + \mathbb{R}_+^d,$$

i.e. the Minkowski sum of a polytope and a positive orthant.

In this paper, we consider the dual construction of a cone partition of a positive orthant with respect to the set of cut vectors for all non-empty cuts $X \subset V$ in the complete graph $K_n = (V, E)$:

$$K_{\max}^+(X) = \{\mathbf{c} \in \mathbb{R}^d,\ \mathbf{c} \geq \mathbf{0} :\ \langle \mathbf{c}, \mathbf{v}(X) \rangle \geq \langle \mathbf{c}, \mathbf{v}(Y) \rangle,\ \forall Y \subset V\},$$
$$K_{\min}^+(X) = \{\mathbf{c} \in \mathbb{R}^d,\ \mathbf{c} \geq \mathbf{0} :\ \langle \mathbf{c}, \mathbf{v}(X) \rangle \leq \langle \mathbf{c}, \mathbf{v}(Y) \rangle,\ \forall Y \subset V\}.$$

Firstly, these cone partitions and their graphs were introduced in [9] and later studied in [6]. In particular, it was established that the clique number of a graph of a cone partition is linear for the polynomially solvable min-cut problem and exponential for the NP-hard max-cut problem. Similar results for the cut polyhedron and the minimum cut problem are considered in [14,31].

Note that we exclude the empty cut from consideration and identify each cut $X \subset V$ and its complement $\bar{X} = V \backslash X$. Thus, the total number of cuts is $2^{|V|-1} - 1$.

3   Vertex Adjacency

The adjacency criteria in the graphs of cone partitions for cut problems with non-negative edges were introduced in [9]. In this section, we present simpler alternative versions of the criteria with new proofs based on crossing sets terminology and the submodularity of the cut function.

Two subsets $A, B \subset V$ are called *crossing* if

$$A \cap B \neq \emptyset,\ \text{and}\ A \backslash B \neq \emptyset,\ \text{and}\ B \backslash A \neq \emptyset,\ \text{and}\ V \backslash (A \cup B) \neq \emptyset.$$

An example of crossing sets is shown in Fig. 3.

Theorem 2. *The cones $K_{\min}^+(X)$ and $K_{\min}^+(Y)$ are adjacent if and only if the cuts X and Y are not crossing.*

Proof. Suppose that the cuts X and Y are crossing, but the cones $K_{\min}^+(X)$ and $K_{\min}^+(Y)$ are adjacent. The adjacency of cones means that there exists a non-negative vector $\mathbf{c}$ that belongs to both cones $K_{\min}^+(X)$ and $K_{\min}^+(Y)$ but does

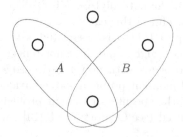

Fig. 3. An example of crossing sets

not belong to any other cone from the partition $K_{\min}^+$:

$$\exists \mathbf{c} \in \mathbb{R}^d \; (\mathbf{c} \geq \mathbf{0}): \; \langle \mathbf{c}, \mathbf{v}(X) \rangle = \langle \mathbf{c}, \mathbf{v}(Y) \rangle < \langle \mathbf{c}, \mathbf{v}(Z) \rangle, \; \forall Z \subset V \; (Z \neq X, Y). \quad (1)$$

An important property of the cut function is *submodularity* (see Schrijver [29]):

$$\langle \mathbf{c}, \mathbf{v}(X) \rangle + \langle \mathbf{c}, \mathbf{v}(Y) \rangle \geq \langle \mathbf{c}, \mathbf{v}(X \cup Y) \rangle + \langle \mathbf{c}, \mathbf{v}(X \cap Y) \rangle.$$

Since the cuts X and Y are crossing, both cuts $X \cup Y$ and $X \cap Y$ exist and are not empty. Moreover, the value of at least one of them does not exceed $\langle \mathbf{c}, \mathbf{v}(X) \rangle$ and $\langle \mathbf{c}, \mathbf{v}(Y) \rangle$ due to the submodularity of the cut function. Therefore, the inequality (1) is violated, and the cones $K_{\min}^+(X)$ and $K_{\min}^+(Y)$ are not adjacent.

Now suppose that the cuts X and Y are not crossing. It is easy to check that in this case at least one of the cuts or its complement is a subset of another cut or its complement. Without loss of generality, we assume that $X \subset Y$.

We consider the following vector $\mathbf{c}$ of edge weights (Fig. 4):

- the total weight of the edges between X and $Y \backslash X$, and between $\bar{Y}$ and $Y \backslash X$ are both equal to 2;
- edges between X and $\bar{Y}$ have total weight 1;
- all other edges have weight 4.

By construction, the values of the cuts X and Y are both equal to 3, the value of the cut $Y \backslash X$ is 4, and all other cuts contain at least one edge of the weight 4. Thus, by the inequality (1), the cones $K_{\min}^+(X)$ and $K_{\min}^+(Y)$ are adjacent.

Theorem 3. *The cones $K_{\max}^+(X)$ and $K_{\max}^+(Y)$ are adjacent if and only if one of the following conditions is satisfied:*

- *cuts X and Y are crossing;*
- *the symmetric difference between cuts X and Y contains exactly one element*

$$|X \ominus Y| = 1, \; or \; |\bar{X} \ominus Y| = 1, \; or \; |X \ominus \bar{Y}| = 1, \; or \; |\bar{X} \ominus \bar{Y}| = 1.$$

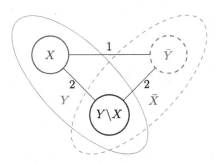

Fig. 4. Cuts of X and Y in the case of $X \subset Y$

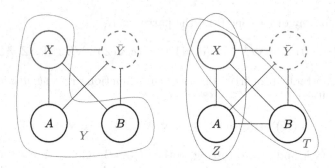

Fig. 5. Cuts X, Y, Z, and T

Proof. Adjacency of the cones $K_{\max}^+(X)$ and $K_{\max}^+(Y)$ means that:

$$\exists \mathbf{c} \in \mathbb{R}^d \ (\mathbf{c} \geq \mathbf{0}): \ \langle \mathbf{c}, \mathbf{v}(X) \rangle = \langle \mathbf{c}, \mathbf{v}(Y) \rangle > \langle \mathbf{c}, \mathbf{v}(Z) \rangle, \ \forall Z \subset V \ (Z \neq X, Y). \quad (2)$$

Let the cuts X and Y be not crossing and contain more than one element in the symmetric difference. Without loss of generality, we examine the case $X \subset Y$. Since $|X \ominus Y| > 1$, the set $Y \setminus X$ contains at least two elements and can be divided into two non-empty subsets A and B. We consider two additional cuts $Z = X \cup A$ and $T = X \cup B$ (Fig. 5). By the submodularity of the cut function, we obtain that

$$\langle \mathbf{c}, \mathbf{v}(Z) \rangle + \langle \mathbf{c}, \mathbf{v}(T) \rangle \geq \langle \mathbf{c}, \mathbf{v}(X = Z \cap T) \rangle + \langle \mathbf{c}, \mathbf{v}(Y = Z \cup T) \rangle.$$

The value of at least one of the cuts Z or T cannot be less than the value of X and Y. Thus, by (2), the cones $K_{\max}^+(X)$ and $K_{\max}^+(Y)$ are not adjacent.

Now suppose that the cuts X and Y are crossing. Again we consider the special vector $\mathbf{c}$ of edge weights (Fig. 6):

- all edges between $X \cap Y$ and $\bar{X} \cap \bar{Y}$, and between $X \cap \bar{Y}$ and $\bar{X} \cap Y$ have positive weights with a total sum both equal to 1;
- the weights of all other edges are zero.

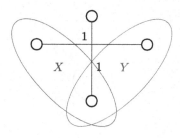

Fig. 6. The case of crossing cuts X and Y

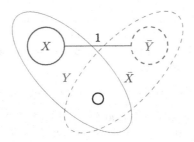

Fig. 7. Case of cuts X and Y with one element in the symmetric difference

Both cuts X and Y have a value of 2 equal to the total sum of all edges in the graph. Any other cut skips at least one non-zero edge, and its value is less than 2. Therefore, the cones $K_{\max}^+(X)$ and $K_{\max}^+(Y)$ are adjacent.

It remains to consider the last configuration when the cuts X and Y do not cross and contain only one element in the symmetric difference. Without loss of generality, we assume $X \subset Y$ and $|X \ominus Y| = 1$. We consider the following vector **c** of edge weights (Fig. 7):

- each edge between X and $\bar{Y}$ has a positive weight and their total sum is equal to 1;
- the weights of all other edges are equal to 0.

Again, the total sum of the weights of all edges in the graph and the values of the cuts X and Y are equal to 1. Any other cut skips at least one non-zero edge, and its value is less than 1. Thus, the cones $K_{\max}^+(X)$ and $K_{\max}^+(Y)$ are adjacent.

Note that if $X, Y \subset V$ are two cuts, then verifying whether the sets are crossing and the corresponding cones are adjacent can be done in linear time $O(V)$.

4   Graph Diameter

In this section, we present new results on the diameter of the graphs of cone partitions for the min-cut and max-cut problems with non-negative edges.

Theorem 4. *The diameter $d(K_{\min}^+)$ of the graph of cone partition for the min-cut problem with non-negative edges is equal to 2 for all $|V| \geq 4$.*

Proof. Cases $|V| \leq 3$ are trivial: in a graph on two vertices there is only one non-empty cut, and in a graph on three vertices, the cones of all three non-empty cuts are pairwise adjacent.

Recall that *the eccentricity* $\epsilon(\mathbf{v})$ of a graph vertex $\mathbf{v}$ is the greatest distance between $\mathbf{v}$ and any other vertex of a graph. Let us show that the graph of the cone partition $K_{\min}^+$ contains vertices with eccentricity 1, i.e. vertices that are adjacent to all others. We choose a cut X separating exactly one element ($|X| = 1$ or $|\bar{X}| = |V| - 1$). Consider an arbitrary cut Y different from X:

– if $X \subset Y$, then $X \backslash Y = \emptyset$, the cuts are not crossing, and the corresponding cones $K_{\min}^{+}(X)$ and $K_{\min}^{+}(Y)$ are adjacent;
– if $X \not\subset Y$, then $X \cap Y = \emptyset$, the cuts are not crossing, and the cones $K_{\min}^{+}(X)$ and $K_{\min}^{+}(Y)$ are also adjacent.

Thus, $\epsilon(K_{\min}^{+}(X)) = 1$, hence the graph diameter equals 2.

Note that for other cuts Y such that $2 \leq |Y| \leq |V| - 2$, we have $\epsilon(K_{\min}^{+}(Y)) = 2$. Indeed, for each Y there exists a crossing cut Z such that the cones $K_{\min}^{+}(Y)$ and $K_{\min}^{+}(Z)$ are not adjacent, but there is a path between them in the graph of cone partition through $K_{\min}^{+}(X)$, where $|X| = 1$.

Theorem 5. *The diameter $d(K_{\max}^{+})$ of the graph of cone partition for the max-cut problem with non-negative edges is equal to 2 for all $|V| \geq 4$.*

Proof. Again, cases $|V| \leq 3$ are trivial. Besides, if $|V| = 4$, then for any cut X, where $|X| = 2$, the cone $K_{\max}^{+}(X)$ has eccentricity 1. Indeed, any other cut on two elements is crossing with X, and for any cut on 1 or 3 elements, the symmetric difference with X or $V \backslash X$ contains exactly 1 element.

Let us show that if $|V| \geq 5$, then the eccentricity of each vertex in the graph of a cone partition equals 2. We consider two arbitrary cuts X and Y whose cones are not adjacent and construct a cut Z such that the cone $K_{\max}^{+}(Z)$ is adjacent to both $K_{\max}^{+}(X)$ and $K_{\max}^{+}(Y)$.

By Theorem 3, the cones $K_{\max}^{+}(X)$ and $K_{\max}^{+}(Y)$ are not adjacent if and only if the cuts X and Y are not crossing, and the symmetric difference between cuts contains more than one element. In this case, at least one of the cuts or its complement is a subset of another cut or its complement. Without loss of generality, we assume that $X \subset Y$ and $|X \ominus Y| \geq 2$. Let us construct a cut Z according to the following rules.

1. If $|X| > 1$, then $Z = \bar{Y} \cup \{x\}$, where x is one of the elements of X (Fig. 8). The cuts X and Z are crossing, hence the cones $K_{\max}^{+}(X)$ and $K_{\max}^{+}(Z)$ are adjacent. The cut $\bar{Y}$ is a subset of Z and the symmetric difference is exactly one element x. Therefore, the cones $K_{\max}^{+}(Y)$ and $K_{\max}^{+}(Z)$ are also adjacent.
2. If $|\bar{Y}| > 1$, then $Z = X \cup \{y\}$, where y is one of the elements of $\bar{Y}$ (Fig. 9). Similarly to the previous case, the cuts $\bar{Y}$ and Z are crossing, and X is a subset of Z with exactly one element in the symmetric difference. Therefore, the cones $K_{\max}^{+}(X)$ and $K_{\max}^{+}(Y)$ are adjacent to the cone $K_{\max}^{+}(Z)$.
3. If $|X| = |\bar{Y}| = 1$, then $Z = X \cup \bar{Y}$ (Fig. 10). The cuts X and $\bar{Y}$ are subsets of Z and differ from Z by exactly one element. Therefore, the cones $K_{\max}^{+}(X)$ and $K_{\max}^{+}(Y)$ are adjacent to the cone $K_{\max}^{+}(Z)$.

Therefore, for any cuts X and Y there exists a cut Z whose cone $K_{\max}^{+}(Z)$ is adjacent both to $K_{\max}^{+}(X)$ and $K_{\max}^{+}(Y)$. On the other hand, if $|V| \geq 5$, then for any cut X there is a cut Y obtained from X by adding or removing 2 elements, such that the cones $K_{\max}^{+}(X)$ and $K_{\max}^{+}(Y)$ are not adjacent. Thus, the eccentricity of each vertex of the graph of a cone partition equals 2, whence the diameter of the graph is also 2.

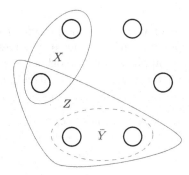

Fig. 8. Case $|X| > 1$ and $Z = \bar{Y} \cup \{x\}$, where x is one of the elements of X

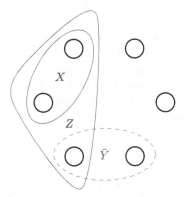

Fig. 9. Case $|\bar{Y}| > 1$, then $Z = X \cup \{y\}$, where y is one of the elements of $\bar{Y}$

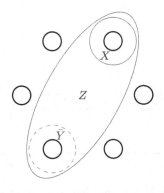

Fig. 10. Case $|X| = |\bar{Y}| = 1$ and $Z = X \cup \bar{Y}$

5   Vertex Degrees

Now we study the degrees of vertices in the graphs of cone partitions for the min-cut and max-cut problems with non-negative edges. They are of interest if the adjacency criteria are applied as neighborhood structures for simplex-like algorithms. Let's call the *cardinality of the cut S* the minimum of the cardinalities $|S|$ and $|V \backslash S|$.

Theorem 6. *Let $|V| = n$ and the cardinality of the cut $X \subset V$ equals k, then the degree of the vertex $K_{\min}^+(X)$ in the graph of the cone partition for the min-cut problem with non-negative edges is equal to*

$$2^{n-k} + 2^k - 4.$$

Proof. We separately consider two cases by the cardinality of the cut. If $k = 1$, then, by Theorem 2, the corresponding cone is adjacent to the cones of all other cuts in the graph. Therefore, the degree of a vertex in the graph of cone partition is equal to $2^{n-1} - 2$.

Now we examining the case $k > 1$. Consider some cut X and its complement $\bar{X} = V \backslash X$ (see Fig. 11). By definition, a cut Y crossing with X must contain some elements from X and some elements from $\bar{X}$. The number of subsets of a finite set of k elements, excluding the empty set and the entire set, is $2^k - 2$. Consider all combinations of admissible subsets in X and $\bar{X}$, divided by 2 to exclude complements, and we get that the total number of cuts that cross X is

$$\frac{(2^k - 2) \cdot (2^{n-k} - 2)}{2} = 2^{n-1} - 2^k - 2^{n-k} + 2.$$

The degree of the vertex in the graph of the cone partition corresponding to the cut X can be obtained by subtracting the number of crossing cuts from the total number of cuts, different from X:

$$\deg(K_{\min}^+(X)) = 2^{n-1} - 2 - \left(2^{n-1} - 2^k - 2^{n-k} + 2\right)$$
$$= 2^{n-k} + 2^k - 4.$$

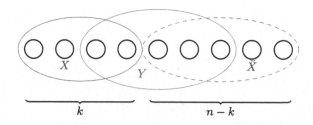

Fig. 11. The cut X, its complement $\bar{X}$, and the crossing cut Y

Corollary 1. *The degree of a vertex in the graph of the cone partition for the min-cut problem with non-negative edges is bounded above and below by*

$$2^{\lceil \frac{n}{2} \rceil} + 2^{\lfloor \frac{n}{2} \rfloor} - 4 \leq \deg(K_{\min}^{+}(X)) \leq 2^{n-1} - 2.$$

We now turn to the max-cut problem.

Theorem 7. *Let $|V| = n$ and the cardinality of the cut $X \subset V$ equals k, then the degree of the vertex $K_{\max}^{+}(X)$ in the graph of the cone partition for the max-cut problem with non-negative edges is equal to*

$$\begin{cases} n - 1, & \text{if } k = 1, \\ 2^{n-1} - 2^{k} - 2^{n-k} + 2 + n, & \text{otherwise.} \end{cases}$$

Proof. Similarly, we consider separately the cases $k = 1$ and $k > 1$. Let $k = 1$. A unit cut X cannot cross with any other cut. Therefore, the cone $K_{\max}^{+}(X)$ is adjacent only to the cones $K_{\max}^{+}(Y)$ for which $|X \ominus Y| = 1$, i.e. $X \subset Y$ and $|Y \backslash X| = 1$. There are exactly $n - 1$ such cuts in total.

Now consider some cut X of cardinality $k > 1$. As previously stated, the number of cuts that are crossing with X is equal to

$$2^{n-1} - 2^{k} - 2^{n-k} + 2.$$

The number of cuts Y for which $|X \ominus Y| = 1$ is equal to n since there are k ways to subtract one element from the cut X and $n - k$ ways to add one element to X (see Fig. 12). Therefore, the degree of the vertex corresponding to the cut X is equal to

$$\deg(K_{\max}^{+}(X)) = 2^{n-1} - 2^{k} - 2^{n-k} + 2 + n.$$

Corollary 2. *The degree of a vertex in the graph of the cone partition for the max-cut problem with non-negative edges is bounded above and below by*

$$n - 1 \leq \deg(K_{\max}^{+}(X)) \leq 2^{n-1} - 2^{\lceil \frac{n}{2} \rceil} - 2^{\lfloor \frac{n}{2} \rfloor} + 2 + n.$$

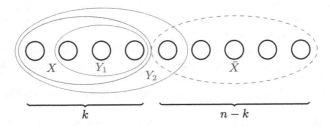

Fig. 12. An example of cuts X, Y_1, Y_2, such that $|X \ominus Y_1| = |X \ominus Y_2| = 1$

6   Conclusion

The results of the research are summarized in the pivot Table 1. The cut problem with arbitrary edges and the max-cut problem with non-negative edges are NP-hard. On the other hand, the min-cut problem with non-negative edges is polynomially solvable. Studying the 1-skeleton and the graphs of cone partitions associated with the cut problem, we see that in all cases verifying the adjacency is a simple problem, the diameter of the polyhedral graph does not exceed 2, and the degrees of vertices are exponential. Of all the characteristics of a 1-skeleton, only the clique number correlates with the complexity of the problem: linear for the polynomially solvable min-cut problem with non-negative edges, and superpolynomial for the NP-hard max-cut problem with non-negative edges and cut problems with arbitrary edges. Besides, the adjacency criteria and the structure of the graphs of cone partitions can be applied to develop and analyze simplex-like combinatorial algorithms for cut problems with non-negative edges.

Acknowledgements. We are very grateful to the anonymous reviewers for their comments and suggestions which helped to improve the presentation of the results in this paper.

References

1. Balas, E., Padberg, M.: On the set-covering problem: Ii. an algorithm for set partitioning. Operations Research **23**(1), 74–90 (1975). https://doi.org/10.1287/opre.23.1.74
2. Balinski, M.L.: Signature methods for the assignment problem. Oper. Res. **33**(3), 527–536 (1985). https://doi.org/10.1287/opre.33.3.527
3. Barahona, F., Grötschel, M., Jünger, M., Reinelt, M.: An application of combinatorial optimization to statistical physics and circuit layout design. Oper. Res. **36**(3), 493–513 (1988). https://doi.org/10.1287/opre.36.3.493
4. Barahona, F., Mahjoub, A.R.: On the cut polytope. Math. Program. **36**(2), 157–173 (1986). https://doi.org/10.1007/BF02592023
5. Barahona, F.: The max-cut problem on graphs not contractible to K_5. Oper. Res. Lett. **2**, 107–111 (1983). https://doi.org/10.1016/0167-6377(83)90016-0
6. Bondarenko, V., Nikolaev, A.: On graphs of the cone decompositions for the min-cut and max-cut problems. Int. J. Math. Math. Sci. **2016**, 7863650 (2016). https://doi.org/10.1155/2016/7863650
7. Bondarenko, V.A.: Nonpolynomial lower bounds for the complexity of the traveling salesman problem in a class of algorithms. Autom. Remote. Control. **44**(9), 1137–1142 (1983)
8. Bondarenko, V.A.: Estimating the complexity of problems on combinatorial optimization in one class of algorithms. Phys.-Dokl. **38**(1), 6–7 (1993)
9. Bondarenko, V.A., Nikolaev, A.V.: Combinatorial and geometric properties of the max-cut and min-cut problems. Dokl. Math. **88**(2), 516–517 (2013). https://doi.org/10.1134/S1064562413050062
10. Bondarenko, V.A., Nikolaev, A.V.: On the skeleton of the polytope of pyramidal tours. J. Appl. Ind. Math. **12**(1), 9–18 (2018). https://doi.org/10.1134/S1990478918010027

11. Bondarenko, V.A., Nikolaev, A.V., Shovgenov, D.A.: Polyhedral characteristics of balanced and unbalanced bipartite subgraph problems. Autom. Control. Comput. Sci. **51**(7), 576–585 (2017). https://doi.org/10.3103/S0146411617070276

12. Boros, E., Hammer, P.L.: On clustering problems with connected optima in Euclidean spaces. Discret. Math. **75**(1), 81–88 (1989). https://doi.org/10.1016/0012-365X(89)90080-0

13. Chen, R.W., Kajitani, Y., Chan, S.P.: A graph-theoretic via minimization algorithm for two-layer printed circuit boards. IEEE Trans. Circuits Syst. **30**(5), 284–299 (1983). https://doi.org/10.1109/TCS.1983.1085357

14. Conforti, M., Rinaldi, G., Wolsey, L.: On the cut polyhedron. Discrete Math. **277**(1), 279–285 (2004). https://doi.org/10.1016/j.disc.2002.12.001

15. Dantzig, G.B.: Linear Programming and Extensions. RAND Corporation, Santa Monica, CA (1963). https://doi.org/10.7249/R366

16. Edmonds, J., Karp, R.M.: Theoretical improvements in algorithmic efficiency for network flow problems. J. ACM **19**(2), 248–264 (1972). https://doi.org/10.1145/321694.321699

17. Floudas, C.A., Pardalos, P.M.: Encyclopedia of Optimization. Springer, New York (2009). https://doi.org/10.1007/978-0-387-74759-0

18. Ford, L.R., Fulkerson, D.R.: Maximal flow through a network. Can. J. Math. **8**, 399–404 (1956). https://doi.org/10.4153/CJM-1956-045-5

19. Frieze, A.M., Teng, S.H.: On the complexity of computing the diameter of a polytope. Comput. Complex. **4**(3), 207–219 (1994). https://doi.org/10.1007/BF01206636

20. Garey, M.R., Johnson, D.S.: Computers and Intractability: A Guide to the Theory of NP-Completeness (Series of Books in the Mathematical Sciences). Freeman, W. H (1979)

21. Ikura, Y., Nemhauser, G.L.: Simplex pivots on the set packing polytope. Math. Program. **33**, 123–138 (1985). https://doi.org/10.1007/BF01582240

22. Karp, R.M.: Reducibility among combinatorial problems. In: Miller, R.E., Thatcher, J.W., Bohlinger, J.D. (eds.) Complexity of Computer Computations. The IBM Research Symposia Series, pp. 85–103. Springer, Boston (1972). https://doi.org/10.1007/978-1-4684-2001-2_9

23. Lun, D.S., Médard, M., Koetter, R., Effros, M.: On coding for reliable communication over packet networks. Phys. Commun. **1**(1), 3–20 (2008). https://doi.org/10.1016/j.phycom.2008.01.006

24. Maksimenko, A.: Combinatorial properties of the polyhedron associated with the shortest path problem. Comput. Math. Math. Phys. **44**, 1611–1614 (2004)

25. Nikolaev, A.V.: On 1-skeleton of the polytope of pyramidal tours with step-backs. Siberian Electron. Math. Rep. **19**, 674–687 (2022). https://doi.org/10.33048/semi.2022.19.056

26. Padberg, M.: The Boolean quadric polytope: some characteristics, facets and relatives. Math. Program. **45**(1), 139–172 (1989). https://doi.org/10.1007/BF01589101

27. Pardalos, P.M., Du, D.Z., Graham, R.L.: Handbook of Combinatorial Optimization. Springer, New York (2013). https://doi.org/10.1007/978-1-4419-7997-1

28. Picard, J.C., Queyranne, M.: On the structure of all minimum cuts in a network and applications. In: Rayward-Smith, V.J. (ed.) Combinatorial Optimization II. Mathematical Programming Studies, vol. 13, pp. 8–16. Springer, Heidelberg (1980). https://doi.org/10.1007/BFb0120902

29. Schrijver, A.: Combinatorial Optimization: Polyhedra and Efficiency. Springer, Heidelberg (2003)

30. Simanchev, R.Y.: On the vertex adjacency in a polytope of connected k-factors. Trudy Inst. Mat. i Mekh. UrO RAN **24**, 235–242 (2018). https://doi.org/10.21538/0134-4889-2018-24-2-235-242

31. Skutella, M., Weber, A.: On the dominant of the s-t-cut polytope: vertices, facets, and adjacency. Math. Program. **124**(1), 441–454 (2010). https://doi.org/10.1007/s10107-010-0373-7

32. de Sousa, S., Haxhimusa, Y., Kropatsch, W.G.: Estimation of distribution algorithm for the max-cut problem. In: Kropatsch, W.G., Artner, N.M., Haxhimusa, Y., Jiang, X. (eds.) GbRPR 2013. LNCS, vol. 7877, pp. 244–253. Springer, Heidelberg (2013). https://doi.org/10.1007/978-3-642-38221-5_26

33. Stoer, M., Wagner, F.: A simple min-cut algorithm. J. ACM **44**(4), 585–591 (1997). https://doi.org/10.1145/263867.263872

34. Ziegler, G.: Lectures on Polytopes. Graduate Texts in Mathematics. Springer, New York (1995). https://doi.org/10.1007/978-1-4613-8431-1

A Pattern-Based Heuristic for a Temporal Bin Packing Problem with Conflicts

A. Ratushnyi[(✉)][iD]

Sobolev Institute of Mathematics SB RAS, Novosibirsk, Russia
alexeyratushny@gmail.com

Abstract. We introduce a new temporal bin packing problem with conflicts that originated from cloud data centers. For each item (virtual machine), we know an arrival time, a finish time, and two weights (CPU and RAM) which are defined by a corresponding type. Each bin (server) has two capacities and is divided into several non-identical parts, called NUMA nodes. Some items are large and have to be split into two predefined parts. Our goal is to pack all items into the minimum number of bins during the known time horizon. For this problem, we design a heuristic based on a column generation approach. We consider a static problem at one moment with a heavy load to obtain a lower bound. The resulting patterns (columns) are then reused to construct a feasible solution to the entire temporal problem. However, the distribution of types may vary at different moments. To address this, we propose to generate more versatile sets of patterns that take into account several moments. We perform computational experiments on semi-synthetic instances with ten types, 10000 items for bins with two NUMA nodes, and 20000 items for bins with four NUMA nodes. Every instance has a 10000-long time horizon. Computational results demonstrate the effectiveness of the approach and a small average gap.

Keywords: bin packing problem · knapsack problem · column generation · temporal · conflicts

1 Introduction

Cloud computing raises a wide variety of challenging problems. One of them is the optimal distribution of client requests for resource allocation. This necessarily brings us to the temporal bin packing problem. It is a well-known and widely studied optimization problem by many researchers. This paper considers a new variation of the temporal bin packing problem. It has type conflicts and the NUMA (Non-uniform memory access) architecture [10] of bins. We also consider large and small sizes of items. Size affects the packing procedure. The problem is NP-hard since it is a generalization of the bin packing problem [11].

A variation of the temporal bin packing problem considered in the conditions of the data center is proposed in [3]. Authors propose MIP and both a

M. Khachay et al. (Eds.): MOTOR 2023, CCIS 1881, pp. 161–175, 2023.
https://doi.org/10.1007/978-3-031-43257-6_13

lower bound and an upper bound. One of the conclusions is that this problem is rather hard for solvers even despite a breaking symmetry technique. The work [7] reviews the column generation approach for a temporal case. Yet, the columns take into account the entire horizon and all items, while a procedure for reducing time moments is proposed to decrease the runtime. Authors of the paper [12] propose reduction methods for a temporal case with a counted number of server activations (or so-called fire-ups). The number of server activations is also an important indicator because it consumes a significant amount of electricity. Minimizing this objective can significantly reduce financial costs. A branch-and-price algorithm, based on column generation, for a temporal case, is implemented in [4]. It is able to prove optimality for instances with up to 500 items. The paper [18] shows an approach with a predefined set of patterns. The works [5,6] give an overview of both online and offline variations of the bin packing problem with conflicts. The bin packing problem with clique-graph conflicts is considered in [1]. It finds an application in security in cloud computing. The authors propose an asymptotic polynomial time approximation scheme and a special approach for packing small items. The publication [16] provides a description of a branch-and-price approach. It considers a special case when the conflict graph is an interval graph. The dependence of the complexity of the problem on the density of the graph conflicts is also explored. The authors of the works [9,15] make some basic analysis of different heuristics. They compare First Fit, First Fit Decreasing, Next Fit, Best Fit, A genetic algorithm, approaches based on solving a sequence of a subset-sum problem, and some others. In the work [13], a team from Microsoft proposes new geometric heuristics for vector bin packing problem and compare them with First Fit Decreasing. It also considers bad instances for different approaches. A couple of works [2,17] consider multi-objective versions of the problem.

In our previous work [14], we considered a temporal bin packing problem without any conflicts and with an even distribution of resources across NUMA nodes. To obtain a lower bound, we used a column generation procedure to solve a linear relaxation of a static bin packing problem (a subproblem that considers only virtual machines from a single moment). At the same time, the upper bound was obtained by extending an integer solution of a static problem using the First Fit heuristic. This approach provided close lower and upper bounds.

In this paper, we extend our previous work and show how to cope with type conflicts and unequal NUMA nodes. To compute lower bounds, we use the same idea and obtain a solution for a single moment by the column generation approach, where the pricing problem takes into account type conflicts. Then, we run a couple of pattern-based heuristics to extend an integer static solution over the entire horizon and obtain a feasible solution for the problem. In the process of extension, these heuristics utilize the packing patterns (columns) obtained during the column generation. Both of the heuristics aim at maintaining packing patterns on servers, that is, packing virtual machines according to patterns throughout the horizon. However, the distribution of types of virtual machines sometimes varies at different moments. Thus, some patterns obtained at one moment may be poorly suited for another. To this end, one of the heuristics tries to use a more versatile set of patterns. Our computational experiments show the effectiveness of these heuristics on large-scale problems.

The paper is organized as follows. Section 2 gives the details and notation of the problem. We present the models for the master problem and the pricing problem in Sect. 3. They are used in obtaining the lower bound. We also provide the details on the column generation procedure. The developed heuristics for a temporal case are given in Sect. 4. Section 5 contains a description of the data sets. It also provides computational results. In Sect. 6, we conclude.

2 Formulation of the Problem

We are given a set of identical servers S, each one has a set N of NUMA nodes. Every node $n \in N$ has a capacity C_r^n of a resource $r \in R = \{\text{CPU}, \text{RAM}\}$. The goal is to place every virtual machine (VM) from the set M on any server, in such a way to use the least number of servers. Every VM $m \in M$ has a start time α_m and a finish time ω_m. Thus, it exists only in time moments $t \in T$ such that $\alpha_m \leq t < \omega_m$. In addition, there is a set of types L. It is assumed that $|L| \ll |M|$. Each VM m has a certain type $l \in L$ which defines the required amount of CPU and RAM for the virtual machine m. For example, a type may indicate that some VM requires 1 CPU and 4 RAM or 32 CPU and 64 RAM. More examples can be found at [8]. A type also specifies whether the corresponding VM is large or small. A large VM is divided into two fixed parts that are placed on the same server, but on two different NUMA nodes and processed simultaneously for the entire lifetime of the VM. The part $p \in P = \{1, 2\}$ of any large virtual machine of type l takes d_{lr}^p amount of resource $r \in R$. At the same time, a small virtual machine of type l has only one part, which occupies d_{lr}^p (defined only for $p = 1$) amount of the resource $r \in R$, and should be completely placed on one node.

In addition, the problem input has a set of conflicts K^L. Each conflict is a pair of types (l_1, l_2), where $l_1, l_2 \in L$. If $(l_1, l_2) \in K^L$ then any VM m_1 with the type l_1 and any VM m_2 with the type l_2 cannot be placed on the same server.

To present the full model, we introduce additional notation. We denote by M^{small} and M^{large} ($M = M^{small} \cup M^{large}$) the subsets of small and large virtual machines. The same can be done for the set $L = L^{small} \cup L^{large}$. The set $M_t \subseteq M$ consists of all the VMs existing at the time moment $t \in T$. Similarly, the set M_t also can be split into two subsets $M_t = M_t^{small} \cup M_t^{large}$. The set of conflicts K^L for types L can be mapped into the equivalent set of conflicts K^M for virtual machines. If $(l_1, l_2) \in K^L$, then $(m_1, m_2) \in K^M$ for every virtual machine m_1 that has the type l_1 and every virtual machine m_2 that has the type l_2.

The decision variables are as follows: $x_{msn}, m \in M, s \in S, n \in N$ equals 1 if the small virtual machine (or the first part of the large virtual machine) m is placed on the node n of the server s and 0 otherwise; $y_{msn}, m \in M^{large}, s \in S$ equals 1 if the second part of the large virtual machine m is placed on the node n of the server s and 0 otherwise; $z_s, s \in S$ equals 1 if the server s was active at any moment (at least one) and 0 otherwise; F is the number of servers used.

It is worth noting that since virtual machines cannot be moved from one server to another, the variables x_{msn}, y_{msn} do not require any index t. In addition, we should mention that the total number of servers F used may be greater than the number of active servers at any moment $t \in T$. While it is the maximum number of servers over all the moments that has been used in our work [14], different objective function (1) is used in this paper.

$$\min_{(x_{msn}),(y_{ms}),(z_s)} F \tag{1}$$

$$\text{s.t. } F = \sum_{s \in S} z_s, \tag{2}$$

$$\sum_{s \in S} \sum_{n \in N} x_{msn} = 1, \ m \in M, \tag{3}$$

$$\sum_{n} x_{msn} \leq \sum_{n} y_{msn}, \ m \in M^{large}, s \in S, \tag{4}$$

$$x_{msn} + y_{msn} \leq 1, \ m \in M^{large}, n \in N, s \in S, \tag{5}$$

$$\sum_{n \in N} x_{msn} \leq z_s, \ m \in M, s \in S, \tag{6}$$

$$\sum_{n} x_{m_1 sn} + \sum_{n} x_{m_2 sn} \leq 1, \ (m_1, m_2) \in K^M, s \in S, n \in N, \tag{7}$$

$$\sum_{m \in M_t} d_{mr}^1 x_{msn} + \sum_{m \in M_t^{large}} d_{mr}^2 y_{msn} \leq C_r^n, \ t \in T, s \in S, n \in N, r \in R, \tag{8}$$

$$x_{msn}, y_{msn}, z_s \in \{0, 1\}. \tag{9}$$

The objective (1) is to minimize the number of servers. Equality (2) defines the number of servers used. Constraints (3–4) enforce every virtual machine to be placed. Inequalities (5) prohibit both parts of any large VM to be placed on the same node. Inequalities (6) show that the server s is active when at least one virtual machine was placed there. Constraints (7) manage the conflicts. Inequalities (8) limit the number of resources on each server at every moment.

The model (1)–(9) is extremely hard to solve, even having only two NUMA nodes. Therefore, it is necessary to develop heuristic approaches capable of quickly obtaining a solution. This is especially important for data centers with a huge number of servers because they usually have very large instances. And the runtime is important for analyzing various situations and making decisions or applying the solution to other problems.

3   Column Generation

Based on the results of the previous work [14] we assume that using only a subset of virtual machines can be enough to get a good lower bound. We also assume that the column generation is capable to provide a better result than other approaches. It is proposed to select only one moment $\hat{t}$ and solve the linear relaxation of the bin packing problem with conflicts for all the virtual machines from the set $M_{\hat{t}}$. The result will become the lower bound for the problem considered. A good example of such a moment $\hat{t}$ can be a moment with the highest load on one of the resources. The load can be computed using the following formula $load_r = (\sum_m \in M_{\hat{t}}^{small} d_{lr}^1 + \sum_m \in M_{\hat{t}}^{large} \sum_{p \in P} d_{lr}^p)/\sum_{n \in N} C_r^n$, where l (of d_{lr}^p) is the type of the corresponding virtual machine $m \in M_{\hat{t}}$. Notice that if we remove the temporality and consider only one moment, then virtual machines of the same type have no difference. Thus, in this section, we will operate with the set of types L instead of the set M.

To describe the applied column generation procedure, we first have to introduce several additional models. The first model corresponds to the bin packing problem or so-called master problem. Here J is the set of all possible patterns for one server. The value a_{lj} denotes the integer number of VMs of the type $l \in L$ in the column $j \in J$. The $J' \subset J$ is usually a small subset of patterns that must be large enough for the problem to have a feasible solution. One possible option is to take columns j_l with the only non-zero element $a_{lj_l} > 0$ for every $l \in L$. The value h_l represents the number of requests for a virtual machine of the type l. The variable x_j is equal to the number of servers packed according to the pattern j. The master problem has the following form.

$$\min \sum_{j \in J'} x_j \tag{10}$$

$$\text{s.t.} \quad \sum_{j \in J'} a_{lj} x_j \geq h_l, \; l \in L, \tag{11}$$

$$x_j \geq 0, \; j \in J'. \tag{12}$$

The objective (10) is the number of bins. Inequality (11) guarantees that all the virtual machines are packed.

The dual problem looks like the following.

$$\max \sum_{l \in L} h_l \lambda_l \tag{13}$$

$$\text{s.t.} \quad \sum_{l \in L} a_{lj} \lambda_l \leq 1, \; j \in J', \tag{14}$$

$$\lambda_l \geq 0, \; l \in L. \tag{15}$$

Let λ_l^* be the dual variables of the current optimal solution. The pricing problem is used to get new columns that can improve the objective function of the master problem. It is a modification of the knapsack problem and can be represented by the model (16)–(23). Here, the variable z_l^n shows how many small VMs of type $l \in L$ are placed on the node n. The variable y_l^{nk} shows the number of large VMs of type $l \in L$, with the first part placed on node n and the second part placed on node k. The χ_l equals 1 if at least one virtual machine of the type l is placed.

$$\max \alpha = \sum_{n \in N} \sum_{l \in L^{small}} \lambda_l^* z_l^n + \sum_{n \in N} \sum_{\substack{k \in N \\ k \neq n}} \sum_{l \in L^{large}} \lambda_l^* y_l^{nk} \tag{16}$$

$$\sum_{l \in L^{small}} d_{lr}^1 z_l^n + \sum_{\substack{k \in N \\ k \neq n}} \sum_{l \in L^{large}} (d_{lr}^1 y_l^{nk} + d_{lr}^2 y_l^{kn}) \leq C_r^n, \ r \in R, n \in N, \tag{17}$$

$$\chi_l \geq \sum_{n \in N} \frac{z_l^n}{M}, \ l \in L^{small}, \tag{18}$$

$$\chi_l \geq \sum_{n \in N} \sum_{\substack{k \in N \\ k \neq n}} \frac{y_l^{nk}}{M}, \ l \in L^{large}, \tag{19}$$

$$\chi_{l_1} + \chi_{l_2} \leq 1, \ (l_1, l_2) \in K^L, \tag{20}$$

$$\chi_l \in \{0, 1\}, l \in L, \tag{21}$$

$$y_l^{nk} \geq 0, \ integer, \ l \in L^{large}, n, k \in N, \tag{22}$$

$$z_l^n \geq 0, \ integer, \ l \in L^{small}, n \in N. \tag{23}$$

The objective (16) is the weighted number of packed virtual machines. Inequalities (17) limit the capacity of the server. Constraints (18)–(20) add conflicts. The value M is a fixed large number greater than any h_l.

It is also possible to use another model (24)–(32) for the pricing problem with a less number of variables. Let z_l^n show a number of small VMs (or a number of first parts of large VMs) of the type $l \in L$ placed on the node n. And let y_l^n show a number of second parts of large VMs of the type $l \in L^{large}$ placed on the node n.

$$\max \alpha = \sum_{n \in N} \sum_{l \in L} \lambda_l^* z_l^n \tag{24}$$

$$z_l^n \leq \sum_{\substack{k \in N \\ k \neq n}} y_l^k, \ l \in L^{large}, n \in N, \tag{25}$$

$$\sum_{n \in N} z_l^n = \sum_{n \in N} y_l^n, \ l \in L^{large}, \tag{26}$$

$$\sum_{l \in L} d_{lr}^1 z_l^n + \sum_{l \in L^{large}} d_{lr}^2 y_l^n \leq C_r^n, \ r \in R, n \in N, \tag{27}$$

$$\chi_l \geq \sum_{n \in N} \frac{z_l^n}{M}, \ l \in L, \tag{28}$$

$$\chi_{l_1} + \chi_{l_2} \leq 1, \ (l_1, l_2) \in K^L, \tag{29}$$

$$\chi_l \in \{0, 1\}, l \in L, \tag{30}$$

$$y_l^n \geq 0, \ integer, \ l \in L^{large}, n \in N, \tag{31}$$

$$z_l^n \geq 0, \ integer, \ l \in L, n \in N. \tag{32}$$

Constraints (25) ensure that two parts of any large virtual machine are placed and that they are located on the different NUMA nodes. This set of constraints is sufficient for the case $|N| = 2$. We add the additional set of constraints (26) for the case $|N| > 2$. It ensures that no two first parts of any large virtual machine use the same second part. For example, consider the case when the first two parts (of two different large VMs) are placed on the first and second NUMA nodes, and at the same time, only one second part is located on the third NUMA node. Other constraints are taken from the model (16)–(23).

The Algorithm 1 gives the pseudocode of the column generation procedure with the proposed models. We run it at two moments that have the highest load for each resource (CPU and RAM). After that, we choose the largest result as a lower bound.

Algorithm 1: The column generation procedure

Input: An initial subset of columns J', A set of virtual machines $M_{\hat{t}}$;
Output: A lower bound LB, A supplemented set J';
1 **Initialization:** $\alpha^* \leftarrow \infty$;
2 **while** $\alpha^* > 1$ **do**
3     $\lambda_l^* \leftarrow$ solve (13)-(15);
4     $j_{new}, \alpha^* \leftarrow$ solve (24)-(32);
5   Add the new column j_{new} to the set J';
6 **end**
7 $LB \leftarrow \lceil \sum_{l \in L} h_l \lambda_l^* \rceil$;
8 **return:** LB, J';

This approach can be improved in the case when a complement of a conflict graph splits into connected components. Assume we have a graph $G = (V, E)$, where $V = L$ and $(i, j) \in E$ if there is no conflict between i and j. Then, any connected component G_k of the graph G can be mapped to a corresponding set of servers of a static solution (with a corresponding set of virtual machines). Any VM from such a set can not be moved to any other set of servers. Thus, we can compute a lower bound independently for each of these components. Consequently, a lower bound becomes $LB = \sum_k \lceil LB_{G_k} \rceil$.

4 Upper Bounds

We obtain a set of useful packing patterns in the process of column generation. All of them can be adopted for an upper bound. We propose to reuse a set of columns J' that we get as an output of the Algorithm 1. It helps us to construct a static integer solution at time moment $\hat{t}$. This solution is partial and can be extended to the whole horizon. A set of virtual machines M can be divided into three disjoint subsets $M_{middle} = \{m \in M | \alpha_m \leq \hat{t} < \omega_m\}$, $M_{right} = \{m \in M | \alpha_m > \hat{t}\}$, $M_{left} = \{m \in M | \hat{t} \geq \omega_m\}$. We run a different packing procedure for each of these subsets.

First, we solve a bin packing problem (10)–(12) for the set M_{middle} with integer variables x_j. Here we use the columns J' obtained in the Algorithm 1.

This solution shows how many servers ($= x_j$) with each pattern $j \in J'$ we need to use at the moment $\hat{t}$. However, this solution is in terms of types, and it is not in terms of virtual machines. Thus, a decoding procedure is required. After we created the servers and assigned a pattern to each one, we begin to pack the virtual machines from the set M_{middle}. For each VM m of a type $l \in L$ we select the first server that still has a place in its pattern j ($a_{lj} > 0$) and update the pattern ($a_{lj} = a_{lj} - 1$). Note that we update only the pattern associated with the current server and do not touch any others.

Algorithm 2: Pattern-based heuristic

Input: A set of virtual machines M_{right}, A set of servers S with corresponding patterns;

1 $sort(M_{right})$;
   // Sort the VMs in non-decreasing order of start times α_m
2 **for** every $\hat{m} \in M_{right}$ **do**
3    $isPlaced \leftarrow false$;
4   **for** every server $s \in S$ **do**
5   **for** every $m \in s$ **do**
6   **if** $\omega_m \leq \alpha_{\hat{m}}$ **then**
7   remove m from the server s;
8             $a_{lj} \leftarrow a_{lj} + 1$; ;   // if server s has a pattern
9   **end**
10   **end**
11   **if** $a_{\hat{l}j} > 0$ **or** s has enough space **then**
12          $isPlaced \leftarrow true$
13          $a_{\hat{l}j} \leftarrow a_{\hat{l}j} - 1$;
14   stop looking for a server;
15   **end**
16   **end**
17   **if** $isPlaced \neq true$ **then**
18   create a new server s;
19   place $\hat{m}$ on the server s;
20       $isPlaced \leftarrow true$;
21   **end**
22 **end**

The sets M_{left} and M_{right} have similar packing procedures. As presented in the Algorithm 2 we use a heuristic based on First Fit that preserves server patterns. Firstly, the set M_{right} is sorted in non-decreasing order of start times α_m. Then every virtual machine $\hat{m} \in M_{right}$ is packed on the first server that has a vacant place in its pattern ($a_{lj} > 0$) or has enough resources (for servers without patterns). If there is no suitable server, then a new server is created without any pattern. Such a server can take any VM if it doesn't break any conflict. Also, before searching for a suitable server, we are to remove every VM m that has already finished ($\omega_m \leq \alpha_{\hat{m}}$). The same procedure is repeated for the set M_{left}. Yet, we sort it in non-increasing order of finish times ω_m. And we remove virtual machines that have $\alpha_m \geq \omega_{\hat{m}}$. This approach is memory-efficient and fast enough since it only requires keeping one time moment in memory. This moment equals α_m of the current virtual machine for the set M_{right} and ω_m for the set M_{left}.

During the packing process, along with a server, we also have to select one or two NUMA nodes. On each server, the nodes with the highest free space have priority. This way, the loading of all nodes remains close to uniform. It helps to pack large virtual machines since they require resources not from one, but from two NUMA nodes at once.

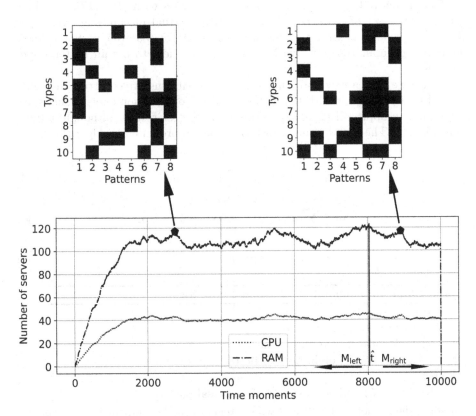

Fig. 1. A representation of the difference between sets of patterns. We run column generation at two different time moments to compare sets of unique patterns. Every pattern is depicted as a column of black and white cells. Each black cell shows that the corresponding type is used in the pattern. At both time moments we have exactly eight used patterns and they are represented in the order of generation.

The approach described above is strongly tied to patterns. Yet, different time moments may have significantly different subsets of virtual machines. Running the Algorithm 1 in such moments would lead to getting different sets J' (see Fig. 1). Thus, the proposed heuristic requires additional improvements.

One of the possible ways to improve it is to use a more versatile set of patterns. Suppose we have a solution for the temporal problem obtained by the approach described above. We can select several difficult time moments $\{t_1, ..., t_k\}$. These moments are the ones with the highest load. To increase diversity, we divide the entire horizon with a step h and choose one moment for each resource $r \in R$ in each interval. Every $t_i \in \{t_1, ..., t_k\}$ can be associated with a corresponding set of patterns $J_i \in \{J_1, ..., J_k\}$. To select a universal subset of patterns J^{un} from the combined set $\hat{J} = \cup_{i=1}^{k} J_i$ we solve the following model (33)-(35). The patterns from the set $J^{un} \subseteq \hat{J}$ correspond to the variables $x_j > 0$.

The model (33)-(35) almost completely coincides with (10)-(12). Yet, here we look for a set of patterns that can pack all virtual machines at every $t_i \in \{t_1, ..., t_k\}$ independently. A solution to the problem (33)-(35) is not a lower bound in general. However, we hope that it can help to obtain better upper bounds.

When we have a new, more universal set of patterns, we construct an upper bound for the moment with the highest load $\hat{t}$. Computation experiments show that such an approach is capable of providing better results, although it takes much longer. The pseudocode of this procedure is presented in Algorithm 3.

$$\min \sum_{j \in \hat{J}} x_j \tag{33}$$

$$\text{s.t.} \quad \sum_{j \in \hat{J}} a_{lj} x_j \geq \max_{t_i \in \{t_1, ..., t_k\}} h_l^{t_i}, \ l \in L, \tag{34}$$

$$x_j \geq 0, integer, \ j \in \hat{J}. \tag{35}$$

Algorithm 3: Temporal pattern-based heuristic

Input: A set of time moments $\{t_1, ..., t_k\}$;

1 **for every** $t_i \in \{t_1, ..., t_k\}$ **do**
2 $\quad$ run Algorithm 1 at the moment t_i;
3 $\quad$ save the corresponding set of patterns J_i obtained;
4 **end**
5 solve (33)-(35) and get a subset of universal patterns J^{un};
6 solve the model (10)-(12) at $\hat{t}$ with integer variables and the set J^{un} for the set M_{middle};
7 run Algorithm 2 for the set M_{right};
8 run Algorithm 2 for the set M_{left};

---

5   Computational Experiments

For our experiments, we generated a couple of semi-synthetic data sets. The requirements of each type $l \in L$ (CPU and RAM) and server configurations are provided by one cloud computing company. Hereby, we have ten types and two server configurations, first one is with two different NUMA nodes and another one is with four different NUMA nodes. Each set contains fifty instances with 10000 VMs for servers with two NUMA nodes and fifty instances with 20000 VMs for servers with four NUMA nodes. The length of the time horizon always equals 10000. In total, we have 200 instances. The algorithms are implemented in python, and the experiments are carried out on a computer with Intel Core i5-11400F and 16 GB of RAM. Gurobi 10.0 is used as a solver.

We compare the following approaches:

- **First Fit (FF).** We pack the sets M_{right} and M_{left} by the First Fit algorithm, with no trying to keep patterns. That is, a virtual machine gets to any server that has a sufficient amount of resources, and conflicts are not violated.
- **Pattern-based (PB).** We pack the sets M_{right} and M_{left} in correspondence with patterns obtained at some moment $\hat{t}$. If there are not enough servers, then we create a new server without any pattern.
- **Temporal pattern-based (TPB).** It works according to Algorithm 3. Thus, it is identical to the approach **PB**, but uses a different set of patterns.

All the approaches are run once for every moment $t_r = argmax(load_r), r \in R$. If the moments with the highest load for both resources coincide, then all approaches are run only once. When selecting difficult moments $\{t_1, ..., t_k\}$, a constant step h and ten intervals are used. We repeat it for every resource $r \in R$. Considering only two resources, we have $10 \leq k \leq 20$, because some time moments may coincide.

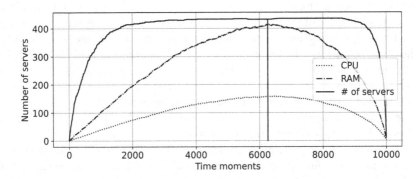

Fig. 2. Example of an instance for the first data set. Start time, finish time, and a type being sampled from the uniform distribution, the resource demand looks like a hill with a pronounced extremum. Different time moments differ rather by the number of virtual machines and not by the ratio of types. CPU and RAM are evaluated in terms of servers (the amount is divided by the server capacity).

The result tables below have the following rows and columns. The columns **FF, PB, TPB** refer to the different approaches listed above. The rows **avg. GAP** and **std. GAP** show the average and standard deviation of the relative

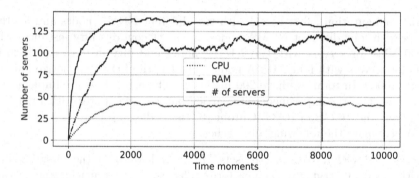

Fig. 3. Example of an instance for the second set. All virtual machines have a relatively short lifetime. The resource demand has several time points that are close to the extremum. The ratio of types at different times differ (see Fig. 1), and as a result, the sets of patterns also differ. CPU and RAM are evaluated in terms of servers (the amount is divided by the server capacity).

percentage deviation of the column generation lower bound and the corresponding upper bound. The rows **MIN_GAP** and **MAX_GAP** show the minimum and maximum deviations. The row **N_SERVERS** shows an average absolute difference between the lower and a corresponding upper bound. The final row **TIME** is an average runtime in seconds.

The first set of instances is generated as follows. The start time α_m is selected from the discrete uniform distribution $U_{[1,9999]}$. The finish time ω_m is selected from the discrete uniform distribution $U_{[\alpha_m+1,10000]}$. The type l is randomly selected from the whole set L using the discrete uniform distribution $U_{[1,10]}$. Conflicts are randomly chosen for each instance from the whole set of conflicts by the discrete uniform distribution $U_{[0,45]}$. Such generation leads to an approximately equal ratio of types all over the horizon. As a consequence, the same set of patterns can be obtained even at sufficiently distant moments. For an example of such an instance, see Fig. 2. The results are presented in Table 1.

The second set is generated in a more complicated way. The start time α_m is selected from the discrete uniform distribution $U_{[1,9999]}$. The finish time is computed as follows $\omega_m = \min(\alpha_m + U_{[1,2000]}, 10000)$. Thereby, the lifetimes of all virtual machines are quite short, and a static solution at any moment t has less effect on the whole temporal solution. Before choosing a type, we select a subset L_5 with five different types. And then we choose a type for the current VM from the set L_5 using the discrete uniform distribution $U_{[1,5]}$. Conflicts are generated in the same way as for the first data set. With such generation, the ratio of types differs at different time moments. As a result, the sets of patterns obtained by column generation at different time moments also differ (see Fig. 1). For an example of such an instance, see Fig. 3. The results are presented in Table 2.

We can draw the following conclusions based on Table 1 and Table 2. The First Fit approach works noticeably worse than both pattern-based heuristics

Table 1. Results for the first set of instances

	2 NUMA nodes			4 NUMA nodes		
	FF	PB	TPB	FF	PB	TPB
avg. GAP	3.22	1.41	1.35	2.83	1.06	0.97
std. GAP	1.45	0.39	0.39	1.52	0.32	0.31
MIN_GAP	0.80	0.68	0.67	0.55	0.37	0.18
MAX_GAP	6.04	2.63	2.63	6.73	2.25	1.80
N_SERVERS	14.78	6.40	6.14	13.12	4.90	4.50
TIME	26.24	28.19	226.29	174.96	204.82	1125.09

Table 2. Results for the second set of instances

	2 NUMA nodes			4 NUMA nodes		
	FF	PB	TPB	FF	PB	TPB
avg. GAP	10.12	6.79	5.07	14.71	5.82	4.77
std. GAP	2.49	1.41	1.52	4.51	1.29	1.22
MIN_GAP	4.72	2.85	2.85	3.51	3.07	2.0
MAX_GAP	14.55	9.27	8.78	22.15	8.27	7.24
N_SERVERS	15.12	9.82	7.22	22.96	8.28	6.72
TIME	35.64	37.34	237.75	204.76	219.88	1315.04

on both data sets. In the worst case, it shows the GAP equal to 22.96%. The Temporal pattern-based idea shows insignificantly better results on the first data set compared to the Pattern-based heuristic. We could expect this because the pattern sets are too similar. For the second data set, the TPB approach can noticeably improve the results of the PB approach. Although it requires significantly more time due to the need to run several column generations. Instances with four NUMA nodes seem to be harder. We noticed that this is caused by a pricing problem (24)–(32) that requires more time for four NUMA nodes. At the same time, the master problem requires insignificant time compared to the pricing problem. The column generation takes most of the time, thus FF and PB have almost identical runtimes.

6 Conclusions

In this work, we presented the new temporal bin packing problem with conflicts and NUMA architecture. We developed a couple of pattern-based heuristics. Each uses a column generation procedure to select proper patterns. And one of them utilizes more universal columns. Two models for the modified knapsack problem were introduced. We carried out computational experiments on two different data sets. One data set is obtained using a fairly primitive generation procedure, while the other simulates complex conditions. Strong results were demonstrated for both data sets. The proposed methods can be generalized to other variations of the bin packing problem.

Acknowledgement. The study was carried out within the framework of the state contract of the Sobolev Institute of Mathematics (project FWNF-2022-0019).

References

1. Arad, I.D., Shachnai, H.: An APTAS for bin packing with clique-graph conflicts. CoRR abs/2011.04273 (2020). https://arxiv.org/abs/2011.04273
2. Aydin, N., Muter, b., Birbil, I.: Multi-objective temporal bin packing problem: an application in cloud computing. Comput. Oper. Res. **121**, 104959 (2020). https://doi.org/10.1016/j.cor.2020.104959
3. D De Cauwer, M., Mehta, D., O'Sullivan, B.: The temporal bin packing problem: an application to workload management in data centres. In: 2016 IEEE 28th International Conference on Tools with Artificial Intelligence (ICTAI), pp. 157–164 (2016). https://doi.org/10.1109/ICTAI.2016.0033
4. Dell'Amico, M., Furini, F., Iori, M.: A branch-and-price algorithm for the temporal bin packing problem (2019)
5. Ekici, A.: Bin packing problem with conflicts and item fragmentation. Comput. Oper. Res. **126**, 105113 (2021). https://doi.org/10.1016/j.cor.2020.105113
6. Epstein, L., Levin, A.: On bin packing with conflicts. In: Erlebach, T., Kaklamanis, C. (eds.) WAOA 2006. Lecture Notes in Computer Science, vol. 4368, pp. 160–173. Springer, Berlin (2007). https://doi.org/10.1007/11970125_13
7. Furini, F., Shen, X.: Matheuristics for the temporal bin packing problem. In: Amodeo, L., Talbi, E.-G., Yalaoui, F. (eds.) Recent Developments in Metaheuristics. ORSIS, vol. 62, pp. 333–345. Springer, Cham (2018). https://doi.org/10.1007/978-3-319-58253-5_19
8. Haider, W., Iqbal, W., Bokhari, F.S., Bukhari, F.: On providing response time guarantees to a cloud-hosted telemedicine web service. In: Zhang, Y., Peng, L., Youn, C.-H. (eds.) CloudComp 2015. LNICST, vol. 167, pp. 234–243. Springer, Cham (2016). https://doi.org/10.1007/978-3-319-38904-2_24
9. Haouari, M., Serairi, M.: Heuristics for the variable sized bin-packing problem. Comput. Oper. Res. **36**(10), 2877–2884 (2009). https://doi.org/10.1016/j.cor.2008.12.016
10. Majo, Z., Gross, T.R.: (Mis)understanding the NUMA memory system performance of multithreaded workloads. In: 2013 IEEE International Symposium on Workload Characterization (IISWC), pp. 11–22 (2013). https://doi.org/10.1109/IISWC.2013.6704666
11. Martello, S., Toth, P.: Knapsack Problems: Algorithms and Computer Implementations. John Wiley & Sons Inc, USA (1990)
12. Martinovic, J., Strasdat, N., Valério de Carvalho, J., Furini, F.: Variable and constraint reduction techniques for the temporal bin packing problem with fire-ups. Optim. Lett., 1–26 (2021). https://doi.org/10.1007/s11590-021-01825-x
13. Panigrahy, R., Talwar, K., Uyeda, L., Wieder, U.: Heuristics for vector bin packing (2011). https://www.microsoft.com/en-us/research/publication/heuristics-for-vector-bin-packing/

14. Ratushnyi, A., Kochetov, Y.: A column generation based heuristic for a temporal bin packing problem. In: Pardalos, P., Khachay, M., Kazakov, A. (eds.) MOTOR 2021. Lecture Notes in Computer Science, vol. 12755, pp. 96–110. Springer, Cham (2021). https://doi.org/10.1007/978-3-030-77876-7_7

15. Rieck, B.A.: Basic analysis of bin-packing heuristics. ArXiv: abs/2104.12235 (2021)

16. Sadykov, R., Vanderbeck, F.: Bin packing with conflicts: a generic branch-and-price algorithm. INFORMS J. Comput. **25**, 244–255 (2012). https://doi.org/10.1287/ijoc.1120.0499

17. Shi, F., Lin, J., Li, Q.: Virtual machine resource allocation optimization in cloud computing based on multiobjective genetic algorithm. Intell. Neurosci. 2022 (2022). https://doi.org/10.1155/2022/7873131

18. Shi, J., Luo, J., Dong, F., Jin, J., Shen, J.: Fast multi-resource allocation with patterns in large scale cloud data center. J. Comput. Sci. **26**, 389–401 (2018). https://doi.org/10.1016/j.jocs.2017.05.005

Integer Models for the Total Weighted Tardiness Problem on a Single Machine

R. Yu. Simanchev[1,2](✉) [ID] and I.V. Urazova[1] [ID]

[1] Omsk State University, Omsk, Russia
osiman@rambler.ru
[2] Omsk Scientific Center of SB RAS, Omsk, Russia

Abstract. The paper deals with the following scheduling problem. Jobs are served by a single machine. Each job is characterized by a positive weight, a release date, a due date. Servicing times are the same for all jobs. Preemptions are allowed. We investigate the case of the discrete time. The goal is to find a schedule for servicing jobs that minimizes the weighted sum of tardiness. The complexity status of this problem is unknown today. Our paper proposes two models of boolean linear programming for this problem. A comparative analysis of the models is carried out, and results of the computational experiment are described.

Keywords: Schedules · Tardiness · Polyhedral approach · Valid inequalities · (0,1)-programming

1 Introduction

The total weighted tardiness problem on a single machine is considered. The problem is formulated as follows. Let V, $|V| = n$, be a set of jobs that must be served by a single machine. Each job $i \in V$ is characterized by a positive weight ($w_i > 0$), a release date ($r_i \geq 0$), and a due date ($d_i > 0$). Servicing times ($p_i > 0$) are the same for all jobs, that is $p_i = p$ for all $i \in V$. Preemptions are allowed. All of these characteristics are integer. We investigate the case of the discrete time. The goal is to find a schedule for processing jobs that minimizes the weighted sum of tardiness. A jobs schedule is a $(0,1)$-vector x with coordinates x_{ik}, $i \in V$, $k = 1, 2, \ldots$, which are determined by the rules: $x_{ik} = 1$, if at time k the job i is being served; otherwise $x_{ik} = 0$.

Let us introduce the value $C_i(x) = \max\{k \in D \mid x_{ik} = 1\}$, the last moment of processing job i in the schedule x. The tardiness of a job $i \in V$ in the schedule x is the value $T_i(x) = \max\{0, C_i(x) - d_i\}$. The goal is to find a schedule that minimizes the weighted sum of tardiness.

In the notation adopted in scheduling theory, this problem can be written as $1|pmtn; p_i = p; r_i| \sum w_i T_i$. Depending on the type of constraints, the computational complexity of the problem varies. With equal job weights, the problem is polynomially solvable both with and without preemptions ($1|pmtn; p_i =$

$p; r_i | \sum T_i$ and $1 | p_i = p; r_i | \sum T_i)$ [1–3]. In case when preemptions are disabled, the processing times for jobs are different, but the times of their arrival the same, the problem is NP-hard both in weighted and unweighted versions $(1 || \sum T_i$ and $1 || \sum w_i T_i$) [3–5]. As for the considered problem, that is, $1 | pmtn; p_i = p; r_i | \sum w_i T_i$, its complexity status is currently unknown [3].

Since the number of calls is finite, we can introduce some general deadline d, long enough for all jobs to be processed. The set of all schedules on the set V that fit into a given deadline d will be denoted by Σ_d. In addition, we will use the notation $D = \{1, 2, \ldots, d\}$. With such an arrangement, the schedule $x \in \Sigma_d$ is conveniently presented as a table of dimension $n \times d$, in which each row corresponds to the job $i \in V$, and each column corresponds to the time $k \in D$. Any column of this table that does not contain ones is called empty. It is clear that the schedules introduced above are integer solutions of the following system of constraints:

$$\sum_{k \in D} x_{ik} = p, \quad i \in V; \tag{1}$$

$$\sum_{i \in V} x_{ik} \le 1, \quad k \in D; \tag{2}$$

$$x_{ik} \ge 0, \quad i \in V, \ k = r_i + 1, r_i + 2, \ldots, d; \tag{3}$$

$$x_{ik} = 0, \quad i \in V, \ k = 1, 2, \ldots, r_i; \tag{4}$$

The polyhedron specified by conditions (1)–(4) is, in fact, a polyhedron of the transportation problem with integer right-hand sides and therefore it is integer.

As it has been noted, scheduling theory considers a large number of problems on the set of schedules Σ_d. These problems differ in the structure of objective functions. We consider the problem of minimizing the function

$$f(x) = \sum_{i \in V} w_i T_i(x) = \sum_{i \in V} w_i \max\{0, C_i(x) - d_i\}, \tag{5}$$

on the set of vertices of the polyhedron (1)–(4). Note that the problem is not an integer linear programming problem, since its objective function is non-linear.

Review [6] presents a large number of integer models for various problems in scheduling theory, including for a single machine. In the present study, a new integer model is proposed for the problem under consideration. The model has a visual interpretation in terms of graph theory. In addition, this model is easily projected onto a number of other scheduling problems on a single machine with preemptions, among which there are polynomially solvable, NP-hard problems and problems with an open complexity status, for example: $1 | pmtn; p_i = p; r_i | \sum w_i C_i$, $1 | pmtn; r_i | \sum C_i$, $1 | pmtn; p_i = p; r_i | \sum T_i$, $1 | pmtn; r_i | \sum w_i C_i$, $1 | pmtn; r_i | \sum U_i$, $1 | pmtn; p_i = p; r_i | \sum w_i U_i$, $1 | pmtn; r_i | \sum w_i U_i$. All these problems are mentioned in [3].

The current paper explores the polyhedral properties of the problem. To do this, we need the following concepts and notation.

Let G be an ordinary graph. If i and j are adjacent vertices of the graph G, then we will use the notation ij to denote the edge between them. The degree of a vertex i of the graph G is denoted by $d_G(i)$. A graph H is a subgraph of graph G ($H \subset G$) if the vertex set (edge set) of graph H is a subset of vertex set (edge set) of graph G. For the graphs G_1 and G_2 the notation $G_1 \cup G_2$ means the graph obtained by the union of the vertex sets and the edge sets of these graphs. A polytope in a finite-dimensional Euclidean space is a convex hull of a finite set of points, a polyhedron is a solution set of a finite system of linear equations and inequalities if it is bounded. The polyhedron containing the given polytope will be called the polyhedral relaxation of the polytope.

Let M be a subset of the t-dimensional Euclidean space R^t. A linear inequality in R^t is called valid with respect to M if any point from M satisfies it. A valid inequality is called a support one if there is a point in M that satisfies it as an equality. Support inequalities are important in constructing convex hulls of sets and in cutting plane algorithms.

In this paper, we consider a special approach to constructing a linear objective function for the problem $1|pmtn; p_i = p; r_i| \sum w_i T_i$, which makes it possible to use the apparatus of integer linear programming to solve it. Section 2 proves the monotonicity property of the objective function, which makes it possible to determine the minimum general deadline for processing of all jobs. The polynomial resolvability of the problem of finding this minimum deadline is shown. In Sect. 3, we construct a special graph that allows us to consider the set of all schedules as a family of its subgraphs. In terms of this graph, we formalize the last moments of processing jobs. In Sect. 4, the mentioned graph is associated with the Euclidean space, in terms of which the Boolean linear programming model of the problem under consideration is described. In addition, three classes of valid inequalities for the scheduling polytope are found. Each of these classes consists of a polynomial number of constraints, resulting in several more polyhedral relaxations of the scheduling polytope. In Sect. 5 these classes of inequalities are described. Finally, Sect. 6 presents the computational results of testing the constructed models and comparing the proposed polyhedral relaxations.

2   On a Minimal General Deadline

Note that the sets Σ_d for different d have the monotonicity property, that means that for $d < d'$ there is an inclusion $\Sigma_d \subset \Sigma_{d'}$. In this view, we can talk about the value d_{min}, which is determined by the conditions: $\Sigma_{d_{min}} \neq \emptyset$, but $\Sigma_{d_{min}-1} = \emptyset$. In this section, we will show that this monotonicity is also preserved with respect to the optimal value of the objective function (5).

Theorem 1. *Let $d \geq d_{min}$. For any schedule $x \in \Sigma_{d+1}$ there exists a schedule $x' \in \Sigma_d$ such that $f(x') \leq f(x)$.*

Proof. Let $x \in \Sigma_{d+1}$. Let us order the jobs by no increase of characteristics r_i, $i \in V$, i.e. $r_1 \leq r_2 \leq \ldots \leq r_{d+1}$. Let $r(x) = \max\{k \mid x_{ik} = 0, \ i \in V\}$ be the rightmost empty column of the table corresponding to schedule x. If $r(x) = d+1$, then everything is proved since the column $d+1$ in the schedule x can be simply deleted, and the rest of the table will be the required schedule $x' \in \Sigma_d$. Let $r_{s-1} < r(x) \leq r_s \leq d$. Let us show that there exists a job $i \leq s$ such that $C_i(x) > r(x)$. Indeed, suppose that $C_i(x) < r(x)$ for each $i = 1, 2, \ldots, s - 1$. If $r(x) < r_s$, then the columns $r(x) + 1, r(x) + 2, \ldots, r_s$ are also empty and, without loss of generality, we can assume that $r(x) = r_s$. This means that the table fragment for schedule x, formed by rows $s, s + 1, \ldots, n$ and columns $r_s + 1, r_s + 2, \ldots, d + 1$, has no empty columns and satisfies requirements (1) and (2). But then $d + 1 = d_{min}$, which contradicts the condition of the theorem.

So, there is a job $i \leq s$ such that $C_i(x) > r(x)$. Consider a new schedule $\tilde{x}$ obtained from the schedule x by moving the unit from the cell $(i, C_i(x))$ to the cell $(i, r(x))$. As a result, we get $C_i(\tilde{x}) < C(x)$ and $r(\tilde{x}) = C_i(x) > r(x)$. Since the number of columns is finite, repeating this procedure the needed number of times, we will get $r(x') = d + 1$, i.e. the column $d + 1$ will be empty. At the same time, the value of the objective function did not increase at each step since the transfer of unity was always carried out in the direction of decreasing column numbers.

The theorem is proved.

It follows from Theorem 1 that the set $\Sigma_{d_{min}}$ contains optimal schedules. This fact and the intention to reduce the dimensionality of the problem makes the search for the minimum general deadline necessary. The polyhedron specified by conditions (1)–(4) is, in fact, a polyhedron of the transportation problem with integer right-hand sides and, therefore, is integer. To search for d_{min}, a polynomial dichotomy procedure is proposed. In this procedure we use the integrality of the polyhedron (1)–(4). Let us denote this polyhedron by M_d.

The dichotomy scheme for searching for d_{min}.

We set $A = 1$, $B = np + \max\{r_i \mid i \in V\}$.

i-th iteration

Step 1. Calculate $d = \lceil \frac{A+B}{2} \rceil$.

Step 2. If $M_d = \emptyset$, then set $A := d$, $B := B$ and go to step 3. If $M_d \neq \emptyset$, then we set $A := A$, $B := d$ and go to Step 3.

Step 3. If $B - A = 1$, then output: $d_{min} = B$. Otherwise, go to $i + 1$-th iteration.

Checking the condition $M_d = \emptyset$ (or, equivalently, $M_d \neq \emptyset$) at step 2 can be done by solving LP problem with an arbitrary linear objective function on the polyhedron M_d. Since the polyhedron is integer, the condition $M_d = \emptyset$ is equivalent to no schedules for given d.

Further, everywhere we will assume that the general deadline for all jobs is minimal, that is, $d = d_{min}$.

3   The Description of a Schedule Set and Jobs Completion Times in Graph Theory Terms

Problem (5) is not an integer linear programming problem since the objective function is not linear. This difficulty is typical for most scheduling problems when the apparatus of integer linear programming are used.

Let us proceed to the description of the set of schedules in graph theory terms. Let us consider three sets of vertices: V is the set of jobs, D is the set of time moments at which the jobs are processed, Y is a duplicate of the set V (this is an auxiliary set that will be used to calculate the completion times of jobs). Moreover, for the sets V and Y the order of the elements is not essential, the set $D = \{1, 2, \ldots d\}$ is ordered. Since the sets V and Y are identical, in cases where the element i belongs to the set V and the set Y at the same time, we will use the notation $i \in V (= Y)$. Let $G_1 = (V, D; E_1)$ be a complete bipartite graph with parts V and D and edge set $E_1 = \{ik \mid i \in V, k \in D\}$; $G_2 = (Y, D; E_2)$ is a complete bipartite graph with parts Y and D and edge set $E_2 = \{ik \mid i \in Y, k \in D\}$. Let us form the graph $G = G_1 \cup G_2$. In the graph G, we define a family of subgraphs $\mathcal{H}$ according to the following rule. The subgraph $H \subset G$, $H = H_1 \cup H_2$, $H_1 \subset G_1$, $H_2 \subset G_2$ belongs to the family $\mathcal{H}$ (called a schedule) if the following three conditions hold for H

$$d_{H_1}(i) = p, \quad i \in V; \tag{6}$$

$$d_{H_1}(k) = 0 \text{ for } k = 1, 2, \ldots, r_i, \text{ and } d_{H_1}(k) \leq 1 \text{ for } k = r_i + 1, r_i + 2, \ldots, d; \tag{7}$$

for $i \in V (= Y)$ i $k \in D$ the inclusion $ik \in EH_2$

occurs if and only if there is $l \in D$, $l \geq k$, such that $il \in EH_1$. $\tag{8}$

Figure 1 shows a fragment of such a graph $H \in \mathcal{H}$ for the vertex $i = 3$.

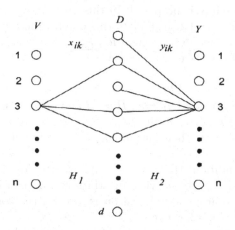

Fig. 1. Fragment of graph $H \in \mathcal{H}$ for $i = 3$.

If $H = H_1 \cup H_2 \in \mathcal{H}$, then by virtue of (6) and (7) the inclusion $ik \in EH_1$ means that at time $k \in D$ the job $i \in V$ and only it is being serviced, and condition (8) guarantees that the vertex $i \in Y$ will be connected by edges to all vertices of the set $\{1, 2, \ldots C_i\} \subset D$. Thus, the subgraph $H_1 \subset H$ defines the schedule and the degrees of the vertices $i \in Y$ with respect to the subgraph $H_2 \subset H$ are the jobs completion times.

This construction allows us to formulate the problem under consideration in the form of:

$$\min\{\sum_{i \in Y} w_i \max\{0, d_{H_2}(i) - d_i\} \mid H = (H_1, H_2) \in \mathcal{H}\} \tag{9}$$

4   The Polyhedral Formulation of the Problem

Let us construct a Boolean linear programming model for problem (9). Let $R^E = R^{E_1} \times R^{E_2}$ be the space associated with the edge set of the graph G. In other words, R^E is the space of column vectors whose components are indexed by elements of the set $E = E_1 \cup E_2$. The coordinate axes in R^{E_1} will be denoted x_{ik}, in $R^{E_2} - y_{ik}$. Let $H \in \mathcal{H}$. The incidence vector of the subgraph $H = H_1 \cup H_2$ is the vector $(x^{H_1}, y^{H_2}) \in R^{E_1} \times R^{E_2}$ with coordinates

$$\text{for } i \in V, k \in D \quad x_{ik}^{H_1} = \begin{cases} 1, \text{ if } ik \in EH_1, \\ 0, \text{ otherwise.} \end{cases}$$

$$\text{for } i \in Y, k \in D \quad y_{ik}^{H_2} = \begin{cases} 1, \text{ if } ik \in EH_2, \\ 0, \text{ otherwise.} \end{cases}$$

Now the set of schedules $\mathcal{H}$ can be associated with a polytope

$$P_{\mathcal{H}} = conv\{(x^{H_1}, y^{H_2}) \in R^E \mid H = H_1 \cup H_2 \in \mathcal{H}\}.$$

Theorem 2. *A subgraph $H = H_1 \cup H_2 \in G$ belongs to the family $\mathcal{H}$ (is a schedule) if and only if its incidence vector $(x^{H_1}, y^{H_2}) \in R^E$ is an integer point of the polyhedron which is formed by constraints (1)–(4) and constraints*

$$\sum_{l=k}^{d} x_{il} \leq p y_{ik}, \quad i \in V(=Y), \quad k \in D; \tag{10}$$

$$y_{ik} \leq \sum_{l=k}^{d} x_{il}, \quad i \in V(=Y), \quad k \in D; \tag{11}$$

$$y_{ik} \leq 1, \quad i \in Y, \quad k \in D. \tag{12}$$

Proof. Let $H = H_1 \cup H_2 \in \mathcal{H}$. The fulfillment of constraints (1)–(4) and (12) for the point (x^{H_1}, y^{H_2}) is obvious. Let us consider constraints (10) and (11). Let $i \in V(= Y)$, $k \in D$. Suppose $\sum_{l=k}^{d} x_{il}^{H_1} > p y_{ik}^{H_2}$. If $y_{ik}^{H_2} = 1$, then $p < \sum_{l=k}^{d} x_{il}^{H_1} \leq \sum_{l=1}^{d} x_{il}^{H_1} = d_{H_1}(i)$. It contradicts condition (6) from the definition of the family $\mathcal{H}$. If $y_{ik}^{H_2} = 0$, then $\sum_{l=k}^{d} x_{il}^{H_1} > 0$. Therefore, there are $l \geq k$ such that $il \in EH_1$. Then, from condition (8) we obtain that $ik \in EH_2$ or, which is the same, $y_{ik}^{H_2} = 1$. The fulfillment of constraints (10) for the point (x^{H_1}, y^{H_2}) is proved.

Now, let us supppose that $y_{ik}^{H_2} > \sum_{l=k}^{d} x_{il}^{H_1}$. If $y_{ik}^{H_2} = 0$, then we obtain a contradiction with the non-negativity of the vector x^{H_1}. If $y_{ik}^{H_2} = 1$, then

$ik \in EH_2$ and from (8) it follows that there is a number $s \geq k$ such that $is \in EH_1$ or, which is the same, $x_{is}^{H_1} = 1$. Therefore, $1 > \sum_{l=k}^{d} x_{il}^{H_1} \geq x_{is}^{H_1} = 1$, which does not make sense. This proves the fulfillment of constraints (11) for the point (x^{H_1}, y^{H_2}). Now, let us prove the reverse implication. Let $(\bar{x}, \bar{y})$ be an integer solution of the system $((1)–(4), (10)–(11))$. Note that for any $i \in V(= Y)$ and $k \in D$, due to (2) and (3) we have $\bar{x}_{ik} \leq 1$, and due to (3) and (10) we have $\bar{y}_{ik} \geq 0$. Let $H_1 \subset G_1$ and $H_2 \subset G_2$ be such that $ik \in EH_1$ if and only if $\bar{x}_{ik} = 1$, and $ik \in EH_2$ if and only if $\bar{y}_{ik} = 1$.

Let us show that $H_1 \cup H_2 \in \mathcal{H}$. The fulfillment of conditions (6) and (7) from the definition of the family $\mathcal{H}$ immediately follows from constraints (1) and (2). Let us prove condition (8). If $ik \in EH_2$, then $\bar{y}_{ik} = 1$. Then from (11) we have $1 < \sum_{l=k}^{d} \bar{x}_{il}$. This means that there exists $l \geq k$ such that $\bar{x}_{il} = 1$ or, equivalently, $il \in EH_1$. If $ik \notin EH_2$, then from (3) and (10) we get $\sum_{l=k}^{d} \bar{x}_{il} = 0$. This means that $\bar{x}_{il} = 0$ for any $l \geq k$ or, equivalently, $il \notin EH_1$.

The theorem is proved.

The polyhedron defined by the constraints $((1)–(4), (10)–(12))$ will be denoted by $M_{\mathcal{H}}$. It is obvious that $P_{\mathcal{H}} \subseteq M_{\mathcal{H}}$, i.e. $M_{\mathcal{H}}$ is a polyhedral relaxation of the polytope $P_{\mathcal{H}}$. Moreover, since $M_{\mathcal{H}}$ and $P_{\mathcal{H}}$ lie in the unit cube, we can state the corollary of Theorem 2.

Corollary 1. *The vertex set of the polytope $P_{\mathcal{H}}$ coincides with the set of integer points in the space R^E that belong to the polyhedron $M_{\mathcal{H}}$.*

The above description of the set of schedules in terms of the graph G allows us to linearize the objective function of problem (9). As already noted, for the schedule $H = H_1 \cup H_2 \in \mathcal{H}$ the completion time of the job $i \in V$ can be expressed as $C_i(x^{H_1}) \equiv C_i(x^{H_1}, y^{H_2}) = d_{H_2}(i) = \sum_{k \in D} y_{ik}$. Let us show that $\max\{0, C_i(x^{H_1}) - d_i\} = \sum_{k=d_i+1}^{d} y_{ik}$. Indeed, if $C_i(x^{H_1}) \leq d_i$, then $y_{ik} = 0$ for all $k > d_i$ and, consequently, $\sum_{k=d_i+1}^{d} y_{ik} = 0$. If $C_i(x^{H_1}) > d_i$, then $y_{ik} = 1$ for all $k = d_i + 1, d_i + 2, \ldots, C_i(x^{H_1})$ and hence $\sum_{k=d_i+1}^{d} y_{ik} = \sum_{k=d_i+1}^{C_i(x^{H_1})} y_{ik} = C_i(x^{H_1}) - d_i$.

Now problem (9) can be formulated as a Boolean programming problem of the form

$$\min\{\sum_{i \in Y} \sum_{k=d_i+1}^{d} w_i y_{ik} \mid (x, y) \in M_{\mathcal{H}} \cap Z^E\}, \tag{13}$$

where Z^E is an integer lattice of the space R^E.

The form of the objective function in problem (13) implies the presence of the condition $d_i < d$, although, generally speaking, the situation $d_i \geq d$ is possible. However, in this case for $x \in \Sigma_d$ we have $\max\{0, C_i(x) - d_i\} = 0$ and consequently the jobs $i \in V$ for which $d_i \geq d$, we can simply not include it in the objective function. Therefore, further, without loss of generality, we will assume that $d_i < d$ for all $i \in V$.

The following example (Fig. 2) show that the polyhedron $M_{\mathcal{H}}$ also has non-integer vertices, i.e. the inclusion $P_{\mathcal{H}} \subset M_{\mathcal{H}}$ is strict. We present the vertex of the $M_{\mathcal{H}}$ polyhedron in graph form. Here $n = 3$, $p = 4$, $d = 12$, $r = (0, 2, 2)$. The solid line shows the variables for which the point $(x, y) \in M_{\mathcal{H}}$ has the value 1, the dotted line shows non-integer variables. In addition, Table 1 shows the values for each variable y_{ik}, $i \in Y$, $k \in D$. This example was obtained by applying the simplex method to the polyhedron $((1)–(4), (10)–(12))$ (the form of the objective function does not matter).

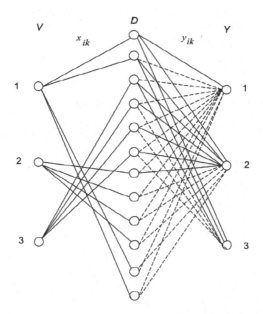

Fig. 2. Non-integer point of the polyhedron $M_{\mathcal{H}}$.

Table 1. The values of the variables y_{ik} for the point from Fig. 2.

y_{ik}	1	2	3	4	5	6	7	8	9	10	11	12
1	1	0.75	0.5	0.5	0.5	0.5	0.5	0.5	0.5	0.5	0.5	0.25
2	1	1	1	1	1	1	1	0.75	0.5	0.25	0	0
3	1	1	1	0.75	0.5	0.25	0	0	0	0	0	0

5   Classes of Valid Inequalities

This section describes the classes of inequalities valid with respect to the polytope $P_{\mathcal{H}}$. The inequalities of these classes strengthen the constraints of the polyhedron $M_{\mathcal{H}}$ in the sense that $M_{\mathcal{H}}$ has points cut off by them. We propose three classes of valid inequalities.

The first class consists of the set of hyperplanes that entirely contain the polytope $P_{\mathcal{H}}$. Since the processing of the job i cannot be completed before the moment $r_i + p$, so by condition (8) from the definition of the family $\mathcal{H}$ we have

Proposition 1. *The polytope $P_{\mathcal{H}}$ lies entirely in each of the hyperplanes given by the equations*

$$\sum_{k=1}^{r_i+p} y_{ik} = r_i + p, \qquad i \in V.$$

The class of these hyperplanes will be denoted by $S1$. Note that since the number of these constraints is equal to n, it is expedient to include them in the initial polyhedral relaxation of the polytope $P_{\mathcal{H}}$.

Equally simple and directly following from condition (8) is the second class of valid inequalities.

Proposition 2. *For any $i \in V$ and $k \in D$ the inequality*

$$x_{ik} \leq y_{ik}$$

is valid with respect to the polytope $P_{\mathcal{H}}$.

This class of inequalities will be denoted by $S2$.
And finally, the third class.

Proposition 3. *For any $k \in D$ the inequality*

$$\sum_{i \in V} y_{ik} \geq n - \lfloor \frac{k-1}{p} \rfloor$$

is valid with respect to the polytope $P_{\mathcal{H}}$.

Indeed, according to the construction of the model (6)–(8) and its polyhedral description ((1)–(4), (10)–(12)) we have the following: if $1 - y_{ik} = 0$, then job $i \in V$ is in processing at time $k \in D$; if $1 - y_{ik} = 1$, then job i has been complete processing at previous moments of times. Therefore, $\sum_{i \in Y}(1 - y_{ik})$ is the number of jobs whose processing is completed before k, that is, in the period from 1 to $(k - 1)$ inclusive. During this time, no more than $\lfloor \frac{k-1}{p} \rfloor$ jobs can be completely processed. So $\sum_{i \in Y}(1 - y_{ik}) \leq \lfloor \frac{k-1}{p} \rfloor$ or, equivalently, $\sum_{i \in Y} y_{ik} \geq n - \lfloor \frac{k-1}{p} \rfloor$.

Note that for $k \geq pn+1$ we have $n - \lfloor \frac{k-1}{p} \rfloor \leq 0$ and the inequality from Proposition 3 becomes trivial. Therefore, when using these inequalities, we assume that $k \leq pn$. This class of inequalities will be denoted by $S3$.

It is easy to check directly that the point from Fig. 2 is cut off, for example, by the inequalities $x_{36} \leq y_{36}$, $x_{28} \leq y_{28}$, and so on. These inequalities belong to the class $S2$. The same point is cut off by the inequality $y_{18} + y_{28} \geq 3 - \lfloor \frac{8-1}{4} \rfloor$ belonging to the class $S3$. This means that the inequalities constructed in this section can serve as cutting planes in the corresponding algorithms.

6   Probation of Models and Experimental Comparison of Relaxations

In this work, a computing experiment was conducted. The goal of experiment was to compare the models that were built in Sects. 4 and 5, namely:
- **model** $M1$, formed by constraints (1)–(4), (10)–(12) and by objective function

$$\min \sum_{i \in y} \sum_{k=d_i+1}^{d} w_i y_{ik};$$

- **model** $M2$, formed by constraints (1)–(4), (10)–(12) and constraints of $S1$, $S2$, $S3$ and the same objective function.

The classes $S1$, $S2$ and $S3$ were described in Sect. 5. They are valid for a polytope $P_{\mathcal{H}}$. In addition, examples of fractional points were given, cut off by constructed inequalities. Therefore these inequalities can be used as cutting planes in the corresponding algorithms. Note that the capacities of these classes are polynomial in n. Accordingly, these inequalities can be simultaneously added to the constraints on the problem (13), not fearing that this will entail the exponential increase in the number of constraints.

To conduct an experiment, a program was developed in java. To solve the integer linear problems, the programs from the IBM ILOG CPLEX 20.1.0 package are used.

The experiment was conducted on a computer with a dual-core Intel Pentium P6200 processor with a frequency of 2.13 GHz. When calculating the time of solving the problem, the processor time was taken into account, that is, the time that the processor spent directly on solving the problem.

During the experiment, a series of problems with pre-set values of n and p were generated. The w_i parameters were randomly selected from the interval from 1 to 100. The values of r_i were selected so that for each task $d = d_{min} = pn$, the values of d_i were selected with the condition of $d_i < d$.

The generated problems were launched alternately on both models. The problem was considered unresolved if the solution time exceeded 60 min.

Table 2 presents the average values of the results of launching problems of $M1$ and $M2$ on various input data. The number of jobs of n changed from 10 to 120, servicing time $p = 3.4.5$. For each n, 10 instances with various parameters of p, r_i, d_i, w_i were solved. An empty cell means that none of the instances with these parameters was solved in 60 min. As can be seen from Table 2, the addition of the valid inequalities $S1$, $S2$ and $S3$ to the constraints of polyhedron $M_{\mathcal{H}}$ can significantly increase the dimension of the problems solved.

Table 2. The results of solving the instances using $M1$ and $M2$ on various input data.

n	p	$M1$ (sec)	$M2$ (sec)	n	p	$M1$ (sec)	$M2$ (sec)
10	3	0,42	0,24	50	3	210,45	50,10
	4	1,37	0,62		4		192,31
	5	11,62	2,79		5		611,40
20	3	5,90	6,07	60	3	54,15	35,60
	4	75,08	84,09		4		178,00
	5	856,25	417,65		5		309,60
30	3	7,06	3,03	70	3		634,40
	4	774,20	6,24	80	3		1180,00
	5		12,07	100	3		477,60
40	3	5,98	5,43	120	3		2203,00
	4	1798,00	18,38				
	5		27,71				

7   Conclusion

In this work in the graph theory terms, new models that formalize the problem of finding a schedule on one machine that minimizes the weighted sum of tardiness are built.

The main feature of the models proposed in this work is that here the schedule set are interpreted as a family of subgraphs of the some special graph. As a result, we get a rich arsenal of possibilities for analyzing the combinatorial structure of the problem. These possibilities is significantly expanding when using the polyhedral properties of the corresponding polytope.

The numerical experiment was carried out in order to evaluate the efficiency of additional classes of valid inequalities S1, S2 and S3. The model $M2$ with additional constraints behaves much better than the $M1$ model in terms of solution time. At the same time, the model $M2$ contains a significantly more constraints than the model $M1$. Nevertheless, as can be seen from the experiment, that usage of these inequalities leads to an increase of the dimension of problems that we can solve. We did not set ourselves the task of obtaining records in terms of dimension. This goal will be relevant to our next work on this topic.

Acknowledgement. This work was carrited out within the governmental order for Omsk Scientific Center SB RAS (project registration number 121022000112-2).

References

1. Tian, Z., Ng, C.T., Cheng, T.C.E.: An $O(n^2)$ algorithm for scheduling equal-length preemptive jobs on a single machine to minimize total tardiness. J. Sched. **9**(4), 343–364 (2006)
2. Baptiste, P.: Scheduling equal-length jobs on identical parallel machines. Discrete Appl. Math. **103**(1), 21–32 (2000)
3. Brucker, P., Knust, S.: Complexity results for scheduling problems. https://www.mathematik.uni-osnabrueck.de/research/OR/class
4. Du, J., Leung, J.Y.-T.: Minimizing total tardiness on one machine is NP-hard. Math. Oper. Res. **15**(3), 483–495 (1990)
5. Lenstra, J.K., Kan, A.R., Brucker, P.: Complexity of machine scheduling problems. Ann. Discrete Math. **1**, 343–362 (1977)
6. Blazewicz, J., Dror, M., Weglarz, J.: Mathematical programming formulations for machine scheduling: a survey. Eur. J. Oper. Res. **51**, 283–300 (1991)

Solving Maximin Location Problems on Networks with Different Metrics and Restrictions

Gennady G. Zabudsky[(⊠)] [iD]

Sobolev Institute of Mathematics SB RAS, Novosibirsk 630090, Russia
zabudsky@ofim.oscsbras.ru

Abstract. Several optimal location problems of an obnoxious facility on a network of roads connecting settlements are considered. It is necessary to find such location of the facility so that a minimum distance to a nearest settlement is as large as possible taking into account the resident population. Such facility can be, for example, a nuclear power plant, a waste recycling plant. An overview of various formulations, the properties of the problems and algorithms for solving are given. The main focus is on the problem taking into account a restriction on transportation costs for servicing the settlements by the facility. The cost of servicing the settlements by the facility is determined using the shortest paths in the network. The objective function uses Euclidean metric. Exact algorithm for solving of this problem is proposed.

Keywords: Euclidean metric · Maxmin criterion · Obnoxious facility · Shortest paths · Voronoi diagram

1 Introduction

One of the actively developing areas of operations research is the analysis and solving of facilities location problems. Such problems have great applied importance. They need to be solved in various fields of activity: location of service points, technological equipment in workshops, automated design of electronic devices [1–3].

In general, a facilities location problem is formulated as follows: there is an area with facilities fixed in it and new facilities that need to be located in the area. Specified restrictions on location of new facilities are to be met, and some criterion of a quality of location is to be optimal. The various formulations of such problems are defined by sizes of the facilities, areas in which they should be located (line, plane, network), various restrictions and types of criterion and so on [4–6].

The criterion of optimality in the location problems can be different, and it is depending on the specifics of the facilities and on what functions they

The research was funded in accordance with the state task of the IM SB RAS, project FWNF-2022-0020.

perform. A term "facility" can be interpreted quite widely. Minimization of a maximum weighted distance between facilities often is used when considering an optimal location of hospitals, police stations, fire stations. Minimization a weighted sum of distances is used when choosing a location of switches in a telephone network, warehouses, substations in a power grid [7–9]. In addition to these criterions, a maximin and maximum criterions are applied according to which a minimum distance between facilities or a minimum weighted sum of distances is maximized, respectively. These criterions are applied when dealing with obnoxious (undesirable) production facilities that have adverse effects to people or environment. Note that the adverse effects of such facilities decreases with increasing distance to them. Therefore it is easiest to locate them as far away as possible from populated areas.

Most often the maximin criterion in the location problems on networks is applied in two variants. In the first variant, distances are measured by the shortest paths in the networks. This metric is used in significant part of the location problems on networks with the maximin criterion [7,11–13]. In the second variant, several metrics are used. For example, Euclidean metric is used to measure a negative effect of the obnoxious facility, and the shortest paths metric is used to calculate the cost of servicing customers by the facility [6,14].

In this paper, the location problems on various networks with a maximin criterion are considered. An overview of the formulations and algorithms for solving of the problems in which the distances in the objective function are measured along the shortest paths in the network is given. The main attention is paid to the problem of optimal location of an obnoxious facility on a transport network connecting some settlements. It is necessary to find such a location of the facility so that a minimum distance to a nearest settlement is as large as possible and at the same time a budget for transportation costs for servicing the settlements by the facility was not violated. The problem uses Euclidean metric to determine the magnitude of the adverse effects of the obnoxious facility to the settlements. The shortest paths in network are used to determine the transportation costs for servicing the settlements by the facility. A polynomial algorithm for exact solving of this problem is proposed.

Section 2 provides an overview of maximin location problems on arbitrary and special networks when distances are measured by shortest paths.

In Sect. 3, a formulation of a maximin problem with different metrics is given. Euclidean metric is used in the objective function of the problem. Shortest paths metric is used in a restriction on transportation costs. Several properties of the problem and a polynomial algorithm for solving this problem are presented.

2   Shortest Paths Metric

The Section deals with a location problem of a facility on arbitrary and special networks with the maximin criterion. The distances between the vertices of the network and between the vertices and points on the edges are determined using the shortest paths [7,11–13]. The maximin location problem is NP-hard for an arbitrary number of facilities, even if the network has one edge [15]. Problem for one facility on general network is polynomially solvable. An overview of some properties of the problem and algorithms for its solving are given.

2.1 General Network

To formulate a mathematical model of the problem, we introduce the following notations [11,12]. Let $G = (V, E)$ be a connected, undirected network with a set of vertices $V = \{v_1, v_2, \ldots, v_n\}$ and a set of edges $(v_i, v_j) \in E$, $v_i, v_j \in V$, $i < j$, $i, j \in N = \{1, \ldots, n\}$. Each edge connecting v_i and v_j has positive weight (length) $c(v_i, v_j)$. Denote by $d(v_i, v_j)$ the length of a shortest path between vertices v_i and v_j. For any $(v_i, v_j) \in E$, $c(v_i, v_j) \geq d(v_i, v_j)$. We also denote by α_i a parameter (weight) of the vertex $v_i, i \in N$, which is always positive.

It is necessary to locate a facility on the network. Therefore, the set of candidate points is the set Z consisting of the vertices and the infinite set of points on the edges. The location of a point on an edge is determined using the distances from the vertices of the edge. For example, z is located on the edge (v_i, v_j) at a distance of $c(v_i, z) = \lambda c(v_i, v_j)$ from the vertex v_i and at a distance $c(v_j, z) = (1 - \lambda)c(v_i, v_j)$ from the vertex v_j, where $0 \leq \lambda \leq 1$. Denote by $d(v_i, z)$ the length of a shortest path from vertex v_i to $z, z \in Z$. Our objective will be to maximize

$$r(z) = \min_{i \in N} \alpha_i d(v_i, z), \tag{1}$$

where $z \in Z$.

If a point z^* solves the problem (1), then it is said to be a maximin location with the optimal value of the objective function $r(z^*)$.

For network distance function $d(v_i, z)$, $z \in (v_p, v_q)$, $(v_p, v_q) \in E$, the following properties hold [8,12]:

$$d(v_i, z) = \min\{d(v_i, v_p) + c(v_i, z), \ d(v_i, v_q) + c(v_p, v_q) - c(v_i, z)\}$$

where $i = 1, \ldots, n$, and $0 \leq c(v_i, z) \leq c(v_p, v_q)$.

Fuction $d(v_i, z)$ is continuous and concave on segment $[0, c(v_p, v_q)]$ and one of the conditions is met

(a) linearly increases with slope 1 in the edge;
(b) linearly decreases with a slope -1 on the edge:
(c) linearly increases with slope 1 on the segment $[0, z_i(p, q)]$ and linearly decreases with slope -1 on the segment $[z_i(p, q), c(v_p, v_q)]$, where

$$z_i(p, q) = (d(v_i, v_q) + c(v_p, v_q) - d(v_i, v_p))/2.$$

In [12], the problem (1) on general network was investigated. The paper presents certain properties of the problem, which allow to find a solution to the problem. Although the problem is non-convex its solution space G can be divided into edges and resulting subproblems can be solved more easily than the original problem.

In [12], it is shown that the problem (1) on the edge (v_p, v_q) is equivalent to the following linear programming problem:

$$\max g(z)$$

$$g(z) = \min_{1 \leq i \leq 2n, 0 \leq c(v_p, z) \leq c(v_p, v_q)} \{B_i z + C_i\},$$

where $B_i = \alpha_i$, $B_{n+i} = -\alpha_i$, $i = 1, \ldots, n$ and $C_i = \alpha_i d(v_i, v_p)$, $C_{n+i} = \alpha_i(d(v_i, v_q) + c(v_p, v_q))$, $i \in N$.

Thus, for any edge $(v_p, v_q) \in E$, the function $g(z)$, $z \in (v_p, v_q)$ is continuous, piecewise linear and concave on the segment $[0, c(v_p, v_q)]$, consisting of at most $2n$ strictly monotone segments.

In [12], it is proved that there is any unique maximin location on each edge. As a consequence, there are no more than m local maximums on the network. A combinatorial algorithm for solving of the problem (1) with a time complexity $O(mn)$ is proposed.

In [7], a linear programming model to solve the problem (1) was used. The problem is solved for each edge and the maximum value among the set of values of the objective function is selected. Let the obnoxious facility be placed on the edge $(v_p, v_q) \in E$. The shortest path from the facility to the vertex v_i is $\min\{d(v_p, v_i) + c(v_p, z); d(v_q, v_i) + c(v_p, v_q) - c(v_p, z)\}$. Then the model has the following form:

$$y \to \max,$$

$$\alpha_i(d(v_p, v_i) + c(v_p, z)) \geq y,$$

$$\alpha_i(d(v_q, v_i) + c(v_p, v_q) - c(v_p, z)) \geq y,$$

$$c(v_p, z) \leq c(v_p, v_q),$$

$$y, c(v_p, z) \geq 0.$$

2.2   Tree Networks

Often in the literature, the maximin location problems on networks of a special type are considered. Network structure allows to find useful properties of the problem and determine new ways to solve it.

If a network is a path, then we can put n vertices on a real line and identify them with real numbers such that

$$0 = x_1 < x_2 < \ldots < x_n$$

and $d(x_i, x_j) = |x_i - x_j|$. Then objective function $r(z)$ is the following expression

$$r(z) = \min\{\alpha_i|z - x_i| : i = 1, \ldots, n\} = \min\{r^+(z), r^-(z)\},$$

where

$$r^+(z) = \min\{\alpha_i(z - x_i) : x_i \leq z\},$$
$$r^-(z) = \min\{\alpha_i(x_i - z) : x_i \geq z\}.$$

In [16], a linear-time algorithm is given that finds a maximum value of the objective function $r(z)$ along the path.

The star is a tree consisting of a central vertex v_0 which has edges with n remaining vertices $\{v_1, \ldots, v_n\}$ [16]. By S_i, denote a problem which consists in determining a local optimal solution on an edge (v_0, v_i). It is equivalent to the problem on the path as follows: we place all vertices on the real line in such a way that v_0 in the point $x_0 = 0$, vertex v_i is to the right of v_0 at a distance of $c(v_0, v_i) = x_i$, and all other vertices v_j $(j \neq i)$ are to the left of v_0 with a distance of $c(v_0, v_j) = x_j$. In problem S_i, it is necessary to maximize the following function:

$$r_i(z) = \min_{j=0,\ldots,n} \{\min \alpha_j(x_j + z), \ \alpha_i(x_i - z)\}$$

for $0 \leq z \leq x_i$.

Let a maximum be reached in $z^{(i)}$. Point $z^{(i)}$ is a solution following linear programming problem with two variables y and z:

$$\min\{z : y \leq \alpha_j(x_j + z), \ j = 0, \ldots, n, \ y \geq \alpha_i(x_i - z)\}.$$

The constraints that are common to all problems S_i can be written in the form $y \leq h(z)$ where

$$h(z) = \min_{j=0,\ldots,n} \alpha_j(x_j + z).$$

The function $h(z)$ is piecewise linear and increasing. Point $z^{(i)}$ is an intersection of $h(z)$ with $\alpha_i(x_i - z)$. Due to the monotonicity of $h(z)$ it is required to calculate $z^* = \max z^{(i)}$. The value of z^* can be obtained as an optimal solution to following linear programming problem:

$$\min\{z : y \leq \alpha_j(x_j + z), y \geq \alpha_i(x_i - z), i, j = 0, \ldots, n\} \qquad (2)$$

Consider an optimal solution (z^*, y^*) of the problem (2). In [17], it is proved that optimal locations of the obnoxious facility are points on all edges (v_0, v_i) at the distance z^* from the central vertex v_0 for which $y^* = \alpha_i(x_i - z^*)$.

The problem (2) has $2n$ constraints and 2 variables. Using an algorithm from [4] for solving linear programming problem with 2 variables, problem (2) can be solved in linear time. Thus, the optimal solution of problem (1) on the star can be found in linear time.

Consider the maximin location problem on weighted trees. To solve the problem on an arbitrary tree with n vertices, two algorithms are proposed [9,15] with time complexity $O(n \log_2 n)$ and $O(kn \log_2 n)$, respectively. The parameter k depends on the structure of the tree. For paths and stars, $k = O(1)$. For a balanced tree, $k = O(\log n)$ but there are trees such that $k = \Theta(n)$. For an unweighted tree, a linear algorithm is proposed [17].

3   Euclidean and Shortest Paths Metrics

Section deals with the location problem of an obnoxious facility on an arbitrary network with maximin criterion in Euclidean metric and a restriction on transportation costs. The transportation costs are determined using the shortest paths metric. Model that takes into account the natural geometry of transport networks is proposed. The properties of the problem are found on the basis of which an exact algorithm for finding a solution of the problem is proposed. The analysis of the computational complexity of the algorithm is carried out.

3.1   Problem Formulation

There is a network of roads on a plane connecting some settlements. The population size in each settlement and the distances between them are known. It is necessary to place, for example, a waste recycling plant (facility) on the network. There is a budget of transportation costs for the transportation of waste from settlements to the facility. The facility has the adverse effects to the population and therefore it should be located as far away from the settlements. It is needed to place the facility on the network as far as possible from the settlements, taking into account the population size of the settlements and that transportation costs do not exceed the budget.

To formulate a mathematical model of the problem we use some previously introduced notations and introduce new ones. Let $G = (V, E)$ be an undirected network representing the roads and settlements on a plane. The vertices of the network correspond to the settlements and the edges correspond to the roads. Two positive parameters (α_i, w_i) are assigned to each vertex v_i, $i \in N$. Parameter α_i reflects a degree of undesirability of placing the facility near the settlement corresponding v_i. Parameter w_i, on the contrary, reflects a requirement to place the facility as close as possible to the settlement corresponding v_i. Parameter w_i, for example, is the number of peoples in the settlement correspond to vertex v_i. Parameter α_i is the inverse value of the number population in the settlement correspond to vertex v_i. Each vertex v_i has coordinates (a_i, b_i) $i \in N$ on the plane. The edges of the network are segments of straight lines on the plane with known lengths $c(v_i, v_j) > 0, (v_i, v_j) \in E, i, j \in N$. As in Sect. 2, we denote by Z the infinite set of points on the network G. The adverse effects of the facility to the settlements will be measured in Euclidean metric $\rho(v_i, z), z \in Z, i \in N$. We will use the shortest paths $d(v_i, z), z \in Z, i \in N$ in G for determining transportation costs of servicing the settlements. Let T be an available budget for the transportation of waste from settlements to the facility. The mathematical model has form:

$$r_1(z) = \min_{i \in N} \alpha_i \rho(v_i, z) \to \max_z, \qquad (3)$$

$$\sum_{i \in N} w_i d(v_i, z) \le T, \qquad (4)$$

$$z \in Z. \qquad (5)$$

Expression (3) means maximizing of the minimum weighted distance from the facility to the vertices. Condition (4) guarantees the fulfillment of the restriction on transportation costs. The left part of the inequality (4) is a sum of the weighted shortest paths from some point $z \in Z$ to all vertices of the network.

Condition (5) means, that the facility is located on infinite set of points on all edges of G, including all vertices. Problem (3)–(5) is a nonlinear, nonconvex problem with one or more local maximum.

Note that a linear approximation can be used to model real road sections. This will lead to the additional introduction of dummy vertices and edges to the network. This method is used, for example, in [6]. In this paper, we do not consider any method of approximating real roads by straight-line segments.

In [18], an algorithm for finding an approximate solution to the problem is proposed. The algorithm considers a finite set of points on an arbitrary edge of the network. The values of the objective function are calculated at the points, where the budget restriction is met. The point with a maximum value of the objective function determines a local maximum on the edge. An approximate solution to the problem (3)–(5) is the best of the point. An experiment was conducted to solve the problem with the proposed algorithm on the network of the main railway lines of France.

Problems close to the formulated one without taking into account the budget constraint were studied, for example, in [11, 17].

In [19–21], a problem of locating an obnoxious facility in a polygonal area on a plane (polygon) is considered. The problem is finding a point in the polygon, which maximizes a minimum weighted Euclidean distance to given set points in the polygon. It does not matter whether the polygon is convex or not. In fact, this is the problem of maximization function (3) in the polygonal area on the plane. It is proved, that an optimal solution is either in the convex hull of the vertices or on the boundary (segments) of the polygon. In the case of searching the solution on the boundary segment (local optimum) it is necessary to solve a system of two equations: the linear equation of the boundary segment and the nonlinear equation of Euclidean distances. If the local optimum is not located on the boundary of the polygon, then it is necessary to solve a system of three equations of Euclidean distances.

3.2   Transportation Costs Function

Let's analyze some properties the sum function of weighted shortest paths from some point $z \in Z$ on arbitrary edge (v_p, v_q) to all vertices of the network G.

$$\sigma(z) = \sum_{i \in N} w_i d(z, v_i). \tag{6}$$

In [11] a concept of edge bottleneck points is introduced. Let a point x be located on the edge (v_p, v_q). The point x is an edge bottleneck point with respect to vertex v_k if there is a vertex v_k such that

$$d(v_k, v_p) + c(v_p, x) = d(v_k, v_q) + c(v_q, x),$$

where $c(v_p, x), c(v_q, x) > 0$ are the distances from v_p and v_q to point x on the edge (v_p, v_q). The equality $c(v_p, x) + d(v_q, x) = c(v_p, v_q)$ is correct.

Note that a bottleneck point on edge (v_p, v_q) with respect to vertex v_k is associated with a cycle formed by the shortest path from vertex v_k to vertex v_p, edge (v_p, v_q), and the shortest path from vertex v_q to back to vertex v_k. This is illustrated in Fig. 1., where the shortest paths from vertex v_k to v_p and v_q are shown in the form of curved lines. The cycle contains the bottleneck point x.

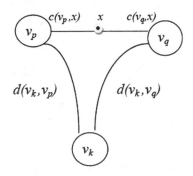

Fig. 1. Cycle contains the bottleneck point x.

Edge (v_p, v_q) contains an edge bottleneck point with respect to vertex v_k if and only if following inequality holds

$$|d(v_k, v_p) - d(v_k, v_q)| < c(v_p, v_q). \tag{7}$$

This means that there is no shortest path from v_k to v_p and from v_k to v_q containing the edge (v_p, v_q).

The papers [1, 11] show that $\sigma(z)$ on an edge is continuous, piecewise linear, concave function with critical points (intersection of linear functions) only at the bottleneck points of the edge. If G is a tree, then in this case there are no edge bottlenecks points as there are no cycles. Then the following proposition is true.

Proposition 1. *If G is a tree, then function $\sigma(z)$ is linear on arbitrary edge.*

Proof. Let $(v_p, v_q) \in E$ be an edge of the G. The set of vertices of the network G by the edge (v_p, v_q) is divided into two sets: V_L and V_R. The set V_L is such set of vertices of the network G that the paths from them to the v_q pass through the vertex v_p. By N_L denote a set of indexes of such vertices. The set V_R is such set of vertices of the network G that the paths from them to the v_p pass through the vertex v_q. By N_R denote a set of indexes of such vertices. There are relations $V_L \cap V_R = \emptyset$ and $V_L \cup V_R = V$. The following equalities take place.

$$d(z, v_i) = d(v_i, v_p) + c(v_p, z), i \in N_L.$$
$$d(z, v_i) = d(v_i, v_q) + c(v_p, v_q) - c(v_p, z), i \in N_R.$$

$$\sigma(z) = \sum_{i \in N} w_i d(z, v_i) = \sum_{i \in N_L} w_i d(z, v_i) + \sum_{i \in N_R} w_i d(z, v_i) =$$

$$\sum_{i \in N_L} w_i [d(v_i, v_p) + c(v_p, z)] + \sum_{i \in N_R} w_i [d(v_i, v_q) + c(v_p, v_q) - c(v_p, z)] =$$

$$\sum_{i \in N_L} w_i d(v_i, v_p) + c(v_p, z) \sum_{i \in N_L} w_i + \sum_{i \in N_R} w_i d(v_i, v_q) +$$

$$+ c(v_p, v_q) \sum_{i \in N_R} w_i - c(v_p, z) \sum_{i \in N_R} w_i =$$

$$\sum_{i \in N_L} w_i d(v_i, v_p) + \sum_{i \in N_R} w_i d(v_i, v_q) + c(v_p, v_q) \sum_{i \in N_R} w_i +$$

$$+ c(v_p, z) \left(\sum_{i \in N_L} w_i - \sum_{i \in N_R} w_i \right).$$

Let 's put

$$a = \sum_{i \in N_L} w_i d(v_i, v_p) + \sum_{i \in N_R} w_i d(v_i, v_q) + c(v_p, v_q) \sum_{i \in N_R} w_i,$$

$$b = \sum_{i \in N_L} w_i - \sum_{i \in N_R} w_i.$$

We get the equation of a straight line $\sigma(z) = a + bc(v_p, z)$. Moreover, b it can be negative or positive. Note that this form will have function $\sigma(z)$ on any edge of a network, if the edge is a bridge.

3.3   Domain of Admissible Solutions

Let's look some properties of the problem (3)–(5) that allow to find all local optimums and exact solution to the problem. A domain of admissible solutions of the problem is only some part of the network G due to the restriction of transportation costs (budget). All edge segments, on which the value of transportation costs do not exceed the value of the budget T form the domain of admissible solutions of the problem (3)–(5). By D denote the domain. If the budget restriction is violated for the vertices of an edge, then the edge does not belong to the domain D. It follows from the concavity property of the transportation costs function.

Before constructing the domain of admissible solutions of the problem (3)–(5) on an edge (v_p, v_q), it is necessary to find all edge bottleneck points. The following algorithm for finding such points on the edge is proposed. Consistently consider all vertices of the network G. For a current vertex v_k we check performing of the inequality (7). If the inequality (7) does not hold, then the edge (v_p, v_q) does not contain an edge bottleneck point relative to v_k, and we move to another vertex G. If the inequality holds, then the edge (v_p, v_q) contains an edge bottleneck point relative to the vertex v_k. Distance s from the vertex v_p to the point is equal to

$$s = \frac{|d(v_k, v_p) - d(v_k, v_q)| + c(v_p, v_q)}{2}.$$

After constructing all the bottleneck points on the edge, we determine the domain of admissible solutions on the edge. To do this, we arrange the bottleneck points in ascending order $s_1 < s_2 < \dots s_n$ from the vertex v_p, $p < q$. Sequentially calculate following values

$$T_1(s_i) = \sum_{k \in N} w_k \min\{d(v_k, v_p) + s_i; d(v_k, v_q) + c(v_p, v_q) - s_i\}.$$

The following variants are possible.

1) $T_1(v_p) \leq T$ and $T_1(v_q) \leq T$.
1) We look through the bottleneck points sequentially. There is a pair of points s_t and s_{t+1} such that $T_1(s_t) \leq T$ and $T_1(s_{t+1}) > T$. If $T_1(s_t) < T$, then we construct the equation of the line $l(s_t, s_{t+1})$ through the points $(s_t, T_1(s_t))$ and $(s_{t+1}, T_1(s_{t+1}))$. Solving the equation $l(s_t, s_{t+1}) = T$, and we find a point z_1, for which $T_1(z_1) = T$. If $T_1(s_t) = T$, then the value of z_1 is defined as $z_1 = s_t$. Next, we calculate a value of the function T_1 at the points s_{t+2}, s_{t+3} etc. There is such k, that $T_1(s_k) > T$ and $T_1(s_{k+1}) \leq T$. If $T_1(s_{k+1}) < T$, then we construct the equation of the line $l(s_k, s_{k+1})$ through the specified pair of the points. Solving the equation $l(s_k, s_{k+1}) = T$, and we find a point z_2, for which $T_1(z_2) = T$. If $T_1(s_{k+1}) = T$, then the value of z_2 is defined as $z_2 = s_{k+1}$. Thus two points are found on the edge (v_p, v_q), for which $T_1(z_1) = T_1(z_2) = T$. The domain of admissible solutions on the edge (v_p, v_q) represents two segments: $[v_p, z_1]$ and $[z_2, v_q]$, as shown in Fig. 2.
If there is no such pair of points s_t and s_{t+1}, for which $T_1(s_t) \leq T$ and $T_1(s_{t+1}) > T$, then all points of the edge (v_p, v_q) belong to the domain of admissible solutions.
2) $T_1(v_p) \leq T$ and $T_1(v_q) > T$.
There is a pair of points s_t and s_{t+1} such that $T_1(s_t) \leq T$ and $T_1(s_{t+1}) > T$. If $T_1(s_t) < T$, then we construct the equation of the line $l(s_t, s_{t+1})$ through the points $(s_t, T_1(s_t))$ and $(s_{t+1}, T_1(s_{t+1}))$. Solving the equation $l(s_t, s_{t+1}) = T$, and we find the point z_1, for which $T_1(z_1) = T$. If $T_1(s_t) = T$, then the value of z_1 is defined as $z_1 = s_t$. The domain of admissible solutions on the edge (v_p, v_q) represents the segment $[v_p, z_1]$.
3) $T_1(v_p) > T$ and $T_1(v_q) \leq T$.
There is a pair of points s_t and s_{t+1} such that $T_1(s_t) > T$ and $T_1(s_{t+1}) \leq T$. If $T_1(s_{t+1}) < T$, then we construct the equation of the line $l(s_t, s_{t+1})$ through the points $(s_t, T_1(s_t))$ and $(s_{t+1}, T_1(s_{t+1}))$. Solving the equation $l(s_t, s_{t+1}) = T$, and we find the point z_2, for which $T_1(z_2) = T$. If $T_1(s_{t+1}) = T$, then the value of z_2 is defined as $z_2 = s_{t+1}$. The domain of admissible solutions on the edge (v_p, v_q) represents a segment $[z_2, v_q]$.

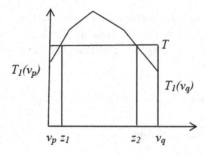

Fig. 2. Function $\sigma(z)$ and Domain of admissible solutions on edge (v_p, v_q).

As a result, the domain of admissible solutions of the problem (3)–(5) on an arbitrary edge (v_p, v_q) has no more than two segments. One of the ends of these segments will be vertex v_p or v_q. Thus, to solve problem (3)–(5) on the domain D, this is solving a series of the problems on edge segments of the network G.

If G is a tree, then there is no need to find bottleneck points. If $T_1(v_p) \leq T$ and $T_1(v_q) > T$ ($T_1(v_p) > T$ and $T_1(v_q) \leq T$), we find an intersection point z_1 (z_2) of the straight line $l(v_p, v_q)$ with the straight line $y = T$. Thus, we get the domain of admissible solutions on edge (v_p, v_q) segment $[v_p, z_1]$ ($[z_2, v_q]$).

3.4   Algorithm

At the first step of the algorithm for solving the problem (3)–(5) it is necessary to check, that the domain D is not empty. To do this, we solve a problem of finding 1-median on the network in the shortest paths metric. If the solution of the problem has the value of the objective function no more than T, then the domain of admissible solutions of problem (3)–(5) is not empty.

When searching a local optimum on an edge (v_p, v_q) belonging to the admissible domain, it is necessary to construct a weighted Voronoi diagram for the vertices of the network G [3,5]. Next, we find the intersection points of the edges of the Voronoi diagram with the segments admissible domain D on the edge (v_p, v_q). As a result, the edge segments will be divided into pieces by the intersection points. Each the piece will be located inside some locus of the Voronoi diagram. The function r_1 is convex on each such piece [6]. Consequently, the function r_1 reaches an optimal value at the intersection points of the edges of the Voronoi diagram with the edge segments (v_p, v_q) belonging to the domain D. Choosing a point with a maximum value of the function r_1, we find a local optimum of the problem (3)–(5) on the edge (v_p, v_q).

Looking through all the edges of the network G, the segments of which belong to the domain of admissible solutions D, we get a set of local optimums. Choosing a maximum value from them, we obtain a global solution to the problem (3)–(5).

The algorithm for finding a local optimum of the problem (3)–(5) on an edge can be briefly presented as follows.

Step 1. Solve the 1-median problem on network G. If optimal value of objective function of the problem is greater T, then the problem (3)–(5) has no solution.

Step 2. Construct the domain of admissible solutions D.

Step 3. Construct the weighted Voronoi diagram of vertices of the network G.

Step 4. Find intersection points of all Voronoi edges with the segments of the edge, belonging to the domain D.

Step 5. Calculate the values of the function (3) at the specified intersection points. The point, at which the maximum value of the objective function is reached, will be a local optimum of the problem on the edge.

There are $O(n^2)$ vertices and $O(n^2)$ edges in a weighted Voronoi diagram, which can be generated in $O(n^2)$ time. There are $O(mn^2)$ intersection points the edges of Voronoi diagram with the edges of network G, since each Voronoi edge can be intersect an edge of network G at most twice [3]. Therefore a complexity of the algorithm is $O(mn^2)$.

4   Conclusion

The problems of optimal location of an obnoxious facility on a networks of roads connecting settlements are considered. It is necessary to find such location of the facility so that the minimum distance to the nearest settlement is as large as possible. An overview of the properties and algorithms for solving the problems using the shortest paths metric is given.

The main attention is paid to the problem taking into account a restriction on transportation costs for servicing the settlements by the facility. The objective function uses Euclidean metric. The restriction take into account the shortest paths in the network. A polynomial algorithm for finding all local maximums on the edges of the network is proposed. The choice of the optimal solution among the specified optimums can be made by the decision-maker from any additional conditions. Considered model and proposed algorithm can be applied for solving, for example, a problem of location a waste processing plant to reduce the adverse effects to the population.

One of the conditions of the problem that is the assumption, that roads are line segments on a plane. Real roads can be approximated by straight-line segments. Furthermore, an interesting continuation the study of the problem is it solution without the use of Voronoi diagram. For example, using an approach to solving a maximin problem within a bounded region on a plane with Euclidean metric and applying Karuch-Kuhn-Tucker optimality conditions.

References

1. Eiselt, H.A., Marianov, V.: Foundations of Location Analysis. Springer, New York (2011). https://doi.org/10.1007/978-1-4419-7572-0
2. Glukov, V.I., Varepo, L.G., Shalay, V.V., Simonenko, R.V., Belyave, P.S.: Coordinate structure geometrical location specifications. J. Phys.: Conf. Ser. **1901** (2021). https://doi.org/10.1088/1742-6596/1901/1/012008

3. Melachrinoudis, E., Smith, J.: An $O(mn^2)$ algorithm for the maximin problem in E^2. Oper. Res. Lett. **18**, 25–30 (1995)
4. Megiddo, N.: Linear-time algorithms for linear programming in R^3 and related problems. SIAM J. Comput. **12**, 759–776 (1983). https://doi.org/10.1137/0212052
5. Preparata, P., Shamos, M.: Computational Geometry: An Introduction. Springer-Verlag, New York (1985)
6. Heydari, R., Melachrinoudis, E.: Location of a semi-obnoxious facility with elliptic maximin and network minisum objectives. Eur. J. Oper. Res. **223**, 452–460 (2012). https://doi.org/10.1016/j.ejor.2012.06.039
7. Berman, O., Drezner, Z.: A note on the location of an obnoxious facility on a network. Eur. J. Oper. Res. **120**, 215–217 (2000)
8. Hakimi, S.: Optimum locations of switching centers and the absolute centers and medians of a graph. Oper. Res. **11**, 450–459 (1964). https://doi.org/10.1287/opre.12.3.450
9. Tamir, A.: Improved complexity bounds for center location problems on networks by using dynamic data structures. SIAM J. Discr. Math. **1**, 377–396 (1988). https://doi.org/10.1137/0401038
10. Moon, I.: Maximin center of pendant vertices in a tree network. Transp. Sci. **23**, 213–216 (1989). https://doi.org/10.1287/trsc.23.3.213
11. Church, R.L., Garfinkel, R.S.: Location an obnoxious facility on a network. Transp. Sci. **12**, 107–118 (1978). https://doi.org/10.1287/trsc.12.2.107
12. Melachrinoudis, E., Zhang, G.: An O(mn) algorithm for the 1-maximin problem on a network. Comput. Oper. Res. **26**, 849–869 (1999). https://doi.org/10.1016/S0305-0548(98)00099-9
13. Berman, O., Drezner, Z., Wang, J., Wesolowsky, G.O.: The minimax and maximin location problems on a network with uniform distributed weights. IIE Trans. **35**, 1017–1025 (2003). https://doi.org/10.1080/07408170304397
14. Melaehrinoudis, E., Yavuz, E., Heydari, R.: An $O(m^2 + mn^2)$ algorithm for the bi-objective location problem on a network with mixed metrics. Int. J. Oper. Res. **23**, 427–450 (2015). https://doi.org/10.1504/IJOR.2015.070144
15. Tamir, A.: Obnoxious facility location on graphs. SIAM J. Discr. Math. **4**, 550–567 (1991). https://doi.org/10.1137/0404048
16. Burkard, R., Dollani, H., Lin, Y., Rote, G.: The obnoxious center problem on a tree. SIAM J. Discr. Math. **14**, 498–509 (2001). https://doi.org/10.1137/S0895480198340967
17. Burkard, R., Dollani, H.: Center problems with pos/neg weights on trees. Eur. J. Oper. Res. **145**, 483–495 (2003). https://doi.org/10.1016/S0377-2217(02)00211-4
18. Zabudsky, G.G., Lisina, M.S.: Approximately algorithm for maximin location problem on network. In: IEEE Conference 2018 Dynamics of Systems, Mechanisms and Machines, Omsk (2018). https://doi.org/10.1109/Dynamics.2018.8601502
19. Melachrinoudis, E., Cullinane, T.P.: An locating an obnoxious facility within a polygonal region. Ann. Oper. Res. **6**, 137–145 (1986). https://doi.org/10.1007/BF02026821
20. Dasarathy, B., White, L.J.: A maximin location problem. Oper. Res. **28**, 1385–1401 (1980). https://doi.org/10.1287/opre.28.6.1385
21. Drezner, Z., Wesolowsky, G.O.: A maximin location problem with maximum distance constraints. AIIE Trans. **12**, 249–252 (1980). https://doi.org/10.1080/05695558008974513

Operations Research

On Probability Shaping for 5G MIMO Wireless Channel with Realistic LDPC Codes

Evgeny Bobrov[1,2]([✉]) [iD] and Adyan Dordzhiev[3]

[1] Moscow Research Center, Huawei Technologies, Moscow, Russia
eugenbobrov@ya.ru
[2] M. V. Lomonosov Moscow State University, Moscow, Russia
[3] National Research University Higher School of Economics, Moscow, Russia
adyandordzhiev@yandex.ru

Abstract. Probability Shaping (PS) is a method to improve a Modulation and Coding Scheme (MCS) in order to increase reliability of data transmission. It is already implemented in some modern radio broadcasting and optic systems, but not yet in wireless communication systems. Here we adapt PS for the 5G wireless protocol, namely, for relatively small transport block size, strict complexity requirements and actual low-density parity-check codes (LDPC). We support our proposal by a numerical experiment results in Sionna simulator, showing 0.6 dB gain of PS based MCS versus commonly used MCS.

Keywords: QAM · MCS · OFDM · 5G · PS · FEC · BICM · LDPC

Abbreviations

5G	Fifth Generation
AWGN	Additive White Gaussian Noise
BICM	Bit-interleaved Coded Modulation
BLER	Block Error Rate
CM	Coded Modulation
ESS	Enumerative Sphere Shaping
FEC	Forward Error Correction
GS	Geometric Shaping
IC	Inter-Carrier
IS	Inter-Symbol
LCM	Least Common Multiple
LDPC	Low-Density Parity-Check Code
LLR	Log-Likelihood Ratio
LMMSE	Linear Minimum Mean Squared Error
LOS	Line-of-Sight

E. Bobrov and A. Dordzhiev—Equal contribution.

LS	Least Squares
LTE	Long-Term Evolution
MCS	Modulation and Coding Scheme
MIMO	Multiple-input multiple-output
NLOS	Non-Line-of-Sight
NUC	Non-uniform Constellations
OFDM	Orthogonal Frequency-Division Multiplexing
PS	Probability Shaping
PSCM	Probability Shaped Coded Modulation
QAM	Quadrature Amplitude Modulation
SNR	Signal-to-Noise Ratio

1   Introduction

In the 5G New Radio downlink procedure, the user equipment proposes the serving base station for use in the next signal transmission of the optimal modulation and coding scheme (MCS) [1] based on quadrature amplitude modulation (QAM). In order to achieve the capacity of the additive white Gaussian noise (AWGN) channel, the transmit signal must be Gaussian distributed. The use of uniformly distributed QAM symbols with optimal coded modulation (CM) leads to a shaping loss of up to 1.53 dB for high order constellations [4]. Bit-interleaved coded modulation (BICM) with parallel bit-wise demapping as currently employed in LTE leads to an additional loss.

Non-uniform constellations (NUC) and geometric shaping (GS) have been recently adopted for the next-generation terrestrial broadcast standard [12]. The QAM constellations are optimized for each target signal-to-noise ratio (SNR) by maximizing the BICM capacity for uniformly distributed bits. Note that, in contrast to standard Gray-labeled QAM (Fig. 1), the Non-Uniform constellations do not allow for a simple independent demapping of the real and imaginary part. Therefore, one-dimensional NUCs for each real dimension were also studied in [12], which provide a reduced shaping gain. The performance of BICM can be improved by using non-uniform constellations (NUC), but there remains a gap to the capacity with Gaussian transmit signal.

In the traditional approach of data transmission, each point in a particular constellation has an equal chance of being transmitted. While this technique gives the highest bit rate for a given constellation size, it ignores the energy cost of the individual constellation points. So, as an alternative to GS, it is also possible to adjust the probabilities of the constellation points such that they follow an approximate discrete Gaussian distribution, using the probability shaping (PS) method [11]. Probabilistically shaped coded modulation (PSCM) enables the BICM system to close the gap to the capacity with Gaussian transmit signal. PS is a CM strategy that combines constellation shaping and channel coding.

In the literature, Gallager's error exponent approach has been used to study the achievable information rates of PS [5, Ch. 5]. In particular, it was shown that the PS method has achievable capacities for additive white Gaussian noise channels [2]. In [6], the authors revisit the capacity achieving property of PS. The concept of selecting constellation points using a nonuniform Maxwell-Boltzmann PS is investigated in the study [11]. Nonuniform PS signaling scheme reduces the entropy of the transmitter output and, as a result, the average bit rate. However, if low-energy points are picked more frequently than high-energy points, energy savings may (more than) compensate for the bit rate reduction. Authors of [16] proposed a new PS distribution that outperforms Maxwell-Boltzmann is studied for the nonlinear fiber channel. In [9] the authors successfully tested the suitability of PS constellations in a German nationwide fiber ring of Deutsche Telekom's R&D field test network. In [3] the PS method is implemented in 64-QAM coherent optical transmission system. In [10], a proposed extension to the 5G New Radio polar coding chain is the introduction of a shaping encoder in front of the polar encoder, which will improve the performance with higher order modulation using this PS scheme.

The main objectives of the study:

- In this paper, we investigate the PS Enumerative Sphere Shaping (ESS) [7] method known in the literature with respect to a realistic MIMO OFDM wireless channel with LDPC at a given coderate.
- We provide numerical experiments on the modern Sionna [8] simulation platform and find local optimal parameters for the ESS method, minimizing BLER and providing a gain of up to 0.6 dB over the QAM-16 baseline.
- This study could be interesting from a scientific point of view, since there are almost no published papers on PS that consider such realistic and contemporary scenarios, while considering only theoretical distributions [15].

The basic principle of PS method is presented in Fig. 2. We change the probability of constellation points, which allows us to scale their coordinate with preserving of the mathematical expectation of constellation power.

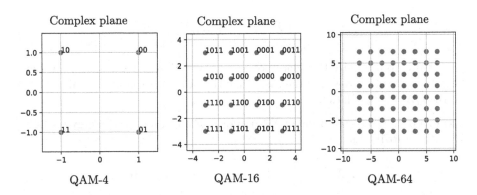

Fig. 1. The base station selects the appropriate QAM scheme for use in the next data transmission. With the increasing of the system quality, the higher QAM can be used.

2   System Model

A block diagram of the proposed PSCM transmitter and receiver is shown in Fig. 6. The main difference to conventional BICM is the distribution matcher that maps the uniformly distributed data bits to bit streams with a desired distribution, which determine the amplitudes of the transmitted QAM symbols. The forward error correction (FEC) encoder generates additional parity bits, which are uniformly distributed and determine the signs of the transmitted QAM symbols. This results in an approximately Gaussian distributed transmit signal using the same constellation mapping as in Long-Term Evolution (LTE).

At the receiver side, the QAM demapper calculates the bit-wise log-likelihood ratios (LLRs) based on the observed receive signal, taking the non-uniform transmit symbol distribution into account. These LLRs are fed to the FEC decoder as in conventional BICM, and the decoder output is finally mapped back to data bits by the distribution deshaper. Note that both the distribution matcher and deshaper correspond to simple one-to-one mappings, which can be efficiently implemented.

3   Optimal Distribution for Probability Shaping

Let $P_X = (p_1, \ldots, p_m)$ be the vector of probabilities of each constellation point and $X = (x_1, \ldots, x_m)$ is its random variable — transmitted points on the constellation, where m is a number of the constellation points.

Let Y be a random variable — received points:

$$Y = X + N_0, \qquad N_0 \sim \mathcal{CN}(0, \sigma^2), \qquad 0 < \sigma^2 < \infty, \tag{1}$$

where N_0 is a Gaussian random variable — noise of a channel.

The energy of the constellation is equal to the expectation of $|X|^2$, i.e.

$$\mathbb{E}[|X|^2] = \sum_{i=1}^{m} p_i |x_i|^2.$$

Our goal is to minimize the energy of the constellation to reduce errors in symbols. In this paper, we study the case when the random variable X is distributed on the QAM constellation.

Example. Initial distribution of X is uniform. For instance, the energy of QAM-16 is equal to 10 since

$$\mathbb{E}[|X|^2] = \frac{1}{16} \cdot (4 \cdot 2 + 8 \cdot 10 + 4 \cdot 18) = 10.$$

Now if we change the distribution of X in such a way that

- four points with coordinates $(\pm 1, \pm 1)$ have probability 0.125
- eight points with coordinates $(\pm 1, \pm 3), (\pm 3, \pm 1)$ have probability 0.0375

– four points of coordinates $(\pm 3, \pm 3)$ have probability 0.05.

In this case, the energy will be equal to 7.6 since

$$\mathbb{E}[|X|^2] = 0.125 \cdot 4 \cdot 2 + 0.0375 \cdot 8 \cdot 10 + 0.05 \cdot 4 \cdot 18 = 7.6,$$

and if we shift the points by multiplying them by the square root of ratio of the energies of the constellations $\sqrt{\frac{10}{7.6}}$, then the energy again become equal to 10. It follows that the points of constellation are further apart, and the variance is the same. So, the probability of error are less.

3.1   Problem Statement

The physical meaning of the problem (2) is to minimize the constellation energy at a fixed constellation entropy. The entropy $H(X)$ means the amount of information transmitted by the constellation, and the energy $\mathbb{E}[|X|^2]$ means the power the transmitter has to expend in transmitting the data.

$$
\begin{cases}
\mathbb{E}[|X|^2] = \displaystyle\sum_{i=1}^{m} p_i \cdot |x_i|^2 \to \min_{P_X} \\[2ex]
\displaystyle\sum_{i=1}^{m} p_i = 1 \\[2ex]
H(X) = -\displaystyle\sum_{i=1}^{m} p_i \cdot \log_2 p_i = const
\end{cases}
\tag{2}
$$

There is no analytical expression for the problem (2), and so the constellation points are assumed to have a Maxwell-Boltzmann distribution (3) since it is close to the optimal distribution [11] and maximizes the entropy of the constellation with a constraint on its energy:

$$\widehat{p}_i = \frac{e^{-\mu|x_i|^2}}{\sum_{j=1}^{m} e^{-\mu|x_j|^2}}, \quad i = 1, \ldots, m \tag{3}$$

Parameter $\mu = \mu(\widehat{p}_1, \ldots \widehat{p}_m)$ — is a scaling of constellation points:

$$\sum_{i=1}^{m} p_i \cdot |x_i|^2 = \sum_{i=1}^{m} \widehat{p}_i \cdot |\mu x_i|^2 \quad \Longleftrightarrow \quad \mu^2 = \frac{\sum_{i=1}^{m} p_i \cdot |x_i|^2}{\sum_{i=1}^{m} \widehat{p}_i \cdot |x_i|^2} \tag{4}$$

where p_i is the uniform distribution, $\widehat{p}_i$ is the optimal distribution, and x_i is the complex coordinates of the points.

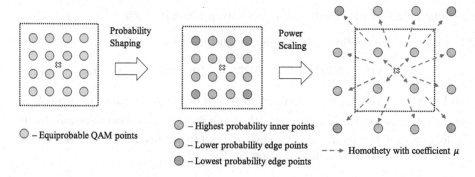

Fig. 2. The probability shaping method increases system performance by scaling constellation points, which is allowed while preserving the constellation power — the mathematical expectation of the modulus of the complex points.

4   Coded Modulation Design for QAM-16

According to the labelling procedure, we can notice that the first two bits in the binary representation of the constellation points are responsible for symmetry about the coordinate axes, and the last two bits are responsible for absolute value. (Fig. 3). In what follows, we will refer to the first two bits as sign bits, and the last two bits as amplitude bits. Thus, amplitude bit zero corresponds to points with coordinates ± 1, and amplitude bit corresponds to points with coordinates ± 3.

4.1   Constellation Energy Minimisation

For the practical finite block-length codes, it is required to implement the Enumerative Sphere Shaping (ESS) method [7].

The energy minimization process is fairly straightforward. We take the constellation points with the smallest absolute value with a higher probability, and the points with the largest absolute value with a lower probability. Thus, we are more interested in constellation points that have more zeros than ones at the amplitude bit positions in the binary representation, and then it is sufficient to maximize the probability of zero at the amplitude bit positions. We also assume that the sign bits are uniformly distributed, i.e. the probability of zero and one of the first two bits in the binary representation of each constellation point is equal to $\frac{1}{2}$.

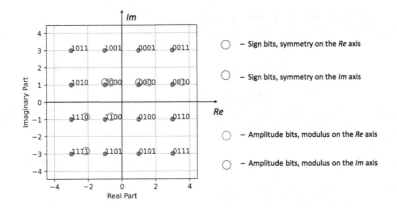

Fig. 3. Gray labeling of the sign and amplitude bits. (Color figure online)

As noted above, amplitude bit one corresponds to more distant points from the origin, and amplitude bit zero corresponds to closer points. Thus, we can assume that the *energy* of a sequence of n amplitude bits consisting of k ones and $n - k$ zeros, is equal to

$$\underbrace{1^2 + \ldots + 1^2}_{n-k} + \underbrace{3^2 + \ldots + 3^2}_{k}.$$

It can be seen that the nearest points to the origin have the lowest energy.

For a given number of input amplitude bits k and block length n, the most efficient way to change probabilities is to map all possible 2^k realisations to the 2^k sequences of n amplitude bits with minimal energy. After that, we can calculate the probability of one $p_a(1)$ and probability of zero $p_a(0)$ in a set of blocks of length n.

Now if we know the distribution of the amplitude bits, then we can find the probability of the constellation points. Each constellation point contains two sign bits and two amplitude bits, so the probability of a point is equal to

$$\widehat{p_i} = (\frac{1}{2})^2 \cdot p_a(0)^k \cdot (1 - p_a(0))^{1-k}, \quad i = 1, \ldots, 16 \tag{5}$$

where k is the number of zero amplitude bits in the bit representation of the constellation point.

After we have changed the distribution of constellation points, we can calculate the scaling parameter μ as the ratio of the initial energy to the received energy:

$$\mu^2 = \frac{\mathbb{E}[|X|^2]}{\mathbb{E}[|\widehat{X}|^2]} = \frac{10}{\sum_{i=1}^{16} \widehat{p_i} \cdot |x_i|^2}, \tag{6}$$

where $\widehat{X}$ is a new random variable with distribution 5. Finally, we shift points of the constellation by multiplying them by the parameter μ, thereby reducing the probability of error.

4.2   Amplitude Shaper and Sign Delay

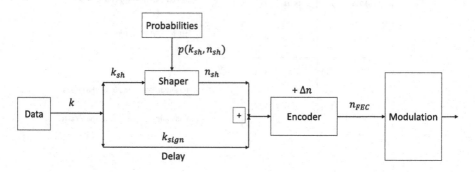

Fig. 4. Data flow through the amplitude probability shaper and encoder to modulation.

In this subsection we describe the model provided in Fig. 4. Initially, the input is k informational bits with a uniform distribution. These bits are divided into two groups, one of which will be the amplitude bits, and the other group will be part of the sign bits. Amplitude bits are transmitted through the shaper block, which works according to the algorithm described above. The shaper output is a block of a different length, in which the amplitude bits are already distributed according to the algorithm. We will denote the number of bits in the first group by k_{sh}, the number of bits in the second group by k_{sign} and the number of bits at the shaper output by n_{sh}.

After that, k_{sign} bits and n_{sh} amplitudes bits are concatenated and encoded using the LDPC procedure. The LDPC procedure, in turn, generates additional Δn check bits, which are also considered to be uniformly distributed. We will denote the number of bits at the encoder output as $n_{FEC} = k_{sign} + n_{sh} + \Delta n$. Note that the $k_{sign} + \Delta n$ bits are sign bits, which have a uniform distribution, while the amplitude bits n_{sh} are distributed according to the algorithm. The number of sign k_{sign} and error correction bits Δn is equal to the number of shaper output bits, i.e. $n_{sh} = k_{sign} + \Delta n$.

For this procedure, coderate $R \in (0,1]$ is fixed, while shaper input size k_{sh} and shaper output n_{sh} vary. The values of k_{sign} and n_{FEC} can be calculated using the code rate formulas.

Extra bits are now shared between Shaper with rate $R_{sh} = \frac{k_{sh}}{n_{sh}}$ and Encoder with rate $R_{FEC} = \frac{n_{sh} + k_{sign}}{n_{FEC}}$, afterall $R = \frac{1}{2}(R_{sh} + 2R_{FEC} - 1)$.

The problem is to find the optimal proportion between R_{sh} and R_{FEC}.

4.3   Example of Generated Probabilities Using ESS

In Tables 1, 2 examples of generated probabilities for QAM-16 and QAM-64 (Fig. 1) using the ESS method [7] are given. Note that for QAM-64 and above the probabilities of zeros and ones depend on each other, so joint probabilities need to be determined.

Table 1. Example of amplitude probabilities for QAM-16, $n_{sh} = 256$.

k_{sh}	40	80	120	160	200	240	256
$p_a(0)$	0.97	0.94	0.90	0.84	0.76	0.64	0.5
$p_a(1)$	0.03	0.06	0.10	0.16	0.24	0.36	0.5

Table 2. Example of amplitude probabilities for QAM-64, $n_{sh} = 1024$.

k_{sh}	300	400	500	600	700	800	900	1000	1024
$p_a(00)$	0.87	0.80	0.73	0.66	0.59	0.52	0.44	0.32	0.25
$p_a(01)$	0.13	0.18	0.23	0.27	0.30	0.32	0.31	0.28	0.25
$p_a(10)$	0.00	0.01	0.03	0.06	0.09	0.13	0.18	0.23	0.25
$p_a(11)$	0.00	0.00	0.00	0.01	0.01	0.03	0.07	0.16	0.25

4.4 Probability Shaping Mapping

For mapping purposes, we form a special PS matrix (Fig. 5) with uniform sign bits and non-uniform amplitude bits, following the data flow scheme (Fig. 4). We generate k_{sign} sign bits with equal probability of zeros and ones $p_s = \frac{1}{2}$, and n_{sh} amplitude bits with unequal probability of zeros and ones: $p_a \neq \frac{1}{2}$ such that $p_a(0) > p_a(1)$. The probability of amplitude bits can be determined by the proper values of k_{sh} and n_{sh} using the ESS method [7].

Finally, after the PS matrix is constructed, the mapping to the QAM is performed. With mapping procedure, bits are converted to the constellation points (or symbols) using the mapping table, which gives a specific coordinate on the complex plane for each unique sequence of bits. Notice that, given bit probabilities, there is a one-to-one correspondence to symbol probabilities.

After mapping procedure is done the symbols go through the MIMO channel, demodulation and decoding, probability deshaping and BLER calculation, which are described in Sect. 2 and Fig. 6. The demodulation and deshaping methods are the same procedures described earlier and are performed in reverse order. The decoding procedure is a complex process, which uses loopy belief propagation [14] to iteratively recover the correct bits (LLRs).

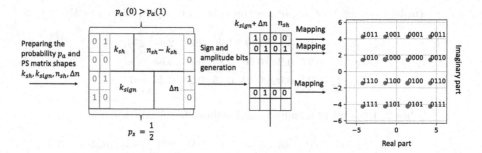

Fig. 5. Creating a code block virtual matrix, generating sign k_{sign} bits, LDPC Δn bits and amplitude n_{sh} bits with final symbol mapping. The red dots represent the mapped constellation points from the generated binary sequence.

4.5 Arrangement of Finite Code Block Shapes

To consistent all the shapes k_{sh}, k_{sign} and Δn, and form the PS matrix (Fig. 5) we solve the system of integer equations (7) finding LCM. Hereafter, the values of N_{fr}^{sh} and N_{fr}^{FEC} define the multiplicative constants balancing these equations. The values of N_A and N_S define the number of amplitude and sign bits in the code block. It is implicitly assumed that everywhere in Figs. (4), (5) the values are $k_{sh} := N_{fr}^{sh} k_{sh}$, $k_{sign} := N_{fr}^{sh} k_{sign}$, $\Delta n := N_{fr}^{FEC} \Delta n$, $n_{FEC} := N_{fr}^{FEC} n_{FEC}$:

$$\begin{cases} N_A = n_{sh} N_{fr}^{sh} \\ C N_S = N_A \\ N_S + N_A = N_{fr}^{FEC} n_{FEC}, \end{cases} \tag{7}$$

where the value of C defines the constellation system, i.e. $C = 1$ — QAM-16, $C = 2$ — QAM-64, $C = 3$ — QAM-256 and so on.

5   Numerical Experiments

In the experiments, Coded BLER is the average error of transmitted block of bits before the LDPC encoder and after the decoding in Fig. 6, which takes into account the realistic coding-encoding procedure.

5.1   Energy per Bit and Noise Ratio

The energy per bit to noise ratio E_b/N_0 is a normalized SNR measure, also known as SNR per bit. The E_b/N_0 measure can be used to express the relationship between signal power and noise power.

The energy per bit measure E_b is the energy we use to transmit one bit of information with the total power P and the LDPC coderate R:

$$E_b = \frac{P}{R},$$

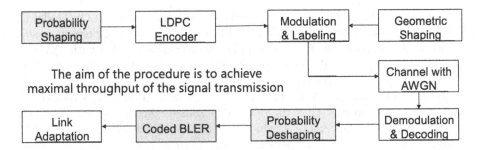

Fig. 6. Block diagram of probability shaping transmitter and receiver.

The noise measure N_0 is the noise variance per real and imaginary parts:

$$N_0 = 2\sigma^2$$

Thus, E_b/N_0 can be expressed in terms of SNR:

$$E_b/N_0 = \frac{P}{R}\frac{1}{2\sigma^2} = \frac{P}{\sigma^2}\frac{1}{2R} = \frac{\text{SNR}}{2R}$$

In decibel, E_b/N_0 is

$$E_b/N_0 \text{ in dB} = 10\log_{10}(E_b/N_0) = 10\log_{10}\left(\frac{P}{\sigma^2}\frac{1}{2R}\right) \tag{8}$$

We use the value of E_b/N_0 in the Monte Carlo experiments. For a given value of E_b/N_0 with the power P and coderate R the variable noise power σ^2 disturbs the symbols transmitted over the channel (1).

5.2 Realistic Simulations Using Sionna

This study considers OFDM MIMO with a base station and a user equipped with multiple cross-polarised antennas. We provide simulations in Sionna [8] on the OFDM channel using 5G LDPC codes. The architecture of the system consists of LDPC, Bit Interleaver, Resource Grid Mapper, LS Channel Estimator, Nearest Neighbor Demapper, LMMSE Equalizer [13], OFDM Modulator and presented in Fig. 6. Optimization variables are constellation type, a bit order, coderate, BLER, SNR, code block sizes and 5G model (LOS D, NLOS A).

The system uses soft estimates of LLRs for the decoder. Channel model is chosen to be OFDM 5G 2.6 GHz with delay spread of 40ns. The block size is 1536 with 10^5 number of Monte-Carlo trials in the simulations and so in total $1.536 \cdot 10^8$ bits were processed for each point of E_b/N_0. For all simulation, 20 iterations of LDPC have been used. The code source is random binary tensors. The system use 3GPP wireless both Line of Sight (LOS) and Non Line of Sight (NLOS) channel models D and A. The model is simulated in real time domain considering inter-symbol (IS) and inter-carrier (IC) interferences. In Table 3 we provide simulation parameters for Sionna.

In Figs. 7 and 8, we provide an experiment for both QAM-16 LOS Model D and NLOS Model A probability shaped (PS) constellations. We present experiments Coded BLER (see Fig. 6) with Gaussian transmit signal with QAM16 Baseline and amplitude PS with different shaping parameters, where coderate is r, block size is n and parameters of PS are n_{sh} and k_{sh}.

There is a local optimum for $k_{sh} = 192$ PS QAM-16 in both LOS and NLOS models. It is noteworthy that the optimal parameter k_{sh} is the same for the different LOS and NLOS models, which tells us that the chosen parameters are stable. Note that with wrong parameter settings, e.g. a strong shaping factor $k_{sh} = 128$, the quality of the PS method is worse than that of baseline QAM.

In Fig. 9 present experiments with AWGN channel and LDPC, which show a higher gain than for the OFDM channel model.

In Table 4, we provide gains in dB of the proposed PS method at 10% BLER. The PS method achieves 0.56 dB gain in Model A NLOS Uplink compared to the baseline. From the massive experiments, we conclude that the proposed PS method constellations superior the baseline QAM for real 5G Wireless System using FEC LDPC.

The model codes and related experiments can be found in the repository[1].

Table 3. Simulation parameters in Sionna.

Carrier frequency	2.6e9
Delay spread	40e-9
Cyclic prefix length	6
Num guard carriers	[5, 6]
FFT size	44
Num user terminal antennas	2
Num base station antennas	2
Num OFDM symbols	14
Num LDPC iterations	20

Table 4. Gain in dB of the proposed PS method at 10% BLER.

	Model D LOS	Model A NLOS
Uplink	0.52 dB	0.56 dB
Downlink	0.5 dB	0.5 dB

---

https://github.com/eugenbobrov/On-Probabilistic-QAM-Shaping-for-5G-MIMO-Wireless-Channel-with-Realistic-LDPC-Codes.

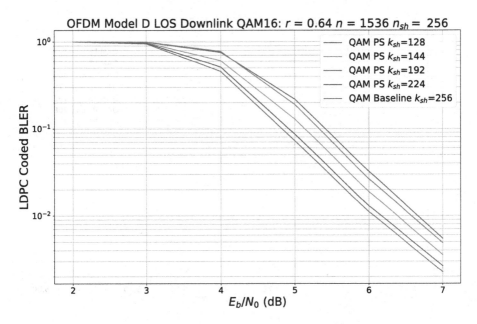

Fig. 7. Model D LOS Downlink channel Coded BLock Error Rate for OFDM QAM16.

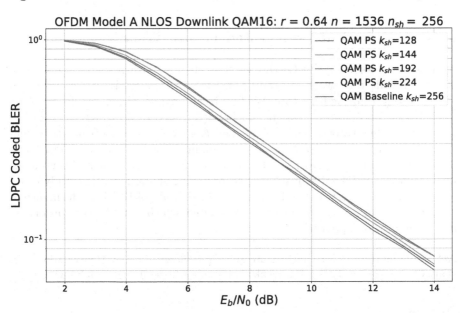

Fig. 8. Model A NLOS Uplink channel Coded BLock Error Rate for OFDM QAM16.

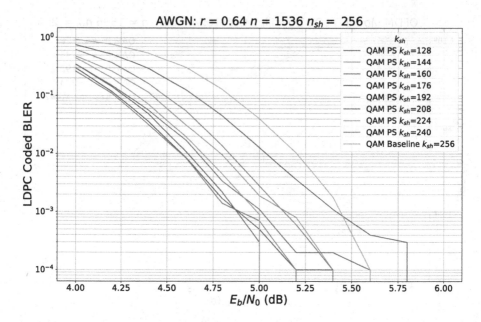

Fig. 9. AWGN channel Coded BLock Error Rate QAM16.

6 Conclusions and Suggested Future Work

In this paper, for a MIMO OFDM wireless channel with realistic LDPC code at a given code rate, we study the PS scheme of Enumerative Sphere Shaping (ESS) known from the literature. We find local optimal parameters for the ESS method that minimise the BLER and provide a gain of up to 0.6 dB over the QAM-16 baseline through numerical experiments on the state-of-the-art Sionna simulation platform, modeling physical communication system level. Since there are almost no published works on PS that consider such realistic and contemporary scenarios, while only considering theoretical distributions, this study could be of scientific interest. In the future, a detailed study of BLER performance of a combination of PS and GS methods is possible, which could be very promising in communication applications.

Acknowledgements. The authors are grateful to Sergey Loktev, Dmitry Minenkov, Dmitry Shmelkin, Sviatoslav Panchenko, Alexandr Khodunin and Ivan Sobolev.

References

1. Bobrov, E., Kropotov, D., Lu, H., Zaev, D.: Massive MIMO adaptive modulation and coding using online deep learning algorithm. IEEE Commun. Lett. **26**(4), 818–822 (2021)
2. Böcherer, G.: Achievable rates for probabilistic shaping. arXiv preprint: arXiv:1707.01134 (2017)

3. Buchali, F., Steiner, F., Böcherer, G., Schmalen, L., Schulte, P., Idler, W.: Rate adaptation and reach increase by probabilistically shaped 64-QAM: an experimental demonstration. J. Lightwave Technol. **34**(7), 1599–1609 (2016)
4. Forney, G., Gallager, R., Lang, G., Longstaff, F., Qureshi, S.: Efficient modulation for band-limited channels. IEEE J. Sel. Areas Commun. **2**(5), 632–647 (1984)
5. Gallager, R.G.: Information Theory and Reliable Communication, vol. 588. Springer, Cham (1968)
6. Gültekin, Y.C., Alvarado, A., Willems, F.M.: Achievable information rates for probabilistic amplitude shaping: an alternative approach via random sign-coding arguments. Entropy **22**(7), 762 (2020)
7. Gültekin, Y.C., Fehenberger, T., Alvarado, A., Willems, F.M.: Probabilistic shaping for finite blocklengths: distribution matching and sphere shaping. Entropy **22**(5), 581 (2020)
8. Hoydis, J., et al.: Sionna: an open-source library for next-generation physical layer research. arXiv preprint: arXiv:2203.11854 (2022)
9. Idler, W., et al.: Field trial of a 1 Tb/s super-channel network using probabilistically shaped constellations. J. Lightwave Technol. **35**(8), 1399–1406 (2017)
10. İşcan, O., Böhnke, R., Xu, W.: Probabilistic shaping using 5G new radio polar codes. IEEE Access **7**, 22579–22587 (2019)
11. Kschischang, F.R., Pasupathy, S.: Optimal nonuniform signaling for Gaussian channels. IEEE Trans. Inf. Theory **39**(3), 913–929 (1993)
12. Loghin, N.S., Zöllner, J., Mouhouche, B., Ansorregui, D., Kim, J., Park, S.I.: Non-uniform constellations for ATSC 3.0. IEEE Trans. Broadcast. **62**(1), 197–203 (2016)
13. Mineev, D., Bobrov, E., Kuznetsov, V.: On the interference cancellation by reduced channel zero forcing class of precodings in massive MIMO systems. Telecommun. Syst., 1–12 (2023)
14. Murphy, K., Weiss, Y., Jordan, M.I.: Loopy belief propagation for approximate inference: an empirical study. arXiv preprint: arXiv:1301.6725 (2013)
15. Neskorniuk, V., Carnio, A., Marsella, D., Turitsyn, S.K., Prilepsky, J.E., Aref, V.: Model-based deep learning of joint probabilistic and geometric shaping for optical communication. In: CLEO: science and Innovations, pp. SW4E-5. Optica Publishing Group (2022)
16. Sillekens, E., et al.: A simple nonlinearity-tailored probabilistic shaping distribution for square QAM. In: 2018 Optical Fiber Communications Conference and Exposition (OFC), pp. 1–3. IEEE (2018)

Additive Routing Problem for a System of High-Priority Tasks

Alexandr G. Chentsov[1,2]([✉]) [ID] and Pavel A. Chentsov[1,2] [ID]

[1] Krasovskii Institute of Mathematics and Mechanics of the Ural Branch of the Russian Academy of Sciences, Yekaterinburg 620108, Russia
chentsov@imm.uran.ru, chentsov.p@mail.ru
[2] Ural Federal University, ul. Mira, 19, Yekaterinburg 620002, Russia

Abstract. It is considered the routing problem for which some fixed tasks must be serviced above all. Other tasks can be serviced only after realization of above-mentioned original tasks. It is supposed that each our task is the megalopolis (nonempty finite set) visiting with fulfilment of some works. In our setting, two partial interconnected routing problems arise. We suppose that, in each partial routing problem, the corresponding precedence conditions are given. Using widely understood dynamic programming (DP), we obtain the optimal composition solution for initial total problem. As an application, we note the known engineering problem connected with sheet cutting by zones on CNC machines. By DP procedure the optimal algorithm realized on PC was constructed.

Keywords: Dynamic programming · Precedence conditions · Route

1 Introduction

The known hard-to-solve traveling salesman problem (TSP) is the natural prototype of considered extremal routing problems with elements of decomposition. In connection with TSP we note [1–3]; moreover, we note [4,5] as investigations connected with dynamic programming (DP) for TSP solution. But, in routing problems oriented to engineering applications, many essential singularities arise. In these connection, first of all, we note different constraints (of course, in problems connected with sheet cutting on CNC machines, many constraints arise). In particular, in practical routing problems, precedence conditions are realized. For problems connected with sheet cutting, conditions excluding thermal deformations of details are very important. For this aim, it is possible to use penalty method; in this case, cost functions with task list dependence arise. We use this approach. In the following, constructions of [6] are used. In connection with applications, we are oriented to monograph [7]. We note some investigations connected with routing for megalopolises: see [8–10]. Moreover, in [11–23], questions connected with sheet cutting are considered (see remarks in Sect. 5). But, in this investigation, new approach to optimization for compositional solutions set out: following [6] and [9], we consider the procedure for construction

M. Khachay et al. (Eds.): MOTOR 2023, CCIS 1881, pp. 218–230, 2023.
https://doi.org/10.1007/978-3-031-43257-6_17

of optimal compositional solutions. Similar results are unknown to the autors (of course, the routing problems with megalopolises visiting under precedence conditions mean). It is important, that for real problem connected with sheet cutting, manages to get a solution in a reasonable time (see Sects. 4 and 5).

2   The Mathematical Setting

We fix a nonempty set X and a nonempty finite subset X^0 of X. Elements of X^0 are considered as starting points. Fix a natural number $\mathbf{n}$, $\mathbf{n} \geq 4$, nonempty finite sets $M_1, \ldots, M_\mathbf{n}$ (megalopolises), each of which is a subset of X, and nonempty relations

$$\mathbb{M}_1, \ldots, \mathbb{M}_\mathbf{n}$$

for which

$$\mathbb{M}_1 \subset M_1 \times M_1, \ldots, \mathbb{M}_\mathbf{n} \subset M_\mathbf{n} \times M_\mathbf{n}. \tag{1}$$

So, elements of $\mathbb{M}_j$, where j is a natural number with $1 \leq j \leq \mathbf{n}$, are ordered pairs (OP). Let $M_1, \ldots, M_\mathbf{n}$ be pairwise disjunctive; $M_1 \cap X^0 = \emptyset, \ldots, M_\mathbf{n} \cap X^0 = \emptyset$. We consider the next movements

$$(x \in X^0) \rightarrow (x_{1,1} \in M_{\gamma(1)} \rightsquigarrow x_{1,2} \in M_{\gamma(1)}) \rightarrow \ldots \tag{2}$$
$$\rightarrow (x_{\mathbf{n},1} \in M_{\gamma(\mathbf{n})} \rightsquigarrow x_{\mathbf{n},2} \in M_{\gamma(\mathbf{n})}),$$

$$(x_{1,1}, x_{1,2}) \in \mathbb{M}_{\gamma(1)}, \ldots, (x_{\mathbf{n},1}, x_{\mathbf{n},2}) \in \mathbb{M}_{\gamma(\mathbf{n})}, \tag{3}$$

where γ is a permutation of indexes $1, \ldots, \mathbf{n}$.

Some General Designations. We suppose that $\mathbb{N} \triangleq \{1; 2; \ldots\}$ ($\triangleq$ is the equality by definition), $\mathbb{N}_0 \triangleq \{0; 1; 2; \ldots\}$ and

$$\overline{p, q} \triangleq \{k \in \mathbb{N}_0 \mid (p \leq k) \& (k \leq q)\} \ \forall p \in \mathbb{N}_0 \ \forall q \in \mathbb{N}_0.$$

Let $\mathbb{R}_+ \triangleq \{\xi \in \mathbb{R} \mid 0 \leq \xi\}$, where $\mathbb{R}$ is the real line. If H is a nonempty set, then by $\mathcal{R}_+[H]$ we denote the set of all functions from H into $\mathbb{R}_+$ (nonnegative real-valued functions on H). For each OP h, by $\mathrm{pr}_1(h)$ and $\mathrm{pr}_2(h)$ we denote the first and the second elements of h respectively; of course, $h = (\mathrm{pr}_1(h), \mathrm{pr}_2(h))$. To a set H, we associate the family $\mathcal{P}(H)$ of all subsets of H and $\mathcal{P}'(H) \triangleq \mathcal{P}(H) \backslash \{\emptyset\}$.

Let $\mathcal{M} \triangleq \{M_i : i \in \overline{1, \mathbf{n}}\}$ and, for fixed $\mathbf{N} \in \overline{2, \mathbf{n} - 2}$,

$$\mathcal{M}_1 \triangleq \{M_i : i \in \overline{1, N}\}, \quad \mathcal{M}_2 \triangleq \mathcal{M} \backslash \mathcal{M}_1 = \{M_i : i \in \overline{N + 1, \mathbf{n}}\};$$

we consider the sets of $\mathcal{M}$ as megalopolises. We take the next requirement: in (2), (3), visiting to megalopolises of $\mathcal{M}_2$ is admissible only after visiting to all megalopolises of $\mathcal{M}_1$. So, we obtain $\mathcal{M}_1$-problem and $\mathcal{M}_2$-problem; these problems are interconnected.

Suppose that $\mathbb{P}$ is the set of all permutations of indexes of $\overline{1,\mathbf{n}}$ and $\mathbb{P}_1$ and $\mathbb{P}_2$ are the sets of all permutations of indexes of $\overline{1,N}$ and $\overline{1,\mathbf{n}-N}$ respectively. We call routes permutations of $\mathbb{P}$, $\mathbb{P}_1$, and $\mathbb{P}_2$. If $\alpha \in \mathbb{P}_1$ and $\beta \in \mathbb{P}_2$, then

$$\alpha \diamond \beta \in \mathbb{P}$$

is defined by the rule

$$((\alpha \diamond \beta)(k) \stackrel{\triangle}{=} \alpha(k) \ \forall k \in \overline{1,N}) \& ((\alpha \diamond \beta)(l) \stackrel{\triangle}{=} \beta(l-N) + N \ \forall l \in \overline{N+1,\mathbf{n}}). (4)$$

So, (4) is the route coalescence. Suppose that the choice of $\mathbb{P}_1$ and $\mathbb{P}_2$ can be subordinated to precedence conditions. In this connection, we fix the sets $\mathbf{K}_1$ and $\mathbf{K}_2$ for which

$$\mathbf{K}_1 \subset \overline{1,N} \times \overline{1,N}, \quad \mathbf{K}_2 \subset \overline{1,\mathbf{n}-N} \times \overline{1,\mathbf{n}-N}.$$

Elements of $\mathbf{K}_1$ and $\mathbf{K}_2$ are called address pairs. If h is an address pair, then $pr_1(h)$ is called sender and $pr_2(h)$ is called receiver of h. We suppose that

$$(\forall \mathbf{K}^0 \in \mathcal{P}'(\mathbf{K}_1) \ \exists z^0 \in \mathbf{K}^0 : pr_1(z^0) \neq pr_2(z) \ \forall z \in \mathbf{K}^0) \tag{5}$$
$$\& (\forall \tilde{\mathbf{K}}^0 \in \mathcal{P}'(\mathbf{K}_2) \ \exists z^0 \in \tilde{\mathbf{K}}^0 : pr_1(z^0) \neq pr_2(z) \ \forall z \in \tilde{\mathbf{K}}^0)$$

(in applied problems, conditions (5) are fulfilled typically). Then, by [8, (2.2.53)]

$$\mathcal{A}_1 \stackrel{\triangle}{=} \{\alpha \in \mathbb{P}_1 \mid \alpha^{-1}(pr_1(z)) < \alpha^{-1}(pr_2(z)) \ \forall z \in \mathbf{K}_1\} \neq \emptyset, \tag{6}$$

$$\mathcal{A}_2 \stackrel{\triangle}{=} \{\alpha \in \mathbb{P}_2 \mid \alpha^{-1}(pr_1(z)) < \alpha^{-1}(pr_2(z)) \ \forall z \in \mathbf{K}_2\} \neq \emptyset, \tag{7}$$

where α^{-1} is the inverse permutation for every permutation α. In (6) and (7), admissible (by precedence) routes are introduced. Using (6) and (7), we obtain that

$$\mathbf{P} \stackrel{\triangle}{=} \{\alpha \diamond \beta : \alpha \in \mathcal{A}_1, \beta \in \mathcal{A}_2\} \neq \emptyset \tag{8}$$

and $\mathbf{P} \subset \mathbb{P}$. So, (8) is considered as the set of all admissible routes in $\mathcal{M}$-problem (the problem about visiting to all megalopolises of $\mathcal{M}$).

Along with routes, we introduce trajectories in $\mathcal{M}$-problem. Let $\mathbb{Z}$ be the set of all mappings from $\overline{0,\mathbf{n}}$ in $X \times X$. Under $x \in X^0$ and $\gamma \in \mathbf{P}$, in the form of

$$\mathcal{Z}_\gamma[x] \stackrel{\triangle}{=} \{(z_t)_{t \in \overline{0,\mathbf{n}}} \in \mathbb{Z} \mid (z_0 = (x,x)) \tag{9}$$
$$\& (z_t \in \mathbb{M}_{\gamma(t)} \ \forall t \in \overline{1,\mathbf{n}})\},$$

we obtain nonempty finite set of all trajectories starting from x and coordinated with route γ. For $x \in X^0$ we obtain that

$$\tilde{\mathbf{D}}[x] \stackrel{\triangle}{=} \{(\gamma, (z_t)_{t \in \overline{0,\mathbf{n}}}) \in \mathbf{P} \times \mathbb{Z} \mid (z_t)_{t \in \overline{0,\mathbf{n}}} \in \mathcal{Z}_\gamma[x]\} \tag{10}$$

is the set of all admissible solutions in the $(\mathcal{M}, x)$-problem ($\mathcal{M}$-problem with starting point x). Finally, we introduce the set

$$\mathbf{D} \stackrel{\triangle}{=} \{(\gamma, (z_t)_{t\in\overline{0,n}}, x) \in \mathbf{P} \times \mathbb{Z} \times X^0 \mid (\gamma, (z_t)_{t\in\overline{0,n}}) \in \tilde{\mathbf{D}}[x]\} \qquad (11)$$

of all (admissible) routing processes. So, $\mathbf{D}$ is a nonempty finite set. We consider elements of (11) as admissible solutions in the basic $\mathcal{M}$-problem for which starting optimization is also admissible.

We suppose that $\mathfrak{N}$ is the family of all nonempty subsets of $\overline{1, \mathbf{n}}$. For $j \in \overline{1, \mathbf{n}}$, suppose that

$$(\mathfrak{M}_j \stackrel{\triangle}{=} \{pr_1(z) : z \in \mathbb{M}_j\}) \& (\mathbf{M}_j \stackrel{\triangle}{=} \{pr_2(z) : z \in \mathbb{M}_j\});$$

moreover, let

$$(\mathbf{X} \stackrel{\triangle}{=} \bigcup_{i=1}^{n} \mathfrak{M}_i) \& (\mathbf{X} \stackrel{\triangle}{=} (\bigcup_{i=1}^{n} \mathbf{M}_i) \cup X^0).$$

As cost functions, we fix

$$\mathbf{c} \in \mathcal{R}_+[\mathbf{X} \times \mathbf{X} \times \mathfrak{N}], c_1 \in \mathcal{R}_+[\mathbb{M}_1 \times \mathfrak{N}], \ldots, c_\mathbf{n} \in \mathcal{R}_+[\mathbb{M}_\mathbf{n} \times \mathfrak{N}], \qquad (12)$$

$$f \in \mathcal{R}_+[\bigcup_{i=1}^{n-N} \mathbf{M}_{N+i}].$$

In terms of functions (12), we introduce additive criterion: for $x \in X^0$, $\gamma \in \mathbf{P}$, and $(z_t)_{t\in\overline{0,n}} \in \mathcal{Z}_\gamma[x]$

$$\mathfrak{C}_\gamma[(z_t)_{t\in\overline{0,n}}] \stackrel{\triangle}{=} \sum_{t=1}^{n} [\mathbf{c}(pr_2(z_{t-1}), pr_1(z_t), \{\gamma(k) : k \in \overline{t, \mathbf{n}}\}) \qquad (13)$$

$$+ c_{\gamma(t)}(z_t, \{\gamma(k) : k \in \overline{t, \mathbf{n}}\})] + f(pr_2(z_\mathbf{n})).$$

In (13), we estimate all steps of movements (2), (3). For $x \in X^0$, the following $(\mathcal{M}, x)$-problem is investigated:

$$\mathfrak{C}_\gamma[(z_t)_{t\in\overline{0,n}}] \to \min, \quad (\gamma, (z_t)_{t\in\overline{0,n}}) \in \tilde{\mathbf{D}}[x]; \qquad (14)$$

we associate with (14) the corresponding extremum $\tilde{V}[x]$ and (nonempty) extremal set $(\text{sol})[x]$:

$$\tilde{V}[x] \stackrel{\triangle}{=} \min_{(\gamma, (z_t)_{t\in\overline{0,n}}) \in \tilde{\mathbf{D}}[x]} \mathfrak{C}_\gamma[(z_t)_{t\in\overline{0,n}}] \in \mathbb{R}_+, \qquad (15)$$

$$(\text{sol})[x] \stackrel{\triangle}{=} \{(\gamma, (z_t)_{t\in\overline{0,n}}) \in \tilde{\mathbf{D}}[x] \mid \mathfrak{C}_\gamma[(z_t)_{t\in\overline{0,n}}] = \tilde{V}[x]\} \neq \emptyset. \qquad (16)$$

But, our basic routing problem ($\mathcal{M}$-problem) is

$$\mathfrak{C}_\gamma[(z_t)_{t\in\overline{0,n}}] \to \min, \quad (\gamma, (z_t)_{t\in\overline{0,n}}, x) \in \mathbf{D}; \qquad (17)$$

for this $\mathcal{M}$-problem, we consider extremum $\mathbb{V}$ and (nonempty) extremal set **SOL**:

$$\mathbb{V} \overset{\triangle}{=} \min_{(\gamma,(z_t)_{t\in\overline{0,\mathbf{n}}},x)\in\mathbf{D}} \mathfrak{C}_\gamma[(z_t)_{t\in\overline{0,\mathbf{n}}}] \qquad (18)$$

$$= \min_{x\in X^0} \min_{(\gamma,(z_t)_{t\in\overline{0,\mathbf{n}}})\in\tilde{\mathbf{D}}[x]} \mathfrak{C}_\gamma[(z_t)_{t\in\overline{0,\mathbf{n}}}] = \min_{x\in X^0} \tilde{V}[x] \in \mathbb{R}_+,$$

$$\mathbf{SOL} \overset{\triangle}{=} \{(\gamma,(z_t)_{t\in\overline{0,\mathbf{n}}},x) \in \mathbf{D} \mid \mathfrak{C}_\gamma[(z_t)_{t\in\overline{0,\mathbf{n}}}] = \mathbb{V}\} \neq \emptyset. \qquad (19)$$

By (15), the extremum function $\tilde{V}[\cdot]$ defined on X^0 is realized. We consider the problem

$$\tilde{V}[x] \to \min, \quad x \in X^0, \qquad (20)$$

of starting point optimization; of course, $\mathbb{V}$ is extremum of (20) and

$$X^0_{\text{opt}} \overset{\triangle}{=} \{x \in X^0 \mid \tilde{V}[x] = \mathbb{V}\} \neq \emptyset \qquad (21)$$

is the corresponding extremal set. Then, by (15), (16), (18), (19), and (21)

$$(\gamma^*,(z_t^*)_{t\in\overline{0,\mathbf{n}}},x^*) \in \mathbf{SOL} \ \forall x^* \in X^0_{\text{opt}} \ \forall(\gamma^*,(z_t^*)_{t\in\overline{0,\mathbf{n}}}) \in (\text{sol})[x^*]. \qquad (22)$$

3   The General Scheme of Algorithm

For solution of $\mathcal{M}$-problem (17), the corresponding (optimal) algorithm was proposed in [6]. This algorithm allows you to solve the problem of tangible dimension in a reasonable time (see examples in [6]). The corresponding values of the counting time are given when considering the examples in Sects. 4 and 5. The basic element of this solution is DP realizable under decomposition of $\mathcal{M}$-problem by the system of $\mathcal{M}_1$-problem and $\mathcal{M}_2$-problem. In addition, formalization of $\mathcal{M}_1$-problem and $\mathcal{M}_2$-problem is analogous to constructions for $\mathcal{M}$-problem (see (9)–(19)) with some transformation of cost functions compared to (12). Now, we confine ourselves to the presentation of the algorithm scheme.

In connection with used DP procedures, we note that here variant ascending to [8, Section 4.9] is realized (see constructions of [10–12]). In addition (under precedence conditions), we don't build the whole array of Bellman function values. Instead of this, we define special layers of this function. For this, layers of position space are first created. In turn, essential lists of tasks are constructed for this. In the last build, precedence conditions are used significantly. In addition, we are achieving significant savings in computing resources. So, we use precedence conditions "in positive direction". We connect every layer of position space with fixed power value for essential lists which are elements of the given layer. Important, that for construction of all layers of Bellman function, a fairly simple recurrent procedure is realized. Of course, this recurrent procedure is based on the Bellman equation. In our case, this Bellman equation takes into account precedence conditions.

We note that, in our case, constructions based on the Bellman equation are realized for $\mathcal{M}_1$-problem and $\mathcal{M}_2$-problem separately. After realization each of two recurrent procedures, we obtain extremum function for corresponding partial problem. This function is defined on the starting point set. But, for $\mathcal{M}_2$-problem, the latter set more to be created. This operation is envisaged in the first step of our algorithm. Now, let's move on to the consideration of the algorithm scheme.

1) Create a set of starting points for $\mathcal{M}_2$-problem in the form

$$X^{00} \triangleq \bigcup_{i \in \overline{1,N} \setminus \tilde{\mathbf{K}}_1} \mathbf{M}_i, \tag{23}$$

where $\tilde{\mathbf{K}}_1 \triangleq \{\mathrm{pr}_1(h) : h \in \mathbf{K}_1\}$. Form $\mathcal{M}_2$-problem as a system of $(\mathcal{M}_2, x)$-problems with $x \in X^{00}$.

2) Define the layers of the Bellman function for $\mathcal{M}_2$-problem. As final layer of the Bellman function, we define the extremum function of $\mathcal{M}_2$-problem.

3) By extremum function of $\mathcal{M}_2$-problem we create terminal component of (additive) criterion in $\mathcal{M}_1$-problem. Form $\mathcal{M}_1$-problem as a system of $(\mathcal{M}_1, x)$-problems, where $x \in X^0$.

4) Define the layers of the Bellman function for $\mathcal{M}_1$-problem. The extremum function of $\mathcal{M}_1$-problem is realized as final layer of the Bellman function. By this layer we define optimal starting point x^0 and extremum as the value of the final layer for this point.

5) Using standard (for DP) procedure, we construct optimal solution of $(\mathcal{M}_1, x^0)$-problem in the form of OP route-trajectory. As a result, we obtain $\mathcal{M}_1$-solution with the start x^0.

6) Fix finish point x^{00} on the trajectory of $\mathcal{M}_1$-solution. Define optimal solution on $(\mathcal{M}_2, x^{00})$-problem in the form of OP route-trajectory that is $\mathcal{M}_2$-solution (this construction is realized by DP procedure).

7) Glue together $\mathcal{M}_1$-solution and $\mathcal{M}_2$-solution. Glue together separately routes and trajectories. As a result, we obtain optimal solution of $(\mathcal{M}, x^0)$-problem. Adding the point x^0 to this solution, we get see (22) the optimal route process that is an element of **SOL** (19).

The detailed proofs of theoretical statements are contained in [6,9]. Now, we note only several important statements about connection of $\mathcal{M}$-problem and $\mathcal{M}_1$-problem. Namely, extremum $\mathbb{V}$ (18) of $\mathcal{M}$-problem coincides with analogous extremum for $\mathcal{M}_1$-problem. And what is more, the extremum function

$$\tilde{V}[\cdot] \triangleq (\tilde{V}[x])_{x \in X^0} \in \mathcal{R}_+[X^0]$$

coincides with analogous extremum function of $\mathcal{M}_1$-problem. Finally, the set (21) coincides with the set of all optimal starting points for $\mathcal{M}_1$-problem. These properties are connected with step 3): terminal component of criterion of $\mathcal{M}_1$-problem is defined by extremum function of $\mathcal{M}_2$-problem. So, step 3) defines the natural connection of $\mathcal{M}_1$-problem and $\mathcal{M}_2$-problem. This circumstance shows the important role of terminal component using as part of the criterion.

4 Computing Experiment: Sheet Cutting Under Two Zones

The optimal algorithm of previous section was implemented as a standard program for the simplest variant of the instrument control under sheet cutting on CNC machines by zones. In addition, the problem solution on a PC was considered. As X, a nondegenerate rectangle in the plane was used: we have cut sheet of metal. Sets $M_1, \ldots, M_n$ are realized by the contour sampling (really: equidistant curves). In addition, two types of points on these sets are used: tie-in-points and switch off points. Each of relations $\mathbb{M}_1, \ldots, \mathbb{M}_n$ consists of OP of above-mentioned type (tie-in-point and switch off point). In the above mentioned simplest variant of the instrument control by zones, $\mathcal{M}_1$- and $\mathcal{M}_2$-problem are realized like Sect. 1. So, a decomposition of total $\mathcal{M}$-problem is realized. In $\mathcal{M}_1$-problem and $\mathcal{M}_1$-problem, there are precedence conditions given by the sets $\mathbf{K}_1$ and $\mathbf{K}_2$ of address OP. Moreover, in $\mathcal{M}$-problem, there are conditions connected with heat removal under thermal cutting. The corresponding exact definitions of cost functions (12) are given in [10, Section 5,6]. So, now, we consider the variant of two zones. For this variant, the optimal algorithm with 1)–7) was realized.

For computations was used computer with Intel i5-11300H CPU, 8 Gb RAM, Windows 11 OS. The program was written in the C++ language with using of Qt interface library.

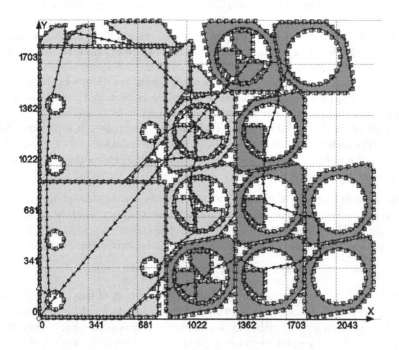

Fig. 1. Sample 1: computation result.

Sample 1. Number of contours is $\mathbf{n} = 50$. Precedence condition pairs number equals 31. It split in two clusters with dimension $|\mathcal{M}_1| = |\mathcal{M}_2| = 25$. The numbers of precedence conditions for these clusters are $|\mathbf{K}_1| = 14$ and $|\mathbf{K}_2| = 17$. The external price function has threshold value of the penalty 0,5, length of the finish cut area 100 mm, width – 25 mm. The penalty value is 1000000. Obtained result is 119,9. So, the result is correct in sense of thermal restrictions set through cost functions. Indeed, this result is significantly less than the penalty constant. Counting time is 25 min 16 s. The obtained result shown on the Fig. 1.

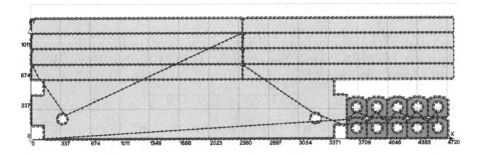

Fig. 2. Sample 2: computation result.

Sample 2. Now, we consider one example for which we will try to take into account the consideration of technological nature (see [7, §1.3.3]). Namely, in the case of thermal cutting, so called long details must be cut first. It's about the details for which the linear dimensions differ to much. Such details are most susceptible to thermal deformation. Therefore, it is advisable to cut them earlier in order to ensure more efficient removal of heat generated during insertion. In this connection, we form megalopolises on $\mathcal{M}_1$-problem including discrete contours of long details (we allow adding some more megalopolises to $\mathcal{M}_1$). The remaining megalopolises are included in $\mathcal{M}_2$.

Number of contours is $\mathbf{n} = 31$. Precedence condition pairs number equals 12. It split in two clusters with dimension $|\mathcal{M}_1| = 11$, $|\mathcal{M}_2| = 20$. The numbers of precedence conditions for these clusters are $|\mathbf{K}_1| = 2$ and $|\mathbf{K}_2| = 10$. The external price function parameters was used like in sample 1. Obtained result is 83,6. It has no penalties, so, the result is correct in sense of thermal restrictions set through cost functions. Counting time is 20 s. The obtained result shown on the Fig. 2.

5   Computing Experiment: Sheet Cutting by Zones in the General Case

Now, we consider natural evolution of algorithm of Sect. 2 for problem connected with cutting by zones. Namely, we consider more general variant: so, we assume that r zones are set, where $r \in \mathbb{N}$ and $r \geq 3$ (suppose that X, X^0, $\mathbf{n}$, $M_1, \ldots, M_\mathbf{n}$, $\mathbb{M}_1, \ldots, \mathbb{M}_\mathbf{n}$ correspond to Sect. 1; analogously, remain cost functions (12) with obvious correction domain of definition for f). Fix numbers

$$N_0 \in \overline{1, \mathbf{n}}, \quad N_1 \in \overline{1, \mathbf{n}}, \ldots, N_r \in \overline{1, \mathbf{n}}$$

for which $N_0 = 0$, $N_r = \mathbf{n}$ and $N_s + 2 \leqslant N_{s+1}$ under $s \in \overline{0, r-1}$. In the following, we suppose that

$$\mathcal{M}_j \triangleq \{M_i : i \in \overline{N_{j-1} + 1, N_j}\} \ \forall j \in \overline{1, r}.$$

Then, $\{\mathcal{M}_j : j \in \overline{1, r}\}$ is a partition of $\mathcal{M} = \{M_j : j \in \overline{1, \mathbf{n}}\}$. With

$$\mathcal{M}_1, \ldots, \mathcal{M}_r$$

we connect r zones (clusters). Suppose that, in the following, $\mathbf{K}_1, \ldots, \mathbf{K}_r$ are sets of address OP: for $j \in \overline{1, r}$,

$$\mathbf{K}_j \subset \overline{1, N_j - N_{j-1}} \times \overline{1, N_j - N_{j-1}}$$

and $\forall \mathbf{K}^0 \in \mathcal{P}'(\mathbf{K}_j) \ \exists z^0 \in \mathbf{K}^0 : \mathrm{pr}_1(z^0) \neq \mathrm{pr}_2(z) \ \forall z \in \mathbf{K}^0$. Under $j \in \overline{1, r}$, we introduce the set $\mathbb{P}_j$ of all permutations of indexes of $\overline{1, N_j - N_{j-1}}$ and the set

$$\mathcal{A}_j \triangleq \{\alpha \in \mathbb{P}_j \mid \alpha^{-1}(\mathrm{pr}_1(z)) < \alpha^{-1}(\mathrm{pr}_2(z)) \ \forall z \in \mathbf{K}_j\} \neq \emptyset.$$

of all admissible (by precedence) permutations of such type. So, we have partial routes. We realize gluing together routes of $\mathcal{A}_1, \ldots, \mathcal{A}_r$ and obtain a nonempty set of admissible routes in the basic $\mathcal{M}$-problem similar to (17). Of course, we have also $\mathcal{M}_j$-problem for each $j \in \overline{1, r}$. Now, we confine ourselves to the presentation of the algorithm scheme (more detailed consideration given in [6, Section 12]).

1) For each $j \in \overline{1, r-1}$, we define the set

$$\hat{\mathbf{K}}_j \triangleq \{\mathrm{pr}_1(z) : z \in \mathbf{K}_j\}.$$

Then, we introduce the set

$$X_j^{00} \triangleq \bigcup_{s \in \overline{1, N_j - N_{j-1}} \backslash \hat{\mathbf{K}}_j} \mathbf{M}_{N_{j-1}+s}$$

and consider X_j^{00} as the set of all possible starting points in $\mathcal{M}_{j+1}$-problem. Moreover, suppose that $X_0^{00} \triangleq X^0$ (recall that X^0 is given by our suppositions).

2) For $\mathcal{M}_r$-problem, we define layers of the Bellman function for $\mathcal{M}_r$-problem. As final layer, the extremum function of $\mathcal{M}_r$-problem is realized.

3) If $j \in \overline{2, r-1}$ and, for $\mathcal{M}_{j+1}$-problem, the extremum function was constructed, we define terminal component of additive criterion of $\mathcal{M}_j$-problem as above-mentioned extremum function for $\mathcal{M}_{j+1}$-problem. So, $\mathcal{M}_j$-problem formed. Now, we define layers of the Bellman function for this $\mathcal{M}_j$-problem. As final layer, the extremum function of $\mathcal{M}_j$-problem is realized.

4) After realization of step 3) for all $\mathcal{M}_j$-problem, where $j \in \overline{2, r-1}$, we use the extremum function of $\mathcal{M}_2$-problem for construction of terminal component of additive criterion for $\mathcal{M}_1$-problem. Then, we sequentially define layers of the Bellman function for $\mathcal{M}_1$-problem. Final layer is defined on X^0 and corresponds to extremum function of this $\mathcal{M}_1$-problem. We find the point $x^0 \in X^0$ realizing minimum of extremum function.

5) Using layers of the Bellman function of $\mathcal{M}_1$-problem, we construct optimal solution of this problem with start x^0. This solution is realized as OP route-trajectory.

6) If $k \in \overline{1, r-1}$ and sequentially optimal solutions of $\mathcal{M}_j$-problems, $j \in \overline{1, k}$, already were constructed, we fix the finish point x^{00} on $\mathcal{M}_k$-trajectory in the form of second element of OP corresponding to trajectory value for final index (for $\mathcal{M}_k$-problem). We use x^{00} as starting point for $\mathcal{M}_{k+1}$-problem after which determine optimal solution of this problem as OP route-trajectory.

7) After building sequentially optimal solutions of all problems $\mathcal{M}_1, \ldots, \mathcal{M}_r$, we realize component-wise gluing of these solutions (routes stick together with routes and trajectories stick together with trajectories). The resulting OP is completed by the point x^0. The obtained triplet is considered as required realization of route process.

Now, we briefly consider example of realization of the above-mentioned algorithm in problem connected with sheet cutting by zones.

Sample 3. Number of contours is **n** $= 100$. Precedence condition pairs number is 70. It split in five clusters with dimension $|\mathcal{M}_1| = 16$, $|\mathcal{M}_2| = 20$, $|\mathcal{M}_3| = 24$, $|\mathcal{M}_4| = 20$, $|\mathcal{M}_5| = 20$. The numbers of precedence conditions for these clusters are $|\mathbf{K}_1| = 9$, $|\mathbf{K}_2| = 14$, $|\mathbf{K}_3| = 18$, $|\mathbf{K}_4| = 15$, $|\mathbf{K}_5| = 14$. The external price function parameters was used like in sample 1. Obtained result is 238,2. It again has no penalties. Counting time is 53 s. The obtained result shown on the Fig. 3.

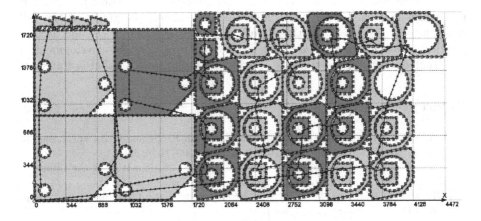

Fig. 3. Sample 3: computation result.

Based on a computational experiment, it can be noted that the maximum number of contours in one cluster (zone) for computing on a personal computer is in the range of 25–35, and depends on the number of nested contours, i.e. on the number of precedence conditions. The dependence on the number of clusters, if the clusters are similar, is close to linear.

In connection with Cutting Path problem (the tool path optimization problem for CNC sheet-cutting machines), we note monograph [7], in which, the detailed statement of this engineering problem is discussed (moreover, see the detailed consideration in [11, 12]). We recall the known "nesting" problem [13, 14] with which the Cutting Path Problem is connected. Moreover, in should be singled investigations [15–19], where questions of routing algorithms constructing are considered. We especially note research by A.A.Petunin (see [20–23]). For more detailed familiarization with routing methods in Cutting Path Problem, we recommend [11, Introduction].

6   Conclusion

In this article, the routing problem with constraints and complicated cost functions is investigated. This problem is connected with realization implementation of visits to nonempty finite sets (megalopolises) with fulfilment some works. Of course, this mathematical problem has many engineering applications. In given investigation, only one application noted; namely, we have focused on issues connected with sheet cutting on CNC machines. In given problem, often zone cutting mode is used. In simplest case, two zones are fixed. Then, all contour set is divided into the sum of two subsets. It is required to first cut the contours of the first set and only after that to cut the contours of the second subset. This statement is considered in given article. For this statement, optimal solution is constructed. This solution is a triplet with the first element as a route, the second element as a trajectory, and the third elements as a starting point. For solution construction, the widely understood DP is used. The corresponding DP procedures are realized for two zones separately. After, the gluing of two DP procedures is realized. So, in this article, a useful connection of DP and decomposition constructions is established. In the final of article, the development of constructed algorithm for the case of arbitrary finite zones number is stated.

References

1. Gutin, G., Punnen, A.P.: The Traveling Salesman Problem and Its Variations, p. 850. Springer, Berlin (2002). https://doi.org/10.1007/b101971
2. Cook, W.J.: In Pursuit of the Traveling Salesman: Mathematics at the Limits of Computation, 228 p. Princeton University Press, Princeton (2012)
3. Gimadi, E., Khachay, M.: Extremal Problems on Sets of Permutations, 220 p. UrFU Publ., Yekaterinburg (2016)

4. Bellman, R.: Dynamic programming treatment of the travelling salesman problem. J. ACM **9**(1), 61–63 (1962)
5. Held, M., Karp, R.M.: A dynamic programming approach to sequencing problems. J. Soc. Ind. Appl. Math. **10**(1), 196–210 (1962)
6. Chentsov, A.G., Chentsov, P.A.: An extremal two-stage routing problem and procedures based on dynamic programming. Trudy Inst. Mat. i Mekh. UrO RAN **28**(2), 215–248 (2022)
7. Petunin, A.A., Chentsov, A.G., Chentsov, P.A.: Optimal Tool Routing of Shaped Sheet Cutting Machines with Numerical Control. Mathematical Models and Algorithms, 247 p. Ural University Press, Yekaterinburg (2020)
8. Chentsov, A.G.: Extremal Problems of Routing and Job Distribution: Questions of Theory, 238 p. Izhevsk Institute for Computer Research, Izhevsk (2008). (in Russian)
9. Chentsov, A.G., Chentsov, P.A.: Dynamic programming in the routing problem: decomposition variant. Russ. Univ. Rep. Math. **27**(137), 95–124 (2022)
10. Chentsov, A.G., Chentsov, P.A.: Routing under constraints: problem of visit to megalopolises. Autom. Remote. Control. **77**(11), 1957–1974 (2016). https://doi.org/10.1134/S0005117916110060
11. Petunin, A.A., Chentsov, A.G., Chentsov, P.A.: Some applications of optimization routing problems with additional constraints. Bull. Udmurt Univ. Math. Mech. Comput. Sci. **32**(2), 187–210 (2022)
12. Chentsov, A.G., Chentsov, P.A., Petunin, A.A., Sesekin, A.N.: Model of megalopolises in the tool path optimisation for CNC plate cutting machines. Int. J. Prod. Res. **56**(14), 4819–4830 (2018)
13. Dowsland, K.A., Dowsland, W.B.: Solution approaches to irregular nesting problems. Eur. J. Oper. Res. **84**, 506–521 (1995)
14. Stoyan, Y., Pankratov, A., Romanova, T.: Placement problems for irregular objects: mathematical modeling, optimization and applications. In: Butenko, S., Pardalos, P.M., Shylo, V. (eds.) Optimization Methods and Applications. SOIA, vol. 130, pp. 521–559. Springer, Cham (2017). https://doi.org/10.1007/978-3-319-68640-0_25
15. Dewil, R., Vansteenwegen, P., Cattrysse, D.: Sheet metal laser cutting tool path generation: dealing with overlooked problem aspects. Key Eng. Mater. **639**, 517–524 (2015)
16. Dewil, R., Vansteenwegen, P., Cattrysse, D.: A review of cutting path algorithms for laser cutters. Int. J. Adv. Manuf. Technol. **87**(5), 1865–1884 (2016). https://doi.org/10.1007/s00170-016-8609-1
17. Dewil, R., Vansteenwegen, P., Cattrysse, D., Laguna, M., Vossen, T.: An improvement heuristic framework for the laser cutting tool path problem. Int. J. Prod. Res. **53**(6), 1761–1776 (2015)
18. Sonawane, S., Patil, P., Bharsakade, R., Gaigole, P.: Optimizing tool path sequence of plasma cutting machine using TSP approach. In: Web of Conferences, vol. 184, p. 01037 (2020)
19. Levichev, N., Rodrigues, G.C., Duflou, J.R.: Real-time monitoring of fiber laser cutting of thick plates by means of photodiodes. Proc. CIRP **94**, 499–504 (2020)
20. Petunin, A.A., Stylios, C.: Optimization models of tool path problem for CNC sheet metal cutting machines. IFAC-PapersOnLine **49**(12), 23–28 (2016)
21. Petunin, A.A., Chentsov, P.A.: Routing in CNC cutting machines: engineering constraints. Acta Polytech. Hung. **17**(8), 165–177 (2020)

22. Petunin, A.A., Polyshuk, E.G., Chentsov, P.A., Ukolov, S.S., Krotov, V.I.: The termal deformation reducing in sheet metal at manufacturing parts by CNC cutting machines. In: IOP Conference Series: Materials Science and Engineering, vol. 613, p. 012041 (2019)
23. Tavaeva, A., Petunin, A., Ukolov, S., Krotov, V.: A cost minimizing at laser cutting of sheet parts on CNC machines. In: Bykadorov, I., Strusevich, V., Tchemisova, T. (eds.) MOTOR 2019. CCIS, vol. 1090, pp. 422–437. Springer, Cham (2019). https://doi.org/10.1007/978-3-030-33394-2_33

Public-Private Partnership Model with a Consortium

Sergey Lavlinskii[✉], Artem Panin, and Alexander Plyasunov

Sobolev Institute of Mathematics, Novosibirsk, Russia
{lavlin,apljas}@math.nsc.ru

Abstract. A model is proposed for generating a mineral raw materials development program in a resource-rich region. The model is based on a special mechanism of public-private partnership with a consortium. The main idea of the partnership model is to cluster mine fields and set up a system of consortia of private investors who jointly implement projects to construct the necessary production infrastructure in the cluster. Such a mechanism is based on the search for a compromise between the interests of the government and private investors, ensuring a Stackelberg equilibrium. In the process of interaction (two periods, sequential choice), the government acts as a leader by setting quotas on the compensations for the consortia's costs of implementing the infrastructure projects. The system of consortia plays the role of a follower by rationally choosing the infrastructure development program that ensures the profitability of the development projects for private investors, taking into account the costs of shared construction and the compensation schedule offered by the government. This approach allows one to form a targeted development plan by solving a bilevel problem of mathematical programming. This plan determines for each consortium a list of implemented infrastructure projects and, for private investors, a schedule of infrastructure costs and their compensations from the budget. It is proven that the problem of the government belongs to the class of Σ_2^P-hard problems associated with the second level of the polynomial hierarchy. The main directions are proposed in searching for efficient solution algorithms based on metaheuristics and enabling the solution of high-dimensional problems.

Keywords: Stackelberg game · bilevel mathematical programming problems · Σ_2^P-hard problems · stochastic local search · strategic planning · public-private partnership · a consortium of subsoil users

1 Introduction

The development of mechanisms for stimulating private investment presents a timely and relevant, as-yet unresolved problem in Russia, primarily in the under-developed resource-rich regions of Siberia and the Far East. The established practice of making this kind of decision in subsoil resource management tends to operate with political arguments and most unsophisticated effectiveness evaluations, which are derived from analysis of technological projects and current raw materials prices.

Practical experience shows that when attempting to stimulate private investment activity, the government does not succeed in applying the traditional support tools either. Using the financial resources of the Investment Fund of Russia, the government builds production infrastructure (see, e.g., the transport infrastructure project for the development of mineral resources in the southeast of the Chita oblast [1], or finances the implementation of a large complex of infrastructure and environmental activities related to building the reservoir of the Boguchanskaya hydro-based power plant within a project of integrated development of the Lower Angara Region [2].

The expert community is skeptical about the actual progress in implementing these projects. Thus, the railway built by the government in Transbaikalia remained unattended over a long period; in the meantime, it was being dismantled vigorously for scrap metal to be sold in China. It was only in 2016 that the railway was restored and put into temporary operation. In the Lower Angara Region, the budget did not receive the revenues announced in the project documentation, and the people who resettled from the flood zone got neither the promised benefits nor additional social infrastructure. In the Yenisei Siberia Megaproject, the government again plans to take over the implementation of large-scale infrastructure projects to build the Elegest-Kyzyl-Kuragino railway and the Beya transport infrastructure without a detailed analysis of the consequences of such an undertaking in terms of balancing the interests of the government and private investors [3].

Behind all these attempts by the government to support business and intensify the development of natural resources, one can clearly see a lack of qualified expert assessment of the integrated large-scale projects, which the government calls – not quite correctly – public-private partnership projects. In the classical model of public-private partnership (PPP), the investor reaches an agreement with the government on a certain list of infrastructure projects that open up the field development projects of interest to the investor, and then he implements these infrastructure projects at his own expense. The government, in turn, compensates for his costs with a certain lag, e.g., starting from the time it begins to receive taxes from the mining operations set up by the private investor [4–8].

Thus, practical experience shows that the a priori confidence in that any combination of business support tools always brings a positive result does not have a leg to stand on if we set the goal of safeguarding the interests of society as a whole as well as private business. That is why working out a mechanism for a public-private partnership, which would determine a program for development of natural resources in an underdeveloped territory, is a timely and relevant problem of great practical importance, which requires the creation of special tools to support managerial decision-making.

Such a program defines a list of measures necessary for the development of a given territory and it is designed to answer several questions that are important for a potential investor.

How to help the investor overcome the barriers associated with the lack of transport and energy infrastructure, which is so typical of most Siberian and Far Eastern regions of Russia? What kind of a private investment stimulation mechanism, which would combine a range of government investment policy measures, should be put at the core of the program for the development of local natural resources? What organizational activities can be performed to harmonize the goals and pool the resources of individual subsoil users?

These problems are the focus of this paper. Having supplemented the classical PPP model with the consortium mechanism, the authors considerably expand the previously studied partnership models [9–15] by supplementing them with horizontal connections between private investors and resource consolidation effects. It is of fundamental importance that the PPP mechanism and the development program thus become as concrete and targeted as possible, which in itself could be directly beneficial for managerial practice.

2   Mathematical Model

We assume that the mine fields within a region form a system of nonoverlapping clusters; such a hypothesis is true for most of the underdeveloped regions of Siberia and the Far East. The transport and energy infrastructure for such a system of clusters is built by a system of consortia localized in these clusters (one consortium per cluster). In each cluster, the government sets up a management company (MC) that organizes and coordinates the shared financing of infrastructure construction by the private investors that are members of the consortium and pays out compensations for the costs incurred of subsoil users from the budget. It is also assumed that the MC takes over all the functions of the government in coordinating the interests of the investor and the government in terms of mining projects (pre project analysis, environmental control, monitoring, etc.). The output of the model is a targeted plan for each consortium to generate a mineral raw materials development program. The plan determines for each consortium (cluster) a list of infrastructure projects to be implemented and for each investor (mine field) a schedule of costs to build the necessary production infrastructure and a schedule of compensation payments from the budget.

Such a scheme is viable and attractive to the potential private investor only if the latter achieves the necessary profitability of the overall project for developing the mine field and building a part of the necessary infrastructure. It is attractive for the government primarily because of the timing of the development costs: in contrast to the current practice, the compensation payments from the budget can be postponed to the period when taxes are beginning to flow. In addition, the government obtains a tool for regulating the distribution of natural resource rent, a tool that is much more flexible than rent taxes, by assigning a larger share of infrastructure costs in the consortium to highly profitable mine fields. The only thing left to do is to determine the parameters of such a partnership (development program) to ensure profitability for private investors and the maximum possible amount of natural resource rent for the government in the form of taxes.

Consortium Model

How can we describe the functioning of a consortium with an MC coordinating and organizing the processes of mining, the shared construction of mining infrastructure, and the subsequent mutual settlements?

In our case, the MC of the consortium acts as a managerial agent for the PPP project by creating the institutional structure of the partnership and coordinating the interaction of the participants in line with the project goals. Here it is vitally important to take into account transaction costs (TCs) [16,17]. The latter act as friction in coordinating the interests of the investor and the government and substantially affects, first of all, the formation and functioning of the consortium. By its nature, such a specialized consortium must solve complex, inherently optimization-based problems. This circumstance distinguishes it from today's directorates of the aforementioned megaprojects and programs, whose managerial decision-making tools do not measure up to the level of tasks assigned to them.

TCs include not only the costs arising from the conclusion of contracts but also those accompanying the interactions between economic agents [18–20]. In the case of large-scale mining projects implemented in remote regions, TCs may be very high [21]. For an MC, representing the interests of the government, they comprise, e.g., control and monitoring (such as technical oversight and environmental monitoring), the costs of improving supportive public institutions, etc. The TCs of an investor participating in the consortium comprise the costs of conducting an environmental impact assessment, the costs of maintaining business departments responsible for interacting with the relevant regulatory bodies, etc.

Here we distinguish between the *ex ante* and *ex post* TCs of implementing a project. The reason is that *ex ante* and *ex post* TCs in field development address essentially different issues. Moreover, depending on an institutional environment, they can be distributed differently between the government and the private investor, and our model should possess a functional that would capture this circumstance.

Model assumptions:

(1) The size of TCs for both the consortium's MC and the investor depends on the volume of project investment.
(2) While coordinating their interests (T_1 years), the MC and the investor bear *ex ante* TCs; during the entire period of project implementation (T_2 years), they bear *ex post* TCs.
(3) *Ex ante* TCs are increasing until the launch of the project; *ex post* TCs are decreasing in the course of implementing the project.

(4) For the investor, the model specifies not only the level of TCs but also the share of these costs that are attributed to project costs.

By introducing these hypotheses, we can formalize the size of TCs and capture, in an aggregate form, their behavior over time.

PPP Formation Model

The problem statement can be represented as follows. We use the following notation:

$T = \{-T_1, \ldots, 0, 1, \ldots, T_2\}$ is the time horizon; T_0 is the time lag of reimbursement of infrastructure costs by the private investor; I is a set of production projects; J is a set of infrastructure development projects; K is a set of consortia. Each production project has its own private investor.

Production project i in year t:

CFP_i^t is the operating cash cashflow (the difference between the incomes and expenses of operating activities in the process of development);

DBP_i^t are the tax revenues of the budget from the project;

$ITCP_i^t$ and $MTCP_i^t$ are, respectively, the TCs of investor i and the TCs incurred by the MC of the investor's "host" consortium, which arise during the preparation (*ex ante*, $t = -T_1, \ldots, 0$) and implementation (*ex post*, $t = 1, \ldots, T_2$) of the production project.

Infrastructure project j in year t:

ZI_j^t is the schedule of investment costs necessary for implementing project;

VDI_j^t are the off-project revenues of the budget from implementing project j, which are associated with the overall development of the local economy;

$MTCI_j^t$ is the schedule of the TCs incurred by the MC of the consortium implementing project j;

$ITCI_j^t$ is the schedule of the total TCs of the investors partaking in the implementation of project j.

Outside the planning horizon ($t = -T_1, \ldots, 0$), the model parameters CFP_i^t, DBP_i^t, ZI_j^t and VDI_j^t are assumed to be zero.

Interproject connection:

μ_{ij} is the indicator of technological cohesion between production and infrastructure projects; $i \in I$, $j \in J$:

$$\mu_{ij} = \begin{cases} 1, \text{ if the implementation of production project } i \\ \quad \text{necessarily requires the implementation of infrastructure project } j, \\ 0 \text{ otherwise.} \end{cases}$$

The sets of production and infrastructure projects are divided in a nonoverlapping (mutually disjoint) manner into NC consortia, based on the location of the field clusters. A single field is a single consortium consisting of one investor.

The topology of the consortia system is defined by the following parameters:

α_{kj} is the parameter indicating whether infrastructure project j is attributed to consortium k; this parameter equals to one if the attribution is valid and zero otherwise.

β_{ki} is the parameter indicating whether the investor of production project i is attributed to consortium k; this parameter equals to one if the attribution is valid and zero otherwise.

The discounts of the government and the investor: DG and DI respectively.

$BudG^t$, $BudI_i^t$ are the budget constraints of, respectively, the government and investors.

We introduce the following integer variables:

$$z_i = \begin{cases} 1, \text{ if the investor } i \text{ launches production project,} \\ 0 \text{ otherwise;} \end{cases}$$

$$c_j = \begin{cases} 1, \text{ if infrastructure project } j \text{ is implemented by one of the consortia,} \\ 0 \text{ otherwise;} \end{cases}$$

Real-valued variables of the model:

$\bar{W}_k^t$ and W_k^t are the compensation schedules, respectively, offered by the leader (government) and actualized in reality; the compensations reimburse the infrastructure costs incurred by the private investors partaking in consortium k and the expenses associated with the functioning of the MC;

R_i^t is the compensation schedule (determined by the consortium's MC) for the costs of private investor i;

D_{ij} is the share of investor i in the costs of implementing infrastructure project j.

The PPP formation model can be represented as the following problem of bilevel mathematical programming.

The upper-level problem $\widetilde{\mathcal{PG}}$ can be formulated as follows:

$$\sum_{t \in T} \Big(\sum_{i \in I} DBP_i^t z_i + \sum_{j \in J} VDI_j^t c_j - \sum_{k \in K} W_k^t \Big) / (1 + DG)^t \to \max_{\bar{W}, z, c, W, R, D} \quad (1)$$

subject to:

$$\sum_{k \in K} \bar{W}_k^t \leq BudG^t; t \in T; \quad (2)$$

$$\bar{W}_k^t \geq 0; t \in T; k \in K; \quad (3)$$

$$(z, c, W, D, R) \in \mathcal{F}^*(\bar{W}). \quad (4)$$

The set $\mathcal{F}^*$ is a set of optimal solutions of the following low-level parametric consortia problem $\widetilde{\mathcal{PC}}(\bar{W})$:

$$\gamma \sum_{t \in T} \Big(\sum_{i \in I} ((CFP_i^t - ITCP_i^t) z_i - \sum_{j \in J} (ZI_j^t + ITCI_j^t) D_{ij} + R_i^t) \Big) / (1 + DI)^t$$

$$+ (1 - \gamma) \sum_{t \in T} \Big(\sum_{i \in I} DBP_i^t z_i + \sum_{j \in J} VDI_j^t c_j - \sum_{k \in K} W_k^t \Big) / (1 + DG)^t \to \max_{z, c, W, R, D}$$

$$(5)$$

subject to:

$$W_k^t \leq \bar{W}_k^t; t \in T; k \in K; \tag{6}$$

$$\sum_{t \in T} \left(\sum_{i \in I} DBP_i^t z_i \beta_{ki} + \sum_{j \in J} VDI_j^t c_j \alpha_{kj} - W_k^t \right) / (1 + DG)^t \geq 0; k \in K; \tag{7}$$

$$R_i^t = 0; -T_1 \leq t \leq T_0; i \in I; \tag{8}$$

$$R_i^t \geq 0; T_0 + 1 \leq t \leq T_2; i \in I; \tag{9}$$

$$\sum_{i \in I} R_i^t \beta_{ki} + \sum_{j \in J} MTCI_j^t c_j \alpha_{kj} + \sum_{i \in I} MTCP_i^t z_i \beta_{ki} \leq W_k^t; k \in K; t \in T; \tag{10}$$

$$0 \leq D_{ij} \leq \mu_{ij}; i \in I; j \in J; \tag{11}$$

$$\sum_{i \in I} D_{ij} \beta_{ki} = \alpha_{kj} c_j; k \in K; j \in J; \tag{12}$$

$$\sum_{t \in T} \left(\sum_{i \in I} (CFP_i^t - ITCP_i^t) z_i - \sum_{j \in J} (ZI_j^t + ITCI_j^t) D_{ij} \right. \tag{13}$$

$$\left. + R_i^t \right) / (1 + DI)^t \geq 0; i \in I;$$

$$c_j \geq \mu_{ij} z_i; i \in I; j \in J; \tag{14}$$

$$- (CFP_i^t - ITCP_i^t) z_i + \sum_{j \in J} (ZI_j^t + ITCI_j^t) D_{ij} \tag{15}$$

$$- R_i^t \leq BudI_i^t; t \in T; i \in I.$$

In the formulated model, the consortium's MC try to find a compromise between the interests of partners. Thus, the MC seeks to distribute infrastructure costs and budgetary compensation in such a way as to harmonize the rent-seeking behavior of investors and the position of the government as the owner of natural resources (5). The parameter γ reflects the degree to which the interests of a private investor are taken into account.

Constraints (11)–(12) describe the procedure for distributing the infrastructure costs (including TCs) among the members of a consortium. The interdependence of production and infrastructure projects is set by constraint (14); i.e., a mining project cannot be launched if the necessary production infrastructure is not present. The budgets of the government (2) and the investors (15) define restrictions on a feasible set of projects.

The consortium's MC begins compensation payments to investors after T_0 years (8), (9), e.g., from the time of receiving the first tax payments. The MC's budget constraint (10) fixes the important role of TCs arising in the process of coordinating interests in mining and in building infrastructure. The compensation schedule should provide the government with a balance of budget revenues and transfers to the consortium (7) as well as compensate for the infrastructure costs of each investor with a discount factor (13).

The formation of consortium k is expedient for the government only if its MC and investors can ensure an increase in the government's objective function ((1) and (7)). This is why the formation of a single consortium for the full set of mine fields ($NC = 1$) may in some cases be less efficient than creating a system of several consortia covering the entire territory and taking into account the spatial locations of the objects of planning.

The output of the model is a targeted mineral raw materials development program $\{c_j, z_i, R_i^t, D_{ij}\}$ that determines a list of launched infrastructure and production projects specifies a mechanism for implementing shared construction, and defines the main parameters of the compensation policy.

3   Computational Complexity

This paper considers a new bilevel model of interaction between the government and private investors is based on a special mechanism of public-private partnership with a consortium. The structural features of the new model do not allow us to take advantage of previously developed approaches to estimating the computational complexity of bilevel PPP models from [9–15]. The first results obtained for this model in the study of its relationships with the polynomial hierarchy are given below. Let us associate with the government problem $\widetilde{\mathcal{PG}}$ the standard decision problem $D(\widetilde{\mathcal{PG}})$, in which the input is the input of the government problem and an arbitrary rational number k. In the problem $D(\widetilde{\mathcal{PG}})$ we have to decide whether or not there exists a feasible solution with the value of the objective function greater than or equal to k. The decision problem $D(\widetilde{\mathcal{PG}})$ belongs to the class Σ_2^P if there exists a non-deterministic Turing oracle machine that recognizes problem L in polynomial time using some language from the class NP as an oracle. Class Σ_2^P refers to the second level of the polynomial hierarchy [22]. An optimization problem belongs to the class NPO if its standard decision problem belongs to the class NP [22]. Similarly, the class $\Sigma_2^P O$ contains optimization problems for which the corresponding standard recognition problem belongs to the class Σ_2^P [23]. Class NPO refers to the first level of the approximation hierarchy and class $\Sigma_2^P O$, respectively, to the second level of this hierarchy.

We now show that the government problem is Σ_2^P-hard provided that parameter γ is 1 and there is no constraint (9) in the lower-level problem by reducing the Σ_2^P-complete Subset-Sum-Interval problem to the problem in question [24, 25].

Theorem 1. *The problem of the government $\widetilde{\mathcal{P}\mathcal{G}}$ is Σ_2^P-hard.*

Proof. (Sketch of the proof)

The Subset-Sum-Interval problem. Given positive integers $q_1, ..., q_l$, R and r, where r does not exceed l. Does an integer S exist that $R \leq S < R + 2^r$ and none of the subsets $I \subseteq \{1, ..., l\}$ satisfies $\Sigma_{i \in I} q_i = S$.

We reduce this problem to the problem of the government. For this purpose let us construct the following input of the government problem. Let there be $l + 2$ production projects and $l + 1$ infrastructure projects. Assume that we have a single consortium, the planning horizon is three years. For the first l production projects, we have $CFP_i^1 = CFP_i^2 = 0$, $CFP_i^3 = 2q_i$. For other projects: $CFP_{l+1}^1 = 0, CFP_{l+1}^2 = -1/2, CFP_{l+1}^3 = 1, CFP_{l+2}^1 = 0, CFP_{l+2}^2 = R, CFP_{l+2}^3 = \Delta$ and $DBP_{l+1}^1 = 0, DBP_{l+1}^2 = 0, DBP_{l+1}^3 = \Delta, DBP_{l+2}^1 = 0, DBP_{l+2}^2 = 0, DBP_{l+2}^3 = 2\Delta$. For the first l infrastructure projects, we have $ZI_i^1 = ZI_i^3 = 0$, $ZI_i^2 = q_i$. For other projects: $ZI_{l+2}^1 = R, ZI_{l+2}^2 = 0, ZI_{l+2}^3 = 0$, where Δ is a fairly large positive number such as $2R + 2^{r+1}$. Then each of the players is willing to incur any expense to make such a profit. Every production project with the number $i \in \{1, ..., l+2\}$, excluding $l+1$, is required to implement an infrastructure project with the same number. It follows from the accepted agreements that the values of α_{1j} and β_{1i} are equal to 1 for all $j \in \{1, ..., l, l+2\}$ and $i \in \{1, ..., l+2\}$. All other parameters of the production and infrastructure projects are set to zero. The government budget in the first year is $R + 2^r - 1$. In the second and third years, the budget is zero. Each investor has zero budget in any year.

Let S be equal to $\bar{W}_{11}$. The attractiveness of the $l + 2$ project and the zero solvency of investors force the government to give no less than R to the consortium. That is, $S \geq R$. In addition, S does not exceed the value of the government budget. That is, $S \leq R + 2^r - 1$. Excluding the constraint (9) from the model allows the $l + 2$ investor to return up to R to the consortium in the second year. Thus, the consortium can distribute among investors up to S for the implementation of the first $l + 1$ production projects. Consortium income is calculated using the formula:

$$\Delta + S + \sum_{1 \leq i \leq l} q_i z_i + 0.5 z_{l+1},$$

subject to constraint (10):

$$S \geq \sum_{1 \leq i \leq l} q_i z_i + 0.5 z_{l+1}.$$

Consequently, the production project $l+1$ needed by the government will be started only when none of the subsets $I \subseteq \{1, ..., l\}$ satisfies $\Sigma_{i \in I} q_i = S$.

4   Results and Discussion

The formulated model (1)–(15) can serve as a basis for developing model tools to support managerial decision-making when generating mineral raw materials development program based on the above-described PPP mechanism with a consortium. Having addressed the task of developing effective methods to solve the corresponding bilevel problem of high-dimensional mathematical programming, we can work out a practical methodology to develop a government investment policy that stimulates the arrival of a private investor by harmonizing the goals and pooling the resources of individual subsoil users.

This study established the following fact:

The government problem belongs to Σ_2^P-hard problems associated with the second level of the polynomial hierarchy (Theorem 1).

This fact indicates that the search for optimal solutions is unlikely to succeed even at relatively small dimensions. Under these conditions, it does not even make sense to search for approximate solutions with a guaranteed accuracy estimate for the relative deviation from the optimal solution over polynomial time.

Today's literature knows examples of efficient algorithms for solving this kind of complex problems, but these algorithms work only at small dimensions or for problems with a special structure [13, 14, 25–29]. In our case, the structure is the most generic, and the dimension of the consortia problem has increased substantially relative to the previous versions of the model.

Under these conditions, in order to solve the problem on the basis of metaheuristics and exact methods, it is necessary to develop a stochastic approximate hybrid algorithm that generates a "good" initial solution for a stochastic local ascent. In the process of searching for the initial solution, one needs to find an upper bound for the optimal value of the government's problem. To do so, it is proposed to find an optimal solution to the HP-relaxation of the government's problem. Now the initial solution is calculated using a procedure at each step of which the problem of the consortia is solved with an additional threshold constraint on the value of the government's objective function. The threshold constraint is generated using the previously obtained upper bound.

Speaking about the stochastic local search algorithm, it is currently planned to realize this algorithm in the form of a stochastic coordinate-wise ascent. However, in order to refine this algorithm and achieve a good performance on real data, it is necessary to conduct additional theoretical and experimental studies. Under these conditions, the stochastic approximate hybrid algorithm is likely to give a good approximate solution to the bilevel problem.

The authors plan to realize the described scenario of searching for effective approximate methods to solve bilevel problems of large dimensions in the model under study using a special model test site with actual information on Transbaikalia. The result of this work will be a toolkit for supporting managerial decisions with a wide scope of applications. The toolkit will be useful in the practice of real management of the mineral raw materials complex in the Siberian and Far Eastern regions, where large-scale investment projects are being launched with the participation of the government, including dozens of project activities to be implemented over a time horizon of 20–30 years.

Acknowledgements. The study was funded by a grant Russian Science Foundation No. 23-28-00849, https://rscf.ru/project/23-28-00849.

References

1. Glazyrina, I.P., Lavlinskii, S.M., Kalgina, I.S.: Public-private partnership in the mineral resources complex of Zabaikalskii krai: problems and prospects. Geogr. Nat. Resour. **35**(4), 359–364 (2014). https://doi.org/10.1134/S1875372814040088
2. Lavlinskii, S.M.: Public-private partnership in a natural resource region: ecological problems, models, and prospects. Stud. Russ. Econ. Dev. **21**(1), 71–79 (2010). https://doi.org/10.1134/S1075700710010089
3. Bryukhanova, E., Efimov, V., Shishatsky, N.: Research on the issues of economic growth centres establishment in the south of the Angara-Yenisei macroregion. J. Sib. Fed. Univ. Humanit. Soc. Sci. **13**(11), 1736–1745 (2020). https://doi.org/10.17516/1997-1370-0679
4. Reznichenko, N.V.: Public-private partnership models. Bull. St. Petersburg Univ. Ser. 8 Manag. **4**, 58–83 (2010). (in Russian)
5. Quiggin, J.: Risk, PPPs and the public sector comparator. Aust. Account. Rev. **14**(33), 51–61 (2004)
6. Grimsey, D., Levis, M.K.: Public Private Partnerships: The Worldwide Revolution in Infrastructure Provision and Project Finance. Edward Elgar, Cheltenham (2004)
7. Lakshmanan, T.R.: The broader economic consequences of transport infrastructure investments. J. Transp. Geogr. **19**(1), 1–12 (2011)
8. Mackie, P., Worsley, T., Eliasson, J.: Transport appraisal revisited. Res. Transp. Econ. **47**, 3–18 (2014)
9. Lavlinskii, S.M., Panin, A.A., Plyasunov, A.V.: A bilevel planning model for public–private partnership. Autom. Remote. Control. **76**(11), 1976–1987 (2015). https://doi.org/10.1134/S0005117915110077
10. Lavlinskii, S., Panin, A., Pliasunov, A.: Comparison of models of planning the public-private partnership. J. Appl. Ind. Math. **10**(3), 1–17 (2016). https://doi.org/10.1134/S1990478916030017
11. Lavlinskii, S., Panin, A.A., Plyasunov, A.V.: Public-private partnership models with tax incentives: numerical analysis of solutions. In: Eremeev, A., Khachay, M., Kochetov, Y., Pardalos, P. (eds.) OPTA 2018. CCIS, vol. 871, pp. 220–234. Springer, Cham (2018). https://doi.org/10.1007/978-3-319-93800-4_18
12. Lavlinskii, S., Panin, A., Plyasunov, A.V.: Stackelberg model and public-private partnerships in the natural resources sector of Russia. In: Khachay, M., Kochetov, Y., Pardalos, P. (eds.) MOTOR 2019. LNCS, vol. 11548, pp. 158–171. Springer, Cham (2019). https://doi.org/10.1007/978-3-030-22629-9_12
13. Lavlinskii, S., Panin, A., Plyasunov, A.: The Stackelberg model in territorial planning. Autom. Remote. Control. **80**(2), 286–296 (2019). https://doi.org/10.1134/S0005117919020073
14. Lavlinskii, S., Panin, A., Plyasunov, A.: Bilevel models for investment policy in resource-rich regions. In: Kochetov, Y., Bykadorov, I., Gruzdeva, T. (eds.) MOTOR 2020. CCIS, vol. 1275, pp. 36–50. Springer, Cham (2020). https://doi.org/10.1007/978-3-030-58657-7_5
15. Lavlinskii, S., Panin, A., Plyasunov, A.: Bilevel models for socially oriented strategic planning in the natural resources sector. In: Strekalovsky, A., Kochetov, Y., Gruzdeva, T., Orlov, A. (eds.) MOTOR 2021. CCIS, vol. 1476, pp. 358–371. Springer, Cham (2021). https://doi.org/10.1007/978-3-030-86433-0_25

16. Marshall, G.R.: Transaction costs, collective action and adaptation in managing socio-economic system. Ecol. Econ. **88**, 185–194 (2013). https://doi.org/10.1016/j.ecolecon.2012.12.030

17. Ostrom, E.: A general framework for analyzing sustainability of social-ecological systems. Science **325**(5939), 419–422 (2009). https://doi.org/10.1126/science.1172133

18. Holmstrom, B., Milgrom, P.: Multitask principal-agent analyses: incentive contracts, asset ownership, and job design. J Law Econ Organ **7**, 24–52 (1991)

19. Hart, O.: Firms, Contracts, and Financial Structure. Clarendon Press, Oxford (1995)

20. Bolton, P., Dewatripont, M.: Contract Theory. MIT Press, Cambridge (2005)

21. Glazyrina, I., Lavlinskii, S.: Transaction costs and problems in the development of the mineral and raw-material base of the resource region. J. New Econ. Assoc. New Econ. Assoc. **38**(2), 121–143 (2018)

22. Ausiello, G., Crescenzi, P., Gambosi, G., et al.: Complexity and Approximation: Combinatorial Optimization Problems and Their Approximability Properties. Springer, Berlin (1999). https://doi.org/10.1007/978-3-642-58412-1

23. Panin, A.A., Pashchenko, M.G., Plyasunov, A.V.: Bilevel competitive facility location and pricing problems. Autom. Remote. Control. **75**(4), 715–727 (2014). https://doi.org/10.1134/S0005117914040110

24. Eggermont, C.E.J., Woeginger, G.J.: Motion planning with pulley, rope, and baskets. Theory Comput. Syst. **53**(4), 569–582 (2013). https://doi.org/10.1007/s00224-013-9445-4

25. Caprara, A., Carvalho, M., Lodi, A., Woeginger, G.J.: A study on the computational complexity of the bilevel knapsack problem. SIAM J. Optim. **24**(2), 823–838 (2014)

26. Beresnev, V.L., Melnikov, A.A.: Computation of an upper bound in the two-stage bilevel competitive location model. J. Appl. Ind. Math. **16**(3), 377–386 (2022). https://doi.org/10.1134/S1990478922030012

27. Dempe, S., Zemkoho, A. (eds.): Bilevel Optimization Advances and Next Challenges. Springer Optimization and Its Applications (SOIA), vol. 161. Springer, Cham (2020). https://doi.org/10.1007/978-3-030-52119-6

28. Talbi, E.-G. (ed.): Metaheuristics for Bi-Level Optimization. Studies in Computational Intelligence, vol. 482. Springer, Heidelberg (2013). https://doi.org/10.1007/978-3-642-37838-6

29. Salhi, S., Boylan, J. (eds.): The Palgrave Handbook of Operations Research. Studies in Computational Intelligence, p. 905. Palgrave, Cham (2022)

Variable Neighborhood Search Approach for the Bi-criteria Competitive Location and Design Problem with Elastic Demand

Tatiana Levanova[1]([✉]) [ID], Alexander Gnusarev[1] [ID], Ekaterina Rubtsova[2] [ID], and Sigaev Vyatcheslav[3] [ID]

[1] Sobolev Institute of Mathematics, Omsk Division Pevtsova str. 13, 644043 Omsk, Russia
levanovat@gmail.com
[2] Dostoevsky Omsk State University, Prospekt Mira 55A, 644077 Omsk, Russia
[3] Avtomatika-Servis LLC, Omsk, Russia

Abstract. In this paper, we develop a bi-criteria approach to solving the competitive location and design problem with elastic demand. The problem involves a new company, its competitor, and its consumers. The competitor has already placed its enterprises. The new company can choose the locations and the design variants for its facilities within the budget. Consumers independently choose service points from the open facilities of the company or competitor based on their preferences. The goal of the new company is to capture the largest possible share of the total demand. This situation is described using a non-linear integer programming model. In real situations, demand data and other parameters may change. In this case, it is necessary to make a decision that would be stable regarding such changes. We consider one of the concepts of robustness and formulate a new bi-criteria statement of the problem under consideration. In addition to the criterion that maximizes the share of the total demand, it also contains a criterion that maximizes the robustness of the solutions obtained. To solve the bi-criteria problem, we propose an algorithm based on variable neighborhood search and a modified version of the SEMO evolutionary algorithm. The features of the problem and the presence of two criteria are taken into account. Experimental studies have been carried out. The quality of the solutions obtained is analyzed, and a comparison with previous developments is discussed.

Keywords: Robustness · Bi-criteria optimization · Integer programming · Competitive location problem · Variable neighborhood search

This research was supported by state task of the IM SB RAS, project FWNF-2022-0020.

M. Khachay et al. (Eds.): MOTOR 2023, CCIS 1881, pp. 243–258, 2023.
https://doi.org/10.1007/978-3-031-43257-6_19

1   Introduction

Often, when constructing mathematical models, it is necessary to exclude some circumstances of the problem from attention and consider its characteristics unchanged. In real-world applications, ignoring the possibility of data variability can make the optimal solution to the problem unacceptable from a practical point of view. In such situations, the Company may prefer to obtain a solution to the problem in which the presence of some uncertainties will be taken into account in advance. Then there is a need to clarify the concept of optimality, subject to possible changes in the data. The literature has suggested various approaches to optimization problems with different kinds of uncertainties, for discrete location problems [5,8,19,28,31,34]. For example, in the article [32], the authors distinguish between the concepts of risk and uncertainty. In risk situations, there are uncertain parameters whose values are known and determined by probability distributions. Such problems are the subject of stochastic optimization; the general goal is to optimize the mathematical expectation of the value of the objective function [8,16,34]. These studies have been conducted for a long time (see, for example, [28,31,34]). The literature on this approach for location problems is quite wide. For example, the study [7] examines a capacitated facility location problem with uncertain demand. It is assumed that the demands are independent and equally distributed random variables with an arbitrary distribution. The article [35] discusses a situation that may arise when preparing supplies in case of a natural disaster. In such cases, the amount of possible demand and the availability of the transport network cannot be accurately determined. The authors describe the problem using the stochastic facility location model and propose a new matheuristic for its solution. In [16], a stochastic model for a discrete location problem is considered. In order to realize random demand, in addition to the usual stage, the second stage of placement is created. The loss quantile is used as the criterion function of the model. Research in this direction can also be found, for example, in [4,29].

In cases where parameters can change unpredictably and information about them is unknown, system performance is optimized in the worst case. Such tasks are referred to as robust optimization. Robustness can be interpreted as a measure of the flexibility of the solution to achieve almost optimal values under conditions of uncertainty. There are various definitions of robustness. In one of them, a set of uncertainties is used to represent a range of parameter changes in robust optimization problems instead of probabilistic information. For example, there are sets of uncertainties bounded by a rectangle or an ellipsoid [5,6]. The robust variant of the single-source capacitated facility location problem with uncertainty in customer demand using the Danzig-Wolfe decomposition was proposed in [33] and a branch and price algorithm was developed. Authors of [4] introduced the almost robust discrete optimization. This methodology allows us to take into account uncertainties in models with Boolean variables.

The concept of a stability radius is closely related to robustness. Perhaps the first descriptive definition of the stability radius for distances in the traveling salesman problem is given in [21]. Here, the analytical expression of the radius is written out, its properties are described, and some estimates and other theoretical results are given. Later, the authors continued their research and presented a general approach to finding the stability radius of solutions to combinatorial optimization problems [14]. In [27], the authors analyze the relationship between the robustness radius for continuous optimization problems and the mixed-integer linear problem. The applications of the radius of robust feasibility of mixed-integer problem for the facility location are described in [10].

In this paper, we consider one of the approaches to determining the robustness of a solution in continuous location problems, also known in the literature as threshold robustness [10,11]. We study the possibility of generalizing this concept for the case of discrete location problems using the example of a competitive facility location and design problem with elastic demand. In our paper, we will call the initial version of the problem the deterministic location problem (DLP). The paper considers the bicriteria version of this discrete problem, in which, in addition to the main criterion, there is an additional criterion for the robustness of the searching solution. The formulation of DLP in the form of a non-linear bicriteria integer programming problem is proposed. As a solution method, we develop the variable neighborhood search approach.

The paper consists of an introduction and three sections. Section 2 presents models of deterministic and robust variants of the Capacitated Location and Design Problem with Elastic Demand. Notation is introduced and a mathematical model of the original problem is written out. A single-criteria formulation of a robust variant of the problem (RLP) is presented. A new mathematical model is proposed for solving the bi-criteria robust capacitated facility location and design problem with elastic demand (BRLP). In Sect. 3, a new original algorithm for local search with alternating neighborhoods and a modified version of the SEMO evolutionary algorithm are presented, taking into account the need to optimize both the reliability criterion and the maximum income. Section 4 presents test series for experimental analysis of the algorithms, describes the results of numerical experiments, and discusses them. Finally, the prospects of the approach are discussed.

2 Bi-criteria Robust Competitive Facility Location and Design Problem with Elastic Demand

We propose a new bi-criteria version for the following facility location and design problem with elastic demand. The situation is considered when a new company tries to penetrate the goods market. Competitor facilities are already operating in the market that serves the demand of customers. The goal of the new company is to capture as much of the customer's demand as possible. To do this, the company can choose the location and design variants for their facilities, considering the necessary costs and the available budget. Customers independently determine the points of satisfaction of demand by choosing a facilities of a new company or of a competitor.

To write out the model, we introduce the following notation. Let N, C, S, R be the set of customers, set of locations of the competitor's facilities, sets of points of possible locations, and design variants of the new company's facilities, respectively. The coefficients k_{ijr} involve the attractiveness of facilities to customers, the distance to them, and sensitivity to this distance, $i \in N, j \in S, r \in R$. Each customer's demand w_i is also known, $i \in N$. The new company can locate its facilities, taking into account the available budget B and the necessary costs c_{ij} for opening facilities $j \in S$ with the design variants $r \in R$.

Let's agree that if the facility is located in point j, then we will say that the facility i is open. Boolean variables x_{jr} indicate whether or not facility j is open with a design variant r, $j \in S, r \in R$. The benefit of the facility i for client j is $u_{ij} = \sum_{r=1}^{R} k_{ijr} x_{jr}$. Denote U_i the total utility from all located facilities for a customer $i \in N$; MS_i is a new company's total share. Given the notation, the mathematical model looks like this:

$$\max \sum_{i \in N} w_i \cdot g(U_i) \cdot MS_i, \tag{1}$$

$$\sum_{j \in S} \sum_{r \in R} c_{jr} x_{jr} \leq B, \tag{2}$$

$$\sum_{r \in R} x_{jr} \leq 1, j \in S, \tag{3}$$

$$x_{jr} \in \{0, 1\}, \quad r \in R, j \in S. \tag{4}$$

The goal of the new company to capture the maximum share is reflected using the function (1). Constraint (2) allows you to take into account the budget. Inequalities (3) allow choosing only one design variant.

Such a model for a capacitated facility location and design problem with elastic demand was proposed by R. Aboolian, O. Berman, and D. Krass (for more details, see [3]). They used rules from marketing to describe the demand function. More specifically the demand function is $g(U_i) = 1 - \exp\left(-\lambda_i U_i\right)$, where λ_i is the characteristic of elastic demand in point i, $\lambda_i > 0$; The total utility for a customer at $i \in N$ from all open facilities is calculated by the formula:

$$U_i = \sum_{j \in S} \sum_{r=1}^{R} k_{ijr} x_{jr} + U_i(C) = U_i(S) + U_i(C).$$

The company's total share of facility $i \in N$ is

$$MS_i = \frac{U_i(S)}{U_i(S) + U_i(C)} = \frac{\sum_{j \in S} \sum_{r=1}^{R} k_{ijr} x_{jr}}{\sum_{j \in S} \sum_{r=1}^{R} k_{ijr} x_{jr} + \sum_{j \in C} u_{ij}}.$$

In terms of the notation introduced, the goal function (1) looks like this:

$$\max \sum_{i \in N} w_i \left(1 - \exp\left(-\lambda_i \left(\sum_{j \in S} \sum_{r=1}^{R} k_{ijr} x_{jr} + U_i(C)\right)\right)\right) \tag{5}$$

$$\times \left(\frac{\sum_{j \in S} \sum_{r=1}^{R} k_{ijr} x_{jr}}{\sum_{j \in S} \sum_{r=1}^{R} k_{ijr} x_{jr} + \sum_{j \in C} u_{ij}} \right).$$

In the model (5), (2)–(4) all parameters are known and unchangeable. In a real economic situation, the demand, the number of consumers, the volume of production, and others change over time. We will consider a situation where demand may decrease. In this case, the company wants to find out before what change in demand it is profitable for it to locate its facilities. Earlier, we formulated a one-criterion robust statement for a problem with elastic demand [22]. To write it out, we will additionally introduce the minimum allowable volume of demand W; variable ρ is robustness (stability radius of demand); variables γ_i are the deviation from the set volume of demand w_i in the point $i \in N$.

The one-criteria robust capacitated facility location and design problem with elastic demand looks as follows [22]:

$$\max \rho, \tag{6}$$

$$\rho \leq \gamma_i, i \in N, \tag{7}$$

$$\sum_{i \in N} (w_i - \gamma_i) \cdot g(U_i) \cdot MS_i \geq W, \tag{8}$$

$$\gamma_i \leq w_i, i \in N, \tag{9}$$

$$\sum_{j \in S} \sum_{r \in R} c_{jr} x_{jr} \leq B, \tag{10}$$

$$\sum_{r \in R} x_{jr} \leq 1, j \in S, \tag{11}$$

$$x_{jr} \in \{0, 1\}, \rho \geq 0, \gamma_i \geq 0, j \in S, r \in R. \tag{12}$$

In this paper, we consider a bi-criteria formulation of the robust capacitated facility location and design problem with elastic demand (BRLP) in which, in addition to the main criterion that maximizes the share of demand served, there is also a criterion that maximizes the robustness of the solutions obtained. Thus, the robust version of the presented discrete location problem can be written as the following bicriteria nonlinear integer programming problem:

$$\max \rho, \tag{13}$$

$$\max \sum_{i \in N} (w_i - \gamma_i) \tag{14}$$

$$\times \left(1 - \exp \left(-\lambda_i \left(\sum_{j \in S} \sum_{r \in R} k_{ijr} x_{jr} + U_i(C) \right) \right) \right)$$

$$\times \left(\frac{\sum_{j \in S} \sum_{r \in R} k_{ijr} x_{jr}}{\sum_{j \in S} \sum_{r \in R} k_{ijr} x_{jr} + U_i(C)} \right),$$

$$\rho \leq \gamma_i, i \in N, \tag{15}$$

$$\gamma_i \leq w_i, i \in N, \tag{16}$$

$$\sum_{j \in S} \sum_{r \in R} c_{jr} x_{jr} \leq B, \tag{17}$$

$$\sum_{r \in R} x_{jr} \leq 1, j \in S, \tag{18}$$

$$x_{jr} \in \{0,1\}, \rho \geq 0, \gamma_i \geq 0, j \in S, r \in R. \tag{19}$$

To solve the robust problem (13)–(19), the original version of the Variable Neighborhood Search (VNS) algorithm is proposed in the next section.

3   Algorithms for the Bi-criteria Location Problem

Since in the general case, the search for the exact value of the criterion W is not possible [24], we construct an approximate Pareto boundary. The aim of the research was to construct an approximation of the boundary in a new way, using the idea of VNS, and compare the results obtained with the well-known algorithm. To do this, we propose an algorithm based on variable neighborhood search and a modified version of the SEMO evolutionary algorithm [20]. Let a criterion vector function (ρ, W) be given on the set of admissible solutions D, given by constraints (15)-(19). At each iteration of the algorithm SEMO, a new non-dominant pair (ρ, W) is searched for. The result is a set of solutions forming an approximation of the Pareto boundary. The RVNS algorithm finds a sequence of solutions, each of which dominates the previous one. Thus, the algorithm gets one pair (ρ, W) in one run, while ρ improves for the selected W. This implementation is an analog of the method of successive concessions when the main criterion is ρ. Let's describe our developments in more detail.

Simple Evolutionary Multiobjective Optimizer. Introduce a dominance relation for approximate values similar to the classical definition. Let's denote it by $\succ'$, and the corresponding Pareto boundary by F'. Next, we will look for an approximation of the Pareto boundary using the Simple Evolutionary Multiobjective Optimizer (SEMO) [20] algorithm. In the process of the SEMO algorithm at each iteration, a parent individual is randomly selected from a population that contains pairwise non-dominant solutions (individuals). Next, we get a descendant from the parent individual through the mutation operation, which we add to the population if there are no dominant individuals or individuals with the same value of the vector criterion in it. All individuals of the population that are dominated by a descendant are removed. The result of the algorithm is a population calculated by the end of its execution. The scheme of the SEMO algorithm is presented below (see Algorithm 1).

Reduced Variable Neighborhood Search Algorithm. Local search algorithms have taken a firm position among modern methods of approximate solutions. Often, to solve a real-world problem, instead of a time-consuming optimal solution, it is enough to find an approximate solution of good quality in a relatively short time. To do this, it is reasonable to use local search algorithms. Tabu search, simulated annealing, greedy randomized adaptive search

---

Algorithm 1: SEMO algorithm

Generate a solution x randomly and put $\Pi := \{x\}$;
while *the stop criterion has not been met* **do**
$\quad \Pi := \Pi \backslash \{z \in \Pi | x' \succ' z\}$;
$\quad$ **if** $\not\exists z \in \Pi$, *such as* $z \succ' x'$ *or* $(V(z), B(z), Q(z)) = (V(x'), B(x'), Q(x'))$
$\quad$ **then**
$\quad\quad \lfloor \; \Pi := \Pi \cup \{x'\}$
return *population* Π

---

procedure, iterated local search, population-based metaheuristics, and others are widely used for decision-making in various actual situations [2,9,23,25,26]. The Variable Neighborhood Search (VNS) approach is one of such modern methods. Mladenović and Hansen offered this approach for the traveling salesman problem [30]. Currently, we can say that this is a well-known method, it is successfully used to solve complex combinatorial problems such as p-median problem, competitive facility location problem, etc.), scheduling problems, and many others [12,13,15,17,18,24,36]. Depending on whether they use all the steps or not, the authors form various variants of the basic algorithm. In this paper, we propose Reduced Variable Neighborhood Search (RVNS) algorithm to solve the bi-criteria robust facility location and design problem with elastic demand.

The neighborhoods system is a basis of a Variable Neighborhood Search approach (VNS). The construction of special-type neighborhoods is a mandatory part of the algorithm development. In this paper, for the VNS algorithm, the neighborhood types N_{move} and N_{add} of a special form are used, taking into account the features of the location and design problem [22,24]. As a result of the transition to a new solution in the neighborhood N_{move}, we transfer one open facility from a certain point to another. As a result of the application of the neighborhood N_{add}, the two facilities have a different design variant.

Let the integer vector $x = (x_{j_r})$ be such that x_{j_r} corresponds to design variant r of facility in point j as follows: $x_{j_r} = r \Leftrightarrow x_{jr} = 1, j \in S, r \in R$. When the location is selected, i.e. the value of r is defined for each j, then we can calculate the current share of consumer demand served by the company using (14):

$$\overline{W} = \sum_{i \in N} (w_i - \gamma_i) \left(1 - \exp\left(-\lambda_i \left(U_i(S) + U_i(C)\right)\right)\right) \left(\frac{U_i(S)}{U_i(S) + U_i(C)}\right).$$

The function $\Delta(x, \gamma) = \overline{W} - W$ shows the deviation of the value of the share of demand $\overline{W}$ obtained by the algorithm from the minimum allowable volume of demand W. The larger the value of $\Delta(x, \gamma)$, the better the solution is in terms of the company's revenue. We will call a Record the best-found value $\Delta(x, \gamma)$ and the corresponding vectors x and γ.

In the Robust Location and Design Problem with elastic demand, there are $(|N| + 1)$ more variables: ρ is robustness, γ_i is a deviation from $w_i, i \in N$.

The robustness ρ depends on γ and is determined as a least from all deviations $\gamma_i, i \in N$. In our previous work [22] we used the following equation to calculate ρ for a one-criterion robust problem (6)–(12): $\rho(x) = \frac{W \cdot \sum_{i \in N} w_i - W}{|N| \cdot W}$.

In this paper, we will allow the components of the vector γ to take different values in order to be able to select the values of the vector x. As the computational experiment below has shown, such tactics have made it possible to improve not only the reliability criterion (13) but also to increase the share of serviced demand (14). To implement this idea in the variable neighborhood search algorithm, we proposed a special neighborhood for γ.

Definition 1 (Neighborhood $N_{k-iflip}$). *A feasible vector $\gamma\prime$ is called a neighbor for the vector γ in the neighborhood of N_{k-flip} if it can be obtained as follows: select K of the various components of the vector γ and change them according to the rules:*

a) *if $\gamma_k = 0$, then put $\gamma_k' := \gamma_k + 1$;*
b) *if $(\gamma_k = 1)\&(w_k > 1)$, then with probability 0.001 put $\gamma_k' := \gamma_k - 1$, with probability 0.999 put $\gamma_k' := \gamma_k + 1$;*
c) *if $(1 < \gamma_k)\&(w_k > 1)$, then with probability 0.1 put $\gamma_k' := \gamma_k + 1$, with probability 0.9 put $\gamma_k' := \gamma_k - 1$;*
d) *if $\gamma_k = w_k$, with probability 0.9 put $\gamma_k' := \gamma_k - 1$.*

To organize the search for a solution to the bi-criteria problem (13)–(19) by the RVNS algorithm, the following sequence of neighborhoods is selected: $N_1 = N_{k-iflip}$, $N_2 = N_{add}$, $N_3 = N_{move}$. In the Reduced Variable Neighborhood Search algorithm (RVNS) the initial values x will be determined as the best-known solution of the problem (1)–(4). The number of iterations will be the stopping criteria. Denote the iteration number for t, and the maximum number of iterations for T, the number of neighborhood variants for V.

The idea of our algorithm is as follows. We are looking for a pair (x, γ). First, we try to increase part of the company's share of customer demand by changing the vector γ. If this is no longer possible, then we are trying to improve the share of customer demand by changing x, using the idea of looking at neighborhoods as in the VNS approach. The scheme of the Reduced Variable Neighborhood Search algorithm is the following (see Algorithm 2). Information about the results of experimental studies can be found in the next section.

4   Computational Experiments

Experimental studies were conducted in order to compare the results of the proposed RVSN with the SEMO algorithm and the commercial LocalSolver. For this purpose, test cases were created based on a series of test instances for a deterministic facility location and design problem with elastic demand. These examples of the DLP turned out to be difficult for well-known solvers CoinBonmin and Baron [24]. The W parameter has been added to the test instances. It was equal to $W = 0.5 \cdot rec, 0.6 \cdot rec, \ldots, 0.9 \cdot rec$, where rec is the

Algorithm 2: Reduced Variable Neighborhood Search for Bi-Criteria Competitive Location and Design Problem with Elastic Demand

---

Initialization.
Select a set of neighborhood structures N_v, $v = 1, \ldots, V$, that will be used in the search;
Find the initial solution of x by solving the problem (1)–(4);
Set $\gamma = 0$, $\rho_{Rec} = 0$, $Rec := \Delta(x, \gamma)$, $t := 1$;
Step 1.
if $t \leq T$ **then**
    ⌊ set neighborhood number $v := 1$

end of RVNS;
Step 2. (Shaking γ.) Generate a point $\gamma' \in N_v$;
Step 3. (Move or not γ.);
if $\Delta(x, \gamma') > Rec$ **then**
    ⌊ $Rec := \Delta(x, \gamma')$, $\gamma := \gamma'$, $\rho_{Rec} = \min_{i \in N} \gamma_i$, go to step 2;

go to step 4;
Step 4. Until $v \leq V$ repeat the following steps;
    Step 4.1. (Shaking x.) Generate a point $x' \in N_v$;
    Step 4.2. (Move or not x.) **if** $\Delta(x', \gamma) > Rec$ **then**
    ⌊ $Rec := \Delta(x', \gamma)$, $x := x'$, $v := 2$.

Move on to the neighborhood with the next number $v := v + 1$. Go to step 4
Step 5. Set $t := t + 1$. Go to step 1.

---

maximum value of the DLP objective function. The *rec* parameter was calculated by LocalSolver [1]. The number of points was equal to 60, 80, 100, 150, 200, and 300. There are three possible design options with opening costs equal to 1, 2, and 3, respectively. Four budget options were considered, its cost is 3, 5, 7, and 9. Thus, a total of 480 test copies were created. All points are located in a rectangle of size 100 by 150, the distances between the points satisfy the triangle inequality. Consumer demand is elastic ($\lambda = 1$), and consumers are highly sensitive to the distance to service points ($\beta = 2$).

The algorithm is tested on the computer Intel (R) Xeon (R) CPU X5675 @ 3.07 GHz with 96.0 Gb RAM. The algorithms were run 50 times on each of the instances.

It is interesting to note that in the test instances under consideration, the Pareto boundary contains a small number of points. The RVNS and SEMO algorithms showed similar results. For example, for m = 60 SEMO founds a solution (1.26.36), and RVNS founds a solution (1.26.09), in other cases (1,51.78) and (1,52.50), respectively. The CPU time of both algorithms does not exceed 1 s.

We conducted experimental studies comparing VNS and LocalSolver in two stages. First, each of the algorithms performed calculations until they met the stopping criterion. The variable neighborhood search stopped working when it couldn't find a better solution than what it had already found. The counting time of the Local Solver was selected taking into account previous experiments for

DLP and was 20 s. In the second part of the experimental studies, the proposed algorithm and the solver were put under the same conditions in terms of counting time.

We use the following notations in the tables: *dimension* is the number of points of demand; ρ_{RVNS} is a result of RVNS; ρ_{LS} is a Local Solver result; $\rho_{0.5}$ is the value of the ρ in case $W = 0.5 \cdot rec$; error of the results calculated by the formula $error\rho = \frac{\rho_{RVNS} - \rho_{LS}}{\rho_{LS}}$. For the values of W, the notation is similar.

Describe the first stage of an experimental study in which the algorithms worked up to the stop criterion. In Table 1 we give a deviation of the objective function calculated by RVNS from the objective function which calculates LocalSolver. The table shows that the closer the threshold value (W) approaches the optimal, the more the deviation of the objective function grows. In the case where $W = 0.9 \cdot rec$ for all examples of the objective function of the RVNS algorithm and the solver was equal to 0, i.e. no chance to change w_i. The closer the W value is to the optimal DLP value, the less w_i can be changed.

Table 1. Average deviation ρ from LocalSolver, up to stop criteria.

dimension	$error\rho_{0.5}$	$error\rho_{0.6}$	$error\rho_{0.7}$	$error\rho_{0.8}$	$error\rho_{0.9}$
60	2.375	2.875	4.250	9.250	0.000
80	5.625	2.875	4.375	8.750	0.000
100	4.375	6.875	5.250	11.750	0.000
150	7.500	8.625	9.000	21.625	0.000
200	10.250	11.125	12.875	26.000	0.000
300	14.125	14.125	0.000	0.000	0.000

As indicated above $\overline{W}$ is the current share of customer demand served by the Company (14). Deviation $\overline{W}$ of the RVNS algorithm from LocalSolver is given in Table 2. It can be seen from the table that the greatest deviations are achieved in the examples $W = 0.5 \cdot rec$ and they gradually grow to $W = 0.9 \cdot rec$. Negative values mean that the $\overline{W}$ calculated by the RVNS was better than the LocalSolver.

Reduced Variable Neighborhood Search algorithm average running time is rather short and is less than 1 s. The LocalSolver solver time was 20 (standard solver time). It can be seen that the running time of the algorithm grows with the growth of the dimension.

The 95% confidence interval for the probability of obtaining the solution of RVNS better than LocalSolver is given in Table 3. It can be seen from the table that, as in the case with the deviation $\overline{W}$, in the examples $W = 0.5 \cdot rec$ the probability of improving the solver's record is the highest and it gradually decreases to $W = 0.9 \cdot rec$. This is also explained by the fact that the values of the objective function of the algorithm and the solver do not differ in the examples of the series $W = 0.9 \cdot rec$.

Table 2. Average deviation $\overline{W}$ from LocalSolver, up to stop criteria.

Dimension	$errorW_{0.5}$	$errorW_{0.6}$	$errorW_{0.7}$	$errorW_{0.8}$	$errorW_{0.9}$
60	-45.981	-24.912	-8.295	-0.883	1.126
80	-49.807	-26.664	-9.473	-0.517	1.923
100	-45.965	-25.367	-8.958	-0.302	2.932
150	-40.719	-23.634	-8.103	0.317	1.533
200	-44.148	-23.486	-8.557	-0.497	2.074
300	-39.910	-16.452	-5.071	0.657	1.410

Table 3. The 95% confidence interval for the probability of obtaining the solution of RVNS better than LocalSolver.

N	$W = 0.5rec$	$W = 0.6rec$	$W = 0.7rec$	$W = 0.8rec$	$W = 0.9rec$
60	$[0.995; 0.995]$	$[0.991; 0.992]$	$[0.878; 0.880]$	$[0.623; 0.625]$	$[0.344; 0.346]$
80	$[0.994; 0.993]$	$[0.996; 0.997]$	$[0.922; 0.923]$	$[0.591; 0.594]$	$[0.097; 0.098]$
100	$[1.000; 1.000]$	$[0.991; 0.992]$	$[0.889; 0.8901]$	$[0.491; 0.494]$	$[0.108; 0.110]$
150	$[0.987; 0.988]$	$[0.990; 0.990]$	$[0.924; 0.9256]$	$[0.450; 0.453]$	$[0.077; 0.078]$
200	$[0.987; 0.988]$	$[0.968; 0.969]$	$[0.918; 0.919]$	$[0.563; 0.565]$	$[0.098; 0.1000]$
300	$[0.975; 0.975]$	$[0.911; 0.912]$	$[0.799; 0.801]$	$[0.354; 0.356]$	$[0.030; 0.031]$

The superiority of the Reduced Variable Neighborhood Search algorithm over the solver can also be seen in the boxplot (Fig. 1 and 2). The numbers on the horizontal axis indicate the dimensions of the problems. The vertical axis shows deviations in the percentage of RVNS results from LocalSolver results. It can be seen from the graph that for the case $W = 0.5 \cdot rec$ all boxes are located below 0, which means that in most cases the solver's record has been improved. At the same time, there are not so many emissions that would be greater than 0. The boxes are not elongated, this means that the deviations are rather densely grouped relative to the median value. For comparison with RVNS for one-criteria robust problem [22], from Fig. 2 it can be seen that for the case $W = 0.5 \cdot rec$, all boxes are located above 0. It means that a new algorithm receives robust results with a better value of a share of customers' demand. For the case $W = 0.9 \cdot rec$, the results show insignificant deviations from the LocalSolver (Fig. 2). The most densely grouped deviations are for dimensions 60, 80, 150, and 300.

Next, we put the algorithms in the same conditions in terms of counting time. Firstly the RVNS algorithm and the Local Solver were given a running time equal to 1 s. In these conditions, the deviation W does not change in comparison with the results that were given earlier in the first part of the experimental studies, but the deviation ρ is getting better. The results show that in this case, the RVNS more often finds the value of the objective function the same as that of the solver. For example, all dimensions at $W = 0.8 \cdot rec$ and $W = 0.9 \cdot rec$ are the same. It should be noted that the algorithm wins more significantly than the solver in

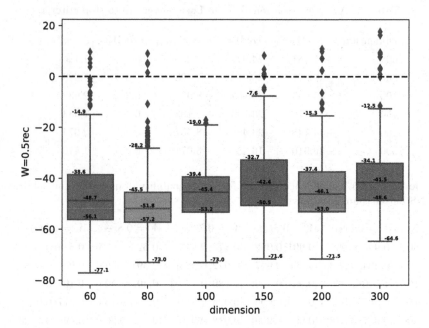

Fig. 1. Statistical analysis for $W = 0.5rec$. for BRLP.

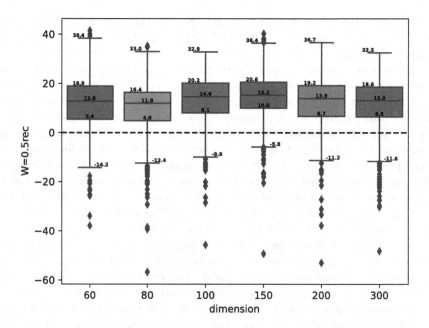

Fig. 2. Statistical analysis for $W = 0.5rec$. for RLP.

terms of the satisfied demand. Then the Reduced Variable Neighborhood Search algorithm and the Local Solver were given a runtime of 20 s. Since in Part 1 of the experiments the greatest difference in the behavior of the algorithms was observed for $W = 0.9 \cdot rec$ and $W = 0.5 \cdot rec$, we will analyze only these test instances. For the case $W = 0.9 \cdot rec$, the deviation of the objective function does not change, for the $W = 0.5 \cdot rec$ it slightly increases. At the same time, on a series of $W = 0.5 \cdot rec$, the deviation of W is not as significant as in Part 2a when the algorithm is running for 1 s, and with $W = 0.9 \cdot rec$, it practically does not differ (see Table 4). For experiments with CPU time at 1 s the deviation of W are the same.

Table 4. Average deviation ρ and $\overline{W}$ from LocalSolver, same time, 20 s.

Dimension	$error\rho_{0.5}$	$errorW_{0.5}$	$error\rho_{0.9}$	$errorW_{0.9}$
60	4.500	−24.912	0.000	1.718
80	5.375	−26.835	0.000	2.169
100	7.250	−25.082	0.000	3.273
150	11.125	−22.445	0.000	0.317
200	16.875	−27.503	0.000	2.063
300	23.000	−31.386	0.000	1.389

In general, computational experiments show that in the cases $W = 0.5 \cdot rec$, $W = 0.6 \cdot rec$, $W = 0.7 \cdot rec$, the algorithm is slightly inferior to the solver in terms of the value of the objective function. At the same time for 1 s, the deviation will be less than if you give them 20 s. This means that this time is not enough for the solver to find good solutions for the value of the objective function. However, if we compare the satisfied demand, then with the examples $W = 0.5 \cdot rec$, $W = 0.6 \cdot rec$, $W = 0.7 \cdot rec$, the algorithm significantly outperforms the solver. Using the examples $W = 0.8 \cdot rec$, $W = 0.9 \cdot rec$, the algorithm, and the solver practically do not differ in the value of the objective function, and the results are comparable in terms of satisfied demand regardless of time. From the point of view of the Company, it can get a solution with the help of the RVNS, if the requirements for the share of demand are relaxed to half of the optimal values of the non-robust problem. If the Company plans to find a stable solution close to the optimal non-robust solution, then the probability of this is small.

5   Conclusions

In this paper, we develop a threshold robustness approach to discrete location problems. We are building a new bi-criteria variant of a competitive location and design problem with elastic demand, in which, in addition to the main criterion, there is a criterion for the reliability of the search solution. We have provided an original version of the variable neighborhood search algorithm with neighborhood systems adjusted to the problem we study. For a comparative analysis of the RVNS results, a modified version of the SEMO evolutionary algorithm was implemented. To conduct numerical experiments, a series of test instances was created by analogy with real data. The values obtained by the well-known LocalSolver were chosen as reference values.

It should be noted that the proposed Reduced Variable Neighborhood Search algorithm shows results comparable to the well-known SEMO evolutionary algorithm. An important characteristic is that the RVNS is quite fast. It finds values better than the LocalSolver given the same amount of time. There is a tendency that the closer the threshold value is to the optimal solution of a non-robust problem, the more difficult it is for the algorithm to find a good solution. This shows an increase in deviations from the results of the Localsolver. Note that in this case, the RVNS exceeds the solver in terms of the share of demand served.

It seems promising to continue research to study the properties of the bicriteria problem and construct other special algorithms for solving it.

References

1. Localsolver. http://www.localsolver.com. Accessed 30 June 2021
2. Local Search in Combinatorial Optimization. Princeton University Press (2003). http://www.jstor.org/stable/j.ctv346t9c
3. Aboolian, R., Berman, O., Krass, D.: Competitive facility location and design problem. Eur. J. Oper. Res. **182**(1), 40–62 (2007)
4. Baron, O., Berman, O., Fazel-Zarandi, M., Roshanaei, V.: Almost robust discrete optimization. Eur. J. Oper. Res. **276**(2), 451–465 (2019). https://doi.org/10.1016/j.ejor.2019.01.043
5. Baron, O., Milner, J., Naseraldin, H.: Facility location: a robust optimization approach. Prod. Oper. Manag. **20**(5), 772–785 (2011)
6. Ben-Tal, A., Nemirovski, A.: Robust convex optimization. Math. Oper. Res. **23**(4), 769–805 (1998). https://doi.org/10.1287/moor.23.4.769
7. Bieniek, M.: A note on the facility location problem with stochastic demands. Omega **55**, 53–60 (2015). https://doi.org/10.1016/j.omega.2015.02.006
8. Birge, J.: State-of-the-art-survey - stochastic programming: computation and applications. INFORMS J. Comput. **9**, 111–133 (1997). https://doi.org/10.1287/ijoc.9.2.111
9. Blum, C., Eremeev, A., Zakharova, Y.: Hybridizations of evolutionary algorithms with large neighborhood search. Comput. Sci. Rev. **46**, 100512 (2022). https://doi.org/10.1016/j.cosrev.2022.100512, https://www.sciencedirect.com/science/article/pii/S1574013722000466
10. Carrizosa, E., Nickel, S.: Robust facility location. Math. Methods Oper. Res. **58**(2), 331–349 (2003). https://doi.org/10.1007/s001860300294
11. Carrizosa, E., Ushakov, A., Vasilyev, I.: Threshold robustness in discrete facility location problems: a bi-objective approach. Optim. Lett. **9**, 1297–1314 (2015)
12. Djenić, A., Radojičić, N., Marić, M., Mladenović, M.: Parallel VNS for bus terminal location problem. Appl. Soft Comput. **42**, 448–458 (2016). https://doi.org/10.1016/j.asoc.2016.02.002

13. Eremeev, A., Gette, A., Hrushev, S., Levanova, T.: Location and design of ground stations for software defined satellite networks. In: 2022 Dynamics of Systems, Mechanisms and Machines (Dynamics), pp. 1–4 (2022). https://doi.org/10.1109/Dynamics56256.2022.10014916
14. Gordeev, E.N., Leontev, V.K.: A general approach to the study of the stability of solutions in discrete optimization problems. Comput. Math. Math. Phys. **1**(36), 66–72 (1996)
15. Hansen, P., Mladenović, N.: Variable neighborhood search: principles and applications. Eur. J. Oper. Res. **130**(3), 449–467 (2001). https://doi.org/10.1016/S0377-2217(00)00100-4
16. Ivanov, S., Akmaeva, V.: Two-stage stochastic facility location model with quantile criterion and choosing reliability level. Vestn. YuUrGU Ser. Mat. Model. Program. **14**(3), 5–17 (2021)
17. Kononova, P., Kochetov, Y.: The variable neighborhood search for the two machine flow shop problem with a passive prefetch. J. Appl. Ind. Math. **7**(1), 54–67 (2013)
18. Kovač, A., Davidović, T., Stanimirović, Z.: Variable neighborhood search methods for the dynamic minimum cost hybrid berth allocation problem. Inf. Technol. Control **47**(3), 471–488 (2018)
19. Laporte, G., Nickel, S., Gama, F.: Location Science. Springer, Cham (2015). https://doi.org/10.1002/9780470258354
20. Laumanns, M., Thiele, L., Zitzler, E., Welzl, E., Deb, K.: Running time analysis of multi-objective evolutionary algorithms on a simple discrete optimization problem. In: Guervós, J.J.M., Adamidis, P., Beyer, H.-G., Schwefel, H.-P., Fernández-Villacañas, J.-L. (eds.) PPSN 2002. LNCS, vol. 2439, pp. 44–53. Springer, Heidelberg (2002). https://doi.org/10.1007/3-540-45712-7_5
21. Leontev, V.K.: Stability of the travelling salesman problem. Comput. Math. Math. Phys. **5**(15), 199–213 (1975)
22. Levanova, T., Gnusarev, A., Rubtsova, E.: On the robust capacitated facility location and design problem with elastic demand. Int. J. Artif. Intell. **21**(2), 93–108 (2023)
23. Levanova, T., Khmara, I.: A local search algorithm for the biclustering problem. In: Burnaev, E., et al. (eds.) AIST 2021. LNCS, vol. 13217, pp. 330–344. Springer, Cham (2022). https://doi.org/10.1007/978-3-031-16500-9_27
24. Levanova, T.V., Gnusarev, A.Y.: Variable neighborhood search algorithms for a competitive location problem with elastic demand. J. Appl. Ind. Math. **14**(4), 693–705 (2020)
25. Levanova, T.V., Belan, S.E.: Local search algorithm for two-stage problem of radio communication systems planning. J. Phys.: Conf. Ser. 1260(8), 082002 (2019). https://doi.org/10.1088/1742-6596/1260/8/082002, http://dx.doi.org/10.1088/1742-6596/1260/8/082002
26. Levanova, T., Gnusarev, A.: Development of threshold algorithms for a location problem with elastic demand. In: Lirkov, I., Margenov, S. (eds.) LSSC 2017. LNCS, vol. 10665, pp. 382–389. Springer, Cham (2018). https://doi.org/10.1007/978-3-319-73441-5_41
27. Liers, F., Schewe, L., Thürauf, J.: Radius of robust feasibility for mixed-integer problems. Inf. J. Comput. (2021). http://www.optimization-online.org/DB_HTML/2019/05/7219.htm
28. Louveaux, F.: Discrete stochastic location models. Ann. Oper. Res. **6**(2), 21–34 (1986)

29. Melnikov, A., Beresnev, V.: Upper bound for the competitive facility location problem with quantile criterion. In: Kochetov, Y., Khachay, M., Beresnev, V., Nurminski, E., Pardalos, P. (eds.) DOOR 2016. LNCS, vol. 9869, pp. 373–387. Springer, Cham (2016). https://doi.org/10.1007/978-3-319-44914-2_30

30. Mladenovic, N., Hansen, P.: Variable neighborhood search. Comput. Oper. Res. **24**, 1097–1100 (1997)

31. Owen, S., Daskin, M.: Strategic facility location: a review. Eur. J. Oper. Res. **111**(3), 423–447 (1998). https://doi.org/10.1016/S0377-2217(98)00186-6

32. Rosenhead, J., Elton, M., Gupta, S.: Robustness and optimality as criteria for strategic decisions. Oper. Res. Quart. **23**(4), 413–431 (1972). https://doi.org/10.1057/JORS.1972.72

33. Ryu, J., Park, S.: A branch-and-price algorithm for the robust single-source capacitated facility location problem under demand uncertainty. EURO J. Transp. Logist. **11**, 100069 (2022). https://doi.org/10.1016/j.ejtl.2021.100069

34. Snyder, L.: Facility location under uncertainty: a review. IIE Trans. **38**(7), 547–564 (2006). https://doi.org/10.1080/07408170500216480

35. Turkeš, R., Sörensen, K., Cuervo, D.: A matheuristic for the stochastic facility location problem. J. Heurist. **27**(4), 649–694 (2021). https://doi.org/10.1007/s10732-021-09468-y

36. Čvokić, D., Kochetov, Y., Plyasunov, A., Savić, A.: A variable neighborhood search algorithm for the $r|p$ hub-centroid problem under the price war. J. Glob. Optim. **83**(3), 405–444 (2022). https://doi.org/10.1007/s10898-021-01036-9

Decomposition Approach
for Simulation-Based Optimization
of Inventory Management

Alexander Yuskov[1]([✉])[iD], Igor Kulachenko[2][iD], Andrey Melnikov[2][iD],
and Yury Kochetov[2][iD]

[1] Novosibirsk State University, Novosibirsk, Russia
a.yuskov@g.nsu.ru
[2] Sobolev Institute of Mathematics of Siberian Branch of Russian Academy
of Sciences, Novosibirsk, Russia
{ink,melnikov,jkochet}@math.nsc.ru

Abstract. We consider a two-echelon inventory management problem,
where customers' requests for spare parts of different types must be fulfilled within a given service level threshold. The supply system is composed of multiple warehouses in the first echelon, where the customers'
requests are processed, and a single second-echelon warehouse, replen-
ishing stocks of the first-echelon warehouses. Replenishment requests of
warehouses are invoked according to inventory policies, which are characterized by one or two numerical parameters and are individual for each
warehouse and each spare part type. The goal is to minimize the total
storage cost for all warehouses at both echelons. System operation is simulated within a black-box function that computes the request satisfaction
rate and inventory holding costs depending on the policy parameters. In
the work, we propose a decomposition approach to adjust these parameters for an industrial-sized supply system. Computational experiments
for up to 1,000 types of items and 100 warehouses are discussed.

Keywords: Grey-box optimization · Multiple-choice knapsack
problem · Local search

1 Introduction

Optimization of the supply chain is crucial for big manufacturers [4,13]. Inventory management aims to deal with the trade-off between customer satisfaction
and inventory holding costs. In practice, the arrival of customer orders is stochastic and difficult to control. Therefore, keeping spare parts in the warehouse is
necessary to satisfy the customers' demands in time but leads to an increase in
holding costs simultaneously.

The multi-echelon model for the first time appeared in the literature in the
paper of Lee [6]. Since then, there have been a number of theoretical studies of
the problem [1] as well as practical algorithms [14]. Some authors try to build a

M. Khachay et al. (Eds.): MOTOR 2023, CCIS 1881, pp. 259–273, 2023.
https://doi.org/10.1007/978-3-031-43257-6_20

mathematical model and solve it using heuristics, exact methods, or stochastic programming [15,16]. However, these algorithms either use greatly simplified models or suffer from computational complexity [10]. There is another method that has several advantages over classical optimization algorithms: a simulation-based approach. It uses a simulator that takes account of all features of the supply chain. In that case, an optimization algorithm uses the simulator as a black-box function and optimizes input parameters. Recent research has demonstrated the potential of this approach in both deterministic [5] and stochastic [3] cases. However, because the properties of the objective function and constraints are now unknown and the algorithm cannot use much problem-specific information, this scheme introduces some new difficulties. There are some algorithms that use only the input and output of the simulator [7,8] or combine simulation with mathematical programming [9]. And the attention to the second approach is increasing [2].

A common way of managing supplies is by using policy parameters at each storage unit that control replenishment requests. Using different policies, warehouses may order new items in a periodic manner or consider the current inventory level. Order sizes can also be fixed or based on incoming demand. A review of different optimization problems related to spare parts inventory management can be found in [17].

In this paper, we propose a new algorithm for the two-echelon inventory management problem. The algorithm iteratively solves black-box sub-problems for each item separately and then combines the solutions into a solution for the whole problem. We perform several computational experiments to evaluate the performance of the developed scheme and compare it with the Nevergrad solver [11] and Multi-Echelon Optimizer which is a specialized algorithm for solving the described problem. The experiments showed the effectiveness of the proposed approach.

The paper is organized in the following way: in Sect. 2 we formulate the two-echelon inventory management problem, in Sect. 3 we describe the proposed approach, particularly in Subsect. 3.2 we describe an upper-level scheme, and in Subsect. 3.3 we describe a local search algorithm used to solve black-box subproblems, in Sect. 4 we present the results of the computational experiments, and Sect. 5 concludes the article.

2   The Inventory Management Problem Formulation

An inventory control policy is to decide when to send the replenishment orders and the order quantity. We consider a two-echelon multi-warehouse inventory system, which contains one upper-level warehouse and multiple local ones.

Local warehouses apply the (s, S) policy. For each individual item, the (s, S) policy contains two positive integer parameters, InvMin (Inventory Minimum or ordering point) and InvMax (Inventory Maximum or order-up-to level), representing the replenishment point and maximum inventory level. When the warehouse meets an external demand for item i, it is immediately satisfied from stock if items are available; otherwise, it is backordered. The inventory position is the amount of stock minus the backorder plus the number of items ordered but not yet received. If the inventory position falls below the InvMin threshold, a request to replenish items up to the InvMax level is sent to the central warehouse.

The central warehouse applies the $(S-1, S)$ policy. The $(S-1, S)$ policy contains a single ROP (Reorder Point) parameter representing a replenishment point. This parameter controls the number of items kept at the warehouse. If the central warehouse receives a replenishment request, it fulfils it from stock and requests the same number of items from an outside supplier. This supplier is assumed to have an unlimited number of items.

When a replenishment order is placed from a downstream unit to an upstream facility in a multi-echelon system, a time period is required for preparing, handling, and delivering goods for that order. The lead time is the duration from the moment an order is placed to the moment the shipment is received.

If there are m local warehouses in total, there are $2m + 1$ parameters to control the inventory levels in all warehouses for a single item. However, there are many items in the spare part warehouses. As a result, the number of variables might be rather large.

The performance indicators of the inventory system are usually related to customer satisfaction; for example, the fulfilment rate is equal to the percentage of orders or demand fulfilled within the due dates. Aside from performance, inventory holding costs are also important to the company. The holding cost is equal to the integral of the inventory level over the time horizon multiplied by the unit holding cost.

The optimization of the spare part inventory system is to decide the value of the inventory control parameters InvMin and InvMax for all the local warehouses and all the items, and the ROP parameters of the central warehouse for all the items, such that the fulfilment rates are not less than a certain value and the inventory holding cost is minimized. Table 1 contains the notation used in the paper.

The inventory management problem (IMP) could be written as follows:

$$\min_{X} \text{InvHoldCost}(X) \tag{1}$$

s.t.

$$\text{LSatRate}(X) \geq \alpha \tag{2}$$

$$\text{CSatRate}(X) \geq \beta \tag{3}$$

$$x_{if}^2 \leq \max(2x_{if}^1, 1), \quad i \in I, f \in F, \tag{4}$$

$$x_{if}^2 \geq x_{if}^1, \quad i \in I, f \in F, \tag{5}$$

$$l(X) \leq X \leq u(X). \tag{6}$$

We consider the case, where we know that total holding cost is a sum of holding costs across all items and total fulfillment rates are a fraction of the satisfied demand to the total demand:

Table 1. Notation

I	set of items
F	set of local warehouses
LDemand	total demand facing by local warehouses
LDemand$_i$	$i \in I$, total demand for item i facing by local warehouses
CDemand$(\cdot)$	total demand facing by central warehouse depending on policy parameters
CDemand$_i(\cdot)$	$i \in I$, total demand for item i facing by central warehouse depending on policy parameters
InvHoldCost$(\cdot)$	inventory holding cost computed within the simulator and depending on policy parameters
InvHoldCost$_i(\cdot)$	$i \in I$, portion of the inventory holding cost associated with item i
LSatDemand$_i(\cdot)$	$i \in I$, total demand for item i satisfied by local warehouses depending on policy parameters
CSatDemand$_i(\cdot)$	$i \in I$, total demand for item i satisfied by central warehouse depending on policy parameters
LSatRate$(\cdot)$	demand satisfaction rate in local warehouses depending on policy parameters
CSatRate$(\cdot)$	demand satisfaction rate in central warehouse depending on policy parameters
x_{if}^1	$i \in I$, $f \in F$, InvMin value for item i in the local warehouse f
x_{if}^2	$i \in I$, $f \in F$, InvMax value for item i in the local warehouse f
x_i^3	$i \in I$, ROP value for item i in the central warehouse
X	variable vector of policy parameter values, composed of all x_{if}^1, x_{if}^2 and x_i^3 values
α	local warehouse satisfaction rate threshold
β	central warehouse satisfaction rate threshold
$l(\cdot)$	lower bound of the variable's domain interval
$u(\cdot)$	upper bound of the variable's domain interval

$$\text{InvHoldCost}(X) = \sum_i \text{InvHoldCost}_i(X),$$

$$\text{LDemand} = \sum_i \text{LDemand}_i,$$

$$\text{CDemand}(X) = \sum_i \text{CDemand}_i(X),$$

$$\text{LSatDemand}(X) = \sum_i \text{LSatDemand}_i(X),$$

$$\text{CSatDemand}(X) = \sum_i \text{CSatDemand}_i(X),$$

$$\text{LSatRate}(X) = \text{LSatDemand}(X)/\text{LDemand},$$

$$\text{CSatRate}(X) = \text{CSatDemand}(X)/\text{CDemand}(X)$$

A black-box simulator is used to obtain LDemand_i, $\text{CDemand}_i(X)$, $\text{InvHoldCost}_i(X)$, $\text{LSatDemand}_i(X)$, and $\text{CSatDemand}_i(X)$ values for solution vector X.

The objective function (1) represents the total inventory holding cost. The satisfaction rates are ensured by constraints (2)–(3). Constraints (4) are introduced due to the business requirements to prohibit the solutions with a non-empty stock at the beginning of the planning period and empty stock at the end of the planning period. Inequalities (5) guarantee that an InvMin level is not greater than the InvMax one. The domain of the variables is defined in (6).

3   Proposed Method

In this section, we describe a two-level reformulation of the IMP that allows us to decompose it into independent subproblems corresponding to individual items. The reason for using such a decomposition is that simulations for different items could be run independently, and it opens the possibility of performing some computations within the model in parallel. Using the decomposition, we propose a multi-agent system, where subproblems are delegated to worker agents coordinated by a guiding agent.

3.1   Two-Level Reformulation

We can notice that the objective function (1) and both constraints (2) and (3) have a sum of independent values across all items. Also, we can compute functions $\text{InvHoldCost}_i(X)$, $\text{CDemand}_i(X)$, $\text{LSatDemand}_i(X)$, and $\text{CSatDemand}_i(X)$ independently for each item. So the idea of the decomposition is that we try to guess the optimal contribution of items into constraints (2) and (3) and then independently find the optimal costs for each item with fixed fulfilment rates requirement.

Along with the notation introduced in Sect. 2, we would use the following ones:

Variables γ_i, $i \in I$ define the lower bound for clients' demand for the i-th item that must be satisfied. Variables δ_i, $i \in I$ define the lower bound for local warehouses' demand for the i-th item, that must be satisfied.

With these notations, the problem could be written as follows:

$$\min_{(\gamma_i),(\delta_i)} \sum_{i \in I} \text{ItemHoldCost}_i(\gamma_i, \delta_i) \tag{7}$$

$$\sum_{i \in I} \gamma_i \geq \alpha \sum_{i \in I} \text{LDemand}_i, \tag{8}$$

$$\sum_{i \in I} \delta_i \geq \beta \sum_{i \in I} \text{CDemand}_i(X_i^*), \tag{9}$$

$$\text{ItemHoldCost}_i(\gamma_i, \delta_i) = \text{InvHoldCost}_i(X_i^*), \quad i \in I, \tag{10}$$

where, for each $i \in I$, X_i^* is a solution of the $\text{OneItem}_i(\gamma_i, \delta_i)$ problem

$$\text{OneItem}_i(\gamma_i, \delta_i) : \quad \min_{X_i} \text{InvHoldCost}_i(X_i) \tag{11}$$

$$\text{LSatDemand}_i(X_i) \geq \gamma_i, \tag{12}$$

$$\text{CSatDemand}_i(X_i) \geq \delta_i, \tag{13}$$

$$x_{if}^2 \leq \max(2x_{if}^1, 1), \quad f \in F, \tag{14}$$

$$x_{if}^2 \geq x_{if}^1, \quad f \in F, \tag{15}$$

$$l(X_i) \leq X_i \leq u(X_i). \tag{16}$$

The upper-level problem (7)–(10) aims to find a threshold of satisfied demand for each individual item such that the resulting satisfaction ratio is at least α for the local warehouses and at least β for the central warehouse. It is worth noting that constraint (9) involve simulation-dependent values CDemand_i, $i \in I$. Moreover, along with common constraints in the form of inequalities and equations, the upper-level problem contains the constraints that, for each $i \in I$, obligate the variables X_i to be a solution to the corresponding one-item problem.

On the lower level, we have $|I|$ optimization problems aiming to minimize the holding cost for each individual item, provided that a certain demand volume must be satisfied. Objective function (11) and constraints (12) and (13) should be computed during simulation.

3.2   Multi-agent Scheme

In this section, we describe a heuristic approach to solving the problem (7)–(16). The approach relies on the fact that one-item problems (11)–(16) could be processed independently for each individual $i \in I$. It allows us to develop a Multi-Agent Scheme (MAS), where several *worker agents* compute in parallel quality solutions by solving one-item problems using a local search algorithm while a *guiding agent* directs the search by requesting solutions for some specific (γ_i) and (δ_i) values and aggregates the solutions of one-item problems into a combined solution of the whole problem.

Solutions Pool and Initialization Step During the computation, we store a pool of best-found solutions $P_i = \{X_i^1, \ldots, X_i^{k_i}\}$ for each problem OneItem_i, $i \in I$ and different (γ, δ)-pairs. Let $K_i = \{1, \ldots, k_i\}$, $i \in I$ denote the indexing set for the corresponding pool of solutions. After the evaluation of the solution X_i^k, $i \in I$, $k \in K_i$, it is attributed with the values $d_{ik}^C = \text{CDemand}_i(X_i^k)$,

$s_{ik}^L = \texttt{LSatDemand}_i(X_i^k)$, $s_{ik}^C = \texttt{CSatDemand}_i(X_i^k)$, and $c_{ik} = \texttt{InvHoldCost}_i(X_i^k)$. Furthermore, the pair $(\gamma_{ik}, \delta_{ik})$ of corresponding values of the $\texttt{OneItem}$ problem parameters is saved.

We say that a solution X_i^j *dominates* a solution X_i^k if $\gamma_{ij} \geq \gamma_{ik}$, $\delta_{ij} \geq \delta_{ik}$, and $c_{ij} \leq c_{ik}$. Given $X_i' \in P_i$, we say that X_i' is *non-dominated* if there is no $X_i'' \in P_i$ such that X_i'' dominates X_i'. Each time a new solution is to be added to the pool, we check to see if it dominates some of the existing ones or is dominated by another one. All the dominant solutions are removed from the pool.

On the initialization step, given $i \in I$, the pool P_i is empty, and the one-item problem solutions, corresponding to pairs (γ_i, δ_i) such that $\delta_i = \gamma_i = \lceil \frac{l}{p} \texttt{LDemand}_i \rceil$ are requested to be computed by worker agents. Here, $p \in \mathbb{Z}^+$ is the algorithm parameter, and l takes integer values from 1 to p. One should notice that a solution obtained for some (γ, δ)-pair is feasible for the one-item problem with parameters (γ', δ'), such that $\gamma \geq \gamma'$ and $\delta \geq \delta'$. Thus, it is reasonable to start filling a solution pool by solving a one-item problem with a (γ, δ)-pair corresponding to the value $l = p$. The obtained solution is used later to initialize the local search when the one-item problem with $l = p - 1$ is being solved. Then the process repeats further by processing the values of l in decreasing order. We use $p = 10$ in our experiments.

If, for an item, it takes less than 10^5 evaluations to fully enumerate all possible policy value configurations, then we apply an exhaustive search to these items and save all non-dominated solutions into the pool.

Guiding Agent. We get a solution to the whole problem (7)–(16) by taking a single solution to each of the one-item problems from the solution pool and combining these solutions together. Given the solution pool fixed $\mathbf{P} = (P_1, \ldots, P_{|I|})$, the best combined solution could be obtained by solving an auxiliary integer programming problem. Let us introduce binary variables (z_{ik}), $i \in I$, $k \in P_i$, where, given i and k, z_{ik} equals one if the solution X_i^k is taken into the combined solution, and zero otherwise. Then, using the newly introduced notation, the auxiliary integer programming problem $\texttt{Master}\,(\mathbf{P})$, which can be regarded as a master problem in our procedure, is formulated as follows:

$$\min_{(z_{ik})} \sum_{i \in I} \sum_{k \in P_i} c_{ik} z_{ik} \tag{17}$$

$$\sum_{k \in K_i} z_{ik} = 1, \quad i \in I, \tag{18}$$

$$\sum_{i \in I} \sum_{k \in K_i} s_{ik}^L z_{ik} \geq \alpha \sum_{i \in I} \texttt{LDemand}_i, \tag{19}$$

$$\sum_{i \in I} \sum_{k \in K_i} s_{ik}^C z_{ik} \geq \beta \sum_{i \in I} \sum_{k \in P_i} d_{ik}^C z_{ik}, \tag{20}$$

$$z_{ik} \in \{0, 1\}, \quad i \in I, k \in K_i. \tag{21}$$

The objective function (21) of the master-problem is the inventory holding cost associated with a combined solution. Inequalities (18) ensure that a single solution, for each of the one-item problems, would be taken into the combined solution, while (19) and (20) guarantee the satisfaction rate level. The master problem could appear to be infeasible due to the improper composition of one-problem solutions $\{X_i^k\}$, $i \in I$, $k \in P_i$ in the pools. But we should notice that the initial state of the pools and the rule to keep non-dominated solutions there, which are provided in Sect. 3.2, ensure that the master-problem is always feasible. Indeed, a solution with fulfilment rates on both local and central warehouses' levels equal to one is always put into a pool during the initialization step when the one-item problem corresponding to $l = p$ is solved. Such a solution could be removed from the pool only if another solution having the same fulfilment rate characteristics is inserted due to the definition of dominance.

We define $Z\left(\tilde{k}\right)$ as a solution to the problem Master $(\mathbf{P})$ with $z_{ik\tilde{k}_i} = 1, z_{ik} = 0$ $\forall i \in I, k \neq \tilde{k}_i$, and the procedure to solve the master-problem will be referred as solveMP$(\mathbf{P})$ thereafter. We call a solution "semi-feasible" if it satisfies constraints (14) and we denote set M_i as a set of indices of all semi-feasible solutions for item i, $\mathbf{M} = \left(M_1, \ldots, M_{|I|}\right)$. Then we can give the following propositions.

Proposition 1. *Problem* IMP *is equivalent to problem* Master $(\mathbf{M})$ *in the sense that if* $(X_i^{k_i^*})$ *is an optimal solution to the problem* IMP, *then* $Z\left(\mathbf{k}^*\right)$ *is an optimal solution to the problem* Master $(\mathbf{M})$, *and vice versa.*

Proposition 2. *If some subset* $K_i \subseteq M_i$ *contains index of an optimal solution to* Master $(\mathbf{M})$, *then any optimal solution to* Master $(\mathbf{K})$ *is also an optimal solution to* Master $(\mathbf{M})$.

Proof. Consider $Z\left(\tilde{k}\right)$ to be any optimal solution for Master $(\mathbf{K})$. It is also a feasible solution for Master $(\mathbf{M})$. Also, because $Z\left(\mathbf{k}^*\right)$ is a feasible solution for Master $(\mathbf{K})$, the objective function of $Z\left(\tilde{k}\right)$ is not greater than the objective function of $Z\left(\mathbf{k}^*\right)$. So $Z\left(\tilde{k}\right)$ is optimal for Master $(\mathbf{M})$. $\qquad\square$

Proposition 2 means that if worker agents occasionally find values corresponding to the optimal solution for the initial problem, then solveMP$(\mathbf{P})$ will choose that solution, and our algorithm will find the optimal solution to IMP.

The workflow of the guiding agent is organized as a sequence of identical steps. On each step, the master-problem is solved for the current state of one-item problems' solution pools. Afterwards, the agent sends requests to compute solutions to one-item problems with fulfilment rates neighbouring those corresponding to the solutions selected in the combined one.

Let (z_{ik}), $i \in I$, $k \in K_i$ be an optimal solution of the master problem on the current step, and let $i \in I$, $k \in K_i$ be such that $z_{ik} = 1$. Then, X_i^k is a part of a combined solution constructed on the current step. Consider the pair $(\gamma_{ik}, \delta_{ik})$ associated with that solution. Before the next step begins, the guiding agent requests to compute the solutions to the problem $\texttt{OneItem}_i$ with slightly shifted parameters.

Let us denote the size of the shift to be applied to the parameters of one-item problems as ε. Initially, it is set to be equal to ε_0 and remains the same as long as the combined solution changes from one step to another. In experiments, we used $\varepsilon_0 = 0.04$. When the combined solution remains the same as it was at the previous step, we set $\varepsilon \leftarrow \varepsilon/2$. Given the current value ε and the current best satisfaction rates γ_i^t, δ_i^t at iteration t, the solutions of the problems $\texttt{OneItem}_i$ with the parameters from the set $A\left(\gamma_i^t, \delta_i^t\right) = \{\gamma_i^t \pm \lceil \varepsilon \texttt{LDemand}_i \rceil\} \times \{\delta_i^t \pm \lceil \varepsilon \texttt{CDemand}_i \rceil\}$ are requested to be computed. When $\lceil \varepsilon \texttt{LDemand}_i \rceil = \lceil \varepsilon \texttt{CDemand}_i \rceil = 1$ for all $i \in I$ and the combined solution does not change during two subsequent steps, the algorithm terminates. The pseudocode of the MAS is represented as Algorithm 1.

Algorithm 1. Multi-agent scheme (MAS)

1: **function** MAS
2:      generate initial set of requested (γ, δ)-pairs: $Q^0 = (Q_i^0), i \in I$.
3:      prepare containers for solution pools: $P_i = \emptyset, i \in I$
4:      **for all** $(\gamma_i, \delta_i) \in Q_i^0$ **do**
5:          $\texttt{oneItemLS}\,(\gamma_i, \delta_i, P_i)$
6:      set $\gamma^0, \delta^0 \leftarrow \texttt{solveMP}\,(P)$
7:      $t \leftarrow 1, \varepsilon \leftarrow \varepsilon_0$
8:      **while** stop criterion is not met **do**
9:          **for** $i \in I$ **do**
10:             $Q_i^t \leftarrow Q_i^{t-1} \cup A\left(\gamma_i^t, \delta_i^t\right)$
11:             **for all** $(\gamma_i, \delta_i) \in Q_i^t \backslash Q_i^{t-1}$ **do**
12:                 $\texttt{oneItemLS}\,(\gamma_i, \delta_i, P_i)$
13:         $\gamma^t, \delta^t \leftarrow \texttt{solveMP}\,(P)$
14:         **if** $\gamma^t = \gamma^{t-1} \wedge \delta^t = \delta^{t-1}$ **then**
15:             $\varepsilon \leftarrow \varepsilon/2$
16:         $t \leftarrow t + 1$
17:     **return** X^*

3.3 Worker Agents and One-Item Problem

When the requests to compute one-item problem solutions are formed by the guiding agent, the worker agents could satisfy them by processing the one-item problems for each individual item independently. To find quality solutions, worker agents perform a randomized local search with additional components that improve their performance.

Given the item $i \in I$ and a pair (γ_i, δ_i), the problem $\texttt{OneItem}_i(\gamma_i, \delta_i)$, defined as (11)–(16), aims to find the parameters X_i of the policies for item i in all the warehouses.

In a case where the pool P_i of non-dominated solutions found so far is not empty, the randomized local search starts with the best suitable solution from the pool. It is a solution $X_i^{k'} \in P_i$, where k' is such that

$$k' = \underset{k \in K_i}{\arg\min} \left\{ \text{InvHoldCost}_i(X_i^k) | \gamma_{ik} \geq \gamma \text{ and } \delta_{ik} \geq \delta \right\}.$$

On the initialization step of the MAS, the solution pool could be empty. According to the order of requests on the initialization step, described in Sect. 3.2, the first solution requested is the one having all the demand satisfied. Such a solution could be constructed by a simple heuristic. This heuristic uses the assumption about the independence of each local warehouse when we have a large ROP value at the central warehouse and need to satisfy the entire demand fully in local warehouses. Using the assumption about the monotonicity of the inventory holding cost in InvMin-InvMax values, we find the best solution by means of a dichotomy algorithm. After that, we adjust the ROP policies of the central warehouse to minimize its holding cost.

The neighbourhood structure used in the randomized local search is formed by semi-feasible vectors (satisfying (14)) that define the policy parameters, which are different for at most two warehouses (both local warehouses and the central warehouse), and the difference is at most two. More formally, given a solution $X_i = ((x_{if}^1), (x_{if}^2), (x_i^3))$, $f \in F$ we define a subset $N'(X_i) = \{\bar{X}_i = ((\bar{x}_{if}^1), (\bar{x}_{if}^2), (\bar{x}_i^3))$ such that $0 < \|X_i - \bar{X}_i\|_\infty \leq 2\}$. Using this notation, the neighborhood $N(X_i)$ of X_i could be defined as

$$N(X_i) = \left\{ \bar{X}_i \in N'(X_i) \mid \bar{x}_i^3 = x_i^3, \#\{f \in F \mid \bar{x}_{if}^1 \neq x_{if}^1 \text{ or } \bar{x}_{if}^2 \neq x_{if}^2\} \leq 2 \right\} \cup$$
$$\left\{ \bar{X}_i \in N'(X_i) | \#\{f \in F \mid \bar{x}_{if}^1 \neq x_{if}^1 \text{ or } \bar{x}_{if}^2 \neq x_{if}^2\} \leq 1 \right\}.$$

A randomized neighbourhood of the solution X_i with a randomization parameter $r \in \mathbb{Z}^*$ would be denoted by $N_r(X_i)$. It consists of r randomly selected elements of the subset $N(X_i)$. In the experiments, we set $r = 21$.

We use a tabu list, where we keep the ten most recently visited solutions, to avoid trapping into a cycle when the set $N(X_i)$ is small and the randomization does not cause leaving the local optima. During the lookup of the neighbourhood $N_r(X_i)$, the algorithm moves to the best neighbour which is not in the tabu list. We use penalty function $\text{InfPenalty}(X_i^k) = \max\left(\gamma_i - \text{LSatDemand}(X_i^k), 0\right) \cdot 10^4 + \max\left(\delta_i - \text{CSatDemand}(X_i^k), 0\right) \cdot 10^4$ to prevent moving to an infeasible solution.

During the computations, visited solutions are compared with the solutions stored in the pool P_i, which is updated when necessary. When the best-found solution does not change during $T = \max\{10, n\}$ iterations, where n is the length of X_i, the search returns to the best solution. The algorithm terminates when the computational budget of 500 simulations is depleted or when the best-found solution does not change during $10T$ iterations. In the MAS algorithm, for $i \in I$ and each pair (γ_i, δ_i), we do not spend more than $2,000$ evaluations in total when solving $\text{OneItem}_i(\gamma_i, \delta_i)$ problems.

The overall scheme of the randomized local search is given by the Algorithm 2.

Algorithm 2. Local search algorithm used for solving one-item problem

---

1: **function** ONEITEMLS(γ_i, δ_i, P_i)
2:     $X_i = \arg\min_{X_i^k \in P_i}\{\text{InvHoldCost}(X_i^k)|\gamma_{ik} \geq \gamma_i \text{ and } \delta_{ik} \geq \delta_i\}$
3:     **while** stop criterion is not met **do**
4:         $X_i \leftarrow \arg\min_{X_i^k \in N_r(X_i)}\{\text{InvHoldCost}(X_i^k) + \text{InfPenalty}(X_i^k, v)\}$
5:         Update pool P_i with solution X_i
6:         $X_i^* \leftarrow \text{Choose}(\{X_i, X_i^*\})$
7:         **if** Intensification criterion is met **then**
8:             $X_i \leftarrow X_i^*$
9:             Clear tabu list
10:     **return** X_i^*

---

4 Computational Results

The MAS algorithm was implemented in Julia. All experiments were performed on a computer equipped with Intel Core i7-8700 CPU running at 3.20 GHz and 32 GB of RAM, running Microsoft Windows 10 Pro operating system.

4.1 Test Instances Description

Test instances were generated similarly to [12]. We use a Poisson distribution to generate client requests for spare parts. We generated two families of instances: small and large. The small instances have 100 items and 10 local warehouses. An item has a 0.025 probability of having requests from a certain warehouse. Taking the central warehouse into consideration, we get about 500 variables per instance. The large instances have 1,000 items, 100 local warehouses, and a 0.045 probability of having requests from a certain warehouse. This gives about 10,000 variables. The levels of the target satisfaction rates were set to $\alpha = 0.95$, $\beta = 0.95$.

4.2 Comparative Analysis

The proposed algorithm was compared with a general-purpose black-box solver, Nevergrad [11], and a Multi-Echelon Optimizer (MEO). For Nevergrad, we use the NGOpt39 optimizer with a $3 \cdot 10^5$ evaluation budget and provide it with an initial solution found by our heuristic for the case of all demands being satisfied. If not provided with an initial solution, Nevergrad needs much more computational budget to show results similar to the ones presented further. Nevergrad and MAS used all available threads, and MEO was run in a single thread. Besides, Nevergrad uses full-model simulations, whereas MAS and MEO use simulations for a one-item model. For MAS, we use the parameters described in the previous sections. The stopping criterion is achieving $\varepsilon = 1$ and $z^t = z^{t-1}$ (Sect. 3). Though it should be mentioned that changing the parameters can shift the balance between computational time and the result quality. In the tables below, we show the inventory holding cost values for the corresponding instances and algorithms.

Table 2 shows the results for small instances. There were 5 runs of the MAS algorithm and 5 runs of Nevergrad (NG) for each instance, and columns "MAS_{min}", "MAS_{avg}", "NG_{min}", and "NG_{avg}" show the best and average results for these runs. Results of MEO are shown in column "MEO". Columns "t_{MAS}, s" and "t_{MEO}, s" show the average running time per one run for MAS and MEO. MAS shows 6% in average better results than MEO in a similar running time, and Nevergrad shows significantly worse results, though its running time was approximately 100 times larger than the one of MAS and MEO. We can see that deviations between the minimal and mean values of the objective function in MAS runs are pretty small. So, we can conclude that the proposed algorithm is consistent.

Table 2. Comparison of MAS, MEO and Nevergrad on small instances

Instance	MAS_{min}	MAS_{avg}	MEO	NG_{min}	NG_{avg}	t_{MAS}, s	t_{MEO}, s
1	**6,486**	6,495	6,912	11,857	12,562	26	20
2	**5,437**	5,480	5,581	11,536	12,181	32	18
3	**5,065**	5,066	5,503	8,413	8,954	19	18
4	**5,590**	5,595	5,868	8,705	9,475	19	18
5	**12,230**	12,238	12,870	16,993	18,540	25	24
6	**7,122**	7,143	7,518	13,493	14,253	24	19
7	**10,846**	**10,846**	11,565	15,747	16,472	22	18
8	**11,577**	11,578	11,882	17,443	18,028	24	17
9	**13,409**	13,414	13,743	20,084	21,290	16	20
10	**8,975**	8,993	9,356	12,999	13,291	22	24

Table 3 shows the results for the large instances. The meaning of the columns is the same as for Table 2. The results achieved by MAS in running time $t = t_{MEO}$ are demonstrated in column "MAS_t". The running time for Nevergrad was approximately 4 h per run. The results in columns "MAS_{avg}" and "MAS_t" are 5% and 4% better than ones of MEO. All solutions found by Nevergrad are much worse than solutions obtained by the other algorithms.

Figure 1 shows the values of the objective function over time for the Multi-Agent Scheme. The horizontal red line shows the value of the objective function found by MEO. The vertical orange line denotes time spent by MEO. The blue ribbon depicts the maximum and minimum values of the objective function obtained from MAS runs. It can be seen from Fig. 1 and Tables 2, 3 that our algorithm performs consistently and there is a narrow range of results from run to run. The maximum relative difference between the best and the worst solution found by MAS was 1.5%. Also, even the worst solution found by MAS is better than the solution found by MEO on all large instances.

Table 3. Comparison of MAS, MEO and Nevergrad on large instances

Instance	MAS_{min}	MAS_{avg}	MAS_t	MEO	NG_{min}	NG_{avg}	t_{MAS}, s	t_{MEO}, s
1	**134,359**	134,469	135,809	140,153	209,120	215,842	1,372	617
2	**128,657**	128,740	129,725	136,991	217,382	219,338	1,454	622
3	**120,136**	120,314	120,654	127,671	204,631	208,680	1,248	892
4	**146,932**	147,193	147,470	154,442	231,940	233,772	1,430	892
5	**125,581**	125,744	126,123	134,837	210,121	213,743	1,397	903
6	**127,484**	127,665	128,579	133,663	217,160	219,995	1,377	620
7	**99,060**	99,260	102,472	104,676	183,820	186,964	1347	356
8	**136,558**	136,691	137,072	141,286	226,715	229,002	1,368	879
9	**129,420**	129,568	133,151	137,233	222,444	224,208	1,547	366
10	**135,778**	135,925	136,435	144,226	238,641	240,459	1,559	909

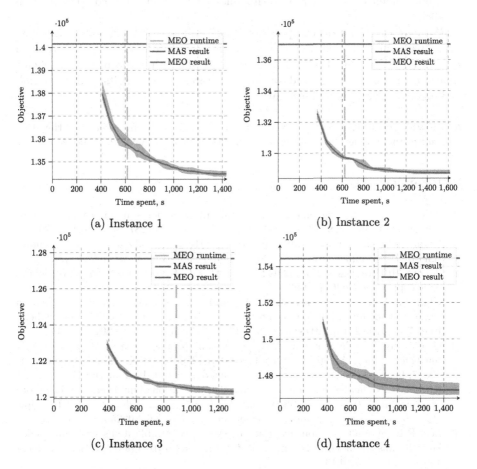

(a) Instance 1

(b) Instance 2

(c) Instance 3

(d) Instance 4

Fig. 1. Comparison of the performances of MAS and MEO (Color figure online)

Auxiliary `Master`(·) problems appeared to be pretty easy for a MIP solver. So its solution takes less than 1 s. Due to the implementation of the simulation algorithm, most computational time is spent on simulations and the fraction of time spent on the algorithm itself is negligible. Also, the whole simulation time is approximately a sum of simulation times for each item.

5   Conclusion

This paper proposes a new algorithm based on a multi-agent grey-box decomposition scheme for a two-echelon spare part inventory problem. The algorithm is distributed between several worker agents and a coordinating agent. Worker agents use local search to solve black-box single-item problems that are set by the coordinating agent. Then the coordinating agent solves a 0–1 linear programming model to combine partial solutions and obtain a solution to the initial problem. The independence of worker problems allows us to run these computations in parallel.

We performed several computational experiments with generated instances to compare our approach with the well-known Nevergrad solver and a provided specialized algorithm MEO. The experiments show that the proposed scheme greatly outperforms Nevergrad and demonstrates better results than a specialized algorithm used as a benchmark.

Acknowledgement. The study was carried out within the framework of the state contract of the Sobolev Institute of Mathematics (project FWNF–2022–0019).

References

1. Axsäter, S.: Modelling emergency lateral transshipments in inventory systems. Manage. Sci. **36**, 1329–1338 (1990). https://doi.org/10.1287/mnsc.36.11.1329
2. Blum, C., Puchinger, J., Raidl, G.R., Roli, A.: Hybrid metaheuristics in combinatorial optimization: a survey. Appl. Soft Comput. **11**(6), 4135–4151 (2011). https://doi.org/10.1016/j.asoc.2011.02.032, https://www.sciencedirect.com/science/article/pii/S1568494611000962
3. Chu, Y., You, F., Wassick, J.M., Agarwal, A.: Simulation-based optimization framework for multi-echelon inventory systems under uncertainty. Comput. Chem. Eng. **73**, 1–16 (2015). https://doi.org/10.1016/j.compchemeng.2014.10.008
4. Grossmann, I.: Enterprise-wide optimization: a new frontier in process systems engineering. AIChE J. **51**(7), 1846–1857 (2005). https://doi.org/10.1002/aic.10617, https://aiche.onlinelibrary.wiley.com/doi/abs/10.1002/aic.10617
5. Köchel, P., Nieländer, U.: Simulation-based optimisation of multi-echelon inventory systems. Int. J. Prod. Econ. **93-94**, 505–513 (2005). https://doi.org/10.1016/j.ijpe.2004.06.046, https://linkinghub.elsevier.com/retrieve/pii/S0925527304002683
6. Lee, H.L.: A multi-echelon inventory model for repairable items with emergency lateral transshipments. Manage. Sci. **33**(10), 1302–1316 (1987). https://doi.org/10.1287/mnsc.33.10.1302

7. Mele, F.D., Guillén, G., Espuña, A., Puigjaner, L.: A simulation-based optimization framework for parameter optimization of supply-chain networks. Ind. Eng. Chem. Res. **45**(9), 3133–3148 (2006). https://doi.org/10.1021/ie051121g

8. Moncayo-Martínez, L.A., Zhang, D.Z.: Multi-objective ant colony optimisation: a meta-heuristic approach to supply chain design. Int. J. Prod. Econ. **131**(1), 407–420 (2011). https://doi.org/10.1016/j.ijpe.2010.11.026, https://www.sciencedirect.com/science/article/pii/S092552731000455X, innsbruck 2008

9. Nikolopoulou, A., Ierapetritou, M.G.: Hybrid simulation based optimization approach for supply chain management. Comput. Chem. Eng. **47**, 183–193 (2012). https://doi.org/10.1016/j.compchemeng.2012.06.045, https://www.sciencedirect.com/science/article/pii/S0098135412002372. fOCAPO 2012

10. Peidro, D., Mula, J., Poler, R., Lario, F.C.: Quantitative models for supply chain planning under uncertainty. Int. J. Adv. Manuf. Technol. **43**, 400–420 (2009). https://doi.org/10.1007/s00170-008-1715-y

11. Rapin, J., Teytaud, O.: Nevergrad - a gradient-free optimization platform (2018). https://GitHub.com/FacebookResearch/Nevergrad

12. Topan, E., Bayındır, Z.P., Tan, T.: Heuristics for multi-item two-echelon spare parts inventory control subject to aggregate and individual service measures. Eur. J. Oper. Res. **256**, 126–138 (2017). https://doi.org/10.1016/j.ejor.2016.06.012

13. Wassick, J.M., Agarwal, A., Akiya, N., Ferrio, J., Bury, S., You, F.: Addressing the operational challenges in the development, manufacture, and supply of advanced materials and performance products. Comput. Chem. Eng. **47**, 157–169 (2012). https://doi.org/10.1016/j.compchemeng.2012.06.041, https://www.sciencedirect.com/science/article/pii/S0098135412002256. fOCAPO 2012

14. Wong, H., van Houtum, G.J., Cattrysse, D., Oudheusden, D.V.: Simple, efficient heuristics for multi-item multi-location spare parts systems with lateral transshipments and waiting time constraints. J. Oper. Res. Soc. **56**, 1419–1430 (2005). https://doi.org/10.1057/palgrave.jors.2601952, https://www.tandfonline.com/doi/full/10.1057/palgrave.jors.2601952

15. Yue, D., You, F.: Planning and scheduling of flexible process networks under uncertainty with stochastic inventory: MINLP models and algorithm. AIChE J. **59**(5), 1511–1532 (2013). https://doi.org/10.1002/aic.13924, https://aiche.onlinelibrary.wiley.com/doi/abs/10.1002/aic.13924

16. Zapata, J.C., Pekny, J., Reklaitis, G.V.: Simulation-optimization in support of tactical and strategic enterprise decisions. In: Kempf, K.G., Keskinocak, P., Uzsoy, R. (eds.) Planning Production and Inventories in the Extended Enterprise. ISORMS, vol. 151, pp. 593–627. Springer, New York (2011). https://doi.org/10.1007/978-1-4419-6485-4_20

17. Zhang, S., Huang, K., Yuan, Y.: Spare parts inventory management: a literature review. Sustain. (Switz.) **13**, 1–23 (2021). https://doi.org/10.3390/su13052460

Optimal Control and Mathematical Economics

The Algorithm for the Construction of a Symbolic Family of Regulators for Nonlinear Discrete Control Systems with Two Small Parameters

Yulia Danik$^{(\boxtimes)}$ and Mikhail Dmitriev

Federal Research Center "Computer Science and Control" of the Russian Academy of Sciences (FRC CSC RAS), Moscow, Russia
yuliadanik@gmail.com
https://www.frccsc.ru/

Abstract. In this paper, the stabilization problem for a discrete weakly nonlinear system with two small positive parameters and a quadratic quality criterion is considered. The parameters are at the nonlinearities in system matrices and can be of different order. We use the asymptotic methods to find a parametric family of solutions in the form of the state feedback. For the construction of a parametric family of feedback control we apply the Discrete State-dependent Riccati equation (D-SDRE) approach, which consists in solving the corresponding discrete matrix algebraic Riccati equation with state-dependent coefficients. An asymptotic expansion of the solution of the corresponding Riccati equation is found in the form of a power series by two parameters. This regular asymptotic series is used for the construction of a one-point matrix Pade approximation by two parameters. The numerical experiments on a grid of parameters demonstrate the stabilization of the closed-loop systems with the proposed regulators. The resulting Pade controllers have interpolation and extrapolation properties, and often significantly improve the approximation accuracy in comparison with controllers based on the regular asymptotic series by one or two parameters.

Keywords: Small parameter · Pade approximation · D-SDRE approach · Asymptotic approximation

1 Introduction

The presence of parameters in mathematical models of applied control problems generates a family of admissible controls and corresponding trajectories for different parameter values, which leads to problems of their approximate analytical description with the goal to reduce the time requires for the calculation of controls.

Asymptotic methods based on the choice of small parameters in models are a powerful tool for obtaining approximate analytical solutions in various applications and, in particular, in control problems. With the help of asymptotics, it is possible to solve "stiff" problems, propose decomposition methods for high-dimensional problems, and construct effective initial approximations for the numerical solution of nonlinear programming and nonlinear optimal control problems. In recent years, the SDRE (State-dependent Riccati equation) approach is often used for the approximate construction of feedback in nonlinear control problems, where the gain matrix is based on the approximate solution of the corresponding matrix Riccati equations and further formal application of the Kalman algorithm for linear-quadratic problems.

Here we consider a quadratic quality criterion, where the first quadratic form with the matrix Q is responsible for the quality of transition process, and the second form with matrix R corresponds to the control costs. The regulators can be constructed by selection of a weighting matrix Q in a form of a power series expansion by parameter, where the terms of the expansion are selected in each approximation. If the norm of the matrix Q is small, then the control costs are firstly minimized and we actually come to the criterion of minimum energy.

At first, the approximate solution of the Riccati equation is found using asymptotic methods and Pade approximations. Here the matrix approximation of Pade [7] is used, which in the scalar case is actively used in many applications in various fields: physics, mechanics [8,9], etc. In the scalar case the Pade method consists in the approximation of a considered function with the help of a ratio of two polynomials, whose coefficients are determined by the special Taylor's series expansion of the function. With the help of Pase constructions, it is often possible to achieve higher approximation accuracy for a wider interval of parameter variation than with the use of asymptotics.

Here matrix Pade approximations (PA) are constructed on the basis of an asymptotic approximation not by one parameter, as in the works [1–3,10], but by two parameters [13]. PA by several parameters in the scalar case are considered in [11,12]. In [11], the expansions by two parameters are considered, and in [12], the coordinates of the state vector of dimension n were used as approximation parameters and a PA of order [3/3] was constructed for the scalar control.

In this paper, on a numerical example it is also shown that the expansion by two parameters gives a better approximation than the expansion by one parameter, that is, the method of reducing of two small parameters to one parameter reduces the accuracy. Note that for the first time in literature the peculiarity of the application of the asymptotic approach with several different small parameters was demonstrated in [5]. In particularly, it is shown there, that in singularly perturbed problems, where there are parameters of different orders of smallness at the derivatives, the algorithm for constructing the system for the zero approximation differs from the algorithm for the case of one small parameter.

2   Problem Statement

We consider a weakly nonlinear discrete control system with two positive parameters and a quadratic quality criterion

$$x(t+1) = A(x,\varepsilon)x(t) + B(x,\mu)u(t)$$
$$= (A_0 + \varepsilon A_1(x))x(t) + (B_0 + \mu B_1(x))u(t), \tag{1}$$

$$x(0) = x_0, \; x(t) \in X \subset R^n, \; u(t) \in R^r, \; t = 0,1,2\ldots, \; 0 < \mu \le \mu_0, \; 0 < \varepsilon \le \varepsilon_0,$$

$$I(u) = \frac{1}{2}\sum_{t=0}^{\infty}(x^T Q(x,\varepsilon,\mu)x + u^T R_0 u) \to \min, \; Q(x,\varepsilon,\mu) > 0, \; R_0 > 0. \tag{2}$$

The D-SDRE (Discrete State-dependent Riccati equation) approach [6], is used for the parametric synthesis construction. This method consists in solving the corresponding discrete matrix algebraic Riccati equation with state-dependent coefficients. Here we choose the control in the form of formally linear feedback

$$u(x,\varepsilon,\mu) = -\tilde{R}(x,\varepsilon,\mu)^{-1}B(x,\mu)^T P(x,\varepsilon,\mu)A(x,\varepsilon)x(t),$$

where $\tilde{R}(x,\varepsilon,\mu) = R_0 + B(x,\mu)^T P(x,\varepsilon,\mu)B(x,\mu)$, is invertible $\forall(x,\varepsilon,\mu)$ and $P(x,\varepsilon,\mu)$ is the solution of the next discrete matrix algebraic state-dependent Riccati equation for all $x \in X, \; 0 < \mu \le \mu_0, \; 0 < \varepsilon \le \varepsilon_0$

$$A^T(x,\varepsilon)P(x,\varepsilon,\mu)A(x,\varepsilon) - P(x,\varepsilon,\mu) - A^T(x,\varepsilon)P(x,\varepsilon,\mu)B(x,\mu)\tilde{R}(x,\varepsilon,\mu)^{-1}$$
$$\times B^T(x,\mu)P(x,\varepsilon,\mu)A(x,\varepsilon) + Q(x,\varepsilon,\mu) = 0, \tag{3}$$

where matrix $Q(x,\varepsilon,\mu)$ is selected in a special way for the solvability of equations for the terms of asymptotics in the form

$$Q(x,\varepsilon,\mu) = Q_0 + \varepsilon Q_{10}(x) + \mu Q_{01}(x) + \varepsilon^2 Q_{20}(x) + \varepsilon\mu Q_{11}(x) + \mu^2 Q_{02}(x).$$

An asymptotic expansion for the solution of the Riccati equation is constructed as a Taylor series expansion in the vicinity of the origin (0,0)

$$P(x,\varepsilon,\mu) = P_0 + \varepsilon P_{10}(x) + \mu P_{01}(x) + \varepsilon^2 P_{20}(x) + \varepsilon\mu P_{11}(x) + \mu^2 P_{02}(x). \tag{4}$$

The coefficients of the matrix expansion $Q(x,\varepsilon,\mu)$ are selected in such a way as to ensure the solvability of the equations for the terms of the asymptotics $P(x,\varepsilon,\mu)$.

Substituting the expression for $P(x,\varepsilon,\mu)$ (4) into the discrete Riccati equation (3) and collecting the terms with the same powers of the parameters and their products, we obtain expressions for the terms of the asymptotic expansion. For P_0 we obtain the matrix algebraic Riccati equation

$$A_0^T P_0 A_0 - P_0 - A_0^T P_0 B_0(R_0 + B_0^T P_0 B_0)^{-1}B_0^T P_0 A_0 + Q_0 = 0.$$

For P_{10} we have a linear equation,

$$-P_{10} + A_0^T P_{10} A_0 - A_0^T P_0 B_0 (R_0 + B_0^T P_0 B_0)^{-1} B_0^T P_{10} A_0$$
$$-A_0^T P_{10} B_0 (R_0 + B_0^T P_0 B_0)^{-1} B_0^T P_0 A_0$$
$$+A_0^T P_0 B_0 (R_0 + B_0^T P_0 B_0)^{-1} B_0^T P_{10} B_0 (R_0 + B_0^T P_0 B_0)^{-1} B_0^T P_0 A_0 + C(x) = 0,$$

where $C(x) = A_0^T P_0 A_1 + A_1^T P_0 A_0 - A_0^T P_0 B_0 (R_0 + B_0^T P_0 B_0)^{-1} B_0^T P_0 A_1 - A_1^T P_0 B_0 (R_0 + B_0^T P_0 B_0)^{-1} B_0^T P_0 A_0 + Q_{10}$.

For P_{01} we have the following linear equation

$$-P_{01} + A_0^T P_{01} A_0 - A_0^T P_0 B_0 (R_0 + B_0^T P_0 B_0)^{-1} B_1^T P_0 A_0$$
$$- A_0^T P_0 B_1 (R_0 + B_0^T P_0 B_0)^{-1} B_0^T P_0 A_0 - A_0^T P_0 B_0 (R_0 + B_0^T P_0 B_0)^{-1} B_0^T P_{01} A_0$$
$$- A_0^T P_{01} B_0 (R_0 + B_0^T P_0 B_0)^{-1} B_0^T P_0 A_0 + A_0^T P_0 B_0 (R_0 + B_0^T P_0 B_0)^{-1} B_0 P_{01} B_0^T$$
$$\times (R_0 + B_0^T P_0 B_0)^{-1} B_0^T P_0 A_0 +$$
$$+A_0^T P_0 B_0 (R_0 + B_0^T P_0 B_0)^{-1} B_0 P_0 B_1^T (R_0 + B_0^T P_0 B_0)^{-1} B_0^T P_0 A_0$$
$$+ A_0^T P_0 B_0 (R_0 + B_0^T P_0 B_0)^{-1} B_1 P_0 B_0^T (R_0 + B_0^T P_0 B_0)^{-1} B_0^T P_0 A_0 + Q_{01} = 0$$

and for P_{20} we get a linear equation

$$-P_{20} + A_1^T P_0 A_1 + A_1^T P_{10} A_0 + A_0^T P_{10} A_1 + A_0^T P_{20} A_0$$
$$- A_1^T P_0 B_0 (R_0 + B_0^T P_0 B_0)^{-1} B_0^T P_0 A_1 - A_0^T P_0 B_0 (R_0 + B_0^T P_0 B_0)^{-1} B_0^T P_0 A_0$$
$$- A_1^T P_0 B_0 (R_0 + B_0^T P_0 B_0)^{-1} B_0^T P_{10} A_0 - A_1^T P_{10} B_0 (R_0 + B_0^T P_0 B_0)^{-1} B_0^T P_0 A_0$$
$$- A_0^T P_0 B_0 (R_0 + B_0^T P_0 B_0)^{-1} B_0^T P_{10} A_1 - A_0^T P_{10} B_0 (R_0 + B_0^T P_0 B_0)^{-1} B_0^T P_0 A_1$$
$$- A_0^T P_0 B_0 (R_0 + B_0^T P_0 B_0)^{-1} B_0^T P_{20} A_0 - A_0^T P_{20} B_0 (R_0 + B_0^T P_0 B_0)^{-1} B_0^T P_0 A_0$$
$$+ A_0^T P_0 B_0 (R_0 + B_0^T P_0 B_0)^{-1} B_0 P_{10} B_0^T (R_0 + B_0^T P_0 B_0)^{-1} B_0^T P_{10} A_0$$
$$+ A_0^T P_{10} B_0 (R_0 + B_0^T P_0 B_0)^{-1} B_0 P_{10} B_0^T (R_0 + B_0^T P_0 B_0)^{-1} B_0^T P_0 A_0$$
$$+ A_1^T P_0 B_0 (R_0 + B_0^T P_0 B_0)^{-1} B_0 P_{10} B_0^T (R_0 + B_0^T P_0 B_0)^{-1} B_0^T P_0 A_0$$
$$+ A_0^T P_0 B_0 (R_0 + B_0^T P_0 B_0)^{-1} B_0 P_{10} B_0^T (R_0 + B_0^T P_0 B_0)^{-1} B_0^T P_0 A_1$$
$$+ A_0^T P_0 B_0 (R_0 + B_0^T P_0 B_0)^{-1} B_0 P_{20} B_0^T (R_0 + B_0^T P_0 B_0)^{-1} B_0^T P_0 A_0 + Q_{20} = 0.$$

The linear equation for P_{02} has the form

$$-P_{02} + A_0^T P_{02} A_0 - A_0^T P_{01} B_0 (R_0 + B_0^T P_0 B_0)^{-1} B_0^T P_{01} A_0$$
$$- A_0^T P_0 B_0 (R_0 + B_0^T P_0 B_0)^{-1} B_0^T P_{02} A_0 - A_0^T P_0 B_0 (R_0 + B_0^T P_0 B_0)^{-1} B_1^T P_{01} A_0$$
$$- A_0^T P_0 B_1 (R_0 + B_0^T P_0 B_0)^{-1} B_0^T P_{01} A_0 - A_0^T P_{02} B_0 (R_0 + B_0^T P_0 B_0)^{-1} B_0^T P_0 A_0$$
$$- A_0^T P_{01} B_0 (R_0 + B_0^T P_0 B_0)^{-1} B_1^T P_0 A_0 - A_0^T P_{01} B_1 (R_0 + B_0^T P_0 B_0)^{-1} B_0^T P_0 A_0$$
$$- A_0^T P_0 B_1 (R_0 + B_0^T P_0 B_0)^{-1} B_1^T P_0 A_0 + A_0^T P_0 B_0 (R_0 + B_0^T P_0 B_0)^{-1} B_0^T P_{01} B_0$$
$$\times (R_0 + B_0^T P_0 B_0)^{-1} B_0^T P_{01} A_0 + A_0^T P_0 B_0 (R_0 + B_0^T P_0 B_0)^{-1} B_1^T P_0 B_0 (R_0 + B_0^T P_0 B_0)^{-1}$$
$$\times B_0^T P_{01} A_0 + A_0^T P_0 B_0 (R_0 + B_0^T P_0 B_0)^{-1} B_0^T P_0 B_1 (R_0 + B_0^T P_0 B_0)^{-1} B_0^T P_{01} A_0$$
$$+ A_0^T P_{01} B_0 (R_0 + B_0^T P_0 B_0)^{-1} B_0^T P_{01} B_0 (R_0 + B_0^T P_0 B_0)^{-1} B_0^T P_0 A_0$$
$$+ A_0^T P_0 B_0 (R_0 + B_0^T P_0 B_0)^{-1} B_0^T P_{02} B_0 (R_0 + B_0^T P_0 B_0)^{-1} B_0^T P_0 A_0$$
$$+ A_0^T P_0 B_0 (R_0 + B_0^T P_0 B_0)^{-1} B_0^T P_{01} B_0 (R_0 + B_0^T P_0 B_0)^{-1} B_1^T P_0 A_0$$
$$+ A_0^T P_0 B_1 (R_0 + B_0^T P_0 B_0)^{-1} B_0^T P_{01} B_0 (R_0 + B_0^T P_0 B_0)^{-1} B_0^T P_0 A_0$$
$$+ A_0^T P_{01} B_0 (R_0 + B_0^T P_0 B_0)^{-1} B_0^T P_0 B_0 (R_0 + B_0^T P_0 B_0)^{-1} B_0^T P_0 A_0$$
$$+ A_0^T P_0 B_0 (R_0 + B_0^T P_0 B_0)^{-1} B_1^T P_0 B_0 (R_0 + B_0^T P_0 B_0)^{-1} B_1^T P_0 A_0$$
$$+ A_0^T P_0 B_1 (R_0 + B_0^T P_0 B_0)^{-1} B_1^T P_0 B_0 (R_0 + B_0^T P_0 B_0)^{-1} B_0^T P_0 A_0$$
$$+ A_0^T P_{01} B_0 (R_0 + B_0^T P_0 B_0)^{-1} B_0^T P_0 B_1 (R_0 + B_0^T P_0 B_0)^{-1} B_0^T P_0 A_0$$
$$+ A_0^T P_0 B_0 (R_0 + B_0^T P_0 B_0)^{-1} B_0^T P_{01} B_1 (R_0 + B_0^T P_0 B_0)^{-1} B_0^T P_0 A_0$$
$$+ A_0^T P_0 B_0 (R_0 + B_0^T P_0 B_0)^{-1} B_0^T P_0 B_1 (R_0 + B_0^T P_0 B_0)^{-1} B_1^T P_0 A_0$$
$$+ A_0^T P_0 B_1 (R_0 + B_0^T P_0 B_0)^{-1} B_0^T P_0 B_1 (R_0 + B_0^T P_0 B_0)^{-1} B_0^T P_0 A_0$$
$$+ A_0^T P_0 B_0 (R_0 + B_0^T P_0 B_0)^{-1} B_1^T P_0 B_1 (R_0 + B_0^T P_0 B_0)^{-1} B_0^T P_0 A_0 + Q_{02} = 0$$

and linear equation for P_{11} is

$$
\begin{aligned}
&-P_{11} + A_1^T P_{01} A_0 + A_0^T P_{01} A_1 + A_0^T P_{11} A_0 - A_1^T P_0 B_0 (R_0 + B_0^T P_0 B_0)^{-1} B_1^T P_0 A_0 \\
&- A_1^T P_0 B_1 (R_0 + B_0^T P_0 B_0)^{-1} B_0^T P_0 A_0 - A_0^T P_0 B_0 (R_0 + B_0^T P_0 B_0)^{-1} B_1^T P_0 A_1 \\
&- A_0^T P_0 B_1 (R_0 + B_0^T P_0 B_0)^{-1} B_0^T P_0 A_1 - A_1^T P_0 B_0 (R_0 + B_0^T P_0 B_0)^{-1} B_0^T P_{01} A_0 \\
&- A_1^T P_{01} B_0 (R_0 + B_0^T P_0 B_0)^{-1} B_0^T P_0 A_0 - A_1^T P_{01} B_0 (R_0 + B_0^T P_0 B_0)^{-1} B_0^T P_0 A_1 \\
&- A_0^T P_0 B_0 (R_0 + B_0^T P_0 B_0)^{-1} B_0^T P_{01} A_1 - A_0^T P_0 B_0 (R_0 + B_0^T P_0 B_0)^{-1} B_0^T P_{11} A_0 \\
&- A_0^T P_{11} B_0 (R_0 + B_0^T P_0 B_0)^{-1} B_0^T P_0 A_0 - A_0^T P_{01} B_0 (R_0 + B_0^T P_0 B_0)^{-1} B_0^T P_{10} A_0 \\
&- A_0^T P_{10} B_0 (R_0 + B_0^T P_0 B_0)^{-1} B_0^T P_{01} A_0 - A_0^T P_{10} B_0 (R_0 + B_0^T P_0 B_0)^{-1} B_1^T P_0 A_0 \\
&- A_0^T P_0 B_0 (R_0 + B_0^T P_0 B_0)^{-1} B_1^T P_{10} A_0 - A_0^T P_0 B_1 (R_0 + B_0^T P_0 B_0)^{-1} B_0^T P_{10} A_0 \\
&- A_0^T P_{10} B_1 (R_0 + B_0^T P_0 B_0)^{-1} B_0^T P_0 A_0 + A_0^T P_0 B_0 (R_0 + B_0^T P_0 B_0)^{-1} B_0^T P_{01} B_0 \\
&\times (R_0 + B_0^T P_0 B_0)^{-1} B_0^T P_{10} A_0 + A_0^T P_0 B_0 (R_0 + B_0^T P_0 B_0)^{-1} B_0^T P_{10} B_0 \\
&\times (R_0 + B_0^T P_0 B_0)^{-1} B_0^T P_{01} A_0 + A_1^T P_0 B_0 (R_0 + B_0^T P_0 B_0)^{-1} B_0^T P_{01} B_0 \\
&\times (R_0 + B_0^T P_0 B_0)^{-1} B_0^T P_0 A_0 + A_0^T P_0 B_0 (R_0 + B_0^T P_0 B_0)^{-1} B_0^T P_{01} B_0 \\
&\times (R_0 + B_0^T P_0 B_0)^{-1} B_0^T P_0 A_1 + A_0^T P_{10} B_0 (R_0 + B_0^T P_0 B_0)^{-1} B_0^T P_{01} B_0 \\
&\times (R_0 + B_0^T P_0 B_0)^{-1} B_0^T P_0 A_0 + A_0^T P_{01} B_0 (R_0 + B_0^T P_0 B_0)^{-1} B_0^T P_{10} B_0 \\
&\times (R_0 + B_0^T P_0 B_0)^{-1} B_0^T P_0 A_0 + A_0^T P_0 B_0 (R_0 + B_0^T P_0 B_0)^{-1} B_0^T P_{11} B_0 \\
&\times (R_0 + B_0^T P_0 B_0)^{-1} B_0^T P_0 A_0 + A_0^T P_0 B_0 (R_0 + B_0^T P_0 B_0)^{-1} B_0^T P_0 B_1 \\
&\times (R_0 + B_0^T P_0 B_0)^{-1} B_0^T P_{10} A_0 + A_1^T P_0 B_0 (R_0 + B_0^T P_0 B_0)^{-1} B_0^T P_0 B_1 \\
&\times (R_0 + B_0^T P_0 B_0)^{-1} B_0^T P_0 A_0 + A_0^T P_0 B_0 (R_0 + B_0^T P_0 B_0)^{-1} B_0^T P_0 B_1 \\
&\times (R_0 + B_0^T P_0 B_0)^{-1} B_0^T P_0 A_1 + A_0^T P_0 B_1 (R_0 + B_0^T P_0 B_0)^{-1} B_0^T P_{10} B_0 \\
&\times (R_0 + B_0^T P_0 B_0)^{-1} B_0^T P_0 A_0 + A_0^T P_{10} B_0 (R_0 + B_0^T P_0 B_0)^{-1} \\
&\times B_0^T P_0 B_1 (R_0 + B_0^T P_0 B_0)^{-1} B_0^T P_0 A_0 + A_0^T P_0 B_0 (R_0 + B_0^T P_0 B_0)^{-1} B_0^T P_{10} B_1 \\
&\times (R_0 + B_0^T P_0 B_0)^{-1} B_0^T P_0 A_0 + A_0^T P_0 B_0 (R_0 + B_0^T P_0 B_0)^{-1} \\
&\times B_1^T P_0 B_0 (R_0 + B_0^T P_0 B_0)^{-1} B_0^T P_{10} A_0 + A_1^T P_0 B_0 (R_0 + B_0^T P_0 B_0)^{-1} B_1^T P_0 B_0 \\
&\times (R_0 + B_0^T P_0 B_0)^{-1} B_0^T P_0 A_0 + A_0^T P_0 B_0 (R_0 + B_0^T P_0 B_0)^{-1} \\
&\times B_1^T P_0 B_0 (R_0 + B_0^T P_0 B_0)^{-1} B_0^T P_0 A_1 + A_0^T P_{10} B_0 \\
&\times (R_0 + B_0^T P_0 B_0)^{-1} B_1^T P_0 B_0 (R_0 + B_0^T P_0 B_0)^{-1} B_0^T P_0 A_0 \\
&+ A_0^T P_0 B_0 (R_0 + B_0^T P_0 B_0)^{-1} B_0^T P_{10} B_0 (R_0 + B_0^T P_0 B_0)^{-1} B_1^T P_0 A_0 \\
&+ A_0^T P_0 B_0 (R_0 + B_0^T P_0 B_0)^{-1} B_1^T P_{10} B_0 (R_0 + B_0^T P_0 B_0)^{-1} B_0^T P_0 A_0 + Q_{11} = 0
\end{aligned}
$$

Note that the terms P_{10} and P_{01} of the asymptotic approximation depend on the matrices $A_1(x)$ and $B_1(x)$, respectively.

Here we assume that

I. *There exist such matrices $Q_0, Q_{10}, Q_{01}, Q_{20}, Q_{11}, Q_{02}$ that the equations are solvable and $P_0, P_{10}, P_{01}, P_{20}, P_{11}, P_{02}$ are positive definite.*

Further, on the basis of the obtained asymptotic expansion for the solution of the Riccati equation (3) for small values of the parameters, a one-point matrix Pade approximation of order [2/2] by two parameters ε, μ is constructed

$$
\begin{aligned}
PA_{[2/2]}(x, \varepsilon, \mu) = {} & (M_0 + \varepsilon M_{10}(x) + \mu M_{01}(x) + \varepsilon^2 M_{20}(x) + \varepsilon\mu M_{11}(x) + \mu^2 M_{02}(x)) \\
& \times (E + \varepsilon N_{10}(x) + \mu N_{01}(x) + \varepsilon^2 N_{20}(x) + \varepsilon\mu N_{11}(x) + \mu^2 N_{02}(x))^{-1},
\end{aligned} \tag{5}
$$

where E is the identity matrix, matrices M, N are square continuously differentiable matrices of dimension $n \times n$.

Unknown matrix coefficients of the Pade approximation are found from a system of equations, which is obtained by equating coefficients with the same degrees of parameters from the equality $PA_{[2/2]}(x, \varepsilon, \mu) = P(x, \varepsilon, \mu)$ and the corresponding expansions of the right and left parts.

The system of equations for 11 unknown matrix Pade coefficients in (5) has the form

$$\varepsilon^0, \mu^0 : \quad M_0 = P_0$$
$$\varepsilon : \quad M_{10} = P_{10} + P_0 N_{10}$$
$$\mu : \quad M_{01} = P_{01} + P_0 N_{01}$$
$$\varepsilon\mu : \quad M_{11} = P_{11} + P_{10}N_{01} + P_{01}N_{10} + P_0 N_{11}$$
$$\varepsilon^2 : \quad M_{20} = P_{20} + P_{10}N_{10} + P_0 N_{20}$$
$$\mu^2 : \quad M_{02} = P_{02} + P_{01}N_{01} + P_0 N_{02}$$
$$\varepsilon\mu^2 : \quad 0 = P_{11}N_{01} + P_{10}N_{02} + P_{02} N_{10} + P_{01}N_{11}$$
$$\varepsilon^2\mu : \quad 0 = P_{20}N_{01} + P_{11}N_{10} + P_{10}N_{11} + P_{01}N_{20}$$
$$\varepsilon^3 : \quad 0 = P_{20}N_{10} + P_{10}N_{20}$$
$$\mu^3 : \quad 0 = P_{02}N_{01} + P_{01}N_{02}$$
$$\varepsilon^2\mu^2 : 0 = P_{20}N_{02} + P_{11} N_{11} + P_{02}N_{20}.$$

The applicability of the algorithm is determined by the solvability of the system of matrix equations for the PA coefficients and the existence of the inverse matrix $(E + \varepsilon N_{10}(x) + \mu N_{01}(x) + \varepsilon^2 N_{20}(x) + \varepsilon\mu N_{11}(x) + \mu^2 N_{02}(x))^{-1}$. The matrices of the "denominator" are found from the system of the last five equations. To get the nonzero coefficients of the "denominator" we add additional optimization parameters $P_{12}, P_{21}, P_{03}, P_{30}, P_{22}$ which are found as positive definite matrices from the minimum of the quality criterion (2).

$$\varepsilon\mu^2 : \quad P_{12} = P_{11}N_{01} + P_{10}N_{02} + P_{02} N_{10} + P_{01}N_{11}$$
$$\varepsilon^2\mu : \quad P_{21} = P_{20}N_{01} + P_{11}N_{10} + P_{10}N_{11} + P_{01}N_{20}$$
$$\varepsilon^3 : \quad P_{30} = P_{20}N_{10} + P_{10}N_{20} \tag{6}$$
$$\mu^3 : \quad P_{03} = P_{02}N_{01} + P_{01}N_{02}$$
$$\varepsilon^2\mu^2 : P_{22} = P_{20}N_{02} + P_{11} N_{11} + P_{02}N_{20}$$

or

$$\begin{pmatrix} P_{02}(x) & P_{11}(x) & P_{01}(x) & 0 & P_{10}(x) \\ P_{11} & P_{20}(x) & P_{10}(x) & P_{01}(x) & 0 \\ P_{20}(x) & 0 & 0 & -P_{10}(x) & 0 \\ 0 & P_{02}(x) & 0 & 0 & P_{01}(x) \\ 0 & 0 & P_{11}(x) & P_{02}(x) & P_{20}(x) \end{pmatrix} \begin{pmatrix} N_{10}(x) \\ N_{01}(x) \\ N_{11}(x) \\ N_{20}(x) \\ N_{02}(x) \end{pmatrix} = \begin{pmatrix} P_{12} \\ P_{21} \\ P_{30} \\ P_{03} \\ P_{22} \end{pmatrix}.$$

The Algorithm of the Pade Regulator Construction

Step 1. An asymptotic approximation of the solution of the matrix discrete state-dependent Riccati equation by powers of two small parameters ε, μ is constructed (4).

Step 2. A one-point matrix PA [2/2] by two parameters for the solution of a matrix discrete Riccati equation is constructed using an asymptotic approximation by two parameters (step 1).

Step 3. Using a one-point matrix PA we get the Pade regulator

$$u(x, \varepsilon, \mu) = -\tilde{R}(x, \varepsilon, \mu)^{-1} B(x, \mu)^T K(x, \varepsilon, \mu) A(x, \varepsilon) x(t), \tag{7}$$

where a symmetric control gain matrix $K(x, \varepsilon, \mu)$ has the form $K(x, \varepsilon, \mu) = 1/2 * (PA_{[2/2]}(x, \varepsilon, \mu) + PA_{[2/2]}(x, \varepsilon, \mu)^T)$.

3   Numerical Experiments

Here, the results of numerical experiments carried out for different parameter values are presented.

Example 1

Let the dynamics of the system be described by the equation

$$x(t+1) = (A_0 + \varepsilon A_1(x))x(t) + (B_0 + \mu B_1(x))u(t),$$

where

$$A_0 = \begin{pmatrix} 1 & 0.1 \\ 0 & 0.5 \end{pmatrix}, \ B_0 = \begin{pmatrix} 0 \\ 0.5 \end{pmatrix}, \ A_1(x) = \begin{pmatrix} 2\frac{\sin(x_1)}{x_1} & 1 \\ \frac{\sin(x_2)}{x_2} & 2 \end{pmatrix}, \ B_1(x) = \begin{pmatrix} 0.1 \\ 0.1 \end{pmatrix}.$$

The results of solving this control problem according to **The Algorithm** are presented in Table. 1 and on the Fig. 1. The coefficients $P_{12}, P_{21}, P_{03}, P_{30}, P_{22}$ in (6) are found by optimisation of criterion (2). It can be seen, that on the selected grid the Pade regulator is close to the D-SDRE regulator by the quality criterion (2) with matrices

$$Q_0 = \begin{pmatrix} 5 & 0 \\ 0 & 5 \end{pmatrix}, \ Q_{10} = Q_{01} = \begin{pmatrix} 15 + 0.01x_1^2 & 0 \\ 0 & 15 + 0.01x_1^2 \end{pmatrix}, \ Q_{11} = \begin{pmatrix} 10 & 0 \\ 0 & 10 \end{pmatrix},$$
$$Q_{20} = Q_{02} = 0, R_0 = 1.$$

Table 1. The comparison of the closed-loop controls by the quality criterion

ε	μ	D-SDRE regulator	Pade [2/2] by two parameters
0,01	0,01	227,43	227,52
0,05	0,01	289,18	295,39
0,1	0,01	335,33	366,55
0,01	0,05	218,01	218,17
0,05	0,05	278,34	284,59
0,1	0,05	325,75	355,09
0,01	0,1	207,38	207,68
0,05	0,1	265,91	272,29
0,1	0,1	314,49	341,63

Closed-loop trajectories for Pade [2/2] regulator and D-SDRE regulator are presented in Fig. 2.

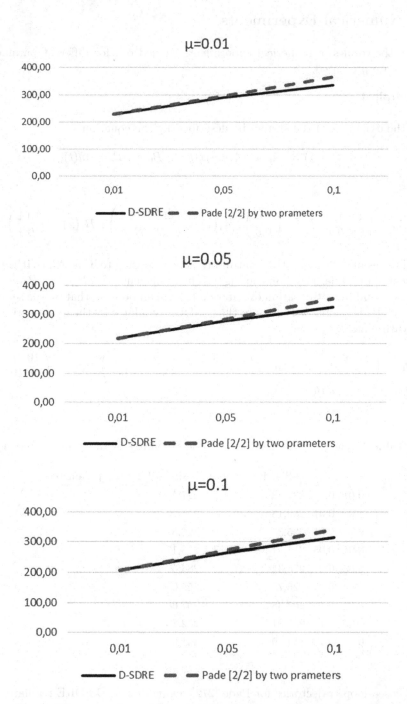

Fig. 1. The comparison of the regulators by the quality criterion values

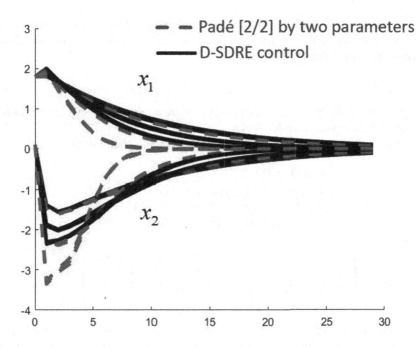

Fig. 2. The trajectories of the closed-loop systems for different parameter values

Example 2

In the next example the problem (1) is reduced to a problem with a single parameter, where it is assumed, for example, that $\mu = c\varepsilon^2$, $c = O(1)$.

By substituting $\mu = c\varepsilon^2$ we get

$$x(t+1) = A(x,\varepsilon)x(t) + B(x,\varepsilon)u(t)$$
$$= (A_0 + \varepsilon A_1(x))x(t) + (B_0 + c\varepsilon^2 B_1(x))x(t),$$
$$x(0) = x_0, \ x(t) \in X \subset R^n, \ u(t) \in R^r, \ t = 0,1,2\ldots,$$

$$I(u) = \frac{1}{2}\sum_{t=0}^{\infty}(x^T Q(x,\varepsilon)x + u^T R_0 u) \to \min,$$

where $X \subset R^n$ is some bounded state-space subset $0 < c$, $0 < \varepsilon \le \varepsilon_0$, $Q(x,\varepsilon) > 0$, $R_0 > 0$. Here we use the second order asymptotics

$$\tilde{P}(x,\varepsilon) = \tilde{P}_0 + \varepsilon\tilde{P}_1(x) + \varepsilon^2\tilde{P}_2(x), \tag{8}$$

using matrices $Q(x,\varepsilon) = Q_0 + \varepsilon Q_1(x) + \varepsilon^2 Q_2(x)$, $Q_1(x) = Q_{10}$, $Q_2(x) = Q_{20}$,

$$Q_0 = \begin{pmatrix} 10 & 1 \\ 1 & 10 \end{pmatrix}, Q_{10} = \begin{pmatrix} 11 + 0.01x_1^2 & 0 \\ 0 & 11 + 0.01x_1^2 \end{pmatrix},$$
$$Q_{01} = Q_{11} = Q_{02} = Q_{20} = 0, R_0 = 1.$$

Now, using second-order asymptotics (8) of the solution of the Riccati equation (3) for small values of the parameter ε, we construct a one-point Pade approximation of order [1/2], namely

$$PA_{[1/2]}(x,\varepsilon) = (M_0(x) + \varepsilon\, M_1(x))\left(E + \varepsilon\, N_1(x) + \varepsilon^2 N_2(x)\right)^{-1}, \qquad (9)$$

where E is an identity matrix, matrices M, N are square quadratic continuously differentiable matrices of order $n \times n$.

A system of equations for the coefficients for Pade coefficients is

$$M_0(x) = \tilde{P}_0(x)$$
$$M_1(x) = \tilde{P}_0(x)N_1(x) + \tilde{P}_1(x)$$
$$0 = \tilde{P}_0(x)N_2(x) + \tilde{P}_1(x)\,N_1(x) + \tilde{P}_2(x)$$
$$0 = \tilde{P}_1(x)N_2(x) + \tilde{P}_2(x)\,N_1(x)$$

or

$$\begin{pmatrix} E & 0 & 0 & 0 \\ 0 & E & -\tilde{P}_0(x) & 0 \\ 0 & 0 & -\tilde{P}_1(x) & -\tilde{P}_0(x) \\ 0 & 0 & -\tilde{P}_2(x) & -\tilde{P}_1(x) \end{pmatrix} \begin{pmatrix} M_0(x) \\ M_1(x) \\ N_1(x) \\ N_2(x) \end{pmatrix} = \begin{pmatrix} \tilde{P}_0(x) \\ \tilde{P}_1(x) \\ \tilde{P}_2(x) \\ 0 \end{pmatrix},$$

We can also construct the Pade approximation by one parameter of order [2/2] using the second order asymptotic approximation. For this we additionally introduce the coefficients $\tilde{P}_3$, $\tilde{P}_4$ and find them with the help of optimisation by the quality criterion (2) to ensure the nondegeneracy of the "denominator" of the Pade approximation.

$$PA_{[2/2]}(x,\varepsilon) = (M_0(x) + \varepsilon\, M_1(x) + \varepsilon^2\, M_2(x))\left(E + \varepsilon\, N_1(x) + \varepsilon^2 N_2(x)\right)^{-1}, \qquad (10)$$

$$M_0(x) = \tilde{P}_0(x)$$
$$M_1(x) = \tilde{P}_0(x)N_1(x) + \tilde{P}_1(x)$$
$$M_2(x) = \tilde{P}_0(x)N_2(x) + \tilde{P}_1(x)\,N_1(x) + \tilde{P}_2(x),$$

$$\begin{pmatrix} -\tilde{P}_2(x) & -\tilde{P}_1(x) \\ -\tilde{P}_3 & -\tilde{P}_2(x) \end{pmatrix} \begin{pmatrix} N_1(x) \\ N_2(x) \end{pmatrix} = \begin{pmatrix} \tilde{P}_3 \\ \tilde{P}_4 \end{pmatrix}.$$

Table 2 and Fig. 3 presents the comparison of controls by quality criteria values for $c = 1$. It is shown that by quality criteria values the asymptotic approximation by two parameters is much closer to the D-SDRE solution and preserve this closeness and qualitative compliance for a large parameter variation interval.

Table 2. The comparison of controls by quality criteria values for c = 1

ε	μ	D-SDRE	Asymptotics by two parameters	Asymptotics by one parameter
0,01	0,0001	392,82	392,9156	392,8548
0,05	0,0025	468,7696	472,8012	650,4096
0,1	0,01	510,0006	538,2427	1613,126
0,2	0,04	555,6039	695,9590	44903,85
0,3	0,09	607,9071	887,2460	15506927

The next table (Table 5) presents the results for Pade regulators constructed on the basis of the asymptotic approximations:

- Pade approximation $PA_{[2/2]}(x, \varepsilon, \mu)$ in (5) by two parameters corresponds to the asymptotic approximation by two parameters from (4);
- Pade approximations by one parameter $PA_{[1/2]}(x, \varepsilon)$ in (9), $PA_{[2/2]}(x, \varepsilon)$ in (10) correspond to the asymptotic approximation by one parameter from (8).

Here matrices $P_{12}, P_{21}, P_{03}, P_{30}, P_{22}$ from (6) are found by optimization for each combination of parameter values and are presented in Table 3. Matrices $\tilde{P}_3$, $\tilde{P}_4$ for $PA_{[2/2]}(x, \varepsilon)$ from (10) by one parameter are presented in Table 4.

Table 3. The optimised coefficients of the Pade system (6)

ε	μ	P_{12}	P_{21}	P_{03}	P_{30}	P_{22}
0,01	0,0001	$\begin{pmatrix} 0,89 & 0,70 \\ 0,70 & 0,87 \end{pmatrix}$	$\begin{pmatrix} 1,41 & 0,91 \\ 0,91 & 0,85 \end{pmatrix}$	$\begin{pmatrix} 1,37 & 0,78 \\ 0,78 & 1,30 \end{pmatrix}$	$\begin{pmatrix} 1,21 & 0,97 \\ 0,97 & 1,32 \end{pmatrix}$	$\begin{pmatrix} 1,07 & 0,67 \\ 0,67 & 1,11 \end{pmatrix}$
0,05	0,0025	$\begin{pmatrix} 1,59 & 0,96 \\ 0,96 & 0,75 \end{pmatrix}$	$\begin{pmatrix} 1,51 & 0,63 \\ 0,63 & 0,57 \end{pmatrix}$	$\begin{pmatrix} 1,93 & 1,20 \\ 1,20 & 1,34 \end{pmatrix}$	$\begin{pmatrix} 1,56 & 0,91 \\ 0,91 & 1,30 \end{pmatrix}$	$\begin{pmatrix} 2,94 & 0,66 \\ 0,66 & 0,41 \end{pmatrix}$
0,1	0,01	$\begin{pmatrix} 1,56 & 0,93 \\ 0,93 & 0,73 \end{pmatrix}$	$\begin{pmatrix} 1,49 & 0,63 \\ 0,63 & 0,57 \end{pmatrix}$	$\begin{pmatrix} 1,80 & 1,25 \\ 1,25 & 1,54 \end{pmatrix}$	$\begin{pmatrix} 1,43 & 0,43 \\ 0,43 & 1,11 \end{pmatrix}$	$\begin{pmatrix} 2,90 & 0,54 \\ 0,54 & 0,38 \end{pmatrix}$
0,2	0,04	$\begin{pmatrix} 2,62 & 0,30 \\ 0,30 & 0,94 \end{pmatrix}$	$\begin{pmatrix} 0,14 & 0,13 \\ 0,13 & 1,78 \end{pmatrix}$	$\begin{pmatrix} 1,87 & 0,00 \\ 0,00 & 3,25 \end{pmatrix}$	$\begin{pmatrix} 7,50 & 0,00 \\ 0,00 & 0,18 \end{pmatrix}$	$\begin{pmatrix} 0,85 & 0,76 \\ 0,76 & 0,02 \end{pmatrix}$
0,3	0,09	$\begin{pmatrix} 0,73 & 1,37 \\ 1,37 & 3,16 \end{pmatrix}$	$\begin{pmatrix} 1,03 & 0,09 \\ 0,09 & 0,76 \end{pmatrix}$	$\begin{pmatrix} 7,34 & 0,45 \\ 0,45 & 0,73 \end{pmatrix}$	$\begin{pmatrix} 0,93 & 0,13 \\ 0,13 & 0,02 \end{pmatrix}$	$\begin{pmatrix} 0,32 & 0,02 \\ 0,02 & 0,13 \end{pmatrix}$

As it can be seen both Pade regulators demonstrate the improvement over the asymptotic approximations (Fig. 3), but the Pade regulator that uses the asymptotic approximation by two parameters works on the larger time interval, than Pade regulator by one parameter and it stabilizes the system for all considered parameter values (see Fig. 4).

Figure 5 emphasizes that the reduction of two small parameters to one is not always successful.

Table 4. The optimised coefficients of the Pade system (10)

ε	μ	$\tilde{P}_3$	$\tilde{P}_4$
0,01	0,0001	$\begin{pmatrix} 1,2984 \ 0,1810 \\ 0,1810 \ 0,7307 \end{pmatrix}$	$\begin{pmatrix} 2,8094 \ 1,6079 \\ 1,6079 \ 1,4387 \end{pmatrix}$
0,05	0,0025	$\begin{pmatrix} 0,4460 \ 0,6395 \\ 0,6395 \ 3,2556 \end{pmatrix}$	$\begin{pmatrix} 3,1227 \ 1,5445 \\ 1,5445 \ 3,1231 \end{pmatrix}$
0,1	0,01	$\begin{pmatrix} 0,4460 \ 0,6395 \\ 0,6395 \ 3,2556 \end{pmatrix}$	$\begin{pmatrix} 3,1227 \ 1,5445 \\ 1,5445 \ 3,1231 \end{pmatrix}$
0,2	0,04	$\begin{pmatrix} 0,4460 \ 0,6395 \\ 0,6395 \ 3,2556 \end{pmatrix}$	$\begin{pmatrix} 3,1227 \ 1,5445 \\ 1,5445 \ 3,1231 \end{pmatrix}$
0,3	0,09	$\begin{pmatrix} 19,9346 \ 0,0125 \\ 0,0125 \ 0,0266 \end{pmatrix}$	$\begin{pmatrix} 1,5964 \ 0,0250 \\ 0,0250 \ 3,5108 \end{pmatrix}$

Table 5. The comparison of Pade controls by quality criteria values

ε	μ	D-SDRE	Pade [2/2] by two parameters	Pade [1/2] by one parameter	Pade [2/2] by one parameter
0,01	0,0001	392,82	392,90	392,7795	392,78
0,05	0,0025	468,7696	474,29	489,9492	511,83
0,1	0,01	510,0006	547,91	2781,8328	1153,78
0,2	0,04	555,6039	656,84	1483,4005	2530,40
0,3	0,09	607,9071	764,96	2000,3707	2511311,34

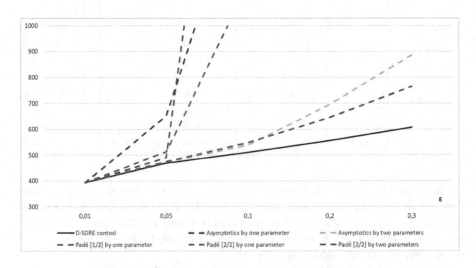

Fig. 3. The comparison of the regulators by the quality criterion values

x_1, x_2

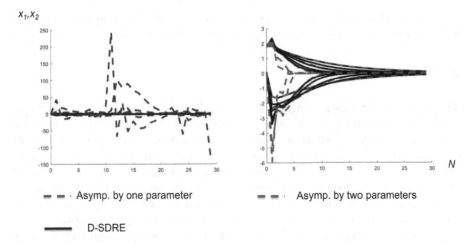

Fig. 4. Trajectories of closed-loop systems for regulators based on second-order asymptotic approximations for different parameter values

x_1, x_2

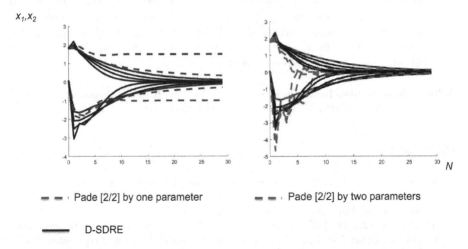

Fig. 5. Trajectories of closed-loop systems for Pade regulators for different parameter values

The asymptotic approximation by two parameters provides stabilization in all the considered cases, in contrast to the asymptotic approximation by one parameter (Fig. 4). As it can be seen, the Pade regulator based on asymptotic approximation by one parameter (8) fails to stabilize the system for some parameter valus (Fig. 5) unlike the Pade regulator by two parameters.

Thus, these experiments demonstrate that the regulators costructed using Pade approximations by two parameters stabilize the closed-loop systems for more cases then regulators that use asymptotic approximations by one parameter. The obtained PA regulators have interpolation and extrapolation properties, which often allow to increase the accuracy of the approximation in comparison with regulators based on pure asymptotic approximations like (4).

4   Conclusion

The algorithm for constructing a symbolic family of Pade regulators for nonlinear control systems with two parameters is proposed. It is based on the asymptotic approximation of the solution of the discrete matrix state-dependent Riccati equation (D-SDRE) in the zero point (0,0) by both parameters simultaniously and the further construction of the matrix Pade approximation in order to extend the admissible parameters' variation intervals. The comparison of the proposed Pade regulators with the controller based on an asymptotic approximation by one parameter demonstrates the advantage of the Pade approximation by two parameters. The experiments show that the trajectories of closed-loop systems calculated along the controls which correspond to the asymptotic approximations by two parameters are closer to the D-SDRE solution for a larger interval of parameter variation, than by one parameter. Moreover, the regulator based on the asymptotic approximation by two parameters is better by the quality criterion values. It is also demonstrated that the use of the matrix Pade approximations gives an improvement over the asymptotic approximations by the quality of stabilization in the considered examples.

Acknowledgements. Research is supported by the Russian Science Foundation (Project No. 21-11-00202), https://rscf.ru/project/21-11-00202/.

References

1. Danik, Yu., Dmitriev, M.: Construction of parametric regulators for nonlinear control systems based on the Pade approximations of the matrix Riccati equation solution. IFAC-PapersOnLine **51**, 815–820 (2018)
2. Danik, Yu.: One D-SDRE regulator for weakly nonlinear discrete state dependent coefficients control systems. In: The 7th International Conference on Control, Decision and Information Technologies (CODIT 2020), pp. 616–621 (2020)
3. Danik, Y., Dmitriev, M.: The construction of stabilizing regulators sets for non-linear control systems with the help of Padé approximations. In: Abramian, A.K., Andrianov, I.V., Gaiko, V.A. (eds.) Nonlinear Dynamics of Discrete and Continuous Systems. ASM, vol. 139, pp. 45–62. Springer, Cham (2021). https://doi.org/10.1007/978-3-030-53006-8_4
4. Gradstein, I.: Differential equations in which the multipliers at the derivatives include various degrees of a small parameter. Dokl. Akad. nauk SSSR **82**(1), 5–8 (1952). (in Russian)
5. Tikhonov, A.N.: Systems of differential equations containing small parameters with derivatives. Mat. Sb. **31**(73), 575–586 (1952). (in Russian)
6. Dutka, A., Ordys, A., Grimble, M.: Optimized discrete-time state dependent Riccati equation regulator. In: Proceedings of the American Control Conference, pp. 2293–2298. IEEE (2005)

7. Baker, G., Graves-Morris, P.: Pade Approximations. Addison-Wesley Publishing (1999)
8. Andrianov, I., Awrejcewicz, J.: Methods of asymptotic analysis and synthesis in the nonlinear dynamics and mechanics of a deformable solid. Izhevsk, IKI (2013). (in Russian)
9. Andrianov, I., Shatrov, A.: Pade approximants, their properties, and applications to hydrodynamic problems. Symmetry **13**(10), 1869–1952 (2021)
10. Danik, Y., Dmitriev, M.: Symbolic regulator sets for a weakly nonlinear discrete control system with a small step. Mathematics **10**, 1–14 (2022)
11. Kitamoto, T.: Multivariate Pad approximation using quantifier elimination. In: Proceedings of the 20th Asian Technology Conference in Mathematics, Leshan, China (2015)
12. Abedinnasab, M., Yoon, Y., Saeedi-Hosseiny, M.: High performance fuzzy-Pade controllers: introduction and comparison to fuzzy controllers. Nonlinear Dyn. **71**(1), 141–157 (2013)
13. Babu, G., Krishnarayalu, M.S.: Suboptimal control of singularly perturbed two parameter discrete control system. Int. Electr. Eng. J. (IEEJ) **5**(11), 1594–1604 (2014)

Analytical Construction of the Singular Set in One Class of Time-Optimal Control Problems in the Presence of Linear Segments of the Boundary of the Target

Lebedev Pavel$^{(\boxtimes)}$ ⓘ and Uspenskii Alexander ⓘ

N.N. Krasovskii Institute of Mathematics and Mechanics of UB RAS,
str. S. Kovalevskaya, 16, Yekaterinburg 620108, Russia
pleb@yandex.ru
https://www.imm.uran.ru/rus/Pages/default.aspx

Abstract. A time-optimal control problem with a spherical velocity vectogram is considered. For one class of non-convex planar target sets with a part of their boundary coinciding with a line segment, conditions are found to construct branches of singular (scattering) curves in analytical form. Explicit formulas are obtained for pseudo-vertices, i.e., singular boundary points of the target set generating branches of the singular set. An analytical relation is revealed between the endpoints of different optimal trajectories with the same initial conditions on the singular set that falls on the target set in a neighborhood of a pseudo-vertex. Formulas are found for the extreme points of the singular set branches. The developed approaches to constructing exact non-smooth solutions for dynamic control problems are illustrated with examples.

Keywords: scattering curve · singular set · pseudo-vertex · mapping · curvature

1 Introduction

In a time-optimal control problem with a spherical velocity vectogram and a closed non-convex target set, the optimal result function is non-smooth regardless of differential properties of the target set boundary. The singularity of the problem solution is due to special points on the target boundary, which are pseudo-vertices. These points generate symmetry sets, known as bisectors. According to conflict management theory, bisectors are scattering curves. Unlike in regular cases, there are two or more optimal trajectories coming from the points of these curves that are directed differently from the singular curve. In

The work was performed as part of research conducted in the Ural Mathematical Center with the financial support of the Ministry of Science and Higher Education of the Russian Federation (Agreement number 075-02-2023-913).

M. Khachay et al. (Eds.): MOTOR 2023, CCIS 1881, pp. 292–307, 2023.
https://doi.org/10.1007/978-3-031-43257-6_22

general, bisectors are constructed by numerical methods. These methods are based on identifying pseudo-vertices of the target set, followed by the construction of bisector branches by solving algebraic or ordinary differential equations. From experience of constructing solutions to such problems to date it is possible to determine conditions for a singular set to be described analytically. This paper analyzes the case with the target set boundary containing line segments. Analytical formulas of the mappings connecting the target set points are obtained. On the one hand, these points are endpoints of optimal trajectories with common initial conditions, while, on the other hand, they are located in the neighborhood of a pseudo-vertex. An example of a set is also found and presented with a curved boundary, i.e. a boundary with no straight line segments, for which, nevertheless, it is possible to construct these mappings in an explicit analytical form.

It should be noted that the specific dynamic problem under consideration reduces studying the singular set of its solution to researching properties of a metric projection onto a closed set of Euclidean space (by geometric methods mainly). At the same time, the paper considers in detail the case with a projection operator with no more than two values on the target set.

2   Problem Statement

The time-optimal control problem on the plane $\mathbf{R}^2$ with a closed non-convex target set $A \subset \mathbf{R}^2$ is considered. The system solution is to be found with current coordinates $\mathbf{x} = (x, y)$ in the optimal time on the target set A. The system dynamics

$$\dot{\mathbf{x}} = \mathbf{v} \tag{1}$$

is defined by vector control $\mathbf{v} = (v_1, v_2)$, with possible values $\mathbf{v} \in O(\mathbf{0}, 1)$, where $O(\mathbf{c}, r) = \{\mathbf{x} \in \mathbf{R}^2 : \|\mathbf{x} - \mathbf{c}\| \leq r\}$ is a circle of radius r centered at point $\mathbf{c}$; $\mathbf{0} = (0, 0)$ is the origin.

In the present problem, the optimal control $\mathbf{v}$ for $\mathbf{x} \notin A$ is a vector of length 1 that is co-directed with the vector originating from point $\mathbf{x}$ to the nearest point $\mathbf{y}$ in the Euclidean metric of the set A boundary. The optimal result function $u(\mathbf{x}) = u(x, y)$ is equal to the Euclidean distance $\rho(\mathbf{x}, A) = \min\{\|\mathbf{x} - \mathbf{y}\| : \mathbf{y} \in A\}$ from the point $\mathbf{x} = (x, y) \in \mathbf{R}^2$ to the set A.

The time-optional dynamic problem (1) and the target set A are tightly connected with the Hamilton-Jacobi differential equations

$$\min_{(v_1, v_2) \in O(\mathbf{0}, 1)} \left(v_1 \frac{\partial u}{\partial x} + v_2 \frac{\partial u}{\partial y} \right) + 1 = 0 \tag{2}$$

and eikonal equations

$$\left(\frac{\partial u}{\partial x} \right)^2 + \left(\frac{\partial u}{\partial y} \right)^2 = 1. \tag{3}$$

The generalized (minimax) solution [1, ch. IV] of the Dirichlet problem for Eq. (2) with a boundary condition

$$u|_{\partial M} = 0 \tag{4}$$

coincides with the optimal result function $u(x, y) = \rho((x, y), A)$ on the set $G = \mathbf{R}^2 \setminus A$ (see [2, Theorem 1]). The fundamental (generalized) solution of the Dirichlet problem for Eq. (3) with a boundary condition (4) (introduced by S. N. Kruzhkov [3]) is equal to the same function to modulo; however, opposite in sign: $u_k(x, y) = -\rho((x, y), A)$.

Let's assume that the boundary $\Gamma = \partial A$ of the target set is a plane curve described by the equation

$$\Gamma = \{\mathbf{x} \in \mathbf{R}^2 : \mathbf{x} = \mathbf{x}(t), t \in \Xi\}. \tag{5}$$

Here $\Xi \subseteq \mathbf{R}$ is a simply connected set, whereas the mapping $\mathbf{x} \colon \Xi \to \mathbf{R}^2$ is and twice differentiable at all internal points Ξ, except an admissible finite number of points. It should be noted that the level surface $\Phi(\tau) = \{(x, y) \in \mathbf{R}^2 : u(x, y) = \tau\}$ of the function $u(x, y) = \rho((x, y), A)$ at the time point $\tau > 0$ coincides with the wavefronts [4] when the wave propagates from the set A for the time τ in case of a spherical velocity vectogram.

If the target set is convex, then the function $u(\mathbf{x}) = \rho(\mathbf{x}, A)$ is also convex and differentiable by $\mathbf{R}^2 \setminus A$ (see [5, ch. II, §8]). If the set A is non-convex, then $u(\mathbf{x})$ has singular sets due to a number of optimal trajectories for a dynamic system (1).

We denote $\Omega_A(\mathbf{y})$ as a union of all points $\mathbf{x} \in A$, closest to $\mathbf{y}$ in the Euclidean metric. It should be noted that this construction is crucial for proving the function $u(x, y) = \rho((x, y), A)$ to be a generalized (minimax) solution of the Dirichlet problem for Eq. (2) with the boundary condition (4). Let's consider the plot $\operatorname{gr} u(x, y)$ of the function $u(x, y)$ restriction to the set $\mathbf{R}^2 \setminus A$. It is a surface in an expanded position space with three coordinates x, y, τ. At least one characteristic passes through its each point [1, ch. I, §1.2] of Eq. (2)—the segment $[(x, y, u(x, y)), (x_p, y_p, 0)]$, where $(x_p, y_p) \in \Omega_A((x, y))$.

Definition 1. *A set*

$$L(A) = \{\mathbf{y} \in \mathbf{R}^2 : \operatorname{card}(\Omega_A(\mathbf{y})) > 1\}$$

is called a bisector $L(A)$ [6] *of a closed set* $A \subset \mathbf{R}^2$.

Here, card card $(\Omega_A(\mathbf{y}))$ is the cardinality of the set $\Omega_A(\mathbf{y})$. It is equal to the number of elements, if the set $\Omega_A(\mathbf{y})$ is finite. However, cases are possible, when $\mathbf{y}$ is the center of a circle $\partial O(\mathbf{x}, r)$, with an arc in Γ. Therefore, all the arc elements can be included into $\Omega_A(\mathbf{y})$. Then, card card $(\Omega_A(\mathbf{y}))$ is the cardinality of an infinite set. If for two different points $\mathbf{x}_1 \in \Gamma$, $\mathbf{x}_2 \in \Gamma$ conditions $\mathbf{x}_1 \in \Omega_A(\mathbf{y})$ and $\mathbf{x}_2 \in \Omega_A(\mathbf{y})$ are met, then they are said to generate a bisector point. According to R. Isaacs's classification for control problems and differential games, the bisector $L(A)$ is characterized by the following property: at least two optimal trajectories—the segment $[\mathbf{y}, \mathbf{x}], \mathbf{x} \in \Omega_A(\mathbf{y})$ proceed from each of its points $\mathbf{y} \in L(A)$ [7, Example 6.10.1]. Sets similar to the bisector are studied in the theory of wavefronts and termed as "conflict set" [8], "symmetry set" [9] and "medial axe" [10]. Is should also be mentioned that in geometric optics, the fundamental solution of the boundary value problem (3), (4) is smooth at points $\mathbf{y} \in G$ if card card $\big(\Omega_A(\mathbf{y})\big) = 1$, and loses its classical differentiability when card card $\big(\Omega_A(\mathbf{y})\big) > 1$ [11]. Another important application of the set $L(A)$ is calculating non-convexity of the set A (for more details see [12]).

Definition 2. *Let us define the point* $\mathbf{x}_0 = \mathbf{x}(t_0)$ *as a pseudo-vertex of the set* A, *whereas* $\widehat{\mathbf{y}}$ *is defined as the bisector extreme point generated by this pseudo-vertex, if there is a sequence* $\widetilde{\mathbf{x}}_n)\}_{n=1}^{\infty} \subset A$ *of point pairs of the set* A *and the sequence* $\{\mathbf{y}_n\}_{n=1}^{\infty} \subset L(A)$ *of bisector points for which the following conditions are met:*

$$\lim_{n \to \infty} (\overline{\mathbf{x}}_n, \widetilde{\mathbf{x}}_n) = (\mathbf{x}_0, \mathbf{x}_0),$$

$$\lim_{n \to \infty} \mathbf{y}_n = \widehat{\mathbf{y}},$$

$$\forall n \in \mathbf{N} \ \{\overline{\mathbf{x}}_n, \widetilde{\mathbf{x}}_n\} \subseteq \Omega_A(\mathbf{y}_n).$$

If the pseudo-vertex lies on a smooth curve section (5), it is crucial to determine the relationships between the values of the parameter t, which define the projections of the bisector points. This enables to construct smooth sections of the singular set $L(A)$. In general, $L(A)$ may contain bifurcation points. However, finding them is of secondary importance. This problem is solved by isolating those elements $\mathbf{y}_i \in L(A)$ for which card card $\big(\Omega_A(\mathbf{y}_i)\big) > 2$. In more detail, singular points of sets similar to the bisector in Euclidean spaces of small dimension were investigated, e.g., by V. D. Sedykh [13,14].

Definition 3. *Let the point* $\mathbf{x}_0 = \mathbf{x}(t_0)$ *be a pseudo-vertex of the set* A. *Then, the continuous function* $t_1 = g(t_2)$, *defined on some right semi-neighborhood* $(t_0, t_0+\varepsilon), \varepsilon > 0$ *is said to be the right-hand bisector mapping in the neighborhood of the pseudo-vertices of* $\mathbf{x}_0$, *if the following conditions are met:*

$$\forall t_2 \in (t_0, t_0 + \varepsilon) \ g(t_2) < t_0, \tag{6}$$

$$\lim_{t_2 \to t_0 + 0} g(t_2) = t_0, \tag{7}$$

$$\forall t_2 \in (t_0, t_0 + \varepsilon) \ \exists \mathbf{y} \in L(A) \colon \big\{\mathbf{x}\big(g(t_2)\big), \mathbf{x}(t_2)\big\} \subseteq \Omega_A(\mathbf{y}). \tag{8}$$

Definition 4. *Let the point* $\mathbf{x}_0 = \mathbf{x}(t_0)$ *be a pseudo-vertex of the set* A. *Then, the continuous function* $t_2 = \overline{g}(t_1)$, *defined on some left semi-neighborhood* $(t_0 - \varepsilon, t_0), \varepsilon > 0$ *is said to be the left-hand bisector mapping in the neighborhood of the pseudo-vertices of* $\mathbf{x}_0$, *if the following conditions are met:*

$$\forall t_1 \in (t_0 - \varepsilon, t_0) \ \overline{g}(t_1) > t_0, \tag{9}$$

$$\lim_{t_1 \to t_0 - 0} \overline{g}(t_1) = t_0, \tag{10}$$

$$\forall t_1 \in (t_0 - \varepsilon, t_0) \ \exists \mathbf{y} \in L(A) \colon \big\{\mathbf{x}(t_1), \mathbf{x}\big(\overline{g}(t_1)\big)\big\} \subseteq \Omega_A(\mathbf{y}). \tag{11}$$

In total, conditions (6) and (7) mean that the chart closure of a continuous function $t_1 = g(t_2)$ has a fixed-point value. Here, $g(t_0) = \lim\limits_{t_2 \to t_0 + 0} g(t_2) = t_0$. Similarly, conditions (9) and (10) mean that the graph closure of a continuous function $t_2 = \bar{g}(t_1)$ has a fixed-point value. Here, $\bar{g}(t_0) = \lim\limits_{t_1 \to t_0 - 0} \bar{g}(t_1) = t_0$. These pairs of supplemented functions provide a continuous parametrization of the curve Γ in the pseudo-vertex neighborhood:

$$\breve{g}(t) = \begin{cases} \bar{g}(t_3), & t_3 \in (t_0 - \varepsilon, t_0), \varepsilon > 0, \\ t_0, & t_3 = t_0, \\ g(t_3), & t_3 \in (t_0, t_0 + \varepsilon^*), \varepsilon^* > 0. \end{cases}$$

For more detail of mapping properties in geometry see, e.g., [15, §4, 1°]. Supplements (8) and (11) show that $\mathbf{x}(t_1)$ and $\mathbf{x}(t_2)$ generate a bisector point. It should be noted that there are pseudo-vertex neighborhoods, where none of the above mentioned mappings are defined. This can be a special point of the curve (5), such that it is a common projection for all points of one of the branches of $L(A)$, as in [16, Example 4]. On the other hand, at some points $\mathbf{x}_0$, the bisector has two different right- or left-hand mappings. The reason is that two different extreme points lying on the normal to Γ in $\mathbf{x}_0$ on different sides of $\mathbf{x}_0$ can correspond to one pseudo-vertex, as in [6, example 4.1].

3   Analytic Formulas for Constructing a Singular Set

The case is investigated with the curve (5) containing an arc that can be presented as a function plot in a Cartesian coordinate system.

Consider sets of functions $f(x)$, of point x_0 and two numbers $\varepsilon_1 > 0, \varepsilon_2 > 0$, for which the following conditions are satisfied:

A1. The function $f(x)$ is differentiable on the interval $(x_0 - \varepsilon_1, x_0 + \varepsilon_2)$.
A2. The function $f(x)$ is twice differentiable on the intervals $(x_0 - \varepsilon_1, x_0)$ and $(x_0, x_0 + \varepsilon_2)$.
A3. The plot $f(x)$ coincides with the arc of the circle at no interval. $(x_1, x_2) \subseteq (x_0 - \varepsilon_1, x_0 + \varepsilon_2)$.

Let's select the curves Γ containing segments of straight lines. We should note that a smooth curve contain straight line segments only if they are smoothly conjugated to each other by arcs of curves. In this case, the abscissa x of the point $\mathbf{x} \in \Gamma$ serves as a parameter in the Eq. (5).

Lemma 1. *Let the boundary Γ of the set A coincide with the plot of an explicitly given function $y = y(x)$. If the following conditions are met:*

1) *for the function $y = y(x)$, the point x_0 and the numbers $\varepsilon_1 > 0, \varepsilon_2 > 0$ the conditions A.1–A.3 are valid;*
2) *the point $\mathbf{x}_0 = (x_0, y(x_0))$, $x_0 \in X$, is a pseudo-vertex of the set A, then there exists a number $r > 0$ such that for all pairs of points (x_1, x_2) and (x_1^*, x_2^*) such that*

$$\exists \mathbf{y} \in L(A) \colon \left\{ (x_1, y(x_1)), (x_2, y(x_2)) \right\} \subseteq \Omega_A(\mathbf{y}), \tag{12}$$

$$\exists \mathbf{y}^* \in L(A) \colon \left\{ (x_1^*, y(x_1^*)), (x_2^*, y(x_2^*)) \right\} \subseteq \Omega_A(\mathbf{y}), \tag{13}$$

$$x_1 \in (x_0 - r, x_0), x_1^* \in (x_0 - r, x_0), \tag{14}$$

$$x_2 \in (x_0, x_0 + r), x_2^* \in (x_0, x_0 + r), \tag{15}$$

the inequation

$$x_1 < x_1^*, \tag{16}$$

is satisfied if and only if

$$x_2 > x_2^*. \tag{17}$$

Proof. Let's denote $\mathbf{x}_i = (x_i, y(x_i))$, $\mathbf{x}_i^* = (x_i^*, y(x_i^*))$, $i = 1, 2$. Without loss of generality, we assume that the function $y = y(x)$ is linear in the left semi-neighborhood $(x_0 - \varepsilon, x_0), \varepsilon > 0$ and the right-hand curvature is $k(x_0 + 0) > 0$. Since a straight line has is zero curvature, the greater limiting curvature is on the right-hand side of the point x_0. Let's consider the part of the bisector $L^0(A)$, lying in a sufficiently small neighborhood $\widehat{\mathbf{y}}$, such that $L^0(A)$ is embedded in the super plot of the function $y(x)$. Since the mapping $\mathbf{y} \mapsto \Omega_A(\mathbf{y})$ is upper semi-continuous, there is a number $r > 0$ such that for all points $\mathbf{x}$ with an abscissa $x \in (x_0 - r, x_0 + r)$, the bisector points $\mathbf{y} \in L(A), \mathbf{x} \in \Omega_A(\mathbf{y})$ generated by these points belong to $L^0(A)$. Let's assume that there are points $x_1, x_2, x_1^*, x_2^* \in (x_0 - r, x_0 + r)$ for which conditions (12)–(16) are met, except for (17). This means that the normal segment π_2 from point $\mathbf{x}_i$ to point $\mathbf{y}$ intersects either with the segment $[\mathbf{x}_1^*, \mathbf{y}^*]$ or with the segment $[\mathbf{x}_2^*, \mathbf{y}^*]$. If on the plot the point $\mathbf{x}_2$ is to the left than $\mathbf{x}_2^*$, the line π_2 is to pass through one of these segments to cross the normal plotted to Γ at point $\mathbf{x}_1$. However, two segments from the points to their set projection are able to only intersect with each other either at the starting point or at the endpoint. According to the Lemma 1, the points $\mathbf{x}_1, \mathbf{x}_2, \mathbf{x}_1^*, \mathbf{x}_2^*$ are pairwise distinguishable; therefore, the segments intersect with each other at the point $\mathbf{y} = \mathbf{y}^*$. In this case, there are four different projections for the point $\mathbf{y} = \mathbf{y}^*$ on A in the neighborhood of the pseudo-vertex $\mathbf{x}_0$. This contradicts the condition that in any sufficiently small neighborhood of the pseudo-vertex, the curve Γ does not coincide with the arc of the circle, which means that at the bisector points $\mathbf{y} \in L^0(A)$ in a sufficiently small neighborhood corresponding to the $\mathbf{x}_0$ of the extreme point $\widehat{\mathbf{y}}$, the card $\operatorname{card}(\Omega_A(\mathbf{y})) = 2$ is fulfilled (for more details, see [17]).

Lemma 2. *Let the set A have a boundary coinciding with the function plot $y = y(x)$. If the following conditions are met*

1) *the set of projections $\Omega_A(\mathbf{y})$ of the bisector point $\mathbf{y}$ includes the points $\mathbf{x}_1 = (x_1, y(x_1))$ and $\mathbf{x}_2 = (x_2, y(x_2))$;*
2) *$x_1 < x_2$;*
3) *the function $y = y(x)$ in the neighborhood of the point $x = x_1$ coincides with the linear function*

$$y = y_0 + a(x - x_0), \tag{18}$$

then, the equation holds true

$$x_1 = x_c - \sqrt{\frac{(x_2 - x_c)^2 + \big(y(x_2) - y_0 - a(x_c - x_0)\big)^2}{1 + a^2}}, \qquad (19)$$

where

$$x_c = \frac{y_0 - ax_0 + y'(x_2)x_2 - y(x_2)}{y'(x_2) - a}. \qquad (20)$$

Proof. Let's denote coordinates of the bisector point for which points x_1 and x_2 are included in the set of projections, as $y = (x^*, y^*)$. By construction, y lies at the intersection of the normals to Γ at x_1 and x_2. Let's denote the intersection point of tangents to Γ at the same points as y_c. The tangent at point x_1 is described with the Eq. (18) (we will denote it as γ_1), and the tangent at point x_2 is described with the equation

$$y - y(x_2) = y'(x_2)(x - x_2)$$

(we will denote it as γ_2). When substituting the ordinate value from (18) into the last equation, we obtain the equation

$$y_0 + a(x_c - x_0) - y(x_2) = y'(x_2)(x_c - x_2)$$

for the abscissa of the point y_c. By explicitly selecting the value x_c from it, the Eq. (20) may be written.

According to the condition of segment equality plotted from the points x_1 and x_2 to the intersection point of the normals y, it follows that the segments plotted from x_1 and x_2 to the intersection point of the tangents y_c are of equal length. Since the first segment lies on the line γ_1, its length is determined by the equation

$$\|x_1 - y_c\| = \frac{|x_1 - x_c|}{\sqrt{a^2 + 1}}. \qquad (21)$$

The length of the second segment (with the intersection abscissa value from (20) and its embedment in the line γ_2 taken into account) is calculated as

$$\|x_2 - y_c\| = \sqrt{(x_2 - x_c)^2 + \big(y(x_2) - y_0 - a(x_c - x_0)\big)^2}. \qquad (22)$$

By condition $x_2 > x_1$, it can be shown that

$$x_c \geq x_1. \qquad (23)$$

Let's assume that $x_c < x_1$. So y_c lies in the semi-neighbourhood Π_-, which is bounded by the line π_1 orthogonal to γ_1 passing through x_1, whereas $x_2 \notin \Pi_-$. Therefore, the angle $\angle(y_c, x_1, x_2)$ in the triangle $\triangle x_1 x_c x_2$ at the vertex x_1 is greater than the right one. However, since $\triangle x_1 y_c x_2$ is an isosceles triangle with vertex x_c, for the angle at its vertex, $\angle(y_c, x_1, x_2) < \pi/2$ is estimated. There is a contradiction.

According to the equality condition of lengths (21) and (22) and the estimate (23) (which makes it possible to expand the expression under the module), the dependence for the abscissas of points x_1, x_c, x_2 is found:

$$\frac{x_c - x_1}{\sqrt{a^2 + 1}} = \sqrt{(x_2 - x_c)^2 + (y(x_2) - y_0 - a(x_c - x_0))^2}.$$

When multiplying both parts of the last equation by $\sqrt{a^2 + 1}$ (this value is obviously different from zero for any real a) and then adding x_1, the equality (19) is derived.

Lemma 3. *Let the set A have a boundary coinciding with the function graph $y = y(x)$. If the following conditions are met*

1) *the set of projections $\Omega_A(\mathbf{y})$ of the bisector point $\mathbf{y}$ includes the points $\mathbf{x}_1 = (x_1, y(x_1))$ and $\mathbf{x}_2 = (x_2, y(x_2))$;*
2) *$x_1 < x_2$,*
3) *the function $y = y(x)$ in the neighborhood of the point $x = x_2$ coincides with the linear function (18);*

the following equation holds true

$$x_2 = \overline{x}_c + \sqrt{\frac{(x_1 - \overline{x}_c)^2 + (y(x_1) - y_0 - a(\overline{x}_c - x_0))^2}{1 + a^2}}, \tag{24}$$

where

$$\overline{x}_c = \frac{y_0 - ax_0 + y'(x_1)x_1 - y(x_1)}{y'(x_1) - a}. \tag{25}$$

Proof is completely identical with the proof of Lemma 2, with the only difference that the intersection point $\overline{\mathbf{y}}_c$ of the tangents to $\mathrm{gr}\, y(x)$ lies to the left of the point $\mathbf{x}_2$. Therefore, in the Eq. (24), the distance between points $\overline{\mathbf{y}}_c$ and $\mathbf{x}_1$ is added to the abscissa $\overline{x}^c$ of the point $\overline{\mathbf{y}}_c$ calculated by Eq. (25).

Theorem 1. *Theorem. Let the boundary Γ of the set A coincide with the function plot $y = y(x)$. If the following conditions are met:*

1) *for the function $y = y(x)$, the point x_0 and the numbers $\varepsilon_1 > 0, \varepsilon_2 > 0$, the conditions A.1–A.3 are valid;*
2) *the point $\mathbf{x}_0 = (x_0, y(x_0)), x_0 \in X$ is a pseudo-vertex of the set A;*
3) *the function $y(x)$ in the left semi-neighborhood $(x_0 - \varepsilon_1, x_0), \varepsilon_1 > 0$, coincides with a linear one,*

then, for some number $\varepsilon > 0$ there is a bisector mapping of the right semi-neighborhood of the pseudo-vertex $\mathbf{x}_0$, given by the equation:

$$x_1(x_2) = x_c^*(x_2) - \sqrt{\frac{(x_2 - x_c^*(x_2))^2 + (y(x_2) - y_0 - y'(x_0)(x_c^*(x_2) - x_0))^2}{1 + y'(x_0)^2}}, \tag{26}$$

$$x_c^*(x_2) = \frac{y(x_0) - y'(x_0)x_0 + y'(x_2)x_2 - y(x_2)}{y'(x_2) - y'(x_0)}. \tag{27}$$

Proof. Since, according to Theorem, the Lemma 1 conditions are fulfilled at the point $\mathbf{x}_0$, in some semi-neighborhood $(x_0 - \varepsilon, x_0)$, the function $x_2 = x_2(x_1)$ connecting the projections abscissas of the bisector points is monotonically decreasing. At the point $x = x_0$, there is a discontinuity in the curvature $k(x)$ of the curve Γ, due to a straight line and a part of Γ that is not a straight line not meeting at this point. In this case, one-sided curvature signs at x_0 coincide on the left and right (more precisely, $k(x_0 - 0) = 0$; therefore, it can be assumed that $k(x_0 - 0)$ has the same sign as $k(x_0 + 0)$). As shown in [18, Theorem 3], the extreme point $\hat{\mathbf{y}}$ corresponding to a pseudo-vertex of this type is unique and coincides with that of the limiting positions of the curvature center $\mathbf{c}(x)$, which is closer to Γ. Since, according to Theorem, in a sufficiently small neighborhood of the pseudo-vertex, Γ does not contain a circle arc, $\Omega_A(\hat{\mathbf{y}}) = \{\mathbf{x}_0\}$. Hence, for all points $\mathbf{y} \in L(A)$ at $\mathbf{y} \to \hat{\mathbf{y}}$, projections pairs $\{\mathbf{x}_1, \mathbf{x}_2\} \subseteq \Omega_A(\mathbf{y})$ converge to the point $\mathbf{x}_0$, while lying on different sides. That is, in the pseudo-vertex neighborhood, the bisector mappings for the point coordinates are defined.

According to Theorem condition, the point $\mathbf{x}_0 = (x_0, y(x_0))$ lies on a smooth plot section of the function $y = y(x)$. Therefore, the derivative $y'(x_0)$ is defined. Since to the left of the pseudo-vertex the curve Γ coincides with the line segment γ, the equation of γ coincides with the equation of the tangent to Γ in the pseudo-vertex:

$$y = y'(x_0)(x - x_0) + y(x_0). \tag{28}$$

By properties of a smooth curve pseudo-vertex in some sufficiently small semi-neighborhoods $[x_0 - \varepsilon_0, x_0]$ and $[x_0, x_0 + \varepsilon_0]$, such that for each point $(x_2, y(x_2))$ at $x_2 \in (x_0, x_0 + \varepsilon_0]$, there is a point $(x_1, y(x_1))$, $x_1 \in [x_0 - \varepsilon_0, x_0)$ such that they generate a bisector point. Let's choose a semi-neighborhood $[x_0 - \varepsilon_0, x_0)$ such that for points $(x, y) = (x_1, y(x_1))$ at $x_1 \in [x_0 - \varepsilon_0, x_0)$, Eq. (28) is fulfilled. Then, according to Lemma 2, for any point $(x_2, y(x_2))$ at $x_2 \in (x_0, x_0 + \varepsilon_0]$, there will be a point with an abscissa x_1 such that these two points generate a bisector point. By substituting the value $y'(x_0)$ as the coefficient a in Eqs. (19), (20), and the value $y'(x_0)$ as the ordinate value y_0, the Eqs. (26) and (27) are obtained.

Corollary 1. *Let the boundary Γ of the set A coincide with the function plot $y = y(x)$. If the following conditions are met:*

1) *for the function $y = y(x)$, the point x_0 and the numbers $\varepsilon_1 > 0, \varepsilon_2 > 0$, the conditions A.1–A.3 are valid;*
2) *the point $\mathbf{x}_0 = (x_0, y(x_0))$ is a pseudo-vertex of the set A;*
3) *the function $y(x)$ in the right semi-neighborhood $(x_0, x_0 + \varepsilon_1)$ coincides with a linear one,*

then, for some number $\varepsilon > 0$ there is a bisector mapping of the left semi-neighborhood of the pseudo-vertex $\mathbf{x}_0$, given by the equation:

$$x_2(x_1) = \overline{x}_c^*(x_1) + \sqrt{\frac{(x_1 - \overline{x}_c^*(x_1))^2 + \big(y(x_1) - y_0 - y'(x_0)(\overline{x}_c^*(x_1) - x_0)\big)^2}{1 + y'(x_0)^2}},$$

$$\tag{29}$$

$$\overline{x}_c^*(x_1) = \frac{y(x_0) - y'(x_0)x_0 + y'(x_1)x_1 - y(x_1)}{y'(x_1) - y'(x_0)}. \tag{30}$$

Proof is completely identical with the proof of Theorem 1; however, it is based on Lemma 3.

Remark 1. The coordinates of the bisector extreme point $\widehat{\mathbf{y}} = (\widehat{x}, \widehat{y})$ for the pseudo-vertex meeting Theorem 1 conditions are found as the limiting position of curvature center [19, ch. III, §25] of the plot $y = y(x)$ at the point $x = x_0$ on the right:

$$\widehat{x} = x_0 - \frac{y'(x_0)^3 + y'(x_0)}{y''(x_0 + 0)}, \tag{31}$$

$$\widehat{y} = y(x_0) - \frac{y'(x_0)^2 + 1}{y''(x_0 + 0)}. \tag{32}$$

Similarly, the bisector extreme point $\widehat{\mathbf{y}} = (\widehat{x}, \widehat{y})$ for the pseudo-vertex corresponding to Corollary 1 conditions coincides with the limiting position of the curvature center at the point $x = x_0$ on the left:

$$\widehat{x} = x_0 - \frac{y'(x_0)^3 + y'(x_0)}{y''(x_0 - 0)}, \tag{33}$$

$$\widehat{y} = y(x_0) - \frac{y'(x_0)^2 + 1}{y''(x_0 - 0)}. \tag{34}$$

Remark 2. Since the pseudo-vertex corresponding to the conditions of Theorem 1 has a discontinuity in the curvature of the boundary of the target set, then the conditions of Theorem 2 from [18] are satisfied in it. Therefore, there is a limit ratio

$$\lim_{x_2 \to x_0, x_2 > x_0} \frac{x_2 - x_0}{x_1(x_2) - x_0} = 0. \tag{35}$$

for the abscissas of the projections of the points of the bisector in the neighborhood of the pseudo-vertex. Similarly, for a pseudo-vertex that satisfies the conditions of Corollary 1, the conditions of Theorem 3 from [18] are satisfied. Therefore, for the coordinates of the points to which the optimal trajectories come in the vicinity of x_0, the equality

$$\lim_{x_1 \to x_0, x_1 < x_0} \frac{x_1 - x_0}{x_2(x_1) - x_0} = 0. \tag{36}$$

holds.

When studying diffeomorphisms in the neighborhood of pseudo-vertices, the question arises on which parameter domain they set the coordinates of the bisector points. In general, to do this, you need to find all the junction points of the smooth branches of the bisector. However, for some special cases of the target set, it is possible to write a system of equations that sets the limits for changing the parameters.

Proposition 1. *Let the set A be a subplot of a smooth function $y = f(x)$ given by gluing three smooth functions*

$$y(x) = \begin{cases} y_1(x), & x \in (-\infty, x^{(1)}], \\ ax + b, & x \in (x^{(1)}, x^{(2)}], \\ y_2(x), & x \in (x^{(2)}, \infty). \end{cases} \tag{37}$$

If the set A has exactly two pseudo-vertices $\left(x_1, f\left(x^{(1)}\right)\right)$ and $\left(x_1, f\left(x^{(2)}\right)\right)$, and the bisector has exactly one bifurcation point $\mathbf{x}^* = (x^*, y^*)$, then the set of projections consists of three points

$$\Omega_* = \{\mathbf{x}_1, \mathbf{x}_2, \mathbf{x}_*\},$$

where

$$\mathbf{x}_1 = \left(x_1, y_1\left(x^{(1)}\right)\right), \mathbf{x}_2 = \left(x_2, y_2\left(x^{(2)}\right)\right), \mathbf{x}_* = (x_*, ax_* + b),$$

for abscissa points, inequalities are fulfilled

$$x_1 < x^{(1)}, x^{(1)} < x_* < x^{(2)}, x_2 > x^{(2)}.$$

At the same time, there are relations

$$\arctan y_1'(x_1) + \arctan y_2'(x_2) = 2\arctan \frac{y_2(x_2) - y_1(x_1)}{x_2 - x_1}, \tag{38}$$

$$x_c - \sqrt{\frac{(x_2 - x_c)^2 + \left(y_2(x_2) - x^{(2)} + a(x_c - x_2)\right)}{1 + a^2}} =$$
$$= \overline{x}_c - \sqrt{\frac{(x_1 - \overline{x}_c)^2 + \left(y_1(x_1) - x^{(1)} + a(\overline{x}_c - x_1)\right)}{1 + a^2}}. \tag{39}$$

Here, the value x_c is found by formula (20), in which $x_0 = x^{(2)}$ is taken as the abscissa x_0, and $y(x) = y_2(x)$ is taken as the function of $y(x)$. Similarly, the value of $\overline{x}_c$ is found by the formula (25), in which the abscissa x_0 is $x_0 = x^{(1)}$ taken, and the function of $y(x)$ is taken $y(x) = y_1(x)$.

Proof. If the bifurcation point is the only one, then it is the junction of the three branches of the bisector. And the set of its projections on A consists of exactly three elements. From the condition that the subplot of the function (37) has two pseudo-vertices, it follows that one of the projections lies between the pseudo-vertices on the graph of the function $f(x)$, and the other two are on opposite sides of the pseudo-vertices. For abscissae $\dot{x}_1 \in (-\infty, x^{(1)})$, $x_2 \in (x^{(2)}, \infty)$ of the points from the segment of normals to the plot gr $f(x)$, deferred to the bifurcation point should be equal. In terms of functions $y_i(x), i = 1, 2$, and their derivatives $y'i(x), i = 1, 2$ this means the equality (38) is fulfilled, see, for example, formula (3.1) from paper [17]. At the same time, according to Theorem 1, the abscissa x_* of the point $\mathbf{x}_*$ and the abscissa x_2 of the point $\mathbf{x}_2$ are connected by equality

(19). According to Corollary 1, the abscissa x_* of the point $\mathbf{x}_*$ and the abscissa x_1 of the point $\mathbf{x}_1$ are connected by equality (26). Hence, for the coordinates x_1 and x_2, the right-hand sides of equalities (19) and (26) coincide (provided that $x^{(1)}$ and $x^{(2)}$, respectively, are taken as the coordinates of the pseudo-vertices in them). Therefore, equality (39) holds true.

4   Example of the Problem Solution

Let's consider time-optimal problem with dynamics (1) solution by selecting a scattering curve if the set A is a function sub plot

$$y(x) = \begin{cases} 2\sec(x+1) - 2, & x \in (-1 - \pi/2, -1], \\ 0, & x \in (-1, 1], \\ \cosh(x-1) - 1, & x \in (1, \infty). \end{cases} \tag{40}$$

The target set boundary analysis demonstrates that there is a pseudo-vertex $\mathbf{x}_0 = \big(x_0, y(x_0)\big) = (1, 0)$, where Theorem 1 conditions are fulfilled. It corresponds to the bisector extreme point, with coordinates found from (31), (32): $\widehat{\mathbf{y}} = (1, 1)$. According to the Eq. (27), the intersection point abscissa of the tangent to the plot of the function (40) at the point $x_2 > x_0$ and the line $y = 0$, which coincides with a part of this plot at $x < 1$, is defined as

$$x_c^*(x_2) = x_2 - \frac{\cosh(x_2 - 1) - 1}{\sinh(x_2 - 1)}.$$

By substituting $\overline{x}_c^*(x_2)$ into (26), a bisector mapping in the right pseudo-vertex semi-neighborhood for the interval $(1, 2.538)$ is obtained:

$$x_1(x_2) = x_2 - \sinh(x_2 - 1). \tag{41}$$

The target set boundary analysis demonstrates that there is another pseudo-vertex $\mathbf{x}_0^* = \big(x_0^*, y(x_0^*)\big) = (-1, 0)$, where Corollary 1 conditions are fulfilled. It corresponds to the bisector extreme point, with coordinates found from (33), (34): $\widehat{\mathbf{y}} = (-1, 0.5)$. According to the Eq. (30), the intersection point abscissa of the tangent to the plot of the function (40) at the point $\overline{x}_1 < \overline{x}_0$ and the line $y = 0$, which coincides with a part of this chart at $x > \overline{x}_0$, is defined as

$$\overline{x}_c^*(x_1) = x_1 - \frac{\cos(x_1 + 1) - \cos^2(x_1 + 1)}{\sin(x_1 + 1)}.$$

By substituting $\overline{x}_c^*(\overline{x}_1)$ into (29), a bisector mapping in the right pseudo-vertex semi-neighborhood for the interval $(-2.059, -1)$, is obtained (on rearrangements):

$$\overline{x}_2(\overline{x}_1) = \overline{x}_1 - \tan\frac{\overline{x}_1 + 1}{2}\left(\sqrt{4\tan^2(\overline{x}_1 + 1) - \cos^2(\overline{x}_1 + 1)} + \cos(\overline{x}_1 + 1)\right). \tag{42}$$

The mapping chart (41) as a red curve and the mapping chart (42) as a blue curve are shown in Fig. 1. From the limit position of the tangent to the $x_1 = x_1(x_2)$ and $\overline{x}_2 = \overline{x}_2(\overline{x}_1)$ charts in the neighborhood of the pseudo-vertices, it can be seen that the limit relations (35) and (36) are satisfied. Figure 1 also shows in green the ploy of the dependence $\breve{x}_2 = \breve{x}_2(\breve{x}_1)$ between the abscissas of the projections of the points of the third branch of the bisector $L(A)$. There is no analytical expression for this dependence, it is found as a numerical solution of a differential equation with $\breve{x}_1$ and $\breve{x}_2$.

The boundary Γ of the target set (green line), the scattering curve $L(A)$ (red curve) and the level lines Φ with a step 0.4 (blue lines) are shown in Fig. 2. Note that the function (40) satisfies the conditions imposed on the function (37) in Proposition 1. Therefore, for the coordinates x_1, x_*, x_2 of the projections of the bifurcation point, the system (38), (39) is valid. The normals to the plot $\operatorname{gr} y(x)$ constructed at these points intersect at the point $(0.3315, 2.413)$. The plot of the optimal result function $u(x, y)$ on a grid with a cell 0.05×0.05 is shown in Fig. 3. The numerical construction of the approximation of the optimal result function was computed using the modernized software package [20].

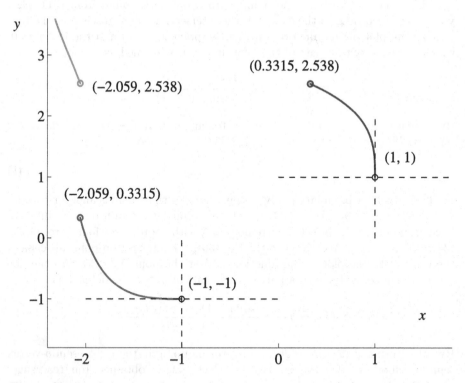

Fig. 1. Charts of the dependencies of the abscissas $x_1 = x_1(x_2)$, $\overline{x}_2 = \overline{x}_2(\overline{x}_1)$ and $\breve{x}_2 = \breve{x}_2(\breve{x}_1)$ of the points to which the optimal trajectories come from the bisector. (Color figure online)

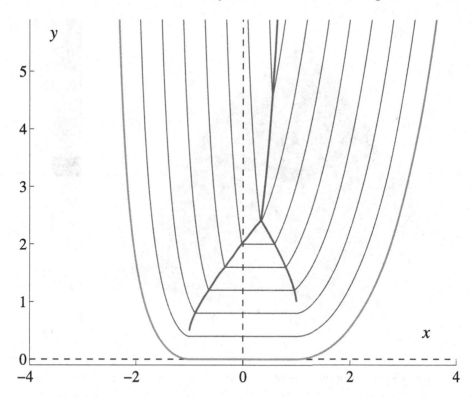

Fig. 2. The boundary Γ of the target set, the bisector $L(A)$ and the level lines Φ of the optimal result function $u(x, y)$. (Color figure online)

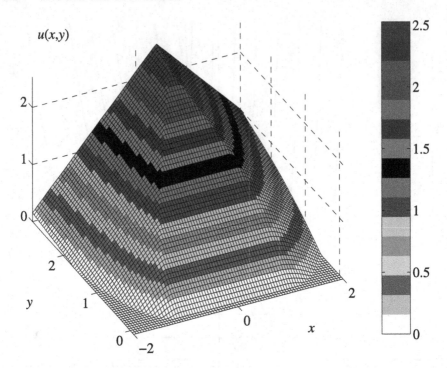

Fig. 3. Plot of the optimal result function $u(x, y)$ in the time-optimal problem.

5 Conclusion

For one class of planar time-optimal control problems for the case of a closed convex set, conditions imposed on the target boundary are determined, thus, enabling to construct a singular set of solutions in an explicit analytical form. Based on analytical equations obtained, algorithms for constructing level lines and a chart of the optimal result function are proposed. Example of control problem with different pseudo vertex of the target set is presented, for which parts of singular set are constructed analytically.

References

1. Subbotin, A.I.: Generalized solutions of first order PDEs: the dynamical optimization perspective. Birkhäuser, Basel (1995). https://doi.org/10.1007/978-1-4612-0847-1
2. Lebedev, P.D., Uspenskii, A.A.: Analytical and numerical construction of the optimal outcome function for a class of time-optimal problems. Comput. Math. Model. **19**(4), 375–386 (2008). https://doi.org/10.1007/s10598-008-9007-9
3. Kruzhkov, S.N.: Generalized solutions of the Hamilton - Jacobi equations of eikonal type. I. Formulation of the problems; existence, uniqueness and stability theorems; some properties of the solutions. Math. USSR-Sbornik **27**(3), 406–446 (1975). https://doi.org/10.1070/SM1975v027n03ABEH002522

4. Arnold, V.I.: Singularities of Caustics and Wave Fronts. Springer, Netherlands (1990). https://doi.org/10.1007/978-94-011-3330-2
5. Dem'yanov, V.F., Vasil'ev, L.V.: Nondifferentiable Optimization. Springer, New York (1985)
6. Lebedev, P.D., Uspenskii, A.A.: Construction of scattering curves in one class of time-optimal control problems with leaps of a target set boundary curvature. Izv. IMI UdGU **55**, 93–112 (2020). https://doi.org/10.35634/2226-3594-2020-55-07
7. Isaacs, R.: Differential Games. Wiley, New York (1965)
8. Siersma, D.: Properties of conflict sets in the plan. Banach Cent. Publ. **50**, 267–276 (1999). https://doi.org/10.4064/-50-1-267-276
9. Giblin, P.J., Reeve, G.: Centre symmetry sets of families of plane curves. Demonstratio Math. **48**(2), 167–192 (2015). https://doi.org/10.1515/dema-201-0016
10. Giblin, P.G.: Symmetry sets and medial axes in two and three dimensions. In: Cipolla, R., Martin, R. (eds.) The Mathematics of Surfaces IX, pp. 306–321. Springer, London (2000). https://doi.org/10.1007/978-1-4471-0495-7_18
11. Alimov, A.R., Tsar'kov, I.G.: Connectedness and solarity in problems of best and near-best approximation. Russ. Math. Surv. **71**(1), 1–77 (2016). https://doi.org/10.1070/RM9698
12. Ushakov, V.N., Ershov, A.A., Pershakov, M.V.: On one addition to evaluation by L.S. Pontryagin of the geometric difference of sets in a plane. Izv. IMI UdGU **54**, 63–73 (2019). https://doi.org/10.20537/2226-3594-2019-54-06
13. Sedykh, V.D.: On the topology of wave fronts in spaces of low dimension. Izv. Math. **76**(2), 375–418 (2012). https://doi.org/10.1070/IM2012v076n02ABEH002588
14. Sedykh, V.D.: Topology of singularities of a stable real caustic germ of type E_6. Izv. Math. **82**(3), 596–611 (2018). https://doi.org/10.1070/IM8643
15. Poznyak, E.G., Shikin, E.V.: Differentsial'naya geometriya: pervoe znakomstvo (Differential geometry: the first acquaintance). MGU, Moscow (1990)
16. Lebedev, P.D.: Calculating the nonconvexity measure of plane sets. Trudy Inst. Mat. Mekh. UrO RAN **13**(3), 84–94 (2007)
17. Lebedev, P.D., Uspenskii, A.A.: Geometry and asymptotics of wavefronts. Russ. Math. (Izv. VUZ Mat.) **52**(3), 24–33 (2008). https://doi.org/10.3103/S1066369X08030031
18. Lebedev, P.D., Uspenskii, A.A.: Geometric singularities of the solution of the Dirichlet boundary problem for Hamilton–Jacobi equation with a low order of smoothness of the border curve. In: Pinelas, S., Kim, A., Vlasov, V. (eds.) CONCORD-90 2018. SPMS, vol. 318, pp. 109–122. Springer, Cham (2020). https://doi.org/10.1007/978-3-030-42176-2_11
19. Rashevskii, P.K.: Kurs differencial'noj geometrii (A Course in Differential Geometry). URSS Publ., Moscow (2003)
20. Lebedev, P.D., Uspenskii, A.A.: Program for constructing wave fronts and functions of the Euclidean distance to a compact nonconvex set. Certificate of state registration of the computer program no. 2017662074 (2017)

On the Existence of Fuzzy Contractual Allocations, Fuzzy Core and Perfect Competition in an Exchange Economy

Valeriy Marakulin[✉][iD]

Sobolev Institute of Mathematics, Russian Academy of Sciences, 4 Acad. Koptyug avenue, Novosibirsk 630090, Russia
marakulv@gmail.com, marakul@math.nsc.ru
https://www.math.nsc.ru/mathecon/marakENG.html

Abstract. The fuzzy core is well-known in theoretical economics, it is widely applied to model the conditions of perfect competition. In contrast, the original author's concept of fuzzy contractual allocation as a specific element of the fuzzy core is not so widely known in the literature, but it also represents a (refined) model of perfect competition. This motivates the study of its validity: the existence of fuzzily contractual allocations in an economic model; it also implies the existence (non-emptiness) of the fuzzy core and develops an approach from [15]. The proof is based on two well-known theorems: Michael's theorem on the existence of a continuous selector for a point-to-set mapping and Brouwer's fixed point theorem. In literature, only the non-emptiness of the fuzzy core was proven under essentially stronger assumptions—typically, it applies replicated economies and Edgeworth equilibria.

Keywords: Fuzzy core · Fuzzy contractual allocation · Edgeworth equilibria · Perfect competition · Existence theorems

1 Introduction

In modern economic theory, the idea of perfect competition is implemented in many ways. Among others one can find the famous Aumann [3] approach based on a model with a non-atomic set of economic agents, non-standard economies according to Brown–Robinson [5] and, of course, the asymptotic Debreu–Scarf Theorem [6], as well as other results, including the contractual approach developed by the author. The history of the idea of perfect competition goes back to Edgeworth and his well-known conjecture [7] that the core (contract curve) shrinks into equilibrium. The proof of this conjecture, based on the idea of replication of economic agents, was proposed in [6]. Later it turned out that the limit

The study was supported by the Program of Basic Scientific Research of the Siberian Branch of the Russian Academy of Sciences (Grant no. FWNF-2022-0019).

M. Khachay et al. (Eds.): MOTOR 2023, CCIS 1881, pp. 308–323, 2023.
https://doi.org/10.1007/978-3-031-43257-6_23

allocations from the core of the replicated economy, named by Aliprantis as Edgeworth equilibria, are elements of the fuzzy core of the economy—a concept introduced in [2]. Edgeworth equilibrium is an attainable allocation whose r-fold repetition belongs to the core of the r-fold replica of the original economy, for any positive integer r.

Due to Debreu–Scarf theorem on the limit coincidence of equilibria and the core for the replicated economy, the fuzzy core was started to be also applied to state the existence of competitive equilibrium. As a result, now the fuzzy core is widely used in theoretical economics, e.g. see [1,8,9]. One can see also [4] as one of the latest results on the existence of fussy core (under essentially stronger assumptions than in [15]). The original author's concept of fuzzy contractual allocation [10,12–14] is not so widely known in the literature, but it also represents an effective model of perfect competition in its simplest form. The idea of fuzzy contractual allocation is that, in the current contractual situation, agents can break contracts *asymmetrically*, without coordination with other individuals and without transferring information about their intentions, i.e. acting in a secret manner. Further, individuals can try to find a new contract, such that this contractual interaction—a break and a signing of a new contract—is beneficial to each of its participants. If this happens, we are talking about fuzzy contractual domination. Allocations that are not dominated in this sense are called fuzzy contractual. They have the highest level of stability and, as it follows from the analysis, every fuzzy contractual allocation belongs to the fuzzy core and presents competitive equilibrium. This allows us to state that it is a model of perfect competition.

Thus, both notions—fuzzy core and fuzzy contractual allocation—play key roles in modern economic theory, and the conditions under which they exist have a high theoretical meaning. The paper examines this problem and states the existence of fuzzy contractual allocations for an economy under very weak conditions[1]. This also implies the non-emptiness of the fuzzy core. Our proof is based on two well-known theorems, they are Michael's theorem on the existence of a continuous selector for a point-to-set mapping and Brouwer's fixed point theorem. A direct proof of the existence of fuzzy contractual allocations is a new result, while the non-emptiness of the fuzzy core is well known (under stronger assumptions). In [15] I suggested the direct proof of fuzzy core non-emptiness, which is efficient and shortest one among others; it also was stimulating our modern study, which develops our approach. As a result, I have produced new results that can be incorporated in proving the existence of Walrasian equilibrium in economies, even with infinite-dimensional commodity spaces, e.g. see [11].

---

[1] A convex model with a compact set $\mathcal{A}(X)$ of feasible allocations and preferences that are continuously extendable to a neighborhood of $\mathcal{A}(X)$.

2 An Economic Model, Fuzzy Core and Contractual Approach

I consider a typical exchange economy in which L denotes the (finite-dimensional) *space of commodities*. Let $\mathcal{I} = \{1, \ldots, n\}$ be a set of agents (traders or consumers). A consumer $i \in \mathcal{I}$ is characterized by a consumption set $X_i \subset L$, an initial endowment $\mathbf{e}_i \in L$, and a preference relation described by a point-to-set mapping $\mathcal{P}_i : X \Rightarrow X_i$ where $X = \prod_{j \in \mathcal{I}} X_j$ and $\mathcal{P}_i(x)$ denotes the set of all consumption bundles strictly preferred by the i-th agent to the bundle x_i relative to allocation $x \in X$. It is also can be applied the notation $y_i \succ_i x_i$ which is equivalent to $y_i \in \mathcal{P}_i(x)$ (to simplify notations; preferences can indirectly depend on other agents consumption $x_j \in X_j$ $j \in \mathcal{I}$, $j \neq i$). So, the pure exchange model may be represented as a triplet

$$\mathcal{E} = \langle \mathcal{I}, L, (X_i, \mathcal{P}_i, \mathbf{e}_i)_{i \in \mathcal{I}} \rangle.$$

Let us denote by $\mathbf{e} = (\mathbf{e}_i)_{i \in \mathcal{I}}$ the vector of initial endowments of all traders of the economy. Denote $X = \prod_{i \in \mathcal{I}} X_i$ and let

$$\mathcal{A}(X) = \{ x \in X \mid \sum_{i \in \mathcal{I}} x_i = \sum_{i \in \mathcal{I}} \mathbf{e}_i \}$$

be the set of all *feasible allocations*. Now let us recall some definitions.

A pair (x, p) is said to be a *quasi-equilibrium* of $\mathcal{E}$ if $x \in \mathcal{A}(X)$ and there exists a linear functional $p \neq 0$ onto L such that

$$\langle p, \mathcal{P}_i(x) \rangle \geq p x_i = p \mathbf{e}_i, \quad \forall i \in \mathcal{I}.$$

A quasi-equilibrium such that $x'_i \in \mathcal{P}_i(x)$ actually implies $p x'_i > p x_i$ is a *Walrasian or competitive equilibrium*.

An allocation $x \in \mathcal{A}(X)$ is said to be dominated (blocked) by a nonempty coalition $S \subseteq \mathcal{I}$ if there exists $y^S \in \prod_{i \in S} X_i$ such that $\sum_{i \in S} y_i^S = \sum_{i \in S} \mathbf{e}_i$ and $y_i^S \in \mathcal{P}_i(x)$ $\forall i \in S$.

The *core* of $\mathcal{E}$, denoted by $\mathcal{C}(\mathcal{E})$, is the set of all $x \in \mathcal{A}(X)$ that are blocked by no (nonempty) coalition.

Everywhere below we assume that model $\mathcal{E}$ satisfies the following assumption.

(**A**) *For each $i \in \mathcal{I}$, $X_i \subset L$ is a convex closed subset, $\mathbf{e}_i \in X_i$ and, for every $x = (x_j)_{j \in \mathcal{I}} \in \mathcal{A}(X)$:*

$$\mathcal{P}_i(x) = \mathrm{co}[\mathcal{P}_i(x) \cup \{x_i\}] \setminus \{x_i\}$$

is a convex set.

Notice that due to (**A**) preferences may be satiated, i.e., $\mathcal{P}_i(x) = \emptyset$ is possible for some agent i and $x \in X$. However, if $\mathcal{P}_i(x) \neq \emptyset$, then preference is *locally non-satiated* at the point x.

For the existence of objects under study, we apply the following (weak) preference continuity assumption.

(C) *For each* $i \in \mathcal{I}$ *there is a point-to-set mapping* $\hat{\mathcal{P}}_i : X \Rightarrow L$ *such that for every* $x \in \mathcal{A}(X)$ *the image* $\hat{\mathcal{P}}_i(x)$ *is convex, open in* L, *implements*

$$\mathcal{P}_i(x) = \hat{\mathcal{P}}_i(x) \cap X_i$$

and for every $y_i \in \hat{\mathcal{P}}_i(x)$ *the set*

$$\hat{\mathcal{P}}_i^{-1}(y_i) = \{z \in X \mid y_i \in \hat{\mathcal{P}}_i(z)\}$$

is open one in X.

Remark 1. Notice that our modern assumptions **(A)** and **(C)** are a bit stronger of applied in [15]. In **(A)** I assume in addition that $\alpha y_i + (1 - \alpha)x_i \in \mathcal{P}_i(x)$ for every $y_i \in \mathcal{P}_i(x)$ and $\alpha \in [1, 0)$. For **(C)** now I assumed also that images $\mathcal{P}_i(x)$, $x \in \mathcal{A}(X)$ can be extended to a neighbourhood of X_i, $i \in \mathcal{I}$.

Also, below without loss of generality to simplify notations, I will assume that X_i is convex and has *full dimension*, i.e. int $X_i \neq \emptyset$ $\forall i \in \mathcal{I}$.

In the framework of model $\mathcal{E}$, a formal mechanism of contracting and recontracting can be introduced. This mechanism reflects the idea that any group of agents can find and realize some (permissible) within-the-group exchanges of commodities referred to as contracts. The mechanism defines the rules of contracting.

2.1   Fuzzy Core and Fuzzy Contractual Allocations

The concept of the fuzzy core is fruitfully working in the theory of economic equilibrium. I recall that any vector

$$t = (t_1, \ldots, t_n) \neq 0, \quad 0 \leq t_i \leq 1, \quad \forall i \in \mathcal{I}$$

maybe identified with a fuzzy coalition, where the real number t_i is interpreted as the measure of agent i participation in the coalition. A coalition t is said to dominate (block) an allocation $x \in \mathcal{A}(X)$ if there exists $y^t \in \prod_{\mathcal{I}} X_i$ such that

$$\sum_{i \in \mathcal{I}} t_i y_i^t = \sum_{i \in \mathcal{I}} t_i \mathbf{e}_i \iff \sum_{i \in \mathcal{I}} t_i(y_i^t - \mathbf{e}_i) = 0 \tag{1}$$

and

$$y_i^t \succ_i x_i, \quad \forall i \in \text{supp}(t) = \{i \in \mathcal{I} \mid t_i > 0\}. \tag{2}$$

The set of all feasible allocations which cannot be dominated by fuzzy coalitions is called the *fuzzy core* of the economy $\mathcal{E}$ and is denoted by $\mathcal{C}^f(\mathcal{E})$.

We begin with a study of the specific properties of the fuzzy core allocations. The elements of fuzzy core are defined via conditions (1), (2) which for non-satiated preferences, i.e., when $\mathcal{P}_i(x) \neq \emptyset$, $\forall i \in \mathcal{I}$, the domination may be equivalently rewritten in the form[2]

$$0 \notin \sum_{i \in \mathcal{I}} t_i(\mathcal{P}_i(x) - \mathbf{e}_i).$$

---

[2] Admitting some inaccuracy in formulas here and below, we identify a vector with a one-element set containing it.

Thus $x \in \mathcal{C}^f(\mathcal{E})$ is now equivalent to[3]

$$0 \notin \mathrm{co}[\underset{\mathcal{I}}{\cup}(\mathcal{P}_i(x) - \mathbf{e}_i)], \tag{3}$$

that after applying separation theorem allows us to conclude that the elements of the fuzzy core are quasi-equilibria. Below we propose other useful in applications characterizations of fuzzy core points presented in "geometrical" terms (introduced in [10]). To this end, let us consider the sets

$$\varUpsilon_i(x) = \mathrm{co}(\mathcal{P}_i(x) \cup \{\mathbf{e}_i\}), \quad i \in \mathcal{I}.$$

Due to the convexity of $\mathcal{P}_i(x)$, for $\mathcal{P}_i(x) \neq \emptyset$, conclude

$$\mathrm{co}(\mathcal{P}_i(x) \cup \{\mathbf{e}_i\}) = \underset{0 \leq \lambda \leq 1}{\cup}[\lambda \mathcal{P}_i(x) + (1-\lambda)\mathbf{e}_i] = \underset{0 \leq \lambda \leq 1}{\cup} \lambda(\mathcal{P}_i(x) - \mathbf{e}_i) + \mathbf{e}_i, \quad i \in \mathcal{I}.$$

This implies that the condition $z + \mathbf{e} \in \prod_\mathcal{I} \varUpsilon_i(x)$, where $\mathbf{e} = (\mathbf{e}_1, \dots, \mathbf{e}_n)$, is equivalent to the existence of $0 \leq \lambda_i \leq 1$ and $[y_i \in \mathcal{P}_i(x) \neq \emptyset$ and $y_i = \mathbf{e}_i$, if $\mathcal{P}_i(x) = \emptyset]$, $i \in \mathcal{I}$ such that

$$z = (\lambda_1(y_1 - \mathbf{e}_1), \dots, \lambda_n(y_n - \mathbf{e}_n)).$$

Hence, due to (1), (2)

$$x \in \mathcal{C}^f(\mathcal{E}) \iff \nexists z \in L^\mathcal{I}, z \neq 0: \ z + \mathbf{e} \in \prod_\mathcal{I} \varUpsilon_i(x) \ \& \ \sum_{i \in \mathcal{I}} z_i = 0 \iff$$

$$\prod_\mathcal{I} \varUpsilon_i(x) \bigcap \mathcal{A}(L^\mathcal{I}) = \{\mathbf{e}\}, \tag{4}$$

where $\mathcal{A}(L^\mathcal{I})$ is a subspace defined by the balance constraints of a pure exchange economy:

$$\mathcal{A}(L^\mathcal{I}) = \{(z_1, \dots, z_n) \in L^\mathcal{I} \mid \sum_{i \in \mathcal{I}} z_i = \sum_{i \in \mathcal{I}} \mathbf{e}_i\}.$$

Notice that characterization (4) is also valid for satiated preferences. In doing so, we have proven the following

Proposition 1. *An allocation $x \in \mathcal{A}(X)$ is the element of fuzzy core if and only if relation (4) is true.*

The direct and effective proof of fuzzy core non-emptiness is based on relation (4) and I suggested it earlier in [15]. In the case of a *2-agent economy*, Fig. 1 presents a graphic illustration of conducted analysis in the Edgeworth's box for a 2-goods economy. In this case, an allocation $x = (x_1, x_2)$ lying in the fuzzy core is equivalent to the convex hulls of $\mathcal{P}_1(x_1) \cup \{\mathbf{e}_1\}$ and of $[\bar{\mathbf{e}} - \mathcal{P}_2(\bar{\mathbf{e}} - x_1)] \cup \{\mathbf{e}_1\}$, $\bar{\mathbf{e}} = \mathbf{e}_1 + \mathbf{e}_2$ having only one point, $\mathbf{e}_1$, in common.

---

[3] Clearly, for a dominating fuzzy coalition t one may always think that $\sum_{i \in \mathcal{I}} t_i = 1$.

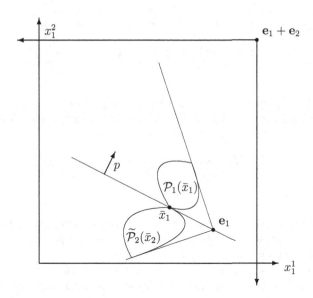

Fig. 1. *Fuzzy core*

One more important notion, which probably still is not good enough qualified in theoretical economy, is the notion of fuzzy contractual allocation. First I recall briefly the conceptual apparatus of the theory of barter contracts, see [10, 12, 13].

Any vector $v = (v_i)_{i \in \mathcal{I}} \in L^{\mathcal{I}}$ satisfying $\sum_{i \in \mathcal{I}} v_i = 0$ is called a barter (exchange) *contract*. Such barter contracts are used in pure exchange economies, as well as in the consumption sector in the economy with production. In what follows, we assume that any barter agreement is valid. With every finite collection V of (permissible) contracts, it can be associated allocation $x(V) = e + \sum_{v \in V} v$, where the vector $e = (e_1, \ldots, e_n) \in X$ is an initial endowments allocation. If $e + \sum_{v \in U} v \in X \; \forall U \subseteq V$, i.e., if any part of the contracts is broken one can get anyway a feasible allocation, then we call V a *web* of contracts. The consideration of webs of contracts allows us to study a huge massive of contractual interactions including different possibilities of contracts breaking (one of them is a fuzzy contractual interaction) for details see [10, 12, 13].

Let V be a web of contracts. For every $v \in V$ we consider and put into correspondence an n-dimension vector

$$t^v = (t_1^v, t_2^v, \ldots, t_n^v), \quad 0 \le t_i^v \le 1, \quad \forall i \in \mathcal{I},$$

and let

$$v^t = (t_1^v v_1, t_2^v v_2, \ldots, t_n^v v_n)$$

be the vector of commodity bundles formed from contract $v = (v_i)_{i \in \mathcal{I}}$ when all agents "break" individual bundles (fragments) of this contract in shares

$(1 - t_i^v)_{i \in \mathcal{I}}$. Denote $T(V) = T = \{t^v \mid v \in V\}$ and introduce

$$V^T = \{v^t \mid v \in V, \ t^v \in T\}, \quad \Delta(V^T) = \sum_{v^t \in V^T} v^t. \tag{5}$$

Definition 1. *An allocation* $x \in \mathcal{A}(X)$ *is called fuzzy contractual if there exists a web V such that $x = x(V)$ and for every $T(V)$ there is no barter contract* $w = (w_1, \ldots, w_n) \in L^{\mathcal{I}}$, $\sum_{i \in \mathcal{I}} w_i = 0$, *such that for*

$$\xi_i = \xi_i(T, V, w) = \mathbf{e}_i + \Delta_i(V^T) + w_i, \quad i \in \mathcal{I} \tag{6}$$

one has

$$\xi_i \succ_i x_i \quad \forall i : \ \xi_i \neq x_i. \tag{7}$$

So, for this kind of allocation the negation of domination means that the implementing web of contracts is *stable* relative to asymmetric *partial breakings of contracts* with or without concluding a new contract.

In economic terms, this notion can be explained in the following way. During recontracting agents may make mistakes, coordination among coalition members may work imperfectly, information can be hiden and so on. As a result, an agent i can (erroneously) think that after the partial breaking of current contracts he/she will have a commodity bundle $x_i^T = \mathbf{e}_i + \Delta_i(V^T)$ and that commodities from x_i^T may be mutually beneficial exchanged so that to dominate the current allocation $x = (x_i)_{\mathcal{I}}$. If allocation $x(V)$ is not fuzzy contractual, then the last possibility may (potentially) destroy agreements and allocation will be changed. Thus fuzzy contractual allocations are protected from this kind of agreement destructions. Notice, that agents also allow only break contracts and do not conclude a new one.

We continue from a preliminary result describing mathematical properties of fuzzy contractual allocations, that is of interest in its own right.

Proposition 2. *An allocation* $x \in \mathcal{A}(X)$ *is fuzzy contractual if and only if*[4]

$$\mathcal{P}_i(x) \cap [x_i, \mathbf{e}_i] = \emptyset \quad \forall i \in \mathcal{I} \tag{8}$$

and

$$\prod_{\mathcal{I}} [(\mathcal{P}_i(x) + \mathrm{co}\{0, \mathbf{e}_i - x_i\}) \cup \{\mathbf{e}_i\}] \bigcap \mathcal{A}(L^{\mathcal{I}}) = \{\mathbf{e}\}. \tag{9}$$

Notice that in this proposition $\mathcal{P}_i(x) = \emptyset$ is possible for some $i \in \mathcal{I}$: by definition $\emptyset + A = \emptyset$ for any $A \subseteq L$. Condition (8) indicates that a partial break of contracts without signing of a new one cannot be beneficial. The requirement (9) denies the existence of a dominating coalition after the partial asymmetric break of the contract $v = (x - \mathbf{e})$.

Figure 2 illustrates Proposition 2 result in the Edgeworth's box. Here $\tilde{\mathcal{P}}_2(x_2) = \bar{\mathbf{e}} - \mathcal{P}_2(\bar{\mathbf{e}} - x_1)$, $\bar{\mathbf{e}} = \mathbf{e}_1 + \mathbf{e}_2$ and one can see that preferred bundles are extended along linear segment with endpoints x_1, $\mathbf{e}_1$.

---

[4] *A linear segment with ends $a, b \in L$ is the set $[a, b] = \mathrm{co}\{a, b\} = \{\lambda a + (1 - \lambda)b \mid 0 \leq \lambda \leq 1\}$.*

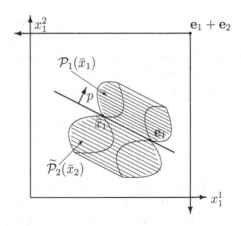

Fig. 2. *Fuzzy contractual allocations*

Proof of Proposition 2. Let x be a fuzzy contractual allocation implemented by a web V, i.e., $x = x(V)$ for some web V, satisfying Definition 1. Then (8) is clearly true one. Suppose that (9) is false and therefore does exist $y = (y_i)_\mathcal{I} \neq \mathbf{e}$ which belongs to the left part of equality (9). Consider coalition $S = \{i \in \mathcal{I} \mid y_i \neq \mathbf{e}_i\}$. Notice $\mathcal{P}_i(x) \neq \emptyset$, $i \in S$ and find $z_i \in \mathcal{P}_i(x)$, $i \in S$ such that $y_i = z_i + \lambda_i(\mathbf{e}_i - x_i)$, for some real $0 \leq \lambda_i \leq 1$, $i \in S$ and $y_i = \mathbf{e}_i$, $i \notin S$. Determine $w_i = y_i - \mathbf{e}_i$, $i \in \mathcal{I}$. Since $\sum_{i \in \mathcal{I}} y_i = \sum_{i \in \mathcal{I}} \mathbf{e}_i$ then $\sum_\mathcal{I} w_i = 0$ and therefore $w = (w_i)_{i \in \mathcal{I}}$ is a contract with supp$(w) = S \neq \emptyset$. One can write

$$z_i = y_i - \mathbf{e}_i + \lambda_i(x_i - \mathbf{e}_i) + \mathbf{e}_i = w_i + \lambda_i \sum_{v \in V} v_i + \mathbf{e}_i, \quad i \in S.$$

Now for all $v \in V$ put $t_i = t_i^v = \lambda_i$, $i \in S$, and $t_i = 1$, $i \notin S$ and apply $T(V) = \{t^v\}_{v \in V}$ for allocation $x = x(V)$. We have $x^T = \mathbf{e} + \Delta(V^T)$, whereby construction $x_i^T = \mathbf{e}_i + t_i(x_i - \mathbf{e}_i)$, $\forall i \in \mathcal{I}$. Therefore, by construction

$$\xi_i = w_i + x_i^T = z_i \in \mathcal{P}_i(x), \quad \forall i \in S = \{j \in \mathcal{I} \mid x_i \neq \xi_i\},$$

that contradicts (7).

Show that if a contractual allocation x satisfies (8) and (9) then it is fuzzy contractual relative to the web $V = \{x - \mathbf{e}\}$. Assume contrary and find $T = \{t\}$ and a contract $w = (w_i)_\mathcal{I}$, supp$(w) = S \neq \emptyset$, such that

$$w_i + t_i(x_i - \mathbf{e}_i) + \mathbf{e}_i \in \mathcal{P}_i(x), \quad \forall i \in S \iff z_i = w_i + \mathbf{e}_i \in \mathcal{P}_i(x) + t_i(\mathbf{e}_i - x_i), \quad \forall i \in S.$$

Let us determine $z_i = \mathbf{e}_i$ for $i \notin S$. Now due to contract's definition conclude $\sum_{i \in \mathcal{I}} z_i = \sum_{i \in \mathcal{I}} \mathbf{e}_i$ that implies the allocation $z \neq \mathbf{e}$ belongs to the left part of (9) and this is a contradiction. ∎

Notice that as soon as for every feasible allocation $x = (x_i)_\mathcal{I}$ we have

$$\mathbf{e}_i \in \Upsilon_i(x) \subset (\mathcal{P}_i(x) + \mathrm{co}\{0, \mathbf{e}_i - x_i\}) \cup \{\mathbf{e}_i\}, \quad \forall i \in \mathcal{I},$$

then due to Propositions 1, 2 every fuzzy contractual allocation belongs to the fuzzy core of economy. However, in general, the property of an allocation to be fuzzy contractual is still a bit stronger than being an element of fuzzy core. The following result clarifies the relationships between two fuzzy notions.

Lemma 1. *Let* $x \in \mathcal{A}(X)$ *and* $\mathcal{P}_i(x) \neq \emptyset$ *for all* $i \in \mathcal{I}$. *Then* $x \in C^f(\mathcal{E})$ *implies:*

$$\prod_{i \in \mathcal{I}} (\mathcal{P}_i(x) + \mathrm{co}\{0, \mathbf{e}_i - x_i\}) \bigcap \mathcal{A}(L^{\mathcal{I}}) = \emptyset. \tag{10}$$

The comparing of formulas (10) and (9) makes clearer the difference between fuzzy core allocation and fuzzy contractual one. One can see that this difference is not too big that allows us in appropriate circumstances to interpret allocations from fuzzy core as fuzzy contractual ones.[5] Moreover, the fact that every element of fuzzy core is a quasi-equilibrium (this is why fuzzy core is so popular in existence theory) can be also easily derived from formula (10). In fact, separating sets in (10) by a (non-zero) linear functional $\pi = (p_1, \ldots, p_n) \in L^{\mathcal{I}}$ one can conclude:

(i) $p_i = p_j = p \neq 0$ for each $i, j \in \mathcal{I}$; this is so because π is bounded on $\mathcal{A}(L^{\mathcal{I}}) = \{(z_1, \ldots, z_n) \in L^{\mathcal{I}} \mid \sum_{i \in \mathcal{I}} z_i = \sum_{i \in \mathcal{I}} \mathbf{e}_i\}$. So, one can take p as a price vector.

(ii) Due to construction and in view of preferences are locally non-satiated at the point $x \in \mathcal{A}(X)$ the points x_i and $\mathbf{e}_i$ belong to the closure of $\mathcal{P}_i(x) + \mathrm{co}\{0, \mathbf{e}_i - x_i\}$. Therefore, via separating property we have

$$\sum_{j \neq i} p\mathbf{e}_j + px_i \geq \sum_{\mathcal{I}} p\mathbf{e}_j \Rightarrow px_i \geq p\mathbf{e}_i \quad \forall i \in \mathcal{I},$$

that is possible only if $px_i = p\mathbf{e}_i \ \forall i \in \mathcal{I}$. So, we obtain budget constraints for consumption bundles.

(iii) By separation property for each i we also have

$$\langle p, \mathcal{P}_i(x) + \mathrm{co}\{0, \mathbf{e}_i - x_i\} \rangle \geq p\mathbf{e}_i,$$

that by (ii) implies $\langle p, \mathcal{P}_i(x) \rangle \geq px_i = p\mathbf{e}_i$. So we proved that p is (quasi)equilibrium prices for allocation $x = (x_i)_{i \in \mathcal{I}}$.

As a result, one can see that if an economic model is such that every quasi-equilibrium is equilibrium, then every fuzzy core allocation is fuzzy contactual one and therefore two fuzzy concepts are equivalent each other. Conditions delivering this fact are well known in literature; for example, it is the case when an economy is irreducible. Moreover there is also a nice possibility to describe fuzzy contractual allocation as an equilibrium with nonstandard prices, see [14].

---

[5] Earlier in literature allocations from fuzzy core were interpreted only as Edgeworth's equilibria and served as a technical tool more than an economic concept.

3   Existence Theorems

Theorem 1. *Let in an exchange economy $\mathcal{A}(X)$ be bounded and assumptions (**A**), (**C**) hold. Then fuzzy contractual allocations do exist.*

Corollary 1. *In Theorem 1 conditions fuzzy core is non-empty, i.e. $\mathcal{C}^f(\mathcal{E}) \neq \emptyset$.*

The existence of contractual core, fuzzy and fuzzy-contractual allocations can be established by applying Brouwer (or Kakutani) fixed point theorem and Michael's [16] continuous selector theorems. The proof of Theorem 1 is presented below; we use characterization described in Proposition 2.

4   Proofs

For the further analysis we need auxiliary lemmas. Let $NS \subset \mathcal{A}(X)$ be an area of all lower unstable contractual allocations, i.e. $x \in \mathcal{A}(X)$ for which (8) is false; also let $\Omega \subset \mathcal{A}(X)$ be a subset consisting the points $x \in \mathcal{A}(X)$ for which (9) is false. Below I study some properties of these sets.

Lemma 2. *If economy $\mathcal{E}$ obeys (**A**) and (**C**), then NS and Ω are open in $\mathcal{A}(X)$.*

Proof. An area of all lower stable contractual allocations is specified as

$$LS = \{x = (x_i)_{i \in \mathcal{I}} \in \mathcal{A}(X) \mid \mathcal{P}_i(x) \cap [x_i, \mathbf{e}_i] = \emptyset \ \forall i \in \mathcal{I}\}$$

and now I consider its supplement $NS = \mathcal{A}(X) \setminus LS$, this is the set of all allocations for which there is an agent interested in a partial break of current contract $v = x - \mathbf{e}$. Suppose $\mathcal{P}_i(x) \cap [x_i, \mathbf{e}_i] \neq \emptyset$ for some $i \in \mathcal{I}$. It means there is $y_i \in \hat{\mathcal{P}}_i(x) \cap [x_i, \mathbf{e}_i]$, $y_i \neq x_i$. Since $\hat{\mathcal{P}}_i(x)$ is assumed to be an open one, there is a *finite* set $A \subset \hat{\mathcal{P}}_i(x)$ such that

$$y_i \in \mathrm{int}(\mathrm{co}A) \quad \Rightarrow \quad x \in \Theta = \bigcap_{a \in A} \hat{\mathcal{P}}_i^{-1}(a),$$

where $\Theta \subset X$ is open in X. Due to $\hat{\mathcal{P}}_i(z)$, $z \in \mathcal{A}(X)$ are also assumed to be convex ones we conclude $y_i \in \mathrm{co}A \subset \hat{\mathcal{P}}_i(z) \ \forall z \in \Theta$. Now if $\varepsilon > 0$ is so that

$$z_i \in L, \ ||z_i - y_i|| < \varepsilon \ \Rightarrow \ z_i \in \mathrm{co}A \ \& \ x' \in \mathcal{A}(X), \ ||x' - x|| < \varepsilon \ \Rightarrow \ x' \in \Theta,$$

then for these allocations one can conclude

$$[x_i', \mathbf{e}_i] \cap \mathrm{co}A \neq \emptyset \quad \Rightarrow \quad [x_i', \mathbf{e}_i] \cap \mathcal{P}_i(x') \neq \emptyset \ \forall x' \in \mathcal{A}(X): \ ||x' - x|| < \varepsilon.$$

As a result we conclude NS is the neighbourhood of every its point and, therefore, is an open subset of $\mathcal{A}(X)$.

Next I consider $\Omega \subset \mathcal{A}(X)$. The reasoning is similar to that presented above: for every $x \in \Omega$ one can find $t = (t_i)_{i \in \mathcal{I}} \in [0,1]^{\mathcal{I}}$ and a contract $w = (w_i)_{i \in \mathcal{I}}$, $\sum_{i \in \mathcal{I}} w_i = 0$, such that

$$y_i = w_i + t_i(x_i - \mathbf{e}_i) + \mathbf{e}_i \in \mathcal{P}_i(x), \quad \forall i : y_i \neq x_i.$$

For these $i \in \mathcal{I}$ there are finite $A_i \subset L$ such that

$$y_i \in \text{int}(\text{co}A_i) \subset \hat{\mathcal{P}}_i(z), \quad \forall z \in \bigcap_{i:y_i \neq x_i} \hat{\mathcal{P}}_i^{-1}(A_i) \cap \mathcal{A}(X) \subset \Omega.$$

It implies there is $\varepsilon > 0$ such that

$$z = (z_i)_{i \in \mathcal{I}} \in \mathcal{A}(X), \; \|z - x\| < \varepsilon \; \Rightarrow \; y_i = w_i + t_i(z_i - \mathbf{e}_i) + \mathbf{e}_i \in \hat{\mathcal{P}}_i(z), \quad \forall i : y_i \neq x_i.$$

As a result $\Omega \subset \mathcal{A}(X)$ is a neighbourhood of every its point and therefore it is an open subset in $\mathcal{A}(X)$, as we wanted to prove. ∎

Let us study other properties of these allocations from Ω. Assuming $x \in \Omega$ we consider the set of contracts $\varphi(x)$ that fuzzily block this allocation:

$$\varphi(x) = \{(v_i, t_i)_{\mathcal{I}} \in (L \times [0,1])^{\mathcal{I}} \mid \sum_{\mathcal{I}} v_i = 0, v \neq 0 : \forall i \notin \text{supp}(v), t_i = 1 \; \&$$

$$\forall i \in \mathcal{I} \; v_i + t_i(x_i - \mathbf{e}_i) + \mathbf{e}_i = g_i(x), \quad \forall i \in \text{supp}(v), \; g_i(x) \succ_i x_i\}. \tag{11}$$

The following lemma presents crucial properties of the point-to-set mapping $\varphi(\cdot)$. First I recall the definition of lower hemicontinuous[6] point-to-set mapping.

Definition 2. *Let Y, Z be topological spaces. A point-to-set mapping $\psi : Y \Rightarrow Z$ is called lower hemicontinuous (l.h.c.) iff*

$$\psi^{-1}(V) = \{y \in Y \mid \psi(y) \cap V \neq \emptyset\}$$

is open for every open $V \subset Z$. For a metric spaces Y, Z a l.h.c. mapping can be equivalently characterized as follows:

For every $y \in Y$, $z \in \psi(y) \subset Z$ and every sequence $y_m \to y$ there is a subsequence $y_{m_k} \in Y$ and a sequence $z_k \in \psi(y_{m_k})$, $m, k \in \mathbb{N}$ such that $z_k \to z$ for $k \to \infty$.

Lemma 3. *If $x \in \Omega$ the set $\varphi(x)$ is convex and non-empty. Moreover, the mapping $\varphi : \Omega \Rightarrow (L \times [0,1])^{\mathcal{I}}$ is lower hemicontinuous one.*

Proof. I first state the convexity of $\varphi(x)$. Let $(w', t'), (w'', t'') \in \varphi(x)$ and $\alpha \in (0,1)$. Then, from the convexity of preferences (**A**), having in mind $t'_i = 1$ for $i \notin \text{supp}(w')$ and, similarly, $t''_i = 1$ for $i \notin \text{supp}(w'')$ we have:

$$\forall i \in \text{supp}(w') \cup \text{supp}(w'')$$

---

[6] According to the modern views, the term semi-continuous mapping is specifically applied for a function—point-to-point map—and hemicontinuous for a correspondence.

$$\alpha(w_i' + t_i'(x_i - \mathbf{e}_i) + \mathbf{e}_i) + (1 - \alpha)(w_i'' + t_i''(x_i - \mathbf{e}_i) + \mathbf{e}_i) \succ_i x_i.$$

Thus, with respect to $t = \alpha t' + (1 - \alpha)t''$, for any $\alpha \in [0, 1]$ the contract

$$w = \alpha w' + (1 - \alpha)w'' : \quad (w, t) \in \varphi(x).$$

Next, we show that due to (**C**) point-to-set mapping $\varphi(\cdot)$, defined in (11) is lower hemicontinuous.

Indeed, let $(v, t) \in \varphi(x)$ be fixed. Now according to (11), for $i \in \mathcal{I}$ such that $v_i \neq 0$ we have

$$g_i(x) = v_i + t_i(x_i - \mathbf{e}_i) + \mathbf{e}_i \in \mathcal{P}_i(x).$$

Clearly, without loss of generality it is enough to study the case $g_i(x) \in \operatorname{int} \mathcal{P}_i(x)$. Now let $A_i \subset \operatorname{int} \mathcal{P}_i(x)$ be a finite subset such that $g_i(x) \in \operatorname{int}(\operatorname{co}A_i)$, i.e. $\operatorname{co}A_i$ is a neighborhood of $g_i(x)$. We specify

$$V_i = \bigcap_{a \in A_i} \mathcal{P}_i^{-1}(a).$$

Due to (**C**) and (**A**) this is an open neighborhood of $x \in \mathcal{A}(X)$ such that $g_i(x) \in \operatorname{co}A_i \subset \mathcal{P}_i(y)$ for every $y \in V_i$. So, if $x^m \in \mathcal{A}(X)$, $x^m \to x$ for the natural $m \to \infty$, then for some $k \in \mathbb{N}$ we have: $\forall m \geq k \; \forall i \in \operatorname{supp}(v)$

$$g_i(x^m) = v_i + t_i(x_i^m - \mathbf{e}_i) + \mathbf{e}_i \in \operatorname{co}A_i \subset \mathcal{P}_i(x^m) \quad \& \quad g_i(x^m) \to g_i(x).$$

As a result, via (11) one concludes $(v^m, t^m) = (v, t) \in \varphi(x^m)$ for all $m \in \mathbb{N}$ big enough. This proves, by definition, $\varphi(\cdot)$ is lower hemicontinuous in $x \in \Omega \subset \mathcal{A}(X)$.   ∎

In the proof of Lemma 4 below I apply the following Michael theorem (see [16] p. 368, Th 3.1''', (c)) on the existence of a continuous selector in its simplified finite-dimensional presentation.[7]

Theorem 2 (Michael, 1956). *Let Y and Z be subsets of finite-dimensional linear spaces. Then every l.h.c. point-to-set mapping $\psi : Y \Rightarrow Z$ having nonempty convex images $\psi(y) \subset Z \; \forall y \in Y$ has a continuous selector.*

Lemma 4. *There is a continuous function $h : \Omega \to \mathcal{A}(X)$ such that for some continuous $\xi_i : \Omega \to X_i$, $\gamma_i : \Omega \to [0, 1]$ such that $\xi_i(x) \in \mathcal{P}_i(x) \cup \{x_i\}$ one has*

$$h_i(x) = \xi_i(x) + \gamma_i(x)(\mathbf{e}_i - x_i), \quad \forall x \in \Omega, \; i \in \mathcal{I}$$

and, moreover, for any $x \in \Omega$ there exists $i \in \mathcal{I}$ such that $h_i(x) = \xi_i(x) \succ_i x_i$, i.e. $\xi_i(x) \in \mathcal{P}_i(x)$ and $\gamma_i(x) = 0$.

---

[7] Note that in original paper item (c) has a typo for the range of $\phi : X \to \mathcal{K}(Y)$. Author denoted $\mathcal{K}(Y)$ as a set of all convex subsets of Y, but speak and prove the result for a narrower class of sets $\mathcal{D}(Y) \subset \mathcal{K}(Y)$, see p. 372. Here I present a less general result, to avoid a cumbersome specification of $\mathcal{D}(Y)$.

Proof. According to assumptions and Lemma 3, the correspondence $\varphi(\cdot)$ specified in (11) obeys all requirements of Michael's theorem on the existence of continuous selector: a lower hemicontinuous correspondence having domain $\Omega \subset \mathcal{A}(X)$, and with convex non-empty images. Thus, there is a continuous mapping satisfying

$$(v, t)(\cdot) : \Omega \to (L \times [0, 1])^{\mathcal{I}} \text{ such that } (v(x), t(x)) \in \varphi(x) \ \forall x \in \Omega.$$

By definition, we have $\sum_{i \in \mathcal{I}} v_i(x) = 0$ and, $t_i(x) = 1$, $g_i(x) = x_i$ for $i \notin \text{supp}(v(x))$ and

$$\forall i \in \text{supp}(v), \quad v_i(x) + t_i(x)(x_i - \mathbf{e}_i) + \mathbf{e}_i = g_i(x) \in \mathcal{P}(x).$$

Therefore, for $f_i(x) = v_i(x) + \mathbf{e}_i = g_i(x) + t_i(x)(\mathbf{e}_i - x_i)$, $i \in \mathcal{I}$ we obtain $\sum_{i \in \mathcal{I}} f_i(x) = \sum_{i \in \mathcal{I}} \mathbf{e}_i$. Thus, we find a continuous mapping $f : \Omega \to \mathcal{A}(L^{\mathcal{I}})$ such that

$$\forall x \in \Omega \ \ \mathbf{e} \neq f(x) \in \prod_{i \in \mathcal{I}} [(\mathcal{P}_i(x_i) + \text{co}\{0, \mathbf{e}_i - x_i\}) \cup \{\mathbf{e}_i\}] \bigcap \mathcal{A}(L^{\mathcal{I}}).$$

Now we define $t^{min}(x) = \min_{j \in \mathcal{I}}(t_j(x))$ and specify

$$h_i(x) = \frac{x_i + g_i(x) + (t_i(x) - t^{min}(x))(\mathbf{e}_i - x_i)}{2}, \quad x \in \Omega, \ i \in \mathcal{I}. \quad (12)$$

So, as

$$\sum_{i \in \mathcal{I}} (g_i(x) + t_i(x)(\mathbf{e}_i - x_i)) = \sum_{i \in \mathcal{I}} \mathbf{e}_i, \quad \sum_{i \in \mathcal{I}} (\mathbf{e}_i - x_i) = 0$$

we conclude

$$\sum_{i \in \mathcal{I}} h_i(x) = \sum_{i \in \mathcal{I}} \mathbf{e}_i.$$

Moreover, for $i \in \mathcal{I}$ such that $t_i(x) = t^{min}(x)$ we have $h_i(x) = \frac{x_i + g_i(x)}{2}$ that due to $x_i \in \text{cl}\,\mathcal{P}_i(x)$ and $g_i(x) \in \mathcal{P}_i(x)$ gives $h_i(x) \in \mathcal{P}_i(x)$. Now we need to show only that $h_i(x) \in X_i \ \forall i \in X_i$.

For $g_i(x) = x_i$ one can put $\frac{(t_i(x) - t^{min}(x))}{2} = \alpha_i \in [0, 1]$ and by (12) conclude

$$h_i(x) = (1 - \alpha_i)x_i + \alpha_i \mathbf{e}_i \in X_i.$$

For $g_i(x) \in \mathcal{P}_i(x) \subset X_i$ via (12) for $\beta_i = t_i(x) - t^{min}(x)$ we have $(1 - \beta_i)x_i + \beta_i \mathbf{e}_i \in X_i$ and therefore

$$h_i(x) = \frac{g_i(x)}{2} + \frac{(1 - \beta_i)x_i + \beta_i \mathbf{e}_i}{2} \in X_i.$$

This proves the map specified in (12) has range $\mathcal{A}(X)$. Now putting

$$\xi_i(x) = \frac{x_i + g_i(x)}{2}, \quad \gamma_i(x) = \frac{t_i(x) - t^{min}(x)}{2}, \quad i \in \mathcal{I},$$

we can redefine the map $h : \Omega \to \mathcal{A}(X)$; it obeys all requirements of Lemma 4. ∎

Proof of Theorem 1. Recall that

$$LS = \{x = (x_i)_{i \in \mathcal{I}} \in \mathcal{A}(X) \mid \mathcal{P}_i(x) \cap [x_i, e_i] = \emptyset \; \forall i \in \mathcal{I}\}$$

is an area of all lower stable contractual allocations and we consider its supplement $NS = \mathcal{A}(X) \backslash LS$, this is the set of all allocations for which there is an agent interested in a partial break of current contract $v = x - \mathbf{e}$. For $x \in NS$ condition (8) is false. Also we specified $\Omega \subset \mathcal{A}(X)$ as a subset consisting the points $x \in \mathcal{A}(X)$ for which (9) is false. Now let us suppose that

$$\Omega \cup NS = \mathcal{A}(X)$$

and show that it is impossible. According to the assumptions and Lemma 2, NS and Ω are an open subsets of $\mathcal{A}(X)$.

We specify $q : NS \to \mathcal{A}(X)$ by formula

$$q(x) = \frac{x + \mathbf{e}}{2}$$

and "glue" this mapping with $h(\cdot)$ defined in Lemma 4, setting

$$f(x) = \alpha(x)q(x) + \beta(x)h(x), \quad x \in \mathcal{A}(X),$$

where $\alpha : \mathcal{A}(X) \to [0, 1]$, $\beta : \mathcal{A}(X) \to [0, 1]$ are continuous functions, such that $\alpha(x) = 1$ for $x \in NS \setminus \Omega$, $\beta(x) = 1$ for $x \in \Omega \setminus NS$, and $\alpha(x) + \beta(x) = 1$ $\forall x \in \mathcal{A}(X)$. For example they can be specified as

$$\alpha(x) = \frac{\rho(x, LS)}{\rho(x, LS) + \rho(x, \mathcal{A}(X) \setminus \Omega)}, \quad \beta(x) = \frac{\rho(x, \mathcal{A}(X) \setminus \Omega)}{\rho(x, LS) + \rho(x, \mathcal{A}(X) \setminus \Omega)}, \quad x \in \mathcal{A}(X),$$

where $\rho(x, S)$ is a distance from the point x to the set $S \subset \mathcal{A}(X)$.[8] Obviously, for $\Omega \cup NS = \mathcal{A}(X)$ the mapping $f : \mathcal{A}(X) \to \mathcal{A}(X)$ is continuous and, by Brouwer's theorem, it must have a fixed point $\bar{x} = f(\bar{x})$. However, where is it?

Suppose $\bar{x} \in NS \setminus \Omega$. Then $f(\bar{x}) = q(x) = \frac{x+\mathbf{e}}{2} \neq \bar{x}$, since otherwise $\bar{x} = \mathbf{e} \notin NS$.

Suppose $\bar{x} \in \Omega \setminus NS$. Then $f(\bar{x}) = h(\bar{x}) = (h_j(\bar{x}))_{j \in \mathcal{I}}$ and by Lemma 4 there is $i \in \mathcal{I}$ such that $h_i(\bar{x}) \in \mathcal{P}_i(\bar{x})$ that is impossible by (**A**).

Suppose $\bar{x} \in NS \cap \Omega$. Now $f(\bar{x}) = \alpha(\bar{x})q(\bar{x}) + \beta(\bar{x})h(\bar{x})$. Clearly $\alpha(\bar{x}) > 0$ and $\beta(\bar{x}) > 0$, since the contrary is impossible. Recall that we also have $h_i(\bar{x}) = \xi_i(\bar{x}) + \gamma_i(\bar{x})(\mathbf{e}_i - \bar{x}_i) \; \forall i \in \mathcal{I}$. Now for $i_0 \in \mathcal{I}$, which is interested in a partial breaking of the contract $\bar{x} - \mathbf{e}$, at the fixed point we have

$$\bar{x}_{i_0} = \alpha \left[\bar{x}_{i_0} + \frac{1}{2}(\mathbf{e}_{i_0} - \bar{x}_{i_0}) \right] + \beta[\xi_{i_0}(\bar{x}) + \gamma_{i_0}(\bar{x})(\mathbf{e}_{i_0} - \bar{x}_{i_0})]$$

$$= \bar{x}_{i_0} + \alpha \frac{1}{2}(\mathbf{e}_{i_0} - \bar{x}_{i_0}) + \beta[\xi_{i_0}(\bar{x}) - \bar{x}_{i_0} + \gamma_{i_0}(\bar{x})(\mathbf{e}_{i_0} - \bar{x}_{i_0})] \quad \Rightarrow$$

$$\left[\frac{1}{2}\alpha + \beta\gamma_{i_0}(\bar{x}) \right] (\mathbf{e}_{i_0} - \bar{x}_{i_0}) + \beta(\xi_{i_0}(\bar{x}) - \bar{x}_{i_0}) = 0. \tag{13}$$

---

[8] It is standardly defined as $\rho(x, S) = \inf_{y \in S} \rho(x, y)$.

Clearly $\xi_{i_0}(\bar{x}) = \bar{x}_{i_0}$ is impossible, otherwise (13) implies $\bar{x}_{i_0} = \mathbf{e}_{i_0}$. Therefore $\xi_{i_0}(\bar{x}) \in \mathcal{P}_{i_0}(\bar{x})$. Also at a fixed point $\bar{x} \in NS$ for some $\lambda \in (0,1]$ we have

$$\lambda \left[\frac{1}{2}\alpha + \beta\gamma_{i_0}(\bar{x}) \right] = \mu > 0, \quad \mu(\mathbf{e}_{i_0} - \bar{x}_{i_0}) + \bar{x}_{i_0} \in \mathcal{P}_{i_0}(\bar{x}).$$

At the same time, due to $(\xi_{i_0}(\bar{x}) - \bar{x}_{i_0}) \in \mathcal{P}_{i_0}(\bar{x}) - \bar{x}_{i_0}$ and $(\mathbf{A})$ we conclude

$$\lambda\beta(\xi_{i_0}(\bar{x}) - \bar{x}_{i_0}) \in \mathcal{P}_{i_0}(\bar{x}) - \bar{x}_{i_0} \quad \Rightarrow \quad \exists \eta_{i_0}(\bar{x}) \in \mathcal{P}_{i_0}(\bar{x}) : \lambda\beta(\xi_{i_0}(\bar{x}) - \bar{x}_{i_0}) = \eta_{i_0}(\bar{x}) - \bar{x}_{i_0}.$$

Now, due to (13) and $(\mathbf{A})$ we have $\mu(\mathbf{e}_{i_0} - \bar{x}_{i_0}) + \eta_{i_0}(\bar{x}) - \bar{x}_{i_0} = 0 \Rightarrow$

$$\frac{\mu(\mathbf{e}_{i_0} - \bar{x}_{i_0}) + \bar{x}_{i_0}}{2} + \frac{\eta_{i_0}(\bar{x})}{2} = \bar{x}_{i_0} \quad \Rightarrow \quad \bar{x}_{i_0} \in \mathrm{co}\mathcal{P}_{i_0}(\bar{x}) = \mathcal{P}_{i_0}(\bar{x}),$$

which is impossible.

Thus, the assumption $\Omega \cup NS = \mathcal{A}(X)$ implies the existence of a continuous mapping $f : \mathcal{A}(X) \to \mathcal{A}(X)$ with no fixed point in $\mathcal{A}(X)$. This contradicts Brouwer's theorem. So, the assumption that there are no fuzzy contractual allocations lead us to a contradiction and it proves the theorem.   ∎

References

1. Aliprantis, C.D., Brown, D.J., Burkinshaw, O.: Existence and Optimality of Competitive Equilibria, p. 284. Springer, Berlin (1989). https://doi.org/10.1007/978-3-642-61521-4
2. Aubin, J.P.: Mathematical Methods of Game and Economic Theory. North-Holland, Amsterdam/New York/Oxford (1979)
3. Aumann, R.: Markets with a continuum of traders. Econometrica **32**(1–2), 39–50 (1964). https://doi.org/10.2307/1913732
4. Allouch, N., Predtetchinski, A.: On the non-emptiness of the fuzzy core. Int. J. Game Theory **37**, 203–10 (2008). https://doi.org/10.1007/s00182-007-0105-2
5. Brown, D.J., Robinson, A.: Nonstandard exchange economies. Econometrica **43**(1), 41–55 (1975)
6. Debreu, G., Scarf, H.: A limit theorem on the core of an economy. Int. Econ. Rev. **4**(3), 235–46 (1963)
7. Edgeworth, F.Y.: Mathematical Psychics: An Essay on the Mathematics to the Moral Sciences. Kegan Paul, London (1881)
8. Florenzano, M.: On the non-emptiness of the core of a coalitional production economy without ordered preferences. J. Math. Anal. Appl. **141**, 484–90 (1989)
9. Florenzano, M.: Edgeworth equilibria, fuzzy core and equilibria of a production economy without ordered preferences. J. Math. Anal. Appl. **153**, 18–36 (1990)
10. Marakulin, V.M.: Contracts and domination in competitive economies. J. New Econ. Assoc. **9**, 10–32 (2011). (in Russian)
11. Marakulin, V.M.: Abstract equilibrium analysis in mathematical economics, p. 348. SB Russian Academy of Science Publisher, Novosibirsk (2012) (in Russian)
12. Marakulin, V.M.: On the Edgeworth conjecture for production economies with public goods: a contract-based approach. J. Math. Econ. **49**(3), 189–200 (2013). ISSN 0304–4068

13. Marakulin, V.M.: On contractual approach for Arrow-Debreu-McKenzie economies. Econ. Math. Methods **50**(1), 61–79 (2014) (in Russian)
14. Marakulin, V.M.: Perfect competition without Slater condition: the equivalence of non-standard and contractual approach. Econ. Math. Methods **54**(1), 69–91 (2018). (in Russian, there is English translation)
15. Marakulin, V.M.: On the existence of a fuzzy core in an exchange economy. In: Pardalos, P., Khachay, M., Mazalov, V. (eds.) MOTOR 2022. Lecture NoteDs in Computer Science, vol. 13367, pp. 210–217. Springer, Cham (2022). https://doi.org/10.1007/978-3-031-09607-5_15
16. Michael, E.: Continuous selections I. Ann. Math. **63**(2), 361–82 (1956)

Linear Interpolation of Program Control with Respect to a Multidimensional Parameter in the Convergence Problem

Vladimir Nikolaevich Ushakov[1] , Aleksandr Anatol'evich Ershov[1,2]($\boxtimes$) ,
Anna Aleksandrovna Ershova[2] , and Aleksandr Vladimirovich Alekseev[3]

[1] N.N. Krasovskii Institute of Mathematics and Mechanics, Yekaterinburg, Russia
ushak@imm.uran.ru, ale10919@yandex.ru
[2] Ural Federal University, Yekaterinburg, Russia
[3] EMDB "Novator", Yekaterinburg, Russia

Abstract. We consider a control system containing a constant three-dimensional vector parameter, the approximate value of which is reported to the control person only at the moment of the movement start. The set of possible values of unknown parameter is known in advance. An convergence problem is posed for this control system. At the same time, it is assumed that in order to construct resolving control, it is impossible to carry out cumbersome calculations based on the pixel representation of reachable sets in real time. Therefore, to solve the convergence problem, we propose to calculate in advance several resolving controls, corresponds to possible parameter values in terms of some grid of nodes. If at the moment of the movement start it turns out that the value of the parameter does not coincide with any of the grid nodes, it is possible to calculate the program control using the linear interpolation formulas. However, this procedure can be effective only if a linear combination of controls corresponding to the same "guide" in the terminology of N.N. Krasovskii's Extreme Aiming Method is used. In order to be able to effectively apply linear interpolation, for each grid cell, we propose to calculate 8 "nodal" resolving controls and use the method of dividing control into basic control and correcting control in addition. Due to the application of the latter method, the calculated solvability set turns out to be somewhat smaller than the actual one. But the increasing of accuracy of the system state transferring to the target set takes place.

Keywords: Control system · Convergence problem · Unknown parameter · Program control · Linear interpolation

1 Introduction

One of the main directions of research in mathematical control theory [1,2] is the reduction of the time required to calculate the resolving optimal control. To

This research was supported by the Russian Science Foundation (grant no. 19-11-00105, https://rscf.ru/en/project/19-11-00105/).

do this, for example, algorithms of the second order of accuracy with respect to the time step are considered in papers [3–7]. The calculation of reachable sets is reduced to the calculation of their boundaries in papers [8–10]. The ways to reduce the number of points required to describe the reachable sets of control systems in article [11]. However, all of the above methods are not fast enough to calculate resolving control in real time. The need for such a fast construction of permissive control may be arised if part of the information about the control system becomes available only at the moment of the start of motion. An example of such situation is described in [12]. According to this article, solving a control problem with an incompletely known initial condition consists of three stages:

1) collecting information about the system,
2) applying this information to eliminate uncertainty,
3) transition to active control.

Firstly, we note, in the conditions of the system movement start, the control person has very short period of time to switch from the second to the third stage of this scheme. Secondly, a significant part of the uncertainties can be reduced to the parametric uncertainty [13, 14, 16].

As a solution to this problem, in paper [15] the advance to construct resolving controls corresponding to several values of an unknown parameter is proposed, and, when the real value of the parameter is obtained at the start moment, we can quickly construct resolving control using specially developed formulas for program control interpolation with respect to a scalar parameter.

This paper is devoted to generalize the results of [15] in case of a multidimensional vector parameter in the framework of the convergence problem for a control system on a finite time interval. In this article we consider a three-dimensional vector parameter as an example of a multidimensional vector parameter. This dimension is large enough to consider the result provable for an arbitrary dimension, while the presentation of the proof for the case of an arbitrary dimension would lead to unnecessary piling up of formulas.

2   Problem Statement

Let the following control system is given on a finite time interval $[t_0, \vartheta]$:

$$\frac{dx}{dt} = f(t, x(t), u(t), \alpha), \quad t \in (t_0, \vartheta),$$
$$x(t_0) = x^{(0)}, \tag{1}$$

where t is the time, $x(t) \in \mathbb{R}^n$ is the phase vector of the system state, $x^{(0)} \in \mathbb{R}^n$ is the initial state, $u(t)$ is the Lebesgue-measurable vector function (vector of controls) with values from the compact set $P \subset \mathbb{R}^p$, n and p are integers, $\alpha \in \mathscr{L}$ is a constant vector parameter, $\mathscr{L}$ is a compact in $\mathbb{R}^3$.

We assume that the following conditions are satisfied.

C1. The vector function $f(t, x, u, \alpha)$ is defined and continuous on $[t_0, \vartheta] \times \mathbb{R}^n \times P \times \mathscr{L}$ and for any bounded and closed domain $\Omega \subset [t_0, \vartheta] \times \mathbb{R}^n$ there is a constant $L = L(\Omega) \in (0, \infty)$ such that

$$\|f(t, x^{(1)}, u, \alpha) - f(t, x^{(2)}, u, \alpha)\| \leq L\|x^{(1)} - x^{(2)}\|,$$

$$(t, x^{(i)}, u, \alpha) \in \Omega \times P \times \mathscr{L}, \ i = 1, 2;$$

here $\| \cdot \|$ is the Euclidean norm of a vector in $\mathbb{R}^n$.

Remark 1. Taking into account the condition C1, we obtain that the continuity modules

$$\omega^{(3)}(\delta) = \max\{\|f(t, x, u_*, \alpha) - f(t, x, u^*, \alpha)\| :$$
$$(t, x, u_*, \alpha), (t, x, u^*, \alpha) \in D \times P \times \mathscr{L}, \ \|u_* - u^*\| \leq \delta\}, \quad \delta \in (0, \infty),$$

$$\omega^{(4)}(\delta) = \max\{\|f(t, x, u, \alpha_*) - f(t, x, u, \alpha^*)\| :$$
$$(t, x, u, \alpha_*), (t, x, u, \alpha^*) \in D \times P \times \mathscr{L}, \ \|\alpha_* - \alpha^*\| \leq \delta\}, \quad \delta \in (0, \infty),$$

are satisfied the limit relations $\omega^{(k)}(\delta) \downarrow 0$ as $\delta \downarrow 0$, $k = 3, 4$.

C2. There is a constant $\gamma \in (0, \infty)$ such that

$$\|f(t, x, u, \alpha)\| \leq \gamma(1 + \|x\|), \quad (t, x, u, \alpha) \in [t_0, \infty) \times \mathbb{R}^n \times P \times \mathscr{L}.$$

Remark 2. By an admissible control $u(t)$, $t \in [t_0, \vartheta]$, we mean a Lebesgue-measurable vector function defined on $[t_0, \vartheta]$ with values from P. The conditions C1 and C2 are sufficient for each admissible control $u(t)$, the motion $x(t)$, which is a solution to the system (1) in the class of absolutely continuous functions, is correspond [17, Sect. 2.1]. In this case, we concider the derivative $\dot{x}(t)$ as a generalization of solution, that satisfied the Newton-Leibniz formula is satisfied for it (see, for example, [18, Ch. 2, Sect. 4]).

Remark 3. By virtue of the condition C2, some sufficiently large domain $\Omega \subset [t_0, \vartheta] \times \mathbb{R}^n$ exists, which certainly contains all possible motions of the system (1) together with all auxiliary constructions for constructing resolving controls. In what follows, we will everywhere use the Lipschitz constant $L = L(\Omega)$ calculated just for this Ω domain.

Denote by $B^k(a, r) = \{\xi \in \mathbb{R}^k : \|\xi - a\| \leq r\}$ the closed ball in the space $\mathbb{R}^k$.
C3. Let the points (t_*, x_*) and (t^*, x^*) are belonged to Ω, where $t^* = t_* + \Delta$, $x^* = x_* + \Delta \cdot f(t_0 + \vartheta - t^*, x^*, \bar{u}, \alpha)$, $\Delta > 0$, $\bar{u} \in \mathring{P}(\rho(\Delta))$, $\alpha \in \mathscr{L}$. In addition, let $\Delta_\alpha > 0$ be also an arbitrary number, but not too large. Then we can define the function $\rho(\Delta)$ in such way to find the correcting vector w from $B^p(\mathbf{0}, \rho(\Delta))$ for solving the following boundary value problem

$$\begin{cases} \dot{x}(t) = f(t, x(t), \bar{u} + w, \tilde{\alpha}), & t \in (t_*, t^*), \\ x(t_*) = x_*, \ x(t^*) = x^* \end{cases}$$

for every value of $\widetilde{\alpha} \in B^3(\alpha, \Delta_\alpha\sqrt{2})$. In this case, the dependence $w = w(\widetilde{\alpha})$ must be from the class $C^2(B^3(\alpha, \Delta_\alpha\sqrt{3}))$ and for all $\widetilde{\alpha} \in B^3(\alpha, \Delta_\alpha\sqrt{3})$ must be satisfied the next inequalities

$$\left\|\frac{\partial^2 w}{\partial\widetilde{\alpha}_i\partial\widetilde{\alpha}_j}\right\| \leq M_2, \quad i,j = \overline{1,3},$$

where the constant $M_2 \geq 0$ is determined by the function $f(\cdot,\cdot,\cdot,\cdot)$, by the domain Ω, and by the values Δ and Δ_α.

Moreover, in addition to conditions C1, C2, C3, we specify *informational conditions*.

We shall assume that at the initial time t_0 the person making the choice of program control $u(t)$ is informed of some approximate value $\alpha^* \in \mathscr{L}$ of the parameter $\alpha \in \mathscr{L}$ with an error not exceeding

$$\|\alpha^* - \alpha\| < \delta_\alpha. \tag{2}$$

In addition, long before the moment t_0 of the movement start, the control person knows the constraint $\mathscr{L} \in \text{comp}(\mathbb{R}^3)$ and approximate position $x^*(t_0)$ of initial point $x(t_0)$ with error

$$\|x^*(t_0) - x(t_0)\| < \delta_x. \tag{3}$$

An additional constraint is the control person cannot perform "heavy" calculations after the moment t_0, when the system starts moving. After this moment, it is necessary to construct an resolving program control in real time, using only some preliminary calculations stored in a limited amount of memory and a priori known $x^*(t_0)$, $\mathscr{L}$.

Thus, we have formulated the information conditions.

Let M be some compact set in $\mathbb{R}^n$ that is the target set for the system (1). Let us formulate the problem of convergence with M for the system (1).

Problem 1. It is required to determine the existence of an admissible program control $u(t)$ that transfers the motion $x(t)$ of the system (1) at the time moment ϑ to a small neighborhood of M, and, if possible, to construct it.

3   Algorithms for Solving the Problem 1

Denote by $\Omega^{(\delta)}(\cdot)$ the mapping that "thinches" the set, i.e. to any bounded set $A \subset \mathbb{R}^k$, $k \in \mathbb{N}$, it associates the finite set $\widetilde{A} = \Omega^{(\delta)}(A)$, consisting possibly from a smaller number of its points and having the property:

$$d(A, \widetilde{A}) \leq \delta,$$

where $d(A, \widetilde{A})$ is the Hausdorff distance between A and $\widetilde{A}$. Methods for constructing such a "thinned out" set $\widetilde{A}$ are given in [20, p. 549].

Denote $\widetilde{P} = \Omega^{(\Delta_u)}(\check{P})$, where $\Delta_u > 0$ is a sufficiently small constant, $\check{P}$ is the restriction of the control from condition C3.

Let us introduce the mapping $X^{(\Delta)} : \mathbb{R} \times \mathbb{R} \times 2^{\Omega} \times \mathscr{L} \mapsto 2^{\Omega}$, which acts according to the rule:

$$X^{(\Delta)}(t^*, t_*, \tilde{X}_*, \alpha) = \bigcup_{x \in \tilde{X}_*} \{x + (t^* - t_*)f(t_*, x, \tilde{P}, \alpha)\}$$

$$= \bigcup_{x \in \tilde{X}_*} \bigcup_{u \in \tilde{P}} \{x + (t^* - t_*)f(t_*, x, u, \alpha)\}.$$

Now, after introducing the necessary notation, we formulate a numerical method for solving the Problem 1 in the form of two algorithms. The first algorithm is intended for calculations that are performed before the system start to move, and the second algorithm is applied directly during the movement.

Algorithm 1

1) *Choose a sufficiently large natural number N and introduce a uniform partition $\Gamma = \{t_0, t_1, t_2, \ldots, t_i, \ldots, t_N = \vartheta\}$ of the time interval $[t_0, \vartheta]$ with diameter $\Delta = \Delta(\Gamma)$, which satisfies to relations $\Delta = t_{i+1} - t_i = N^{-1} \cdot (\vartheta - t_0)$, $i = \overline{0, N-1}$.*

2) *Denote by $\Delta_\alpha > 0$ a sufficiently small constant that satisfies to condition C3 for $\Delta = \Delta(\Gamma)$. Also, the condition C3 defines the function $\rho(\Delta)$ and the control set restriction $\check{P} = \check{P}(\rho(\Delta))$.*

3) *Choose the following set of vectors $\{\alpha^{(j)}\}_{j=1}^{N_\alpha}$, so that any $\alpha \in \mathscr{L}$ is inside "own" cube with eight vertices $\alpha^{(j, \pm, \pm, \pm)} = (\alpha_1^{(j)} \pm \Delta_\alpha/2, \alpha_2^{(j)} \pm \Delta_\alpha/2, \alpha_3^{(j)} \pm \Delta_\alpha/2)$ as a finite subset $\widetilde{\mathscr{L}} \subset \mathscr{L}$.*

4) *Choose a sufficiently small constant $\Delta_x > 0$ and for all $j = \overline{1, N_\alpha}$ define the sets*

$$\tilde{X}_0 = \{x^{(0)}\},$$

$$\tilde{X}_k(\alpha^{(j)}) = \Omega^{(\Delta_x)}(X^{(\Delta)}(t_k, t_{k-1}, \tilde{X}_{k-1}, \alpha^{(j)}), \quad k = \overline{1, N}.$$

When we construct finite sets $\tilde{X}_k(\alpha^{(j)})$, $k = \overline{1, N}$, $j = \overline{1, N_\alpha}$, for each point $\overline{x}^{(k,j)} \in \tilde{X}_k(\alpha^{(j)})$ it is necessary to remember the "parent" point $\overline{x}^{(k-1,j)} \in \tilde{X}_{k-1}(\alpha^{(j)})$ and control $\overline{u}^{(k,j)} = $ const for which the following relation holds:

$$\overline{x}^{(k,j)} = \overline{x}^{(k-1,j)} + \Delta \cdot f(t_{k-1}, \overline{x}^{(k-1,j)}, \overline{u}^{(k,j)}, \alpha^{(j)}).$$

5) *If for all $\alpha^{(j)}$, $j = \overline{1, N_\alpha}$ the distance*

$$\rho(M, \tilde{X}_N(\alpha^{(j)})) = \min\{\|x - y\| : x \in M, y \in \tilde{X}_N(\alpha^{(j)})\} \leq \Delta_x,$$

then we conclude that the Problem 1 is solvable for any $\alpha \in \mathscr{L}$ that implemented in the system and we can continue the algorithm execution.
Otherwise, we are forced to state that we cannot construct a resolving program control for the Problem 1 with acceptable accuracy.

6) *For each* $j = \overline{1, N_\alpha}$ *choose one point* $\overline{x}^{(N,j)} \in \widetilde{X}_N(\alpha^{(j)})$ *that is closest to* M. *We assume that if our algorithm didn't finished at step 5), then* $\rho(\overline{x}^{(N,j)}, M) \leq \Delta_x$, $j = \overline{1, N_\alpha}$. *Further, for each* $j = \overline{1, N_\alpha}$ *we denote by* $\overline{x}^{(k,j)}$ *and* $\overline{u}^{(k,j)}$ *exactly those points and those constant control vectors that "led" us to* $\overline{x}^{(N,j)}$.

7) *For each* $j = \overline{1, N_\alpha}$ *and* $k = \overline{1, N}$ *find 8 constant correcting vectors* $w^{(k,j,\pm,\pm)} \in B^p(\mathbf{0}, \rho(\Delta))$, *which are solutions of the following boundary value problems:*

$$\begin{cases} \dot{x}^{(k,j,\pm,\pm)}(t) = f(t, x^{(k,j,\pm,\pm)}(t), \overline{u}^{(k,j)} + w^{(k,j,\pm,\pm)}, \alpha^{(j,\pm,\pm)}), & t \in (t_{k-1}, t_k), \\ x^{(k,j,\pm,\pm)}(t_{k-1}) = \overline{x}^{(k-1,j)}, \; x^{(k,j,\pm,\pm)}(t_k) = \overline{x}^{(k,j)}. \end{cases}$$

8) *To each* $\alpha^{(j)} \in \widetilde{\mathscr{L}}$ *associate 8 piecewise constant "'nodal"' controls*

$$u^{(j,\pm,\pm,\pm)}(t) = \begin{cases} \overline{u}^{(1,j)} + \overline{w}^{(1,j,\pm,\pm,\pm)}, & t \in [t_0, t_1), \\ \ldots \\ \overline{u}^{(k,j)} + \overline{w}^{(k,j,\pm,\pm,\pm)}, & t \in [t_{k-1}, t_k), \\ \ldots \\ \overline{u}^{(N,j)} + \overline{w}^{(N,j,\pm,\pm,\pm)}, & t \in [t_{N-1}, t_N]. \end{cases} \quad (4)$$

Thus, we have prepared the "'nodal"' program controls and completed the application of the first algorithm. Next, we formulate the second algorithm, which is executed in real time after the moment t_0 of the start of movement.

Algorithm 2

1) *Based on the approximate value of* α^* *obtained at the moment* t_0, *determine* $\alpha^{(j)} \in \widetilde{\mathscr{L}}$, *in whose cube it is in accordance with the partition, which was determined in step 3) of the Algorithm 1.*

2) *Represent the vector* α^* *as a linear combination of vectors* $\alpha^{(j,\pm,\pm,\pm)}$ *as follows:*

$$\alpha^* = \lambda_1 \lambda_2 \lambda_3 \alpha^{(j,-,-,-)} + (1 - \lambda_1)\lambda_2 \lambda_3 \alpha^{(j,+,-,-)} + \lambda_1(1 - \lambda_2)\lambda_3 \alpha^{(j,-,+,-)}$$

$$+ (1 - \lambda_1)(1 - \lambda_2)\lambda_3 \alpha^{(j,+,+,-)} + \lambda_1 \lambda_2 (1 - \lambda_3)\alpha^{(j,-,-,+)} + (1 - \lambda_1)\lambda_2(1 - \lambda_3)\alpha^{(j,+,-,+)}$$

$$+ \lambda_1(1 - \lambda_2)(1 - \lambda_3)\alpha^{(j,-,+,+)} + (1 - \lambda_1)(1 - \lambda_2)(1 - \lambda_3)\alpha^{(j,+,+,+)},$$

where $\lambda_1 + \lambda_2 + \lambda_3 = 0$, $0 \leq \lambda_1 \leq 1$, $0 \leq \lambda_2 \leq 1$, $0 \leq \lambda_3 \leq 1$.

3) *As a desired resolving program control, use the function*

$$\hat{u}(t) = \lambda_1 \lambda_2 \lambda_3 u^{(j,-,-,-)}(t) + (1 - \lambda_1)\lambda_2 \lambda_3 u^{(j,+,-,-)}(t)$$
$$+ \lambda_1(1 - \lambda_2)\lambda_3 u^{(j,-,+,-)}(t) + (1 - \lambda_1)(1 - \lambda_2)\lambda_3 u^{(j,+,+,-)}(t)$$
$$+ \lambda_1 \lambda_2 (1 - \lambda_3)u^{(j,-,-,+)}(t) + (1 - \lambda_1)\lambda_2(1 - \lambda_3)u^{(j,+,-,+)}(t)$$
$$+ \lambda_1(1 - \lambda_2)(1 - \lambda_3)u^{(j,-,+,+)}(t) + (1 - \lambda_1)(1 - \lambda_2)(1 - \lambda_3)u^{(j,+,+,+)}(t).$$
$$(5)$$

4   Error Estimation

Lemma 1. *Let constants* $0 \leq \lambda_1 \leq 1$, $0 \leq \lambda_2 \leq 1$, $0 \leq \lambda_3 \leq 1$, *points* $x = (x_1, x_2, x_3)$ *and* $y = (y_1, y_2, y_3)$ *are from* $\mathbb{R}^3$, *function* $f : \mathbb{R}^3 \to \mathbb{R}^n$, $f \in C^2(\mathbb{R}^3)$, *all its second partial derivatives are bounded by some constant* $m_2 \geq 0$, *i.e.*

$$\left\| \frac{\partial^2 f(x_1, x_2, x_3)}{\partial x_i \partial x_j} \right\| \leq m_2, \quad i, j = \overline{1,3}.$$

Then

$$\big\| f(\lambda_1 x_1 + (1 - \lambda_1)y_1, \lambda_2 x_2 + (1 - \lambda_2)y_2, \lambda_3 x_3 + (1 - \lambda_3)y_3)$$
$$- \lambda_1 \lambda_2 \lambda_3 f(x_1, x_2, x_3) - (1 - \lambda_1)\lambda_2 \lambda_3 f(y_1, x_2, x_3)$$
$$- \lambda_1(1 - \lambda_2)\lambda_3 f(x_1, y_2, x_3) - (1 - \lambda_1)(1 - \lambda_2)\lambda_3 f(y_1, y_2, x_3)$$
$$- \lambda_1 \lambda_2(1 - \lambda_3) f(x_1, x_2, y_3) - (1 - \lambda_1)\lambda_2(1 - \lambda_3) f(y_1, x_2, y_3)$$
$$- \lambda_1(1 - \lambda_2)(1 - \lambda_3) f(x_1, y_2, y_3) - (1 - \lambda_1)(1 - \lambda_2)(1 - \lambda_3) f(y_1, y_2, y_3) \big\|$$
$$\leq \frac{3}{8} m_2 \| x - y \|.$$

Proof. Expanding the function $f(\xi, \eta_2, \eta_3)$ in the first variable at the points $\xi = x_1$ and $\xi = x_2$ into Taylor series with remainder in integral form and substituting $\xi = \lambda_1 x_1 + (1 - \lambda_1)y_1$, we get the next equalities

$$f(\lambda_1 x_1 + (1 - \lambda_1)y_1, \eta_2, \eta_3) = f(x_1 + (1 - \lambda_1)(y_1 - x_1), \eta_2, \eta_3)$$

$$= f(x_1, \eta_2, \eta_3) + (1 - \lambda_1)(y_1 - x_1) \cdot \frac{\partial f(x_1, \eta_2, \eta_3)}{\partial x_1}$$

$$+ \int_{x_1}^{\lambda_1 x_1 + (1 - \lambda_1)y_1} (\lambda_1 x_1 + (1 - \lambda_1)y_1 - t) \frac{\partial^2 f(t, \eta_2, \eta_3)}{\partial t^2} dt,$$

$$f(\lambda_1 x_1 + (1 - \lambda_1)y_1, \eta_2, \eta_3) = f(y_1 + \lambda_1(x_1 - y_1), \eta_2, \eta_3)$$

$$= f(y_1, \eta_2, \eta_3) + \lambda_1(x_1 - y_1) \frac{\partial f(y_1, \eta_2, \eta_3)}{\partial y_1}$$

$$+ \int_{y_1}^{\lambda_1 x_1 + (1 - \lambda_1)y_1} (\lambda_1 x_1 + (1 - \lambda_1)y_1 - t) \frac{\partial^2 f(t, \eta_2, \eta_3)}{\partial t^2} dt,$$

which in turn imply that

$$\big| f(\lambda_1 x_1 + (1 - \lambda_1)y_1, \eta_2, \eta_3) - \lambda_1 f(x_1, \eta_2, \eta_3) - (1 - \lambda_1) f(y_1, \eta_2, \eta_3) \big|$$

$$= \big| \lambda_1 \big(f(x_1 + (1 - \lambda_1)(y_1 - x_1), \eta_2, \eta_3) - f(x_1, \eta_2, \eta_3) \big)$$
$$+ (1 - \lambda_1) \big(f(y_1 + \lambda_1(x_1 - y_1), \eta_2, \eta_3) - f(y_1, \eta_2, \eta_3) \big) \big|$$

$$= \left| \lambda_1 \cdot \left((1 - \lambda_1)(y_1 - x_1)\frac{\partial f(x_1, \eta_2, \eta_3)}{\partial x_1} + \int_{x_1}^{\lambda_1 x_1 + (1 - \lambda)y_1} (\lambda_1 x_1 + (1 - \lambda_1)y_1 - t)\frac{\partial^2 f(t, \eta_2, \eta_3)}{\partial t^2} dt \right) \right.$$

$$\left. + (1 - \lambda_1) \cdot \left(\lambda_1(x_1 - y_1)\frac{\partial f(y_1, \eta_2, \eta_3)}{\partial y_1} + \int_{y_1}^{\lambda_1 x_1 + (1 - \lambda_1)y_1} (\lambda_1 x_1 + (1 - \lambda_1)y_1 - t)\frac{\partial^2 f(t, \eta_2, \eta_3)}{\partial t^2} dt \right) \right|$$

$$= \left| -\lambda_1(1 - \lambda_1)(y_1 - x_1) \int_{x_1}^{y_1} \frac{\partial^2 f(t, \eta_2, \eta_3)}{\partial t^2} dt \right.$$

$$+ \lambda_1 \int_{x_1}^{\lambda_1 x_1 + (1 - c_1)y_1} (\lambda_1 x_1 + (1 - \lambda_1)y_1 - t)\frac{\partial^2 f(t, \eta_2, \eta_3)}{\partial t^2} dt$$

$$\left. + (1 - \lambda_1) \int_{y_1}^{\lambda_1 x_1 + (1 - \lambda_1)y_1} (\lambda_1 x_1 + (1 - \lambda_1)y_1 - t)\frac{\partial^2 f(t, \eta_2, \eta_3)}{\partial t^2} dt \right|$$

$$\leq \lambda_1(1 - \lambda_1)m_2(y_1 - x_1)^2 + \lambda_1 m_2 \frac{(1 - \lambda_1)^2(y_1 - x_1)^2}{2} + (1 - \lambda_1)m_2 \frac{\lambda_1^2(y_1 - x_1)^2}{2}$$

$$= \frac{3}{2}\lambda_1(1 - \lambda_1)m_2(y_1 - x_1)^2 \leq \frac{3}{8}m_2(y_1 - x_1)^2.$$

Substituting $\eta_2 = x_2$, $\eta_2 = y_2$, $\eta_3 = x_3$, and $\eta_3 = y_3$ into the last inequality, we obtain the inequalities

$$\left\| f(\lambda_1 x_1 + (1 - \lambda_1)y_1, x_2, x_3) - \lambda_1 f(x_1, x_2, x_3) - (1 - \lambda_1)f(y_1, x_2, x_3) \right\|$$
$$\leq \frac{3}{8}m_2(y_1 - x_1)^2, \tag{6}$$

$$\left\| f(\lambda_1 x_1 + (1 - \lambda_1)y_1, y_2, x_3) - \lambda_1 f(x_1, y_2, x_3) - (1 - \lambda_1)f(y_1, y_2, x_3) \right\|$$
$$\leq \frac{3}{8}m_2(y_1 - x_1)^2, \tag{7}$$

$$\left\| f(\lambda_1 x_1 + (1 - \lambda_1)y_1, x_2, y_3) - \lambda_1 f(x_1, x_2, y_3) - (1 - \lambda_1)f(y_1, x_2, y_3) \right\|$$
$$\leq \frac{3}{8}m_2(y_1 - x_1)^2, \tag{8}$$

$$\left\| f(\lambda_1 x_1 + (1 - \lambda_1)y_1, y_2, y_3) - \lambda_1 f(x_1, y_2, y_3) - (1 - \lambda_1)f(y_1, y_2, y_3) \right\|$$
$$\leq \frac{3}{8}m_2(y_1 - x_1)^2. \tag{9}$$

Similarly, for any values of η_1, η_2 and η_3 from the domain of the function $f(\cdot, \cdot, \cdot)$, we get the following equalities

$$\left\| f(\eta_1, \lambda_2 x_2 + (1 - \lambda_2)y_2, \eta_3) - \lambda_2 f(\eta_1, x_2, \eta_3) - (1 - \lambda_2)f(\eta_1, y_2, \eta_3) \right\|$$
$$\leq \frac{3}{8}m_2(y_2 - x_2)^2, \tag{10}$$

$$\left\| f(\eta_1, \eta_2, \lambda_3 x_3 + (1 - \lambda_3)y_3) - \lambda_3 f(\eta_1, \eta_2, x_3) - (1 - \lambda_3)f(\eta_1, \eta_2, y_3) \right\|$$
$$\leq \frac{3}{8}m_2(y_3 - x_3)^2. \tag{11}$$

Using the inequalities (6)–(11) and the triangle inequality, we can estimate the difference

$$\big\|f(\lambda_1 x_1 + (1-\lambda_1)y_1, \lambda_2 x_2 + (1-\lambda_2)y_2, \lambda_3 x_3 + (1-\lambda_3)y_3)$$
$$- \lambda_1\lambda_2\lambda_3 f(x_1, x_2, x_3) - (1-\lambda_1)\lambda_2\lambda_3 f(y_1, x_2, x_3)$$
$$- \lambda_1(1-\lambda_2)\lambda_3 f(x_1, y_2, x_3) - (1-\lambda_1)(1-\lambda_2)\lambda_3 f(y_1, y_2, x_3)$$
$$- \lambda_1\lambda_2(1-\lambda_3) f(x_1, x_2, y_3) - (1-\lambda_1)\lambda_2(1-\lambda_3) f(y_1, x_2, y_3)$$
$$- \lambda_1(1-\lambda_2)(1-\lambda_3) f(x_1, y_2, y_3) - (1-\lambda_1)(1-\lambda_2)(1-\lambda_3) f(y_1, y_2, y_3)\big\|$$
$$\leq \big\|f(\lambda_1 x_1 + (1-\lambda_1)y_1, \lambda_2 x_2 + (1-\lambda_2)y_2, \lambda_3 x_3 + (1-\lambda_3)y_3)$$
$$- \lambda_1 f(x_1, \lambda_2 x_2 + (1-\lambda_2)y_2, \lambda_3 x_3 + (1-\lambda_3)y_3)$$
$$- (1-\lambda_1) f(y_1, \lambda_2 x_2 + (1-\lambda_2)y_2, \lambda_3 x_3 + (1-\lambda_3)y_3)\big\|$$
$$+ \big\|\lambda_1 f(x_1, \lambda_2 x_2 + (1-\lambda_2)y_2, \lambda_3 x_3 + (1-\lambda_3)y_3)$$
$$- \lambda_1\lambda_2\lambda_3 f(x_1, x_2, x_3) - \lambda_1(1-\lambda_2)\lambda_3 f(x_1, y_2, x_3)$$
$$- \lambda_1\lambda_2(1-\lambda_3) f(x_1, x_2, y_3) - \lambda_1(1-\lambda_2)(1-\lambda_3) f(x_1, y_2, y_3)\big\|$$
$$+ \big\|(1-\lambda_1) f(y_1, \lambda_2 x_2 + (1-\lambda_2)y_2, \lambda_3 x_3 + (1-\lambda_3)y_3)$$
$$- (1-\lambda_1)\lambda_2\lambda_3 f(y_1, x_2, x_3) - (1-\lambda_1)(1-\lambda_2)\lambda_3 f(y_1, y_2, x_3)$$
$$- (1-\lambda_1)\lambda_2(1-\lambda_3) f(y_1, x_2, y_3) - (1-\lambda_1)(1-\lambda_2)(1-\lambda_3) f(y_1 \cdot y_2 \cdot y_3)\big\|$$
$$\leq \frac{3}{8} m_2 (y_1 - x_1)^2 + \lambda_1 \big\| f(x_1, \lambda_2 x_2 + (1-\lambda_2)y_2, \lambda_3 x_3 + (1-\lambda_3)y_3$$
$$- \lambda_2 f(x_1, x_2, \lambda_3 x_3 + (1-\lambda_3)y_3) - (1-\lambda_2) f(x_1, y_2, \lambda_3 x_3 + (1-\lambda_3)y_3)\big\|$$
$$+ \lambda_1 \big\|\lambda_2 f(x_1, x_2, \lambda_3 x_3 + (1-\lambda_3)y_3) - \lambda_2\lambda_3 f(x_1, x_2, x_3) - \lambda_2(1-\lambda_3) f(x_1, x_2, y_3)\big\|$$
$$+ \lambda_1 \big\|(1-\lambda_2) f(x_1, y_2, \lambda_3 x_3 + (1-\lambda_3)y_3)$$
$$- (1-\lambda_2)\lambda_3 f(x_1, y_2, x_3) - (1-\lambda_2)(1-\lambda_3) f(x_1, y_2, y_3)\big\|$$
$$+ (1-\lambda_1) \big\| f(y_1, \lambda_2 x_2 + (1-\lambda_2)y_2, \lambda_3 x_3 + (1-\lambda_3)y_3)$$
$$- \lambda_2 f(y_1, x_2, \lambda_3 x_3 + (1-\lambda_3)y_3) - (1-\lambda_2) f(y_1, y_2, \lambda_3 x_3 + (1-\lambda_3)y_3)\big\|$$
$$+ (1-\lambda_1) \big\|\lambda_2 f(y_1, x_2, \lambda_3 x_3 + (1-\lambda_3)y_3) - \lambda_2\lambda_3 f(y_1, x_2, x_3) - \lambda_2(1-\lambda_3) f(y_1, x_2, y_3)\big\|$$
$$+ (1-\lambda_1) \big\|(1-\lambda_2) f(y_1, y_2, \lambda_3 x_3 + (1-\lambda_3)y_3)$$
$$- (1-\lambda_2)\lambda_3 f(y_1, y_2, x_3) - (1-\lambda_2)(1-\lambda_3) f(y_1, y_2, y_3)\big\|$$
$$\leq \frac{3}{8} m_2 (y_1 - x_1)^2 + \lambda_1 \cdot \frac{3}{8} m_2 (y_2 - x_2)^2 + \lambda_1\lambda_2 \cdot \frac{3}{8} m_2 (y_3 - x_3)^2$$
$$+ \lambda_1(1-\lambda_2) \cdot \frac{3}{8} m_2 (y_3 - x_3)^2 + (1-\lambda_1) \cdot \frac{3}{8} m_2 (y_2 - x_2)^2$$
$$+ (1-\lambda_1)\lambda_2 \cdot \frac{3}{8} m_2 (y_3 - x_3)^2 + (1-\lambda_1)(1-\lambda_2) \cdot \frac{3}{8} m_2 (y_3 - x_3)^2$$
$$= \frac{3}{8} m_2 (y_1 - x_1)^2 + \frac{3}{8} m_2 (y_2 - x_2)^2 + \frac{3}{8} m_2 (y_3 - x_3)^2 = \frac{3}{8} m_2 \|y - x\|^2.$$

Remark 4. For a single variable scalar function $f : [x_0, x_1] \to \mathbb{R}$ with bounded second derivative (i.e. $|f''(x)| \le m_2$, where the constant $m_2 \ge 0$) from the error estimate for Lagrange's interpolation formula [19, ch. XIV, §14, formula (6)]

$$\left| f(x) - \frac{x - x_1}{x_0 - x_1} f(x_0) - \frac{x - x_0}{x_1 - x_0} f(x_1) \right| \le \frac{\max\limits_{x_0 \le x \le x_1} |f''(x)|}{2} \cdot |(x - x_0)(x - x_1)|$$

and the inequality

$$(x - x_0)(x_1 - x) \le \frac{(x_1 - x_0)^2}{4}, \quad x_0 \le x \le x_1,$$

directly follows the estimaion

$$\left| f(\lambda x + (1 - \lambda)y) - \lambda f(x) - (1 - \lambda)f(y) \right| \le \frac{m_2}{8}(y - x)^2, \quad x, y \in \mathbb{R}, \quad 0 \le \lambda \le 1. \tag{12}$$

However, for a vector function of several variables, the proof of formula (6) given in [19, ch. XIV, §14] will be not correct due to the use of Lagrange's theorem on the finite increment formula, which, as is known, is not applicable to vector functions.

Theorem 1. *Let the system (1) satisfies the conditions C1, C2, C3 and the information conditions described in §2, and system motion $\hat{x}(t)$, $t \in [t_0, \vartheta]$, was generated by the control $\hat{u}(t)$ that generated by the Algorithm 1 and by the Algorithm 2 .*
Then

$$\rho(\hat{x}(\vartheta), M) \le \Delta_x + \delta_x e^{L(\vartheta - t_0)} + \frac{\omega^{(3)}\left(\frac{3}{8} M_2 \Delta_\alpha^2\right) + \omega^{(4)}(\delta_\alpha)}{L} \left(e^{L(\vartheta - t_0)} - 1\right).$$

Proof. According to step 3) of the Algorithm 1, there is an index $j \in \{1, ..., N_\alpha\}$ such that

$$\alpha^* = \lambda_1 \lambda_2 \lambda_3 \alpha^{(j,-,-,-)} + (1 - \lambda_1)\lambda_2 \lambda_3 \alpha^{(j,+,-,-)}$$
$$+ \lambda_1(1 - \lambda_2)\lambda_3 \alpha^{(j,-,+,-)} + (1 - \lambda_1)(1 - \lambda_2)\lambda_3 \alpha^{(j,+,+,-)}$$
$$+ \lambda_1 \lambda_2(1 - \lambda_3)\alpha^{(j,-,-,+)} + (1 - \lambda_1)\lambda_2(1 - \lambda_3)\alpha^{(j,+,-,+)}$$
$$+ \lambda_1(1 - \lambda_2)(1 - \lambda_3)\alpha^{(j,-,+,+)} + (1 - \lambda_1)(1 - \lambda_2)(1 - \lambda_3)\alpha^{(j,+,+,+)},$$

where $0 \le \lambda_1 \le 1$, $0 \le \lambda_2 \le 1$, $0 \le \lambda_3 \le 1$.
The symbol $\hat{x}(t)$ denotes the system (1) motion that corresponding to the control

$$\hat{u}(t) = \lambda_1 \lambda_2 \lambda_3 u^{(j,-,-,-)}(t) + (1 - \lambda_1)\lambda_2 \lambda_3 u^{(j,+,-,-)}(t)$$
$$+ \lambda_1(1 - \lambda_2)\lambda_3 u^{(j,-,+,-)}(t) + (1 - \lambda_1)(1 - \lambda_2)\lambda_3 u^{(j,+,+,-)}(t)$$
$$+ \lambda_1 \lambda_2(1 - \lambda_3)u^{(j,-,-,+)}(t) + (1 - \lambda_1)\lambda_2(1 - \lambda_3)u^{(j,+,-,+)}(t)$$
$$+ \lambda_1(1 - \lambda_2)(1 - \lambda_3)u^{(j,-,+,+)}(t) + (1 - \lambda_1)(1 - \lambda_2)(1 - \lambda_3)u^{(j,+,+,+)}(t),$$

where α is the parameter, and $x(t_0)$ is the initial state. Note that in this notation $\hat{x}(t_0) = x(t_0)$ is the exact initial state of the system, and the estimation (3) can be written as

$$||\hat{x}(t_0) - \overline{x}(t_0)|| = ||x(t_0) - x^*(t_0)|| \leq \delta_x. \tag{13}$$

By construction

$$\rho(\overline{x}(\vartheta), M) \leq \Delta_x. \tag{14}$$

Let us estimate the mismatch $||\hat{x}(\vartheta) - \overline{x}(\vartheta)||$.

By virtue of the condition C3, there is some ideal correcting vector $\overline{w}^{(1,j)} \in B^p(\mathbf{0}, \rho(\Delta))$ such that the system state $\overline{x}(t_0)$ under the action of constant control $\overline{u}^{(1,j)} + \overline{w}^{(1,j)}$ on the interval $[t_0, t_1)$ and, with the parameter α^*, is transferred to the point $\overline{x}(t_1)$ along some trajectory $\overline{x}(t)$.(Therefore, we denote by $\overline{x}(t)$ the entire trajectory of the system (1) passing through the points $\overline{x}(t_0)$, $\overline{x}(t_1)$, ..., $\overline{x}(t_N)$ under the action of the corresponding piecewise constant control $\overline{u}(t) = \overline{u}^{(k,j)} + \overline{w}^{(k,j)}$, $t \in [t_{k-1}, t_k)$, $k = \overline{1, N}$.)

However, according to the Algorithm 2 we use the correcting vector

$$\hat{w}^{(1,j)} = \lambda_1 \lambda_2 \lambda_3 w^{(1,j,-,-,-)} + (1 - \lambda_1)\lambda_2 \lambda_3 w^{(1,j,+,-,-)}$$

$$+ \lambda_1(1 - \lambda_2)\lambda_3 w^{(1,j,-,+,-)} + (1 - \lambda_1)(1 - \lambda_2)\lambda_3 w^{(1,j,+,+,-)}$$

$$+ \lambda_1 \lambda_2(1 - \lambda_3) w^{(1,j,-,-,+)} + (1 - \lambda_1)\lambda_2(1 - \lambda_3) w^{(1,j,+,-,+)}$$

$$+ \lambda_1(1 - \lambda_2)(1 - \lambda_3) w^{(1,j,-,+,+)} + (1 - \lambda_1)(1 - \lambda_2)(1 - \lambda_3) w^{(1,j,+,+,+)}$$

instead of the ideal correcting vector $\overline{w}^{(1,j)}$.

Due to the fact that it is a convex linear combination of the vector set $\{w^{(1,j,\pm,\pm,\pm)}\}$, it also falls into $B^p(\mathbf{0}, \rho(\Delta))$. By virtue of the Lemma 1, the estimation

$$||\hat{w}^{(1,j)} - \overline{w}^{(1,j)}|| \leq \frac{3}{8} M_2 \Delta_\alpha^2$$

is satisfied.

Since, according to the formulas (4) and (5)

$$\hat{u}(t) = \overline{u}^{(1,j)} + \hat{w}^{(1,j)}, \quad t \in [t_0, t_1),$$

then

$$||\hat{u}(t) - \overline{u}(t)|| = ||\hat{w}^{(1,j)} - \overline{w}^{(1,j)}|| \leq \frac{3}{8} M_2 \Delta_\alpha^2, \quad t \in [t_0, t_1).$$

Similarly, for each $k = \overline{1, N}$, by virtue of the condition C3, there is some ideal correcting vector $\overline{w}^{(k,j)} \in B^p(\mathbf{0}, \rho(\Delta))$ such that the system state $\overline{x}(t_{k-1})$ under constant control $\overline{u}(t) = \overline{u}^{(k,j)} + \overline{w}^{(k,j)}$ on the interval $[t_{k-1}, t_k)$ and for $\alpha = \alpha^*$ is transfered to the point $\overline{x}(t_k)$ along the trajectory $\overline{x}(t)$. However, according to the Algorithm 2, on the interval $[t_{k-1}, t_k)$, we use the control

$$\hat{u}(t) = \overline{v}^{(k,j)} + \hat{w}^{(k,j)}$$

$$= \overline{v}^{(k,j)} + \lambda_1 \lambda_2 \lambda_3 w^{(1,j,-,-,-)} + (1 - \lambda_1)\lambda_2 \lambda_3 w^{(1,j,+,-,-)}$$

$$+ \lambda_1(1 - \lambda_2)\lambda_3 w^{(1,j,-,+,-)} + (1 - \lambda_1)(1 - \lambda_2)\lambda_3 w^{(1,j,+,+,-)}$$
$$+ \lambda_1\lambda_2(1 - \lambda_3)w^{(1,j,-,-,+)} + (1 - \lambda_1)\lambda_2(1 - \lambda_3)w^{(1,j,+,-,+)}$$
$$+ \lambda_1(1 - \lambda_2)(1 - \lambda_3)w^{(1,j,-,+,+)} + (1 - \lambda_1)(1 - \lambda_2)(1 - \lambda_3)w^{(1,j,+,+,+)},$$

for which, by virtue of the Lemma 1, he following estimate holds:

$$||\hat{u}(t) - \overline{u}(t)|| = ||\hat{w}^{(k,j)} - \overline{w}^{(k,j)}|| \le \frac{3}{8}M_2\Delta^2, \quad t \in [t_{k-1}, t_k), \quad k = \overline{1, N}. \quad (15)$$

In other words, the estimation (15) is performed on the entire time interval $[t_0, \vartheta]$.

In addition, recall that the value α^* of the parameter α is known with some error δ_α (see (2)).

Thus, we obtain for $t \in [t_0, \vartheta]$ the following integral estimation of the movement mismatch, taking into account (13), (15) and (2):

$$||\hat{x}(t) - \overline{x}(t)|| \le \left|\left| \hat{x}(t_0) + \int_{t_0}^{t} f(\tau, \hat{x}(\tau), \hat{u}(\tau), \alpha)d\tau - \overline{x}(t_0) - \int_{t_0}^{t} f(\tau, \overline{x}(\tau), \overline{u}_j(\tau), \alpha^*)d\tau \right|\right|$$

$$\le ||\hat{x}(t_0) - \overline{x}(t_0)|| + \int_{t_0}^{t} \Big(||f(\tau, \hat{x}(\tau), \hat{u}(\tau), \alpha) - f(\tau, \overline{x}(\tau), \hat{u}(\tau), \alpha) + f(\tau, \overline{x}(\tau), \hat{u}(\tau), \alpha)$$

$$- f(\tau, \overline{x}(\tau), \overline{u}(\tau), \alpha) + f(\tau, \overline{x}(\tau), \overline{u}(\tau), \alpha) - f(\tau, \overline{x}(\tau), \overline{u}(\tau), \alpha^*)||\Big)d\tau$$

$$\le \delta_x + \int_{t_0}^{t} L||\hat{x}(\tau) - \overline{x}(\tau)||d\tau + \int_{t_0}^{t} \omega^{(3)}(\hat{u}(\tau) - \overline{u}(\tau))d\tau + \int_{t_0}^{t} \omega^{(4)}(\alpha - \alpha^*)d\tau$$

$$\le \delta_x + L\int_{t_0}^{t} ||\hat{x}(\tau) - \overline{x}(\tau)||d\tau + (t - t_0) \cdot \omega^{(3)}\left(\frac{3}{8}M_2\Delta_\alpha^2\right) + (t - t_0) \cdot \omega^{(4)}(\delta_\alpha).$$

Hence, by virtue of the strengthened Gronwall lemma [21, ch. 1, §2, p. 26] it follows that

$$||\hat{x}(\vartheta) - \overline{x}(\vartheta)|| \le \delta_x e^{L(\vartheta - t_0)} + \frac{\omega^{(3)}\left(\frac{3}{8}M_2\Delta_\alpha^2\right) + \omega^{(4)}(\delta_\alpha)}{L}(e^{L(\vartheta - t_0)} - 1). \quad (16)$$

The theorem assertion follows from (14) and (16).

5   Conclusion

For simplicity of presentation and clarity, in this article the case of a three-dimensional vector parameter was considered, but it is obvious that the result obtained is valid in the case of any finite dimension of the parameter. No fundamental difficulties are expected in proving the multidimensional analogue of the

Lemma 1 using the method of induction by the number of variables (it is also possible that there is a closer analogue of this auxiliary inequality in the literature than in the specified source [19]. But due to the specific technical character of the Lemma 1, its search is difficult). Note that in the case of an n-dimensional parameter $\alpha = (\alpha_1, ..., \alpha_n)$ the ball $B^3(\alpha, \Delta_\alpha\sqrt{3})$ from the condition C3 should be replaced by the ball $B^n(\alpha, \Delta_\alpha\sqrt{n})$, whose radius is calculated as the length of the diagonal of an n-dimensional cube with side Δ_α. We also note that the condition C3 has a generalized technical nature, and for the practical application of the Algorithms 1 and 2, it is necessary to develop easily verifiable conditions that sufficient for the fulfillment of the condition C3. In comparison with [15], a significant simplification of the algorithm was made: in the new algorithm "reverse" time is not introduced and additional "nodal" resolving controls are not calculated for points inside the partition cells of the possible parameter value set. Also, the number of necessary conditions for the control system is reduced.

Acknowledgements. This research was supported by the Russian Science Foundation (grant no. 19-11-00105, https://rscf.ru/en/project/19-11-00105/).

References

1. Lee, E.B., Markus, L.: Foundations of Optimal Control Theory. Wiley, New York (1967)
2. Sethi, S.P., Thompson, G.L.: Optimal Control Theory: Applications to Management Science and Economics. Springer, New York (2007). https://doi.org/10.1007/0-387-29903-3
3. Ferretti, R.: High-order approximations of linear control systems via Runge-Kutta schemes. Computing **58**(4), 351–364 (1997). https://doi.org/10.1007/BF02684347
4. Veliov, V.M.: Second order discrete approximation to linear differential inclusions. SIAM J. Numer. Anal. **29**(2), 439–451 (1992). https://doi.org/10.1137/0729026
5. Guang, D.H., Mingzhu, L.: Input-to-state stability of Runge-Kutta methods for nonlinear control systems. J. Comput. Appl. Math. **205**(1), 633–639 (2007). https://doi.org/10.1016/j.cam.2006.05.031
6. Baier, R.: Selection strategies for set-valued Runge-Kutta methods. In: Li, Z., Vulkov, L., Waśniewski, J. (eds.) NAA 2004. LNCS, vol. 3401, pp. 149–157. Springer, Heidelberg (2005). https://doi.org/10.1007/978-3-540-31852-1_16
7. Novikova, A.O.: Construction of reachable sets for two-dimensional nonlinear control systems by pixel method. Prikladnaya Matematika i Informatika **50**, 62–82 (2015). [in Russian]
8. Gornov, A.Y., Filkenstein, E.A.: Algorithm for piecewise-linear approximation of the reachable set boundary. Autom. Remote Control **76**(3), 385–393 (2015). https://doi.org/10.1134/S0005117915030030
9. Ershov, A.A., Ushakov, A.V., Ushakov, V.N.: Two game-theoretic problems of approach. Sbornik: Mathematics **212**(9), 1228–1260 (2021). https://doi.org/10.1070/SM9496
10. Wei, L., Taotao, L., Chen, W., Ya, D.: A dimensionality reduction method for computing reachable tubes based on piecewise pseudo-time dependent Hamilton-Jacobi equation. Appl. Math. Comput. **441**(127696), 1–13 (2023). https://doi.org/10.1016/j.amc.2022.127696

11. Filkenstein, E.A., Gornov, A.Yu.: Algorithm for quasi-uniform filling of reachable set of nonlinear control system. Bull. Irkutsk State Univ. Ser. Math. **19**, 217–223 (2017). https://doi.org/10.26516/1997-7670.2017.19.217

12. Nikol'skii, M.S.: A control problem with a partially known initial condition. Comput. Math. Model. **28**(1), 12–17 (2017). https://doi.org/10.1007/s10598-016-9341-2

13. Veliov, V.M.: Parametric and functional uncertainties in dynamic systems local and global relationship. In: Computer Arithmetic and Enclosure Methods, North-Holland, Amsterdam, pp. 1–14 (1992)

14. Kurzhanskii, A.B.: Control and Observation in Conditions of Uncertainty. Naika, Moscow (1977). [in Russian]

15. Ershov, A.A.: Linear parameter interpolation of a program control in the approach problem. J. Math. Sci. **260**(6), 725–737 (2022). https://doi.org/10.1007/s10958-022-05724-z

16. Ushakov, V., Ershov, A., Ushakov, A., Kuvshinov, O.: The problem of guidance the integral funnel to the target set. In: 2022 16th International Conference on Stability and Oscillations of Nonlinear Control Systems (Pyatnitskiy's Conference), pp. 1–2. IEEE, Moscow (2022). https://doi.org/10.1109/STAB54858.2022.9807579

17. Bressan, A., Piccoli, B.: Introduction to the Mathematical Theory of Control. American Institute of Mathematical Sciences, New York (2007)

18. Mikhlin, S.G.: Course of Mathematical Physics. Nauka, Moscow (1968). [in Russian]

19. Demidovich, B.P., Maron, I.A.: Fundamentals of Computational Mathematics. Nauka, Moscow (1966). [in Russian]

20. Ushakov, V.N., Ershov, A.A. On the solution of control problems with fixed terminal time. Vestnik Udmurtskogo Universiteta. Matematika. Mekhanika. Komp'yuternye Nauki **26**(4), 543–564 (2016). [in Russian]

21. Lizorkin, P.I.: Course in Differential and Integral Equations with Additional Chapters of Analysis. Nauka, Moscow (1981). [in Russian]

Behavior of Stabilized Trajectories of a Two Factor Economic Growth Model Under the Changes of a Production Function Parameter

Anastasiia A. Usova[1]([✉]) [iD] and Alexander M. Tarasyev[1,2] [iD]

[1] N.N. Krasovskii Institute of Mathematics and Mechanics UbRAS,
16 S.Kovalevskaya str., Yekaterinburg 620990, Russia
{ausova,tam}@imm.uran.ru
[2] Ural Fedelal University, 19 Mira str., Yekaterinburg 620002, Russia

Abstract. Based on a two-factors economic growth model with a production function of a constant elasticity of substitution, the paper considers a control problem with the infinite time interval and analyzes its stabilized solutions, when the elasticity parameter changes. A qualitative analysis of a Hamiltonian system reveals an existence of a saddle steady state, which continuously depends on the elasticity coefficient. In the domain containing the steady state, the stabilization of a Hamiltonian system is performed, and solutions of the stabilized system are numerically constructed. Varying the elasticity coefficients of CES-production function, these solutions undergo changes. The paper shows that for a limit value of the elasticity parameter, when a production function turns into the Cobb-Douglas production function, corresponding stabilized solutions converge to the limit case associated with the Cobb-Douglas function. Numerical experiments support the theoretical conclusions.

Keywords: Hamiltonian systems · Steady state · Sensitivity analysis · Stabilizer · Production function

1 Introduction

The paper deals with the analysis of growth models with a production function of a constant elasticity of substitution (CES) [9]. Well-known Cobb-Douglas and Leontief production functions are particular (limit) cases of the CES-function. A production function of Cobb-Douglas type is applied in the Solow-Swan exogenous growth model [10]. The Solow-Swan economic model explains long-run economic growth caused by capital accumulation, labor or population growth, and increases in productivity largely driven by technological development. In these

The research of the first author, Anastasiia A. Usova, is supported by the Russian Science Foundation (Project No. 19-11-00105), https://rscf.ru/project/19-11-00105/.

models, an output depends on production factors by means the production function often specified to be of Cobb-Douglas type. Leontief production function is used, for example, in the Harrod-Domar economic growth model [10] that tries to explain an economy's growth rate in terms of the level of saving and of capital, assuming the absence of a natural reason for an economy to have balanced growth. Nowadays, the Leontief function is mainly used to design small-scale production systems and to describe fully automated production systems, while the Cobb-Douglas production function fits for medium-scale production systems with steady and stable performance.

Growth models serve as a basis for optimal control problems aiming at the balanced distribution of investments for achieving the maximum level of a utility function [1,2,4,5,12]. In the cited works, problem analysis was performed for growth models with a production function of Cobb-Douglas type. These models consider production factors, usually one or several of the following: capital, labour, population and resource use, and investigate their influence on the output. However, as was mentioned in [9], there are models where the dependence between the production factors and the output is better described by the CES-function with non-unit elasticity of substitution. It motivates us to consider the corresponding problem and investigate behavior of its stabilized solutions under the changes of elasticity of substitution in order to compare these solutions derived in the case with the Cobb-Douglas production function [12].

The paper has the following structure. The next section introduces a CES-function and its properties, then we describe a two-sector growth model and the corresponding control problem. Based on the Pontryagin maximum principle, the Hamiltonian system is constructed and its qualitative analysis is performed. Forth section deals with the stabilized Hamiltonian system and demonstrates its stabilized solutions for different values of the elasticity parameter. Numerical examples finalize the paper.

2 Problem Statement

2.1 CES Production Function

In the paper, we investigate a growth model with two production factors x_1 and x_2 that are per-capita capital and the labour efficiency respectively. According to the SEDIM model introduced in [14], the labor efficiency is the proportionality coefficient between the labor force and the population. Output y depends on the production factors as a function of the constant elasticity of substitution (CES-function).

$$y = f_\gamma(x_1, x_2) = a \left(\alpha_1 x_1^{-\gamma} + \alpha_2 x_2^{-\gamma} \right)^{-\frac{\nu}{\gamma}}. \tag{1}$$

Positive scale parameter a is the total factor productivity. Coefficients α_1 and α_2 determine the contribution of each factor in the output y and correspond the restrictions $\alpha_{1,2} \geq 0$, $\alpha_1 + \alpha_2 = 1$. Parameter ν ($\nu \in (0, 1]$) is the degree of positive homogeneity of the production function, namely

$$f_\gamma(kx_1, kx_2) = a \left(\alpha_1(kx_1)^{-\gamma} + \alpha_2(kx_2)^{-\gamma}\right)^{-\frac{\nu}{\gamma}} = k^\nu f_\gamma(x_1, x_2).$$

The power parameter γ ($\gamma > -1$) is inversely proportional to the elasticity coefficient σ that satisfies the formula

$$\sigma = -\frac{d\left(\ln x_2/x_1\right)}{d\left(\ln \frac{\partial f(x_1, x_2)}{\partial x_2} \Big/ \frac{\partial f(x_1, x_2)}{\partial x_1}\right)} = \frac{1}{\gamma + 1}.$$

As it was mentioned in the introduction, Leontief and Cobb–Douglas production functions are particular (limit) cases of a CES-function (1). Specifically, Leontief production function is $y = a \min\{x_1^\nu, x_2^\nu\}$ has zero elasticity of substitution (for $\gamma \to +\infty$). The Cobb–Douglas production function $y = ax_1^{\alpha_1\nu} x_2^{\alpha_2\nu} = f_0(x_1, x_2)$ is the CES-function with the unit elasticity ($\sigma = 1$ or $\gamma = 0$). Indeed,

$$\lim_{\gamma \to 0} a \left(\alpha_1 x_1^{-\gamma} + \alpha_2 x_2^{-\gamma}\right)^{-\frac{\nu}{\gamma}} = a \lim_{\gamma \to 0} \exp\left(-\frac{\nu \ln\left(\alpha_1 x_1^{-\gamma} + \alpha_2 x_2^{-\gamma}\right)}{\gamma}\right)$$

$$= a \lim_{\gamma \to 0} \exp\left(\nu \frac{\alpha_1 x_1^{-\gamma} \ln x_1 + \alpha_2 x_2^{-\gamma} \ln x_2}{\alpha_1 x_1^{-\gamma} + \alpha_2 x_2^{-\gamma}}\right) = ax_1^{\alpha_1\nu} x_2^{\alpha_2\nu} = f_0(x_1, x_2). \quad (2)$$

Optimal control problems based on the economic growth models with the Cobb-Douglas production function are comprehensively studied, their optimal and stabilized solutions are also constructed (see [1,2,4,12]).

Before we move to the model description and the problem statement, let us formulate properties of a CES-function related to the parameter γ.

Property 1. First and second order partial derivatives of the CES-function (1) by phase variables tends to the corresponding first and second order partial derivatives of the Cobb-Douglas production function as γ goes to zero, *i.e.*

$$\lim_{\gamma \to 0} \frac{\partial f_\gamma(x_1, x_2)}{\partial x_i} = \frac{\partial f_0(x_1, x_2)}{\partial x_i}, \quad \lim_{\gamma \to 0} \frac{\partial^2 f_\gamma(x_1, x_2)}{\partial x_i \partial x_j} = \frac{\partial^2 f_0(x_1, x_2)}{\partial x_i \partial x_j}, \quad i, j = 1, 2.$$

Proof. To proof the property, we find partial derivatives of the CES-function (1) and express them through the function itself that converges to the Cobb-Douglas function (2) as γ goes to zero. Also, we use here the fact that parameters α_1 and α_2 are weight coefficients, and their sum equals to one.

$$\frac{\partial f_\gamma(x_1, x_2)}{\partial x_i} = \frac{\nu \alpha_i x_i^{-\gamma-1}}{\alpha_1 x_1^{-\gamma} + \alpha_2 x_2^{-\gamma}} f_\gamma(x_1, x_2) \xrightarrow[\gamma \to 0]{} \frac{\nu \alpha_i x_i^{-1}}{\alpha_1 + \alpha_2} f_0(x_1, x_2)$$

$$= \nu \alpha_i \frac{f_0(x_1, x_2)}{x_i} = \frac{\partial f_0(x_1, x_2)}{\partial x_i}, \quad i = 1, 2.$$

Similarly, we deal with the second derivative

$$\frac{\partial^2 f_\gamma(x_1, x_2)}{\partial x_i^2} = \left(-\frac{\nu\alpha_i(\gamma + 1)x_i^{-\gamma-2}}{\alpha_1 x_1^{-\gamma} + \alpha_2 x_2^{-\gamma}} + \frac{\nu\alpha_i\gamma x_i^{-2\gamma-2}}{\left(\alpha_1 x_1^{-\gamma} + \alpha_2 x_2^{-\gamma}\right)^2} \right) f_\gamma(x_1, x_2)$$

$$+ \left(\frac{\nu\alpha_i x_i^{-\gamma-1}}{\alpha_1 x_1^{-\gamma} + \alpha_2 x_2^{-\gamma}} \right)^2 f_\gamma(x_1, x_2)$$

$$\xrightarrow[\gamma\to 0]{} -\nu\alpha_i(1 - \nu\alpha_i)\frac{f_0(x_1, x_2)}{x_i^2} = \frac{\partial^2 f_0(x_1, x_2)}{\partial x_i^2}, \quad i = 1, 2.$$

Finally, we check this property for the second-order mixed derivatives

$$\frac{\partial^2 f_\gamma(x_1, x_2)}{\partial x_1 \partial x_2} = \frac{\nu(\nu + \gamma)\alpha_1\alpha_2 (x_1 x_2)^{-\gamma-1}}{\left(\alpha_1 x_1^{-\gamma} + \alpha_2 x_2^{-\gamma}\right)^2} f_\gamma(x_1, x_2)$$

$$\xrightarrow[\gamma\to 0]{} \nu^2\alpha_1\alpha_2\frac{f_0(x_1, x_2)}{x_1 x_2} = \frac{\partial^2 f_0(x_1, x_2)}{\partial x_1 x_2}.$$

Hence, the property is proven.   □

The present work discusses stabilized solutions of a control problem with the CES-production function and investigates their behaviour when the elasticity parameter σ tends to the unit (or $\gamma \to 0$), in other words, when the CES-production function switches to the Cobb-Douglas function (2).

Next subsection represents the dynamic growth model and poses the control problem.

2.2   Dynamic Growth Model and Control Problem

The paper investigates an economic growth model [12] that describes an output y as a CES-function (1) of two factors $x = (x_1, x_2) \in \mathbb{R}^2_{>0}$. These factors grow due to investments that are the shares u_1 and u_2 of the output y. Investments shares u_1 and u_2 play roles of control parameters and satisfies the restrictions $u = (u_1, u_2) \in [0, \bar{u}_1] \times [0, \bar{u}_2] = \mathcal{U} \subset \mathbb{R}^2$. Negative trends of the factors are caused by the capital depreciation and/or population growth rates denoted by symbols δ and λ respectively. As a result, the dynamic equations of the production factors have the form

$$\begin{cases} \dot{x}_1(t) = f_\gamma(x(t))u_1(t) - (\delta + \lambda)x_1(t), & x_1(0) = x_1^0, \\ \dot{x}_2(t) = bf_\gamma(x(t))u_2(t) - \lambda x_2(t), & x_2(0) = x_2^0. \end{cases} \quad (3)$$

Initial data $x^0 = (x_1^0, x_2^0)$ for the production factors are supposed to be given.

The quality of the control process is estimated by the accumulated consumption index $c(x(t), u(t)) = f_\gamma(x(t))(1 - u_1(t))(1 - u_2(t))$ of a logarithmic type discounted at the infinite time interval

$$J(\cdot) = \int_0^{+\infty} e^{-\rho t} \ln c(x(t), u(t)) dt. \quad (4)$$

The utility function $J(\cdot)$ measures the integrated relative changes of the consumption $c[t] = c(x(t), u(t))$ indeed, using integration by parts, one can get

$$J(\cdot) = \int\limits_0^{+\infty} e^{-\rho t} \ln c[t] dt = \frac{\ln c[0]}{\rho} + \frac{1}{\rho} \int\limits_0^{+\infty} e^{-\rho t} \frac{\dot{c}[t]}{c[t]} dt = \frac{\ln c[0]}{\rho} + \frac{1}{\rho} \int\limits_0^{+\infty} e^{-\rho t} \frac{dc[t]}{c[t]},$$

where $c[0] = f_\gamma(x_1^0, x_2^0)(1 - u_1^0)(1 - u_2^0)$ and u_1^0 and u_2^0 are the initial investment shares. The ratio $dc[t]/c[t]$ can be interpreted as the relative consumption change.

Based on the presented growth model, we formulate the control problem

Problem 1. It is required to synthesize such a control process $\{\mathbf{x}(t), \mathbf{u}(t)\}$ that maximizes the utility function (4), satisfies the dynamic system (3) together with the initial conditions $\mathbf{x}(0) = x^0$ and control restrictions $\mathbf{u} \in \mathcal{U} = [0, \bar{u}_1] \times [0, \bar{u}_2]$.

The problem for the Cobb-Douglas production function is thoroughly investigated in [4,12] (one dimensional case can be found in [1,2]). In this paper, we study this problem for the CES-production function, construct its stabilized solutions and analyze their behaviour for different values of the elasticity parameter γ up to its limit value of zero, when the CES-function turns into the Cobb-Douglas production function.

2.3   Problem Analysis

Problem analysis is conducted using Pontryagin maximum principle [1,12]. In accordance with the Principle, we construct the Hamiltonian function of the form

$$H(x, \psi, u) = \ln c(x, u) + \psi_1(f_\gamma(x)u_1 - (\delta + \lambda)x_1) + \psi_2(bf_\gamma(x)u_2 - \lambda x_2), \quad (5)$$

and derive control parameters $\mathbf{u} = (\mathbf{u}_1, \mathbf{u}_2) \in \mathcal{U}$ providing the maximum to the Hamiltonian function. These controls do exists due to the strict concavity of the Hamiltonian function (5) and the compactness of the control domain $\mathcal{U}$.

$$\mathbf{u}_i = \begin{cases} 0, & (x, \psi) \in \Delta_1^i = \{b^{i-1}\psi_i f_\gamma(x) \leq 1\} \\ 1 - \dfrac{b^{1-i}}{\psi_i f_\gamma(x)}, & (x, \psi) \in \Delta_2^i = \left\{1 \leq b^{i-1}\psi_i f_\gamma(x) \leq \dfrac{1}{1 - \bar{u}_i}\right\} \\ \bar{u}_i, & (x, \psi) \in \Delta_3^i = \left\{b^{i-1}\psi_i f_\gamma(x) \geq \dfrac{1}{1 - \bar{u}_i}\right\}, i = 1, 2. \end{cases} \quad (6)$$

In [12], it is proven that the maximized Hamiltonian

$$\mathbf{H}(x, \psi) = H(x, \psi, \mathbf{u}) = \max_{u \in \mathcal{U}} H(x, \psi, u) \quad (7)$$

is a smooth function of its variables x, ψ and concave in phase variables x.

Further study deals with a Hamiltonian system of the form

$$\dot{x} = \frac{\partial \mathbf{H}(x, \psi)}{\partial \psi}, \quad \dot{\psi} = \rho\psi - \frac{\partial \mathbf{H}(x, \psi)}{\partial x}, \quad (8)$$

and carries out its qualitative analysis, including investigation of the existence and uniqueness of a steady state and its convergence to the steady state of the Hamiltonian system with the Cobb-Douglas production function (see [12]) when the parameter γ tends to 0.

3 Qualitative Analysis

First of all, we search for a steady state and check its location and uniqueness. Due to the system (3) and the structure of the maximizing controls (6), stationary levels of the Hamiltonian system can be located in the domains with non-zero control regimes, i.e. $(x, \psi) \in \Delta_i^1 \cap \Delta_j^2$ $(i, j = 2, 3)$. Therefore, we assume that the steady state is located in the domain $\Delta_2^1 \cap \Delta_2^2$ of the non-constant control regime.

For the convenience, we denote $G(x, \psi) = 1/f_\gamma(x) - \psi_1 - b\psi_2$. In the domain $\Delta_2^1 \cap \Delta_2^2$, the explicit form of the Hamiltonian system (8) has the form

$$\begin{cases} \dot{x}_i = -1/\psi_i + b^{i-1}f_\gamma(x) - ((2-i)\delta + \lambda)x_i, \\ \dot{\psi}_i = ((2-i)\delta + \lambda + \rho)\psi_i + G(x, \psi)\dfrac{\partial f_\gamma(x)}{\partial x_i}, \quad i = 1, 2. \end{cases} \quad (9)$$

Proposition 1. *For the problem with the production CES-function, the Hamiltonian system (9) has a unique steady state. Its coordinates are found explicitly through the solution of the following nonlinear algebraic equation with respect to the parameter* $\xi = \dfrac{x_1}{x_2}$

$$B\xi^{2\gamma+1} + \alpha_1\xi^{\gamma+1} - \alpha_2 K\xi^\gamma - AK = 0, \quad (10)$$

$$K = \frac{\alpha_1(\lambda + \rho)}{b\alpha_2(\delta + \lambda + \rho)}, \quad A = \alpha_1\left(1 - \nu\frac{\delta + \lambda}{\delta + \lambda + \rho}\right), \quad B = \alpha_2\left(1 - \nu\frac{\lambda}{\lambda + \rho}\right).$$

Coordinates $(x_\gamma^*, \psi_\gamma^*)$ *of the steady state satisfy the formulae depending on the parameter* ξ

$$x_{1\gamma}^* = \left(\frac{(\delta + \lambda + \rho)(\alpha_1 + \alpha_2\xi^\gamma)^{\nu/\gamma}(\alpha_1 + B\xi^\gamma)}{a\alpha_1\nu}\right)^{\frac{1}{\nu-1}}, \quad x_{2\gamma}^* = \frac{x_{1\gamma}^*}{\xi}, \quad (11)$$

$$y_\gamma^* = f_\gamma(x_{1\gamma}^*, x_{2\gamma}^*), \quad \psi_{i\gamma}^* = \left(b^{i-1}y_\gamma^* - ((2-i)\delta + \rho)x_{i\gamma}^*\right)^{-1}, \quad (i = 1, 2).$$

Proof of Proposition 1 can be found in [8, Section 3.1].

Proposition 2. *Coordinates of the steady state (11) of the Hamiltonian system (9) converges to the steady state of the Hamiltonian system corresponding to the problem with Cobb-Douglas production function when the parameter* γ *goes to zero.*

$$\lim_{\gamma \to 0} x_{1\gamma}^* = \left(\frac{\delta + \lambda + \rho}{a\alpha_1\nu}\left(1 - \frac{\alpha_2\lambda\nu}{\lambda + \rho}\right)\xi_0^{\alpha_2\nu}\right)^{\frac{1}{\nu-1}} = x_{10}^*, \quad \xi_0 = \frac{A + \alpha_2}{B + \alpha_1}K. \quad (12)$$

Proof. In [12], steady state coordinates for the model with Cobb-Douglas production function is found, and they take the following values

$$\xi_0 = \frac{A + \alpha_2}{B + \alpha_1} K, \quad x_{10}^* = \left(\frac{\delta + \lambda + \rho}{a\alpha_1 \nu} \left(1 - \frac{\alpha_2 \lambda \nu}{\lambda + \rho} \right) \xi_0^{\alpha_2 \nu} \right)^{\frac{1}{\nu-1}}.$$

The Eq. (10) with respect to the parameter ξ for $\gamma = 0$ becomes linear and has the form

$$B\xi + \alpha_1 \xi - \alpha_2 K - AK = 0.$$

Obviously, it has the only solution $\xi = \xi_0$.

Next, we find limit value for the first coordinate $x_{1\gamma}^*$ when γ tends to zero.

$$\lim_{\gamma \to 0} x_{1\gamma}^* = \left(\frac{\delta + \lambda + \rho}{a\alpha_1 \nu} \right)^{\frac{1}{\nu-1}} \lim_{\gamma \to 0} \left((\alpha_1 + \alpha_2 \xi^\gamma)^{\nu/\gamma} (\alpha_1 + B\xi^\gamma) \right)^{\frac{1}{\nu-1}}$$

$$= \left(\frac{(\delta + \lambda + \rho)(\alpha_1 + B)}{a\alpha_1 \nu} \right)^{\frac{1}{\nu-1}} \exp \left(\frac{\nu}{\nu - 1} \lim_{\gamma \to 0} \frac{\ln (\alpha_1 + \alpha_2 \xi^\gamma)}{\gamma} \right)$$

$$= \left(\frac{(\delta + \lambda + \rho)(\alpha_1 + B)}{a\alpha_1 \nu} \right)^{\frac{1}{\nu-1}} \exp \left(\frac{\nu}{\nu - 1} \ln \xi_0^{\alpha_2} \right) = x_{10}^*.$$

All other steady state coordinates are expressed through ξ_0 and x_{10}^*. Thus, the proposition is proven. □

Figure 1 demonstrates phase coordinates $x_{i\gamma}^*$ ($i = 1, 2$) of the steady state as functions of the parameter γ. As is shown, they continuously depend on γ and take values that correspond to the phase coordinates of the steady state in the case with production function of Cobb-Douglas type, when γ equals zero.

Existence of the steady state is required to stabilize the Hamiltonian system. The stabilization algorithm is based on the construction of a stable manifold [6, 7, 12] and is discussed at the next subsection.

4   Stabilized Solutions

In this section, we construct stabilized solutions of the Hamiltonian system (9) and investigate their behaviour when the parameter γ approaches zero.

4.1   Stabilized Solutions for the Model with CES-Function

Construction of a stabilized solution of the Hamiltonian system (9) that converges to the steady state, requires for finding eigenvalues and eigenvectors of Jacobi matrix J of the system (9) at $(x_\gamma^*, \psi_\gamma^*)$. Jacobi matrix satisfies the formula

$$J = \begin{pmatrix} \dfrac{\partial^2 \mathbf{H}(x_\gamma^*, \psi_\gamma^*)}{\partial \psi \partial x} & \dfrac{\partial^2 \mathbf{H}(x_\gamma^*, \psi_\gamma^*)}{\partial \psi^2} \\ -\dfrac{\partial^2 \mathbf{H}(x_\gamma^*, \psi_\gamma^*)}{\partial x^2} & \rho \mathbb{E}_2 - \dfrac{\partial^2 \mathbf{H}(x_\gamma^*, \psi_\gamma^*)}{\partial x \partial \psi} \end{pmatrix} \tag{13}$$

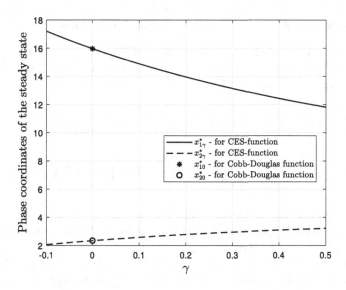

Fig. 1. Phase coordinates $x_{1\gamma}^*$ and $x_{2\gamma}^*$ of the steady state as functions of γ

Due to the properties of the Hamiltonian matrices [3] and their connection with matrix J (13) [6,7], we can claim that the half of the eigenvalues of Jacobi matrix (13) has negative real parts.

Consider matrix V ($V \in \mathbb{R}^{4\times2}$), which columns are composed by the eigenvectors h_1, h_2 corresponding to eigenvalues λ_1, λ_2 having negative reals parts, i.e. $V = (h_1\ h_2)$. The stable manifold $\mathcal{V}$ is a linear subspace generated by these vectors $\mathcal{V} = \langle h_1, h_2 \rangle$.

In order to stabilize the system (9), we project vectors $\begin{pmatrix} x \\ \psi \end{pmatrix}$ onto the subspace $\mathcal{V}$, applying the projector X that is constructed as follows

$$X = V_2 \cdot V_1^{-1}, \quad \text{where } V = \begin{pmatrix} V_1 \\ V_2 \end{pmatrix}, \text{ i.e. } V_1 = \begin{pmatrix} h_{11} & h_{21} \\ h_{12} & f_{22} \end{pmatrix}, V_2 = \begin{pmatrix} h_{13} & h_{23} \\ h_{14} & f_{24} \end{pmatrix}. \quad (14)$$

In the manifold $\mathcal{V}$, the conjugate coordinates $\psi = \widehat{\psi}$ of the state $\begin{pmatrix} x \\ \psi \end{pmatrix}$ are expressed through the phase ones x by the formula $\widehat{\psi} = \psi_\gamma^* + X(x - x_\gamma^*)$. Thus, the stabilized Hamiltonian system has the form (see [7, Theorem1])

$$\begin{cases} \dot{x}_1 = f_\gamma(x) - (\delta + \lambda)x_1 - \widehat{\psi}_1^{-1}(x), \\ \dot{x}_2 = bf_\gamma(x) - \lambda x_2 - \widehat{\psi}_2^{-1}(x), \quad x(0) = \widehat{x}^0 \in \mathcal{V} \cap (\Delta_2^1 \cap \Delta_2^2). \end{cases} \quad (15)$$

Solutions $\widehat{x}_\gamma(t)$ of the system (15) asymptotically converges to the steady state x_γ^* (11) for any admissible values of γ.

4.2 Sensitivity Analysis of the Stabilized Solutions

This section considers behaviour of solutions of the stabilized system (15) when the elasticity parameter γ tends to zero. We start with the analysis of the right-hand parts of the system (15) that can be rewritten in the form

$$\dot{x} = \begin{pmatrix} 1 \\ b \end{pmatrix} f_\gamma(x) - \begin{pmatrix} \delta + \lambda & 0 \\ 0 & \lambda \end{pmatrix} x - \widehat{\Psi}^{-1}(x), \quad \text{where} \tag{16}$$

$$\widehat{\Psi}^{-1}(x) = \begin{pmatrix} \widehat{\psi}_1^{-1}(x) \\ \widehat{\psi}_2^{-1}(x) \end{pmatrix}, \quad \widehat{\psi}(x) = \begin{pmatrix} \widehat{\psi}_1(x) \\ \widehat{\psi}_2(x) \end{pmatrix} = \psi_\gamma^* + X(x - x_\gamma^*).$$

It depends on the production CES-function $f_\gamma(x)$ (1), which is continuous with respect to the parameter γ and converges to the Cobb-Douglas production function when γ goes to zero.

Second term $\widehat{\Psi}^{-1}(x)$ in the right-hand part of the system includes matrix X ($X \in \mathbb{R}^{2 \times 2}$) that is constructed by the coordinates of eigenvectors of Jacobi matrix calculated at the steady state $(x_\gamma^*, \psi_\gamma^*)$. Therefore, first of all we should note, that the steady state continuously depends on γ and approaches values of the steady state (x_0^*, ψ_0^*) derived for the model with Cobb-Douglas production function (see Proposition 2). Next, according on the Hamiltonian system (9), elements of Jacobi matrix (13) contains first and second order partial derivatives of the production CES-function which are continuous with respect to the parameter γ and go to the corresponding derivatives of the Cobb-Douglas production function when γ tends to zero (see Property 1). Thus, elements of Jacobi matrix are continuous functions of the parameter γ.

Eigenvalues and eigenvectors of a matrix continuously depend on matrix components (see [11, 13]). Continuous dependency of the eigenvalues on the matrix elements follows from the representation of the characteristic polynomial, whose coefficients are the sum of all matrix minors of the corresponding order, and, therefore, they continuously depend on the matrix elements. Thus, the required property of the eigenvalues follows from the theorem on the continuous dependence of the roots of a polynomial on its coefficients. Regarding an eigenvector h_J corresponding to an eigenvalue λ_J, we recall that this vector can be found as any non-zero column of the adjoint matrix $\mathrm{adj}(-J + \lambda_J \mathbb{E})$ (see [11]). Therefore, an eigenvector is a continuous function of matrix elements. As a consequence, matrix X (14) constructing by the eigenvectors corresponding to the negative eigenvalues, is a continuous function of parameter γ.

Provided considerations imply the continuity with respect the parameter γ of the right-hand parts of the system (16) and its partial derivatives in phase variables x. Consequently, for any finite-time interval $[0, T]$ ($T < +\infty$) solutions of the stabilized system $\widehat{x}_\gamma(t)$ continuously depend on the parameter γ.

Convergence of the stabilized Hamiltonian system solutions at the infinite time interval is guaranteed by Proposition 2 and the asymptotic stability of the steady state x_γ^* (11) for all admissible values of the parameter γ ($\gamma > -1$). Thus, stabilized trajectories of the problem with the CES-production function tend to the corresponding solutions of the problem with Cobb-Douglas production function when γ approaches zero.

4.3   Numerical Results

For the illustration of the theoretical results, we construct solutions of the stabilized Hamiltonian system for the range of parameter γ,

$$\gamma \in \{0, 0.01, 0.02, 0.05, 0.1, 0.2, 0.3, 0.4, 0.5, 1.0\}.$$

Other model parameters are initialized as follows (see [4,12]): $a = 1$, $\alpha_1 = 0.781$, $\alpha_2 = 0.219$, $\nu = 0.5$, $\rho = 0.05$, $b = 0.31$, $\delta = 0.025$, and $\lambda = 0.005$.

Figures 2 and 3 demonstrate the convergence of the phase coordinates $x_{i\gamma}(t)$ to the $x_{i0}(t)$ $(i = 1, 2)$ when γ goes to zero. As is shown, trajectories have similar shapes and, approximately at $t = 60$, they reach satiration (steady state) levels $x_{i\gamma}^*$, which tend to x_{i0}^* $(i = 1, 2)$ as γ approaches zero.

First phase coordinate $x_{1\gamma}(t)$ grows when the parameter γ decreases. So, the steady state level of $x_{1\gamma}^*$ is higher for smaller γ.

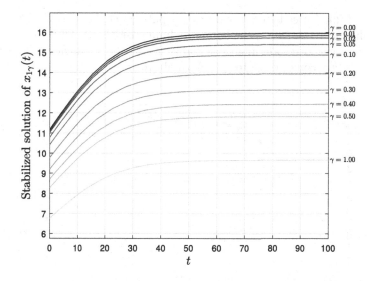

Fig. 2. Stabilized solution for the phase variable $x_{1\gamma}(t)$ when γ goes to zero.

Second phase coordinate $x_{2\gamma}(t)$ has S-shape trends and demonstrates the opposite behavior with respect to γ, *i.e.* it decreases when the parameter γ becomes lesser. So, for the case with Cobb-Douglas function one gets lower saturation level x_{20}^* in comparison with the steady state value $x_{2\gamma}^*$ in the case with CES-function.

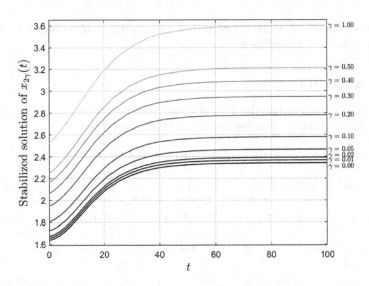

Fig. 3. Stabilized solution for the phase variable $x_{2\gamma}(t)$ when γ goes to zero.

Figure 4 illustrates behaviour of the outputs $y_\gamma(t)$ corresponding to the different values of parameter γ, including its limit case $\gamma = 0$. All trajectories have saturation levels corresponding the equality $y_\gamma^* = f_\gamma(x_\gamma^*)$, which get higher when the parameter γ comes close to zero from above.

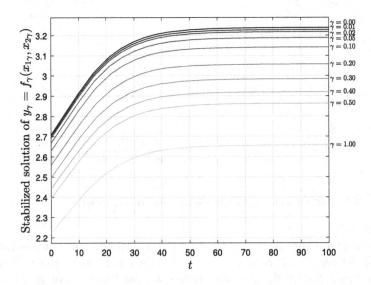

Fig. 4. Stabilized trends of the output $y_\gamma(t)$ when γ goes to zero.

5 Conclusion and Future Work

The paper investigates sensitivity of stabilized solutions of Hamiltonian systems with respect to the elasticity parameter γ of CES-production functions, which is used in the economic growth model to describe dependency of the output on production factors. In the limit case of the unit elasticity (or for parameter $\gamma = 0$), the production function turns into Cobb-Douglas function. It is proven, that the steady state levels of Hamiltonian systems continuously depends on γ. Next, we perform system stabilization and show that the right-hand parts of the stabilized dynamics are continuous functions of its variables including the elasticity parameter γ. Finally, numerical examples calculated for several values of γ illustrate theoretical conclusions on the convergence of the stabilized solutions of the problems with CES-function to the corresponding solutions in the case with Cobb-Douglas production function when γ approaches zero.

In the future research, we plan to analyze the growth model when the elasticity coefficient σ tends to zero or takes values greater than one, that corresponds to the case when parameter γ is negative, $i.e.$ $\gamma \in (-1,0)$. Using proposed approach, we can investigate long term trends of the production factors and describe their qualitative behavior for these ranges of model parameters.

Acknowledgements. The research of the first author, Anastasiia A. Usova, is supported by the Russian Science Foundation (Project No. 19-11-00105), https://rscf.ru/project/19-11-00105/.

References

1. Aseev, S.M., Kryazhimskiy, A.V.: The Pontryagin maximum principle and optimal economic growth problems. Proc. Steklov Inst. Math. **257**, 1–255 (2007). https://doi.org/10.1134/S0081543807020010
2. Krasovskii, A., Kryazhimskiy, A., Tarasyev, A.: Optimal control design in models of economic growth. In: Evolutionary Methods for Design. Optimization and Control, pp. 70–75. CIMNE, Barcelona (2008)
3. Paige, C., Loan, C.V.: A Schur decomposition for Hamiltonian matrices. Linear Algebra Appl. **41**, 11–32 (1981)
4. Tarasyev, A., Usova, A.: Construction of a regulator for the Hamiltonian system in a two-sector economic growth model. Proc. Steklov Inst. Math. **271**, 1–21 (2010)
5. Tarasyev, A.M., Usova, A.A., Wang, W.: Hamiltonian trajectories in a heterogeneous economic growth model for optimization resource productivity. IFAC-PapersOnLine **48**(25), 74–79 (2015). https://doi.org/10.1016/j.ifacol.2015.11.062. ISSN 2405-8963
6. Tarasyev, A.M., Usova, A.A.: Robust methods for stabilization of Hamiltonian systems in economic growth models. IFAC-PapersOnLine **51**(32), 7–12 (2018). https://doi.org/10.1016/j.ifacol.2018.11.344
7. Usova, A.A., Tarasyev, A.M.: Structure of a Stabilizer for the Hamiltonian Systems. In: Tarasyev, A., Maksimov, V., Filippova, T. (eds.) Stability, Control and Differential Games. LNCISP, pp. 357–366. Springer, Cham (2020). https://doi.org/10.1007/978-3-030-42831-0_32

8. Usova, A.A., Tarasyev, A.M.: Analysis of a growth model with a production CES-function. Math. Game Theory Appl. **14**(4), 96–114 (2022). https://doi.org/10. 17076/mgta_2022_4_64. [In Russian]

9. Klump, R., McAdam, P., Willman, A.: Factor substitution and factor augmenting technical progress in the US: a normalized supply-side system approach. ECB Working Paper, vol. 367, 64 p. (2004)

10. Solow, R.M.: A contribution to the theory of economic growth. Q. J. Econ. **70**(1), 65–94 (1956). www.jstor.org/stable/1884513

11. Uteshev, A.Yu.: Notebook. Online resource. http://vmath.ru/vf5/users/au/index. (In Russian)

12. Usova, A.A.: Analysis of properties of Hamiltonian systems and cost functions in dynamic growth models. Thesis, 180 p. (2012)

13. Ostrowski, A.M.: Solution of Equations and System of Equations. Academic Press, New York and London (1960). University of Basel, Switzerland

14. Sanderson, W.C.: The SEDIM model: version 0.1. IIASA Interim Report IR-04-041, 42 p. (2004)

Optimization in Machine Learning

Uncertainty of Graph Clustering in Correlation Block Model

Artem Aroslankin[ID] and Valeriy Kalyagin[(✉)][ID]

Laboratory of Algorithms and Technologies for Network Analysis, HSE University,
Nizhny Novgorod, Russia
vkalyagin@hse.ru
https://nnov.hse.ru/en/latna/

Abstract. Cluster analysis is a powerful tool in network science and it is well developed in many directions. However, the uncertainty analysis of clustering algorithms is still not sufficiently investigated in the literature. To study uncertainty of clustering algorithms we propose to use a new model, which we call correlation block model. We suggest to measure uncertainty of clustering algorithms by error in cluster identification by observations. Uncertainty of different clustering algorithms are compared using proposed methodology. New and interesting phenomena are observed.

Keywords: graph clustering · uncertainty · random variable networks

1 Introduction

Many network models can be represented as simple undirected weighted graph $G = (V, E, W)$, where V is a set of nodes (vertices), E is a set of links (edges), and W is a matrix of weights ($w_{i,j} = 0$ iff $(i,j) \notin E$). In usual setting weight $w_{i,j}$ of the edge (i, j) represents a similarity (degree of connection) between vertices i and j. Clustering or community detection in network is a partition of nodes of network into groups such that nodes in one group are strongly connected, and nodes from different groups are weakly connected. From machine learning prospective clustering is reffered to unsupervised learning. In general there is no ground truth for clustering. In other words, for a given network, communities obtained by one algorithm, have a priori the same value as communities obtained by another algorithm. This explains a large family of clustering algorithms existing in the literature [4,5]. Moreover, it was recently proved that "no free lunch theorem" is valid for clustering too. This theorem is well known in supervised learning [17], and it's non-formal statement for clustering can be formulated as follows (citation from [16]): "For the community detection problem,

Section 1 was prepared within the framework of the Basic Research Program at the National Research University Higher School of Economics (HSE), results of the Sects. 3, 4, 5 and 6 are obtained with a support from RSF grant 22-11-00073.

M. Khachay et al. (Eds.): MOTOR 2023, CCIS 1881, pp. 353–363, 2023.
https://doi.org/10.1007/978-3-031-43257-6_26

with accuracy measured by adjusted mutual information, the uniform average of the accuracy of any method f over all possible community detection problems is a constant that is independent of f".

The "no free lunch theorem" generates the following question: is it possible to compare the quality of clustering algorithms, or in what sense one clustering algorithms can be better than another one? One way to answer this question is to generate a family of graphs with "planted partition" and test clustering algorithms for their ability to recover this planted partition. In random graph theory such generator is known as Stochastic Block Model (SBM) or Planted Partition Model (PPM) [7]. In the simplest version of SBM model set of nodes V is split into k groups $V_1, V_2, \ldots, V_k$ and random graphs are generated with two probabilities: p_{in} (probability to draw an edge between nodes from the same group), and p_{out} (probability to draw an edge between nodes from different groups). For the case where $p_{in} >> p_{out}$, graph is expected to have a prescribed cluster structure and for the case $p_{in} = p_{out}$ the prescribed cluster structure is lost. Planted partition can be considered in this case as a ground truth and one can compare clustering algorithms by their ability to recover this planted partition.

There is an interesting phenomena, called phase transition, observed in clustering with SBM. Let $N = |V|$ be the number of vertices in network. Consider a low density case with 2 clusters: $k = 2$, $|V_1| = |V_2| = N/2$, $p_{in} = c_{in}/N$, $p_{out} = c_{out}/N$. Phase transition phenomena was discovered in [3]. More precisely, it was proven in [3] that there is a threshold in relation between c_{in} and c_{out}, given by $c_{in} - c_{out} = \sqrt{2(c_{in} + c_{out})}$ such that from one side from this threshold it is possible to recover planted cluster structure with some algorithm and from the other side from this threshold there is no algorithm to recover the planted cluster structure. Best algorithm in this sense is known to be a spectral version of modularity maximization [14]. Recent development of clustering with SBM is presented in [1].

In the present paper we develope the study of quality of clustering algorithms for a large class of networks called random variable networks (RVN). This class of networks arises in various network applications: gene expression network, brain connectivity network, climate network, stock market network, and many others. The main problem of interest in random variable networks is the problem of network structure identification by observation [8]. Clustering algorithms applied to random variable networks generate a new and not investigated phenomena - uncertainty of cluster identification by observations. To measure uncertainty of clustering algorithms we follow approach proposed in [8] and define uncertainty of clustering algorithm by a difference between true cluster structure and cluster structure obtained by the algorithm. This difference generates a loss from a false decision. Expected value of this loss will give a measure of uncertainty of a clustering algorithm. Using this uncertainty one can measure the quality of clustering algorithms from a new point of view. To compare the quality of different clustering algorithms we develop a new model, which we call Correlation Block Model (CBM). Using this model we make a comparison of uncertainty of variety of clustering algorithms: from naive threshold method and MST algorithm to advanced Louvain algorithm.

A priori, we have no clear hypothesis what algorithms will show a better performance with respect to uncertainty of identification of cluster structure. The problem is essentially different from clustering in stochastic block model. We will see however, that algorithms with a good reputation in clustering in stochastic block model confirm their abilities in correlation block model too. Moreover we show that phase transition phenomena for SBM is also present in CBM. This phenomena needs to be deeper investigated. Another novelty of our study is related with dependence of uncertainty of clustering on probability distribution behind RVN. We show that for distributions with heavy tails all considered algorithms fail to recover a planted cluster structure in CBM and we discuss how to fix this problem.

The paper is organized as follows. In Sect. 2 we present main definitions and notations. Section 3 is devoted to correlation block model. In Sect. 4 we develop our approach to measure uncertainty of clustering in correlation block model. Results of numerical experiences to compare uncertainty of clustering algorithms are discussed in the Sect. 5. Finally, in Sect. 6 we discuss obtained results and directions for future research.

2   Basic Definitions and Notations

Random variable network is a pair (X, γ), where X is a random vector $X = (X_1, X_2, \ldots, X_N)$, and $\gamma = \gamma(Y, Z)$ is a measure of pairwise dependence (similarity, association,...) between random variables (Y and Z). Random variable network generates a network model - complete weighted graph Γ with N nodes. Node i is associated with the random variable X_i ($i = 1, 2, \ldots, N$). Weight of edge (i, j), $(i \neq j)$ is given by $\gamma_{i,j} = \gamma(X_i, X_j)$. Clustering in random variable network is application of a clustering algorithm to the simple, undirected, weighted graph Γ. However, in practice we have only observations of the random vector X, which we model as independent identically distributed random vectors $X(1), X(2), \ldots, X(n)$ (sample of the size n from the distribution X). The main problem in this case is to identify a cluster structure of the graph Γ by observations. Uncertainty of clustering therefore is related with estimation of how different can be clusters obtained by different observations from the same distribution.

To estimate uncertainty of cluster algorithms we need to measure difference between two cluster partitions. This topic is well studied in the literature (see for example a comprehensive survey on information theoretic measures for clusterings comparison in [18]). For our experiments we choose Adjusted Rand Index (ARI), and Adjusted Mutual Information (AMI). Let $\mathbf{V} = \{V_1, V_2, \ldots, V_k\}$ and $\mathbf{U} = \{U_1, U_2, \ldots, U_l\}$ be two partitions of the set of nodes $V = \{1, 2, 3, \ldots, N\}$. Let $PN_{0,0}$ be the number of pairs of nodes that are in the same clusters in $\mathbf{V}$ and in $\mathbf{U}$, $PN_{0,1}$ be the number of pairs of nodes that are in the same clusters in $\mathbf{V}$ but in different clusters in $\mathbf{U}$, $PN_{1,0}$ be the number of pairs of nodes that are in the same clusters in $\mathbf{U}$ but in different clusters in $\mathbf{V}$, and $PN_{1,1}$ be the number of pairs of nodes that are in different clusters in $\mathbf{V}$ and in $\mathbf{U}$. Then Rand Index (RI) and Adjusted Rand Index (ARI) are defined by

$$RI = \frac{2(PN_{0,0} + PN_{1,1})}{N(N-1)},$$

$$ARI = \frac{2(PN_{0,0}PN_{1,1} - PN_{1,0}PN_{0,1})}{(PN_{0,0} + PN_{0,1})(PN_{0,1} + PN_{1,1}) + (PN_{0,0} + PN_{1,0})(PN_{1,0} + PN_{1,1})}$$

RI is a proportion of pairs of nodes that have the same classification in $\mathbf{V}$ and in $\mathbf{U}$, $RI = 1$ means that two partitions coincide. ARI is adjusted version of RI with advantage to vary in the same interval for all pairs of partitions.

Comparison of two partitions by information based measures uses contingency table between $\mathbf{V}$ and $\mathbf{U}$. Let $N_{i,j} = |V_i \cap U_j|$ (number of nodes that are common for U_i and V_j), $N_{i,*} = \sum_{j=1}^{l} N_{i,j}$, $N_{*,j} = \sum_{i=1}^{k} N_{i,j}$. Entropies H($\mathbf{V}$), H($\mathbf{U}$) and conditional entropies H($\mathbf{V}|\mathbf{U}$), H($\mathbf{U}|\mathbf{V}$) are defined by

$$H(\mathbf{V}) = -\sum_{i=1}^{k} \frac{N_{i,*}}{N} \log \frac{N_{i,*}}{N}, \quad H(\mathbf{U}) = -\sum_{j=1}^{l} \frac{N_{*,j}}{N} \log \frac{N_{*,j}}{N}$$

$$H(\mathbf{V}|\mathbf{U}) = -\sum_{i=1}^{k}\sum_{j=1}^{l} \frac{N_{i,j}}{N} \log \frac{N_{i,j}}{N_{*,j}}, \quad H(\mathbf{U}|\mathbf{V}) = -\sum_{i=1}^{k}\sum_{j=1}^{l} \frac{N_{i,j}}{N} \log \frac{N_{i,j}}{N_{i,*}}$$

Mutual Information (MI) is then defined by

$$MI(\mathbf{V}, \mathbf{U}) = H(\mathbf{V}) - H(\mathbf{V}|\mathbf{U}) = H(\mathbf{U}) - H(\mathbf{U}|\mathbf{V})$$

MI ranges in the interval $[0, \min\{H(\mathbf{V}), H(\mathbf{U})\}]$, and $MI = 0$ means independence between partitions. Adjusted Mutual Information (AMI) is an adjusted version of mutual information (MI) which ranges in $[0, 1]$.

3   Correlation Block Model

We define Correlation Block Model (CBM) as a random variable network model, where the measure of dependence γ is Pearson correlation and the correlation matrix C has a special block structure:

$$C = \begin{bmatrix} C_{1,1} & C_{1,2} & \cdots & C_{1,k} \\ C_{2,1} & C_{2,2} & \cdots & C_{2,k} \\ \cdots & \cdots & \cdots & \cdots \\ C_{k,1} & C_{k,2} & \cdots & C_{k,k} \end{bmatrix} \tag{1}$$

where $C_{i,i}$ ($i = 1, 2, \ldots, k$) are square matrices with 1 on the diagonal and the same value $r_{i,i} \in (-1, 1)$ outside the diagonal, $C_{i,j}$ ($i \neq j$) are rectangular matrices with constant elements equal to $r_{i,j}$, and $C_{j,i} = C_{i,j}^T$. For the case

$k = 2$ (two blocks model) and $r_{1,1} = r_{2,2}$ one has (we denote $r_{in} = r_{1,1} = r_{2,2}$, $r_{out} = r_{1,2} = r_{2,1}$):

$$C = \begin{bmatrix} 1 & r_{in} & \cdots & r_{in} & r_{out} & \cdots & r_{out} \\ r_{in} & 1 & \cdots & r_{in} & r_{out} & \cdots & r_{out} \\ \cdots & \cdots & \cdots & \cdots & \cdots & \cdots & \cdots \\ r_{in} & r_{in} & \cdots & 1 & r_{out} & \cdots & r_{out} \\ r_{out} & r_{out} & \cdots & r_{out} & 1 & \cdots & r_{in} \\ \cdots & \cdots & \cdots & \cdots & \cdots & \cdots & \cdots \\ r_{out} & r_{out} & \cdots & r_{out} & r_{in} & \cdots & 1 \end{bmatrix} \tag{2}$$

Correlation two-block model divides the set of vertices in random variable network in two groups of the size $N_1 = \dim(C_{1,1})$ and $N_2 = \dim(C_{2,2})$ ($N_1 + N_2 = N$). Correlation between random variables inside each group is r_{in} and correlation between random variables from different groups is r_{out}. If $r_{in} \gg r_{out}$ than the network has a clear cluster structure. For $r_{in} = r_{out}$ this cluster structure is obviously lost. Correctness of definition of CBM is related with positive definetness of the matric C in (1). General discussion of block correlation matrices is given in [2]. For the case of two equal blocks we prove the following

Theorem 1. *Let $k = 2$, $N_1 = N_2 = M$, $r_{in} \in (-1, 1)$, $r_{out} \in (-1, 1)$, then the matrix C in (2) is positive definite if and only if*

$$r_{out} < r_{in} + \frac{1}{M}(1 - r_{in}) \text{ and } r_{out} > -r_{in} - \frac{1}{M}(1 - r_{in}).$$

In particular, if $r_{in} > 0$, $r_{out} > 0$ and $r_{out} < r_{in} + \frac{1}{M}(1 - r_{in})$ then matrix C in (2) is positive definite.

Proof. Symmetric matrix is positive definite iff all its eigenvalues are positive. It is not difficult to calculate eigenvalues of the matrix C given by (2) for $N_1 = N_2 = M \geq 2$. Indeed, let $P(\lambda)$ be characteristic polynomial of the matrix C: $P(\lambda) = \det(C - \lambda I_N)$. One has $P(1 - r_{in}) = 0$, because the matrix $(C - (1 - r_{in})I_N)$ has at least two equal rows. For the case $r_{in} \neq r_{out}$, this matrix has rank 2 (only 2 independent rows) and therefore multiplicity of the eigenvalue $\lambda_0 = (1 - r_{in})$ is $(N - 2)$. Two other eigenvalues of the matrix C are $\lambda_1 = 1 + (M - 1)r_{in} + Mr_{out}$ and $\lambda_2 = 1 + (M - 1)r_{in} - Mr_{out}$. First eigenvalue is associated with eigenvector $v_1 = (1, 1, \ldots, 1)^T$, and the second eigenvalue is associated with eigenvector $v_2 = (1, \ldots, 1, -1, \ldots, -1)^T$. For the case $r_{in} = r_{out}$ the matrix C has eigenvalue $\lambda_0 = (1 - r_{in})$ of multiplicity $(N - 1)$ and eigenvalue $\lambda_1 = 1 + (N - 1)r_{in}$ of multiplicity 1. The theorem follows by making positive all eigenvalues of the matrix C: $\lambda_0 > 0$, $\lambda_1 > 0$, $\lambda_2 > 0$.

This result will be used as a theoretical basis of our numerical experiments.

4   Uncertainty of Clustering in Correlation Block Model

In what follows, we consider the simplest case of correlation block model with two blocks equal in size ($k = 2$, $N_1 = N_2 = M$) and $r_{in} \geq r_{out}$. General case can be investigated with the same methodology. To evaluate the quality of different clustering algorithms for correlation block model we assume that the cluster structure of the model is given by these two blocks (the set of vertices in random variable network is split into two groups of the size M). Quality of a clustering algorithm will be measured by it's ability to identify planted cluster structure by observations.

Our methodology is as follows. We fix random vector $X = (X_1, X_2, \ldots, X_N)$ and it's distribution. Given a sample of the size n of observations $x(1), x(2), \ldots x(n)$, $x(t) = (x_1(t), x_2(t), \ldots, x_N(t))$, $t = 1, 2, \ldots, n$ we calculate estimations $\hat{r}_{i,j}$ of correlations $r_{i,j}$ between random variables X_i and X_j ($i, j = 1, 2, \ldots, N$, $i \neq j$) and apply clustering algorithm to the graph $\hat{\Gamma} = (\hat{r}_{i,j})$. Then using Adjusted Rand Index (ARI) and Adjusted Mutual Information (AMI) we measure similarity between obtained and planted partition. Expected values $E(ARI)$ and $E(AMI)$ over all samples will be our measure of quality of clustering algorithm. Uncertainty of clustering algorithms in this setting can be evaluated by $(1 - E(AMI))$ or $(1 - E(ARI))$ (maximal value of AMI and ARI is 1, and it corresponds to the case where there is no error in cluster identification). Less is uncertainty better is ability of algorithm to recover the planted cluster structure.

Uncertainty of a given clustering algorithm for correlation block model is related with two aspects of associated random variable network:

- Distribution of the vector $X = (X_1, X_2, \ldots, X_N)$. In a standard situation, vector X has a multivariate Gaussian distribution. In some applications (stock market network) distributions of X_i have a heavy tails. To model this one can use a large class of elliptical distributions. Multivariate Gaussian and Student distributions are from this class.
- Measure of dependence γ. In standard situation Pearson correlation is used. However, for elliptical distributions there is a connection between Pearson and some other popular (Kendall, Fechner) correlations. It is proved in [8] the following relations between Pearson, Kendall, and Fechner correlations (you can find in this book a detailed description of these correlations and connections between them):

$$\rho^{Kendall} = \rho^{Fechner} = \frac{2}{\pi} \arcsin(\rho^{Pearson}).$$

It implies that correlation bloc model for Pearson correlation is a correlation bloc model for Kendall and Fechner correlations too. Moreover, relation $r_{in} > r_{out}$ for Pearson correlation block model is kept in these new correlation block models. It allows to measure uncertainty of clustering algorithms using these correlations (algorithm is looking to recover the same cluster structure) for elliptical distribution of the vector X. Difference with Pearson correlation network model is in a different way to estimate Kendall and Fechner correlations (see details in [8]). We will see that this can provide a substantial improvement of uncertainty of clustering algorithms for non Gaussian distributions.

5   Comparison of Uncertainty of Clustering Algorithms

In this section we present results of numerical simulations to measure uncertainty of different clustering algorithms for correlation block model. We fix the following parameters of correlation block model: $N = \dim(X) = 40$ (size of network), $N_1 = N_2 = 20$ (cluster size), $n = 40, n = 80$ (sample size). To estimate quality $E(ARI)$ and $E(AMI)$ of clustering algorithms we generate sample for a given distribution $S = 100$ times and calculate average.

We choose four algorithms for a comparison:

- Naive threshold method or MST algorithm. To obtain clusters by the threshold method one can fix a threshold r_0 and delete from the graph all edges with similarity less or equal to r_0. Connected components of the obtained graph are considered as clusters. To obtain a desired number of clusters one needs to choose appropriate value of r_0. It can be shown that the threshold method is equivalent to the construction of the maximum spanning tree of the graph and to cutting a fixed number of weakest (by similarity) edges in this tree (see [13]). MST algorithm is a divisive hierarchical clustering algorithm and using this algorithm one can obtain any number of clusters. For a given undirected weighted graph different experimental techniques are known to choose an appropriate number of clusters. MST algorithm is popular in such applications as computer vision, bio-informatics and many others. Recent discussion of MST algorithms is given in [6].
- Two versions of spectral clustering algorithm: unnormalized spectral clustering and normalized spectral clustering algorithms. Both are heuristics to solve combinatorial optimization problems related with minimization of Ratio Cut and Normalized Cut in weighted graph. Popular source is [11]. In fact, both versions of spectral clustering algorithms can be considered as an embedding of the graph vertices in a vector space of low dimension (space dimension is equal to the prescribed number of clusters). The final step of spectral clustering algorithms is a clustering of points in this vector space. Usually a k-means algorithm is applied at this stage. Important theoretical question, related with spectral clustering algorithms is their consistency i.e. convergence of spectral clustering for growing samples drawn from some underlying probability distribution. From this point of view normalized clustering has a better properties than unnormilized one [12]. Consistency of spectral clustering for Stochastic Block Model is investigated in many sources (see for example [9]).
- Louvain algorithm. This is a heuristic for modularity maximization known to be computationally very effective for large scale networks. Algorithm was presented in 2008 in [10] and since attracted a great attention in the literature.

Experiment 1. $r_{in} = 0,8$, r_{out} is varying from 0 to r_{in}, X has a multivariate Gaussian distribution with zero mean and covariance matrix C given by (2), $n = 40$. The results are presented on the Fig. 1.

One can see on the Fig. 1 that algorithms are different with respect to uncertainty. For both measures ARI and AMI qualitative behavior is the same. Normalized spectral clustering shows the best performance, followed by Louvain, spectral clustering and naive MST algorithms. One can observe some indication to a phase transition phenomena for clustering in correlation block model. In this case the phase transition is displayed by a jump of uncertainty (AMI or ARI) from the region of a good cluster recovery to the region of a bad cluster recovery (clusters obtained by chance).

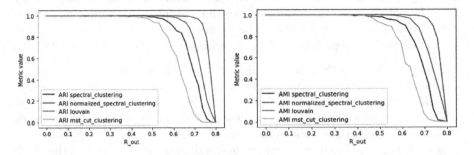

Fig. 1. Quality of clustering algorithms for Gaussian distribution as a function of r_{out} for $r_{in} = 0,8$, $n = 40$

Experiment 2. Degree $d_i = \sum_{j=1}^{N} r_{i,j}$ of each vertex in network is fixed for the same value $d_i = d = 16$, $i = 1, 2, \ldots, N$. In this case r_{in} and r_{out} are related by $19 r_{in} + 20 r_{out} = 16$ and one can vary r_{out} from 0 to its maximal value $(16/39) \approx 0,41$. Fixed degree of vertices is useful when we compare uncertainty for different network dimensions (see similar approach for stochastic block model in [15]). X has a multivariate Gaussian distribution with zero mean and covariance matrix C given by (2), $n = 40, 80$. The results are presented on the Fig. 2.

One can see on the Fig. 2 the same behavior of algorithms uncertainty as for Experiment 1. Similarly, one can observe some indication to a phase transition phenomena for clustering in correlation block model. In this case, similar to Fig. 1, the phase transition is displayed by a jump of uncertainty (AMI or ARI) from the region of a good cluster recovery to the region of a bad cluster recovery (clusters obtained by chance).

Experiment 3. Degree $d_i = \sum_{j=1}^{N} r_{i,j}$ of each vertex in network is fixed for the same value $d_i = d = 16$, $i = 1, 2, \ldots, N$. In this case r_{in} and r_{out} are related by $19 r_{in} + 20 r_{out} = 16$ and one can vary r_{out} from 0 to its maximal value $(16/39) \approx 0,41$. X has a multivariate Student distribution with 2 degree of freedom with zero mean and correlation matrix C given by (2), $n = 40, 80$. The results are presented on the Fig. 3.

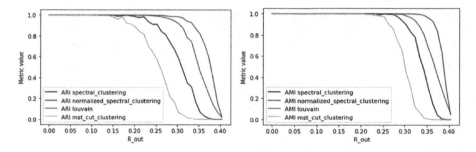

Fig. 2. Quality of clustering algorithms for Gaussian distribution as a function of r_{out} for fixed degree of vertices, $d = 16$, in network. Sample size on the left $n = 40$, sample size on the right $n = 80$

One can see on the Fig. 3 a new and interesting phenomena. All algorithms fail to recover a planted cluster structure for Student distribution. Note, that it is the same cluster structure as for Gaussian distribution with the same correlation matrix C. Quality of algorithms is therefor sensitive to distribution for clustering in correlation block model with Pearson correlation.

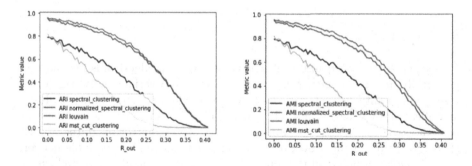

Fig. 3. Quality of clustering algorithms for multivariate Student distribution as a function of r_{out} for fixed degree of vertices, $d = 16$, in network. Sample size $n = 40$ on the left, and sample size $n = 80$ on the right.

Experiment 4. Degree $d_i = \sum_{j=1}^{N} r_{i,j}$ of each vertex in network is fixed for the same value $d_i = d = 16$, $i = 1, 2, \ldots, N$. In this case r_{in} and r_{out} are related by $19r_{in} + 20r_{out} = 16$ and one can vary r_{out} from 0 to its maximal value $(16/39) \approx 0,41$. X has a multivariate Student distribution with 3 degree of freedom with zero mean and correlation matrix C given by (2), $n = 80$. Cluster structure is identified by normalized spectral clustering algorithm in different correlation block models: Pearson, Kendall, and Fechner. The results are presented on the Fig. 4.

One can see on the Fig. 4 another new and interesting phenomena. For correlation block model with Student distribution in random variable network normalized spectral clustering algorithm has a better performance in Kendall and Fechner block models in comparison with Pearson block model. It gives an idea of possibility to construct robust clustering algorithms with uncertainty non dependent on distribution.

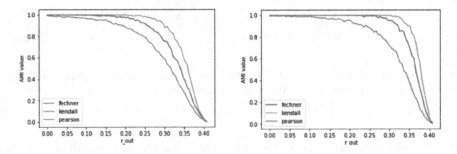

Fig. 4. Quality of normalized spectral clustering algorithms for Student distribution with 3 degree of freedom as a function of r_{out} for fixed degree of vertices, $d = 16$ for three different correlation block models, Pearson, Fechner, and Kendall. Sample size $n = 40$ on the left, and sample size $n = 80$ on the right.

6   Conclusion

The paper deals with quality assessment of clustering algorithms in a new and practically important setting of clustering in random variable networks. In this case a new quality characteristic is shown to be relevant - uncertainty of clusters identification by observations. The paper has two main contributions. First one is methodological. We propose a methodology to assess error magnitude (uncertainty) of a cluster identification by observations in random variable network. Second one has a practical meaning. This is a comparison of different clustering algorithms by their uncertainty of identification of hidden cluster structure in correlation block model. This is also new and gives a new insight into the quality of clustering algorithms. Some new and interesting phenomena are observed. They open a new and promising directions for future investigation of quality of clustering algorithms.

Acknowledgements. Section 1 was prepared within the framework of the Basic Research Program at the National Research University Higher School of Economics (HSE University), results of the Sects. 3, 4, 5 and 6 are obtained with a support from RSF grant 22-11-00073.

References

1. Abbe, E.: Community detection and stochastic block models: recent developments. J. Mach. Learn. Res. **18**, 1–86 (2018)
2. Archakov, I., Hansen, P.R.: A canonical representation of block matrices with applications to covariance and correlation matrices. arXiv:2012.02698v1 [econ.EM] (2020)
3. Decelle, A., Krzakala, F., Moore, C., Zdeborová, L.: Inference and phase transitions in the detection of modules in sparse networks. Phys. Rev. Lett. **107**, 065701 (2011)
4. Fortunato, S.: Community detection in graphs. Phys. Rep. **486**(3), 75–174 (2010)
5. Fortunato, S., Hric, D.: Community detection in networks: a user guide. Phys. Rep. **659**, 1–44 (2016)
6. Gagolewski, M., Cena, A., Bartoszuk, M., Brzozowski, L.: Clustering with minimum spanning trees: how good can it be? arXiv:2303.05679v1 (2023)
7. Holland, P.W., Laskey, K.B., Leinhardt, S.: Stochastic blockmodels: first steps. Soc. Netw. **5**(2), 109–137 (1983)
8. Kalyagin, V.A., Koldanov, A.P., Koldanov, P., Pardalos, P.M.: Statistical Analysis of Graph Structures in Random Variable Networks. Springer, Heidelberg (2020). https://doi.org/10.1007/978-3-030-60293-2
9. Lei, J., Rinaldo, A.: Consistency of spectral clustering in stochastic block models. Ann. Stat. **43**, 215–237 (2015)
10. Blondel, V.D., Guillaume, J.-L., Lambiotte, R., Lefebvre, E.: Fast unfolding of communities in large networks. J. Stat. Mech.: Theory Exp. **10**, P10008 (2008)
11. Luxburg, U.: A tutorial on spectral clustering. Stat. Comput. **17**, 395–416 (2007)
12. Luxburg, U., Belkin, M., Bousquet, O.: Consistency of spectral clustering. Ann. Stat. **36**, 555–586 (2008)
13. Mirkin, B.G.: Core Concepts in Data Analysis: Summarization, Correlation, Visualization (Undergraduate topics in Computer Science). Springer, London (2011). https://doi.org/10.1007/978-0-85729-287-2
14. Nadakuditi, R., Newman, M.E.J.: Graph spectra and the detectability of community structure in networks. Phys. Rev. Lett. **108**, 188701 (2012)
15. Newman M. E. J. and Girvan M.: Finding and evaluating community structure in networks. Phys. Rev. E-Stat. Nonlinear Soft Matt. Phys. **69**(2), 026113 (2004)
16. Peel, L., Larremore, D.B., Clauset, A.: The ground truth about metadata and community detection in networks. Sci. Adv. **3**(5), e1602548 (2017)
17. Wolpert, D.: The lack of a priori distinctions between learning algorithms. Neural Comput. **8**, 1341–1390 (1996)
18. Vinh, N.X., Epps, J., Bailey, J.: Information theoretic measures for clusterings comparison: variants, properties, normalization and correction for chance. J. Mach. Learn. Res. **11**, 2837–2854 (2010)

Multi-target Weakly Supervised Regression Using Manifold Regularization and Wasserstein Metric

Kirill Kalmutskiy[1,2]([✉]) [iD], Lyailya Cherikbayeva[3] [iD], Alexander Litvinenko[4] [iD], and Vladimir Berikov[1,2] [iD]

[1] Sobolev Institute of mathematics, Novosibirsk, Russia
berikov@math.nsc.ru
[2] Novosibirsk State University, Novosibirsk, Russia
k.kalmutskii@g.nsu.ru
[3] Al-Farabi Kazakh National University, Almaty, Kazakhstan
[4] RWTH Aachen, Aachen, Germany
litvinenko@uq.rwth-aachen.de
http://www.uq.rwth-aachen.de

Abstract. In this paper, we consider the weakly supervised multi-target regression problem where the observed data is partially or imprecisely labelled. The model of the multivariate normal distribution over the target vectors represents the uncertainty arising from the labelling process. The proposed solution is based on the combination of a manifold regularisation method, the use of the Wasserstein distance between multivariate distributions, and a cluster ensemble technique. The method uses a low-rank representation of the similarity matrix. An algorithm for constructing a co-association matrix with calculation of the optimal number of clusters in a partition is presented. To increase the stability and quality of the ensemble clustering, we use k-means with different distance metrics. The experimental part presents the results of numerical experiments with the proposed method on artificially generated data and real data sets. The results show the advantages of the proposed method over existing solutions.

Keywords: Weakly supervised learning · Multi-target regression · Manifold regularization · Low-rank matrix representation · Cluster ensemble · Co-association matrix

1 Introduction

Weakly supervised learning is a type of machine learning technique in which a model is trained using incomplete, imprecise, or ambiguous supervision signals, rather than using fully correctly labeled data. Weak supervision often arises in real problems for various reasons. This may be due to an expensive data labeling process, poor accuracy of sensors, insufficient expert qualifications or human error. For example, there is weak supervision in cases where the labeling

M. Khachay et al. (Eds.): MOTOR 2023, CCIS 1881, pp. 364–375, 2023.
https://doi.org/10.1007/978-3-031-43257-6_27

is obtained using crowdsourcing techniques: for each object there is a set of different (possibly inaccurate) labels, the quality of which depends on the skills of the performers. In addition to that, some objects may remain unlabeled if there is not enough budget for them.

Another example is the task of detecting objects in an image [1]. Bounding boxes are a common way to represent the location and extent of objects detected in an image or video frame in object detection tasks. A bounding box is a rectangular box that surrounds the object and is defined by its four corners or coordinates. In some difficult cases, such as detecting objects in medical CT scans, the bounding boxes can be very inaccurate and may highlight unwanted pixels. Moreover, the process of labeling CT images is very time-consuming, so it is not possible to label many objects.

Generally, there are three types of weak supervision: incomplete supervision, inaccurate supervision and inexact supervision [2]. In this work, we focus on the first two types of weak supervision. In particular, we assume that only a small part of the objects have labels, while the labels can be uncertain, and for most of the dataset there are no labels at all.

We propose an algorithm for solving the multi-target weakly supervised regression problem using Wasserstein metric, manifold regularization and a co-association matrix as the similarity matrix. We follow the transductive setting, which means that the objects from test data can be used during training and the task is to find the labels only for these objects. The algorithm for calculating the weighted average co-association matrix is also improved. Finally, we compare the proposed algorithm with existing algorithms of supervised learning and weakly supervised learning on synthetic and real data.

2   Problem Description

Let $X = \{x_1, \ldots, x_n\}$, $x_i = (x_i^1, \ldots, x_i^p)^\top \in \mathbb{R}^p$ are sampled from distribution $\mathcal{P}_X$, where n is the number of objects in the sample and p is the dimensionality of the feature space. In turn, $Y = \{y_1, \ldots, y_n\}$, $y_i = (y_i^1, \ldots, y_i^m)^\top \in \mathbb{R}^m$ are target labels, where m is the dimensionality of the target feature space.

In the semi-supervised transductive learning problem, a dataset $X \times Y = \{(x_1, y_1), \ldots, (x_n, y_n)\}$ is considered, but the target features $\{y_1, \ldots, y_{n_1}\} = Y_1 \subseteq Y$ are only known for a small part of the available data $\{x_1, \ldots, x_{n_1}\} = X_1 \subseteq X$. The rest of the objects $\{x_{n_1+1}, \ldots, x_n\} = X_0 \subseteq X$ are unlabeled. The task is to predict the labels $Y_0 = \{y_{n_1+1}, \ldots, y_n\}$ as accurately as possible according to some criterion.

To model the uncertainty of the observed labels, we use a multivariate normal distribution. We suppose that for each i-th data point, $i = 1, \ldots, n_1$, the value y_i of the target feature is a realization of a random variable Y_i with a cumulative distribution function (cdf) $F_i(y)$ defined on $D_Y \subset \mathbb{R}^m$:

$$Y_i \sim \mathcal{N}(\mu_i, \Sigma_i), \tag{1}$$

where $\mu_i \in \mathbb{R}^m$ is a mean vector, $\Sigma_i \in \mathbb{R}^{m \times m}$ is a covariance matrix, $i = 1, \ldots, n$. The overall degree of uncertainty can be interpreted as $\mathbb{T}_i = |\Sigma_i|$: the larger it is, the greater the uncertainty of the label. Accordingly, for strictly labeled objects, it is expected that $\mathbb{T}_i \approx 0$.

The task is to determine $F_i(y)$ for $i = n_1 + 1, \ldots, n$ following an objective criterion.

3   Related Work

The work [3] provides algorithms WSR-RBF and WSR-LRCM for solving the weakly supervised regression problem in the transductive formulation in the case of a one-dimensional target variable. It uses a univariate normal distribution to model inaccuracy:

$$Y_i \sim \mathcal{N}(a_i, \sigma_i),$$

where σ_i is an indicator of inaccuracy. Then it is proposed to solve the optimization problem by minimizing the distance between the predicted and real distributions using manifold regularization. To approximate the similarity matrix in WSR-LRCM, the co-association matrix is used and to obtain the co-association matrix, the cluster ensemble and the k-means algorithm are used. The WSR-RBF variant uses a weight matrix based on the RBF kernel instead of a low-rank representation:

$$W_{ij} := W(h) = \exp\left(-\frac{h^2}{2\ell^2}\right), \tag{2}$$

where $h = \|x_i - x_j\|$, and ℓ is a parameter.

However, the presented algorithm does not generalize to the multidimensional case. To solve a multi-target regression, it is necessary to train a separate model for each target variable. With this approach, it is possible to effectively solve those problems in which the target variables are independent of each other. If the target variables are not independent, for example, in the problem of object detection [1], these dependencies will be lost during training. These dependencies can be taken into account by using the distance between multivariate distributions, such as the Wasserstein distance [4].

The article [5] presents a detailed analysis of the co-association matrix and the algorithm for its construction. However, it relies on the basic version of the k-means algorithm, which has significant drawbacks, including the use of a single metric option and the uncertainty in choosing the appropriate number of clusters. In [7] the authors analyze the influence of metrics other than Euclidean on the quality of clustering by the k-means algorithm.

4   Proposed Method

Let

- $F^* = \{F_1^*, \ldots, F_{n_1}^*, \ldots, F_n^*\}$ be the set of arbitrary multivariate normal cdf's, each F_i^* is represented by a pair $(a_i, \mathbb{S}_i)$;

– $F = \{F_1, ..., F_{n_1}\}$ be the set of known cdf's, each F_i is represented by a pair (μ_i, Σ_i).

In the following, we assume, both Σ_i and $\mathbb{S}_i$ to be positive-definite matrices. Therefore, they are admitting Cholesky decomposition: $\Sigma_i = \Sigma_i^{1/2} \Sigma_i^{1/2^\top}$, $\mathbb{S}_i = \mathbb{S}_i^{1/2} \mathbb{S}_i^{1/2^\top}$. We denote elements of $\mathbb{S}_i^{1/2}$ as s_{jk}^i, and elements of $\Sigma_i^{1/2}$ as σ_{jk}^i.

4.1 Objective Functional

Consider the following optimization problem:

$$\text{find } F^{**} = \arg \min_{F^*} J(F, F^*), \text{ where}$$

$$J(F, F^*) = \sum_{x_i \in X_1} \mathcal{W}(F_i, F_i^*) + \gamma \sum_{x_i, x_j \in X} \mathcal{W}(F_i^*, F_j^*) W_{ij}$$

where $\mathcal{W}$ is a 2-Wasserstein metric [4] (also known as Kantorovich-Rubenstein distance), $\gamma > 0$ is a parameter, and matrix $W = (W_{ij})$ represents the similarity measures between elements of dataset. For two multivariate Gaussian distributions $N(\mu_0, \Sigma_0)$ and $N(\mu_1, \Sigma_1)$, 2-Wasserstein distance is

$$\mathcal{W}(N(\mu_0, \Sigma_0), N(\mu_1, \Sigma_1)) = ||\mu_0 - \mu_1||_2^2 + ||\Sigma_0^{1/2} - \Sigma_1^{1/2}||_F^2.$$

Following [3], we also add the regularisation term with parameter $\beta > 0$. We can rewrite the objective as

$$\text{find } (a^*, \mathbb{S}^*) = \arg \min_{(a, \mathbb{S})} J(\mu, \Sigma, a, \mathbb{S}), \text{ where}$$

$$J(\mu, \Sigma, a, \mathbb{S}) = \sum_{x_i \in X_1} ||\mu_i - a_i||_2^2 + ||\Sigma_i^{1/2} - \mathbb{S}_i^{1/2}||_F^2$$
$$+ \gamma \sum_{x_i, x_j \in X} W_{ij}(||a_i - a_j||_2^2 + ||\mathbb{S}_i^{1/2} - \mathbb{S}_j^{1/2}||_F^2) \qquad (3)$$
$$+ \beta \sum_{i=1,...,n} ||a_i||_2^2 + ||\mathbb{S}_i||_F^2.$$

4.2 Optimal Solution

To find the optimal solution, we differentiate (3) with respect to elements of a_i and $\mathbb{S}_i^{1/2}$, $i = 1, ..., n$:

$$\frac{\partial J}{\partial a_{ij}} = 2(\mu_{ij} - a_{ij}) + 4\gamma \sum_{l=1,...,n} W_{lj}(a_{lj} - a_{ij}) + 2\beta a_{ij}, \; i = 1, ..., n_1$$

$$\frac{\partial J}{\partial a_{ij}} = 4\gamma \sum_{l=1,...,n} W_{lj}(a_{lj} - a_{ij}) + 2\beta a_{ij}, \; i = n_1, ..., n$$

$$\frac{\partial J}{\partial s_{jk}^i} = 2(s_{jk}^i - \sigma_{jk}^i) + 4\gamma \sum_{l=1,...,n} W_{li}(s_{jk}^l - s_{jk}^i) + 2\beta s_{jk}^i, \; i = 1, ..., n_1$$

$$\frac{\partial J}{\partial s_{jk}^i} = 4\gamma \sum_{l=1,...,n} W_{li}(s_{jk}^l - s_{jk}^i) + 2\beta s_{jk}^i, \; i = n_1, ..., n.$$

Given that the matrices $\Sigma_i^{1/2}$ are lower triangular, we introduce an auxiliary operation $\mathrm{vec}_2 : \mathbb{R}^{m \times m} \to \mathbb{R}^{\frac{m(m+1)}{2}}$ that transforms all elements above the main diagonal (including the main diagonal elements) into a row-by-row vector. Also, a lower triangular matrix can be obtained from a vector using an operation $\mathrm{vec}_2^{-1} : \mathbb{R}^{\frac{m(m+1)}{2}} \to \mathbb{R}^{m \times m}$. Similarly, operation $\mathrm{vec}_3 : \mathbb{R}^{n \times m \times m} \to \mathbb{R}^{n \times \frac{m(m+1)}{2}}$ (as well as $\mathrm{vec}_3^{-1} : \mathbb{R}^{n \times \frac{m(m+1)}{2}} \to \mathbb{R}^{n \times m \times m}$) can be defined for three-dimensional tensors whose elements are lower triangular matrices. Let us denote

$$Y_{1,0} = (\mu_1^\top, ..., \mu_{n_1}^\top, 0, ..., 0) \in \mathbb{R}^{n \times m}$$

$$\Sigma_{1,0} = (\mathrm{vec}_2(\Sigma_1^{1/2})^\top, ..., \mathrm{vec}_2(\Sigma_{n_1}^{1/2})^\top, 0, ..., 0) \in \mathbb{R}^{n \times \frac{m(m+1)}{2}}$$

$$B = \mathrm{diag}(\beta + 1, ..., \beta + 1, \beta, ..., \beta) \in \mathbb{R}^{n \times n}.$$

Then the solution of the optimization problem can be given in the matrix form

$$\begin{aligned} a^* &= (B + 2\gamma L)^{-1} Y_{1,0} \\ \mathbb{S}^* &= \mathrm{vec}_3^{-1}\big((B + 2\gamma L)^{-1}\Sigma_{1,0}\big) \end{aligned} \tag{4}$$

where L is the Laplacian matrix, i.e., $L = D - W$, D is a diagonal matrix with elements $D_{ii} = \sum_j W_{ij}$. If we assume that there is exist $V \in \mathbb{R}^{n \times q}, q \ll n$, such that $W = VV^\top$ then

$$B + 2\gamma L = B + 2\gamma D - 2\gamma VV^\top = G - 2\gamma VV^\top.$$

where $G = B + 2\gamma D$. By using the Woodbury identity [6], the inverse operator $B + 2\gamma L$ in the solution, that takes $O(n^3)$ operations, can be represented as

$$(G - 2\gamma VV^\top)^{-1} = G^{-1} + 2\gamma G^{-1}V(I - 2\gamma V^\top G^{-1}V)^{-1}V^\top G^{-1} \tag{5}$$

where G is diagonal matrix (and therefore can be inverted in linear time), $I - 2\gamma V^\top G^{-1}V \in \mathbb{R}^{q \times q}$. Therefore it takes $O(nq + q^3)$ to perform the inverse, which reduces the computations significantly, since by the assumption $q \ll n$. Finally, we get:

$$\begin{aligned} a^* &= (G^{-1} + 2\gamma G^{-1}V(I - 2\gamma V^\top G^{-1}V)^{-1}V^\top G^{-1})Y_{1,0} \\ \mathbb{S}^* &= \mathrm{vec}_3^{-1}\big((G^{-1} + 2\gamma G^{-1}V(I - 2\gamma V^\top G^{-1}V)^{-1}V^\top G^{-1})\Sigma_{1,0}\big). \end{aligned} \tag{6}$$

In the article [3] it is shown that the weighted average co-association matrix can be used as a similarity matrix. By definition, the weighted average co-association matrix is

$$H = \sum_{l=1}^r \omega_l H_l, \tag{7}$$

where $H_1, ..., H_r$ are the co-association matrices for partitions $P_1, ..., P_r$ with elements indicating whether a pair x_i, x_j belong to the same cluster of this partition or not, $\omega_1, \ldots, \omega_r$ are weights of ensemble elements, $\omega_l \geq 0$, $\sum \omega_l = 1$. This matrix has a low-rank representation:

$$H = VV^\top,$$

where $V = [V_1 V_2 \ldots V_r]$ is a block matrix, $V_l = \sqrt{\omega_l} Z_l$, $Z_l \in \mathbb{R}^{n \times K_l}$ is the cluster assignment matrix for lth partition: $Z_l(i, k) = \mathbb{I}[c(x_i) = k]$, $i = 1, \ldots, n$, $k = 1, \ldots, K_l$ and K_l is the number of clusters in partition P_l, $K_l \ll n$. It is also shown that the Laplacian matrix L for the matrix H can be written in the following form:

$$L = D' - H,$$

$$D' = \mathrm{diag}(D'_{11}, \ldots, D'_{nn}),$$

$$D'_{ii} = \sum_{j=1}^{n} H(i, j) = \sum_{j=1}^{n} \sum_{l=1}^{r} \omega_l \sum_{k=1}^{K_l} Z_l(i, k) Z_l(j, k). \tag{8}$$

Now the optimal solution (5) can be found by using the low-rank representation of the similarity matrix (7) and the diagonal matrix (8).

5   Co-association Matrix: Multimetricity and Optimality

To obtain a low-rank similarity matrix representation, we will use a weighted average co-association matrix as the similarity matrix. However, the standard algorithm for calculating the weighted average co-association matrix [5] has a number of disadvantages:

- The k-means algorithm using the Euclidean metric can only find spherical clusters, so some complex relationships in the data may not be found as a result of clustering;
- The result is strongly influenced by both the choice of the desired number of clusters for the k-means algorithm and the number of different partitions in the ensemble.

To solve these problems, we decided to improve the algorithm for calculating the weighted average co-association matrix. Firstly, we propose to average the co-association matrix over the distance metrics used in the k-means algorithm. Secondly, we propose to use only optimal partitions in terms of cluster validity index in the ensemble in order to reduce the influence of unnecessary partitions and reduce the size of the ensemble.

5.1   Multimetric Weighted Average Co-association Matrix

Let $\{M_t\}_{t=1}^{d}$ be the set of metrics that can be used in the k-means algorithm as the distance between points, for example, the Minkowski distance of order p. Then for each metric from this set, an arbitrary set of partitions variants $\{P_l^{M_t}\}_{l=1}^{r^{M_t}}$ can be obtained using cluster ensemble. Similarly, for each partition, the co-association matrix $H_l^{M_t}$ can be found [3]. Then we define the multimetric weighted average co-association matrix as follows:

$$H = \sum_{t=1}^{d} H^{M_t} = \sum_{t=1}^{d} \sum_{l=1}^{r^{M_t}} \omega_l^{M_t} H_l^{M_t}, \tag{9}$$

where $\omega_1^{M_t}, \ldots, \omega_r^{M_t}$ are weights of ensemble elements, $\omega_l^{M_t} \geq 0$, $\sum_{l=1}^{r^{M_t}} \omega_l^{M_t} = 1$ for each M_t, $t = 1, \ldots, d$.

It should be noted that the clustering quality index, on which partition weights $\omega_l^{M_t}$ depend, should use the selected metric as the distance between points. That is why we assume that $\sum_{l=1}^{r^{M_t}} \omega_l^{M_t} = 1$ for each M_t, $t = 1, \ldots, d$ rather than $\sum_{t=1}^{d} \sum_{l=1}^{r^{M_t}} \omega_l^{M_t} = 1$. As a further improvement, co-association matrices can also be weighted.

Thus, by using different metrics, we can obtain different partitions and reduce the impact of some negative effects arising from the use of the Euclidean distance. For example, in [7] it is shown that using the city blocks metric can reduce the impact of the curse of dimensionality.

5.2   Optimal Weighted Average Co-association Matrix

In general, the number of clusters in each partition is a hyperparameter. For example, in [3] two different set of parameters are used:

- The ensemble size $r = 10$, the number of clusters K_i in i-th partition: $K_i = 2 + i$, $i = 1, \ldots, r$;
- The ensemble size $r = 10$, the number of clusters K_i in i-th partition: $K_i = 100 + i$, $i = 1, \ldots, r$.

However, this choice may not be optimal. So, in the first case, for partitions with a small number of clusters, the weights can be extremely small, which means that their influence on the weighted average co-association matrix will be insignificant. In the second case, in addition to the high computational complexity of finding partitions with a large number of clusters, all resulting partitions can be similar to each other and have almost the same weights. Also, in both cases, it is not guaranteed that at least one optimal partition will be found in terms of any criterion: for example, a partition that achieves a local optimum of the cluster validity index.

We propose another algorithm that calculates weighted average co-association matrix with optimal partitions. The matrix H^* thus obtained is called optimal weighted average co-association matrix. This matrix is optimal in the sense that only optimal partitions according to the cluster validity index are used in its calculation. Below is an algorithm for calculating the optimal weighted average co-association matrix by steps:

Input:
X - dataset.
r - cluster ensemble size.
$k_{\min}$ - minimum number of clusters in a partition.
$k_{\max}$ - maximum number of clusters in a partition.
Output:
H^* - optimal weighted average co-association matrix.
Steps:
1. Find a set of partitions $\{P_k\}_{k=k_{\min}}^{k_{\max}}$ of **X** using the k-means algorithm with different number of clusters k.
2. Calculate a set of cluster validity index values $\{\omega_k\}_{k=k_{\min}}^{k_{\max}}$ for the set of partitions $\{P_k\}_{k=k_{\min}}^{k_{\max}}$.
3. Select **r** largest values $\{\omega_{k_i}\}_{i=1}^{r}$ from a set $\{\omega_k\}_{k=k_{\min}}^{k_{\max}}$ and the corresponding set of partitions $\{P_{k_i}\}_{i=1}^{r}$.
4. Calculate a set of co-association matrices $\{H_{k_i}\}_{i=1}^{r}$ for the set of partitions $\{P_{k_i}\}_{i=1}^{r}$.
5. Calculate optimal weighted average co-association matrix $H^* = \sum\limits_{l=1}^{r} \omega_{k_i} H_{k_i}$
end.

The optimal weighted average co-association matrix thus obtained can be used instead of the original one, including for calculating multimetric weighted average co-association matrix:

$$H^* = \sum_{t=1}^{d} H^{*M_t}. \tag{10}$$

6   C-WSR Algorithm

We formulate three main variants of the Correlated Weakly Supervised Regression (C-WSR) algorithm:

- **RBF:** Radial Basis Function to calculate the similarity matrix is used;
- **LRCM:** a low-rank representation of the weighted average co-association matrix to calculate the similarity matrix is used;
- **LROMCM:** a low-rank representation of the optimal multimetric weighted average co-association matrix (10) to calculate the similarity matrix is used.

Input:
X - dataset with weak supervision, $X_1 \subset$ **X** - labeled sample, $X_2 \subset$ **X** inaccurately labeled sample, $X_3 \subset$ **X** - unlabeled sample.
a_i, Σ_i - mean vectors and covariance matrices of target distributions for each $x_i \in X_1 \cup X_2$
LRCM variant: r, Ω - cluster ensemble size and set of parameters for the k-means for clustering.

LROMCM variant: M - set of metrics for algorithm k-means, r - cluster ensemble size, $k_{\min}$ - minimum number of clusters in a partition, $k_{\max}$ - maximum number of clusters in a partition.

Output:

a^*, $\mathbb{S}^*$ - predicted mean vectors and covariance matrices of target distributions for objects from sample $\mathbf{X}$ (including predictions for the unlabeled sample).

RBF Variant Steps:

Directly calculate predicted mean vectors and covariance matrices of target distributions using (2) and (4).

LRCM Variant Steps:

1. Generate r variants of clustering partition for parameters randomly chosen from Ω; calculate weighted average co-association matrix.
2. Find graph Laplacian in the low-rank representation using (7) and D' in (8).
3. Calculate predicted mean vectors and covariance matrices of target distributions using (6).

LROMCM Variant Steps:

1. Calculate optimal multimetric weighted average co-association matrix with metrics from set M and parameters r, $k_{\min}$, $k_{\max}$ using (9) and (10).
2. Find graph Laplacian in the low-rank representation using (7) and D' in (8).
3. Calculate predicted mean vectors and covariance matrices of target distributions using (6).

end.

7   Experimental Results

In this section, we will compare three variants of the proposed Correlated Weakly Supervised Regression (C-WSR) algorithm. We use the MWD metric when comparing with weakly supervised learning algorithms WSR-RBF and WSR-LRCM from [3] and MAE when comparing with supervised learning algorithms such as Multivariate Linear Regression and gradient boosting from framework XGBoost on real data:

$$\mathrm{MWD}(y, y^*) = \frac{1}{n_{test}} \sum_{x_i \in X_{test}} ||\mu_i - a_i||_2^2 + ||\Sigma_i^{1/2} - \mathbb{S}_i^{1/2}||_F^2,$$

$$\mathrm{MAE}(y, y^*) = \frac{1}{n_{test}} \sum_{x_i \in X_{test}} ||\mu_i - a_i||_2.$$

Since the WSR-RBF and WSR-LRCM algorithms can only be used in a single target scenario, we train a separate model for each target variable. To calculate multimetric weighted average co-association matrix, we use Minkowski metric ρ_p with different $p \in \{1, 2, \infty\}$ and Silhouette as index cluster validity to determine the weights and the optimal number of clusters.

For experiments, we used an AMD Ryzen 9 3850X processor with a clock frequency of 3.5 GHz and 64 GB of RAM.

7.1 Monte-Carlo Simulation

For the Monte Carlo simulation, we generated a dataset of 1000 objects from a mixture of multivariate normal distributions $\mathcal{N}(\mu_k^*, \Sigma_k^*)$, $\mu_k^* = (8k + 1, 8k + 2, ..., 8k + d_x) \in \mathbb{R}^m$, $\Sigma_k^* = \mathrm{diag}(1, ..., 1) \in \mathbb{R}^{d_x \times d_x}$, $d_x = 8$ and $k \in \{1, 2, 3\}$.

For objects generated from the k-th component, we assume that the target function is equal to $Y_k = k + \varepsilon_k$, where ε_k is a random variable with d_y-dimensional normal distribution function $\mathcal{N}(0, D_k D_k^\top)$, D_k is random lower-triangular matrix with elements sampled from normal distribution and $d_y = 4$.

To insure the weak supervision, we assumed 10% of the dataset to be strictly labeled, 20% of the dataset consists of inaccurately labeled objects and the remaining 70% of objects are unlabeled. To model the inaccurate labeling, we use the parameters defined in (1): $\Sigma_i = \Sigma_Y$, where Σ_Y is a covariance matrix of the target function over labeled data. For strictly labeled objects, we assume that the matrix Σ_i is a zero matrix.

For the WSR-LRCM and C-WSR-LRCM algorithms, we used a cluster ensemble of size $r = 30$ and the number of clusters K_i in i-th partition: $K_i = 2+i$, $i = 1, ..., 30$. The C-WSR-LROMCM algorithm uses parameters $r = 10$, $k_{\min} = 2$ and $k_{\max} = 30$. Regularization coefficients $\beta = 0.001$ and $\gamma = 0.001$ are set for all algorithms. The obtained quality metrics were averaged over 100 runs. The results are presented in Table 1.

Table 1. Comparsion on Monte-Carlo simulation.

Supervision type	WSR		C-WSR		
	RBF	LRCM	RBF	LRCM	LROMCM
MWD	0.835	0.760	0.382	0.324	0.227

7.2 CO/NOx Dataset

For CO/NOx dataset [8] we use carbon monoxide (CO) and nitrogen oxides (NOx) emissions for year 2015 as regression targets. This dataset contains 11 features that describe the characteristics of a gas turbine and include 36733 observations.

1% of data is assumed to be strictly labeled, 9% is assumed to be labeled inaccurately, and 90% of data is considered unlabeled. Since the dataset is large, to model the inaccurate labeling, we estimate the mean vectors μ_i and covariance matrices Σ_i by 50 nearest neighbours. For strictly labeled objects, the exact label is used as the mean vector μ_i, and Σ_i is equal to the zero matrix.

As with synthetic data, regularization coefficients $\beta = 0.001$ and $\gamma = 0.001$ are set. For the WSR-LRCM and C-WSR-LRCM algorithms, a cluster ensemble of size $r = 30$ is used with the number of clusters K_i in i-th partition: $K_i = 10 + i$, $i = 1, ..., 30$. The C-WSR-LROMCM algorithm trained with parameters $r = 10$,

$k_{min} = 2$ and $k_{max} = 50$. Supervised learning algorithms (Multivariate Linear Regression (MLR) and XGBoost (XGB)) are trained only on strictly labeled objects. Note that due to the large amount of data in the dataset, finding the inverse matrix in the RBF variant requires a significant amount of computing resources, especially RAM. The results are presented in Table 2.

Table 2. Comparsion on CO/NOx dataset.

Supervision type	WSR		C-WSR			SR	
	RBF	LRCM	RBF	LRCM	LROMCM	MLR	XGB
MWD	72.96	60.07	65.45	52.22	44.74	–	–
MAE	42.11	35.92	38.84	31.92	26.83	38.69	30.48

Thus the results of the experiments show the considerable improvements in the accuracy for the proposed method.

8   Conclusion

In this paper, we considered the problem of multi-target weakly supervised regression with noisy labelling in a transductive setting. Using the multivariate normal distribution, we described an imprecision model in the multi-output case. We also proposed an algorithm for solving the optimisation problem using the Wasserstein metric and manifold regularisation. To speed up the solution of the optimisation problem, we used the cluster ensemble to obtain the co-association matrix and the low-rank representation technique to compress the resulting matrices.

The presented algorithm has shown its advantage over existing machine learning algorithms that cannot use uncertain multidimensional labels during training. We have also made several important improvements to the calculation of the weighted average co-association matrix by introducing an optimal multimetric weighted average co-association matrix. The new approach can significantly improve the quality and stability of the algorithm, and also simplifies the search for optimal hyperparameters to solve each specific problem.

As a further improvement, one can try different distances between distributions in the optimisation problem, and another promising idea would be to use deep learning approaches to find the co-association matrix. It is also worth considering other imprecision models: for example, using different types of multivariate distributions than normal.

Acknowledgements. The work was carried out with the financial support of the Russian Science Foundation, project 22-21-00261. Special thanks to Vladimir Kondratiev for participating in the discussion and experiments.

References

1. Yang, Z., Mahajan, D., Ghadiyaram, D., Nevatia, R., Ramanathan, V.: Activity driven weakly supervised object detection, pp. 2912–2921 (2019). https://doi.org/10.1109/CVPR.2019.00303
2. Zhou, Z.H.: A brief introduction to weakly supervised learning. Natl. Sci. Rev. **5**, 44–53 (2017)
3. Berikov, V., Litvinenko, A.: Weakly supervised regression using manifold regularization and low-rank matrix representation. In: Pardalos, P., Khachay, M., Kazakov, A. (eds.) MOTOR 2021. LNCS, vol. 12755, pp. 447–461. Springer, Cham (2021). https://doi.org/10.1007/978-3-030-77876-7_30
4. Bogachev, V.I., Kolesnikov, A.: The Monge-Kantorovich problem: achievements, connections, and perspectives Russ. Math. Surv. **67**, 785–890 (2012)
5. Berikov, V.B.: Cluster ensemble with averaged co-association matrix maximizing the expected margin. In: International Conference on Discrete Optimization and Operations Research (DOOR 2016), vol. 1623, pp. 489–500. CEUR-WS.org (2016)
6. Higham N.: Accuracy and Stability of Numerical Algorithms. SIAM (2002)
7. Aggarwal, C.C., Hinneburg, A., Keim, D.A.: On the surprising behavior of distance metrics in high dimensional space. In: Van den Bussche, J., Vianu, V. (eds.) ICDT 2001. LNCS, vol. 1973, pp. 420–434. Springer, Heidelberg (2001). https://doi.org/10.1007/3-540-44503-X_27
8. Kaya, H., Tüfekci, P., Uzun, E.: Predicting CO and NOx emissions from gas turbines: novel data and a benchmark PEMS. Turk. J. Electr. Eng. Comput. Sci. **27**, 4783–4796 (2019)

Using General Least Deviations Method for Forecasting of Crops Yields

Tatiana Makarovskikh$^{(\boxtimes)}$, Anatoly Panyukov , and Mostafa Abotaleb

South Ural State University, Chelyabinsk, Russia
{Makarovskikh.T.A,paniukovav}@susu.ru, abotalebmostafa@yandex.ru
http://www.susu.ru

Abstract. Nowadays much attention has been paid to the development of the software, which makes it possible to process and visualize images, and in particular, develop the mathematical models for monitoring and predicting of crops quality to optimize the profit. Modelling the dynamics of vegetation indices in this regard is very relevant and important task especially if it is made with high level of detailing. A number of studies are underway in the agricultural sector to better predict crop yield using machine learning algorithms. In our research we suggest the deterministic approach using general least deviation method to obtain the model for a dynamic process. Using the obtained models data from the previous periods we can compare them with data for the current one and forecast the development of the crops in the current vegetation period. Unlike neural networks, this approach makes it possible to explicitly obtain high-quality quasi-linear difference equations for any field and any region. We use this model for modelling of normalized difference vegetation index dynamics for several years and discuss the opportunities to use our algorithm as a base of software for crop prediction, and detection of the problematic areas of a field.

Keywords: Forecasting · time series · quasilinear model · generalized least deviations method · monitoring crop yields · NDVI dynamics

1 Introduction

Recently, much attention has been paid to the development of the software, which makes it possible to process and visualize satellite and drone information, and in particular, the development of mathematical models for monitoring and predicting of crops quality [1] to optimize the profit. Modelling the dynamics of vegetation indices in this regard is very relevant and important task nowadays, especially if it is made with high level of detailing.

The main objective of the study is to develop a mathematical model based on the available NDVI (Normalized Difference Vegetation Index) statistical data that allows one to describe the dynamic process of development of crops in order

---

TBA later.

© The Author(s), under exclusive license to Springer Nature Switzerland AG 2023
M. Khachay et al. (Eds.): MOTOR 2023, CCIS 1881, pp. 376–390, 2023.
https://doi.org/10.1007/978-3-031-43257-6_28

to assess their productivity. The most of known researches use neural networks [2–5], statistical [6,7] or elementary ordinary differential equations approaches [8].

Our study offers a higher level of detail, which is of particular interest to potential customers. The article [8] considers the average value of NDVI for a single field. There are also known works in which the authors consider the average values of the vegetation index NDVI for rather large territorial objects: (1) for one region [3,6,9], for each administrative region [10], for individual agricultural enterprises, or the average long-term values of this indicator for decades for each region. All these approaches have high errors since there may appear the valuable differences of NDVI even for one field because of forest cover, relief, shading and some other factors. In our research we suggest to divide each considered field to tiles with common properties and NDVI values and research the dynamics of NDVI for each tile separately. It allows us to define and recognize the problematic areas of one field.

In our paper we discuss our methods for identifying the parameters of a single quasilinear difference equation. We use it to solve the problem of regression analysis with mutually dependent observable variables, which allows to implement the general last deviations method (GLDM). Unlike neural networks, this approach makes it possible to explicitly obtain high-quality quasi-linear difference equations (adequately describing the considered process). Our approach is flexible and runs for any initial data for any region and any agriculture, when the neural networks need learning for each considered region, and each agriculture separately. So, the results obtained using neural networks for China, Egypt or even some regions of Russia do not describe the common approach for any data and any region. Our approach also allows to get the model coefficients for all the cases considered earlier: the average NDVI for a region, for administrative region, for one corporation, for one field, and also for part of a field.

2 Data Organizing

A concept of data organizing somewhat similar to the approach of the authors was considered in works [12–14]. The authors propose a method for element-by-element image analysis for the formation of linear contours of objects depicted on aerial photographs. The proposed method makes it possible to recognize objects by their formal features, allows to analyse more details and properties of objects in the process of recognition, and also improves the quality and accuracy of recognition. Unfortunately, in addition to the algorithm itself, the authors do not provide any additional, but very important information. There are no data on the efficiency of using the algorithm, the time of its work with graphic files of large dimensions, the results of computational experiments, analysis of the accuracy of object detection are not given. And these authors, as many others stop on recognition itself, do not concerning the further analysis of state of recognized objects in time.

Our data used for analysis is obtained from the process shown in Fig. 1. Let the object is already recognised, NDVI is already calculated. The NDVI scale

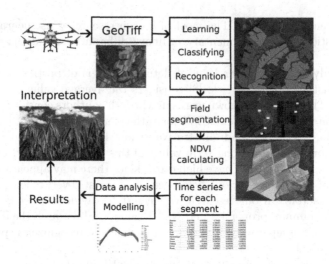

Fig. 1. The scheme of the analysis system executing

$S \in [-1; 1]$. In common negative values $S \leq 0$ show buildings, structures, paved road surfaces, water surfaces, mountains, clouds and snow. Index $S \in [0.1; 0.2)$ usually corresponds to an open soil. In the case of plants, the NDVI index always has positive values $S \in [0.2; 1]$, where $S \in [0.2; 0.4)$ satisfies the weak, sparse vegetation, $S \in [0.4; 0.6)$ means moderate vegetation, and the value $S \geq 0.6$ is an index for healthy, dense vegetation. So, the examined time series consists of values between 0 and 1. And we need to determine the model describing the dynamics of NDVI. In [8] authors showed that NDVI dynamics corresponds to the normal distribution law and defined the model coefficients using the separable differential equation. The parameters of this equation are later obtained using statistical software for one field. In our approach we get all the parameters using only our algorithm. One more peculiarity of approach considered in [8] is that authors use the average NDVI values for the whole field. And in our approach we can split the field for several tiles to discover the dynamics of vegetation index for each of them (see Fig. 2). Each tile is characterized by several peculiarities. For example, the field in Fig. 2 has 6 different tiles:

- T_1 corresponds the area near the pond, it's most likely low;
- T_2 and T_3 satisfy the shadowed area;
- T_4 is the common field without any peculiarities;
- T_5 is the shadowed area near the road;
- T_6 is the area near the road.

For sure, the real examined field can have more areas and more properties defining the different velocity of crops growth. Let us save a vector of the tiles belonging the field. This kind of data organizing may be implemented for the different types of objects such as lakes and islands, crops and forests, etc. If we need

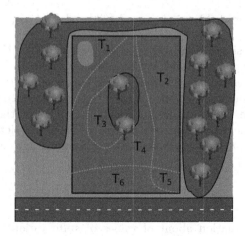

Fig. 2. The example of a field and its division to several tiles

discovering the process in time we can form the 3-dimensional (in common case) matrix as following (see Fig. 3): for each tile T_i, $i = 1, \ldots n$ of recognized image we save the values for time t_1, t_2, ..., t_M for several layers L_j, $j = 1, \ldots k$.

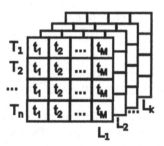

Fig. 3. The matrix of values for each tile in time

The layers may save the information for different bands of multispectral camera, it may be information for different types of objects etc. As soon as the monitoring of crops developing is made in 3–5 days then the number of images M is approximately 100–150 per year if we take images without dependency on a season (for example, for recognition of waterlogged lakes), and 10–40 if we are holding seasonal researches (for example, vegetation of crops, deciduous forest research etc.) The number of tiles N depends on the real size of the recognized objects, and their properties. Hence, to describe the dynamic process for each tile of the recognized object we need to obtain the coefficients for N models. This task is solvable for each agricultural object separately. Now let us consider the common approach of defining the model coefficients.

3   General Least Deviations Method Estimation

Let us consider a single time series for one selected tile. For other tiles, the reasoning is similar up to the defined parameters.

Linear autoregressive models have a small forecasting horizon. The construction of adequate nonlinear models and/or neural networks may not be possible for technical reasons. Quasilinear models allow to increase the forecasting horizon. Let us implement our approach considered in [11] to determine the coefficients $a_1, a_2, a_3 \ldots, a_m \in \mathbb{R}$ of a m-th order quasilinear autoregressive model

$$y_t = \sum_{j=1}^{n(m)} a_j g_j(\{y_{t-k}\}_{k=1}^m) + \varepsilon_t, \quad t = 1, 2, \ldots, T \tag{1}$$

by up-to-date information about of values of state variables $\{y_t \in \mathbb{R}\}_{t=1-m}^T$ at time instants t; here $g_j : (\{y_{t-k}\}_{k=1}^m) \to \mathbb{R}$, j=1, 2, \ldots n(m) are given $n(m)$ functions, and $\{\varepsilon_t \in \mathbb{R}\}_{t=1}^T$ are unknown errors.

The considered approach consists in determining the parameters of the recurrence Eq. (1). The GLDM estimation algorithm [11] gets a time series $\{y_t \in \mathbb{R}\}_{t=-1-m}^T$ of length $T + m \geq (1 + 3m + m^2)$ as an input data and determines the factors $a_1, a_2, a_3 \ldots, a_m \in \mathbb{R}$ by solving the optimization task

$$\sum_{t=1}^T \arctan \left| \sum_{j=1}^{n(m)} a_j g_j(\{y_{t-k}\}_{k=1}^m) - y_t \right| \to \min_{\{a_j\}_{j-1}^{n(m)} \subset \mathbb{R}} \tag{2}$$

The Cauchy distribution

$$F(\xi) = \frac{1}{\pi} \arctan(\xi) + \frac{1}{2}$$

has the maximum entropy among distributions of random variables that have no mathematical expectation and variance. That's why function $\arctan(*)$ is applied as loss function.

Let's consider a m-th order model with quadratic nonlinearity. Then the basic set $g_j(*)$ may contain the following functions

$$g_{(k)}(\{y_{t-k}\}_{k=1}^m) = y_{t-k}, \tag{3}$$
$$g_{(kl)}(\{y_{t-k}\}_{k=1}^m) = y_{t-k} \cdot y_{t-l},$$
$$k = 1, 2, \ldots, m; \; l = k, k+1, \ldots, m.$$

Obviously, in this case $n(m) = 2m + C_m^2 = m(m+3)/2$, and the numbering of $g_{(*)}$ functions can be arbitrary. In particular, for $m = 2$ functions $g_{(*)}$ are the following

$$g_1 = y_1, \quad g_2 = y_2, \quad g_3 = y_1^2, \quad g_4 = y_2^2, \quad g_5 = y_1 \cdot y_2.$$

The model for this case looks like following:

$$y_t = (a_1 y_{t-1} + a_2 y_{t-2}) + (a_3 y_{t-1}^2 + a_4 y_{t-2}^2 + a_5 y_{t-1} y_{t-2}). \tag{4}$$

Predictor forms the indexed by $t = 1, 2, \ldots, T-1, T$ family of the m-th order difference equations

$$\overline{y[t]}_\tau = \sum_{j=1}^{n(m)} a_j^* g_j \left(\{\overline{y[t]}_{\tau-k}\}_{k=1}^m \right),$$

$$\tau = t, t+1, t+2, t+3, \ldots, T-1, T, T+1, \ldots \quad (5)$$

for lattice functions $\overline{y[t]}$ with values $\overline{y[t]}_\tau$ which interpreted as constructed at time moment t the forecasts for y_τ. Let us use the solution of the Cauchy problem for its difference Eq. (5) under the initial conditions

$$\overline{y[t]}_{t-1} = y_{t-1}, \ \overline{y[t]}_{t-2} = y_{t-2}, \ \ldots, \ \overline{y[t]}_{t-m} = y_{t-m} \quad t = 1, 2, \ldots, T-1, T \quad (6)$$

to find the values of the function $\overline{y[t]}$.

So we have the set $\overline{Y}_\tau = \left\{ \overline{y[t]}_\tau \right\}_{t=1}^T$ of possible prediction values of y_τ. Further we use this set to estimate the probabilistic characteristics of the y_τ value.

The task (2), i.e. task of GLDM-estimation, is a concave optimization problem, and entering the additional variables reduces it to the following linear programming task

$$\sum_{t=1}^T p_t z_t \to \min_{\substack{(a_1, a_2, \ldots, a_{n(m)}) \in \mathbb{R}^m, \\ (z_1, z_2, \ldots, z_T) \in \mathbb{R}^T}} \quad (7)$$

$$-z_t \le \sum_{j=1}^{n(m)} [a_j g_j(\{y_{t-k}\}_{k=1}^m)] - y_t \le z_t, \quad t = 1, 2, \ldots, T, \quad (8)$$

$$z_t \ge 0, \quad t = 1, 2, \ldots, T. \quad (9)$$

The task (7)–(9) has a canonical type with variables $n(m) + T$ and $3n$ inequality constraints including the conditions of non-negativity of z_j, $j = 1, 2, \ldots, T$.

The dual to (7) task is

$$\sum_{t=1}^T (u_t - v_t) y_t \to \max_{u, v \in \mathbb{R}^T}, \quad (10)$$

$$\sum_{t=1}^T a_j g_j(\{y_{t-k}\}_{k=1}^m) (u_t - v_t) = 0, \ j = 1, 2, \ldots, n(m), \quad (11)$$

$$u_t + v_t = p_t, \quad u_t, v_t \ge 0, \quad t = 1, 2, \ldots, T. \quad (12)$$

Let us introduce variables $w_t = u_t - v_t$, $t = 1, 2, \ldots, T$. Conditions (12) imply that

$$u_t = \frac{p_t + w_t}{2}, \quad v_t = \frac{p_t - w_t}{2}, \quad -p_t \le w_t \le p_t, \quad t = 1, 2, \ldots, T.$$

So the optimal task (10)–(12) solution is equal to the optimal solution of task

$$\sum_{t=1}^{T} w_t \cdot y_t \rightarrow \max_{w \in \mathbb{R}^T}, \tag{13}$$

$$\sum_{t=1}^{T} g_j(\{y_{t-k}\}_{k=1}^{m}) \cdot w_t = 0, \; j = 1, 2, \ldots, n(m), \tag{14}$$

$$-p_t \le w_t \le p_t, \; t = 1, 2, \ldots, T. \tag{15}$$

Constraints (14) define $(T-n(m))$-dimensional linear variety $\mathcal{L}$ with $(n(m) \times T)$-matrix

$$S = \begin{bmatrix} g_1(\{y_{1-k}\}_{k=1}^{m}) & g_1(\{y_{2-k}\}_{k=1}^{m}) & \cdots & g_1(\{y_{T+1-k}\}_{k=1}^{m}) \\ g_2(\{y_{1-k}\}_{k=1}^{m}) & g_2(\{y_{2-k}\}_{k=1}^{m}) & \cdots & g_2(\{y_{T+1-k}\}_{k=1}^{m}) \\ \vdots & \vdots & \ddots & \vdots \\ g_{n(m)}(\{y_{1-k}\}_{k=1}^{m}) & g_{n(m)}(\{y_{1-k}\}_{k=1}^{m}) & \cdots & g_{n(m)}(\{y_{1-k}\}_{k=1}^{m}) \end{bmatrix}$$

Constraints (15) define T-dimensional parallelepiped $\mathcal{T}$.

The simple structure of the allowed set for task (13)–(15) representing the intersection of $(T - n(m))$-dimensional linear variety $\mathcal{L}$ (14) and T-dimensional parallelepiped $\mathcal{T}$ (15) allows to obtain its solution by algorithm using the gradient projection of the objective function (13) (i.e. vector $\nabla = \{y_t\}_{t=1}^{T}$) on the allowed area $\mathcal{L} \cap \mathcal{T}$ defined by the constraints (14)–(15). The projection matrix on $\mathcal{L}$ is as following

$$S_{\mathcal{L}} = E - S^T \cdot (S \cdot S^T)^{-1} \cdot S,$$

and gradient projection on $\mathcal{L}$ is equal to $\nabla_{\mathcal{L}} = S_{\mathcal{L}} \cdot \nabla$. Moreover, if outer normal on any parallelepiped face forms the sharp corner with gradient projection $\nabla_{\mathcal{L}}$ then movement by this face is equal to zero.

GLDM-estimates are robust to the presence of a correlation of values in $\{y_t \in \mathbb{R}\}_{t=-1-m}^{T}$, and (with appropriate settings) are the best for probability distributions of errors with heavier (than normal distribution) tails (see [15]). The above shows the feasibility of solving the identification problem by algorithm of weighted less deviation method (WLDM) estimation. The established in [16] results allow us to reduce the problem of determining GLDM estimation to an iterative procedure with WLDM estimates [11].

The scheme of algorithm is shown in Fig. 4. Its input data are:

- $S = \{S_t \in \mathbb{R}^N\}_{t \in T}$, the matrix of a linear variety;
- $\nabla_{\mathcal{L}}$, gradient projection of objective function on $\mathcal{L}$;
- weight factors $\{p_t \in \mathbb{R}^+\}_{t=1}^{T}$;
- values of the given state variables $\{y_t \in \mathbb{R}^+\}_{t=1-m}^{T}$.

Algorithm runs as the iteration process for obtaining optimal GLDM solution $A \in \mathbb{R}^{n(m)}$ and the vector of residuals $z \in \mathbb{R}^T$. This process stops when $(A^{(k)} = A^{(k-1)})$. To obtain A and z we run the WLDM estimation algorithm

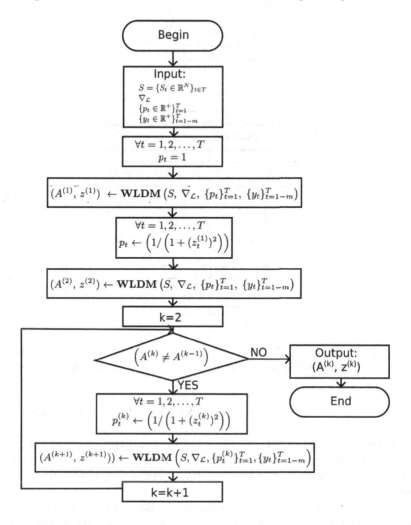

Fig. 4. The scheme of GLDM estimation algorithm

[17] which gets the same input data as GLDM algorithm and calculates the factors

$$a_1, a_2, a_3 \ldots, a_{n(m)} \in \mathbb{R}$$

by solving the optimization task

$$\sum_{t=1}^{T} p_t \cdot \left| \sum_{j=1}^{n(m)} a_j g_j(\{y_{t-k}\}_{k=1}^{m}) - y_t \right| \rightarrow \min_{\{a_j\}_{j=1}^{n(m)} \in \mathbb{R}^{n(m)}} \tag{16}$$

The scheme of this algorithm is shown in Fig. 5. Computational complexity of such algorithm does not exceed $O(T^2)$ due to the simple structure of the admis-

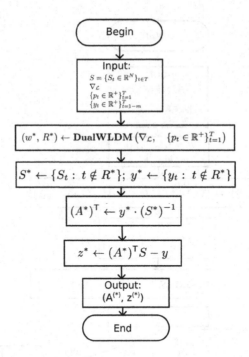

Fig. 5. The scheme of WLDM estimation algorithm

sible set: intersection of T-dimensional cuboid (15) and $(T - n(m))$-dimensional linear variety (14).

Algorithm for dual task (13)–(15) solution begins the search of the optimal solution at 0, moving along direction $\nabla_{\mathcal{L}}$. If the current point falls on the face of brick $\mathcal{T}$, then the corresponding coordinate in the direction of the moving is assumed to be 0.

If (w^*, R^*) is the result of executing the gradient projection algorithm [11], then w^* is the optimal solution to the task (13)-(15), and the optimal solution of the task (10)–(12) is equal to

$$u_t^* = \frac{p_t + w_t^*}{2}, \quad v_t^* = \frac{p_t - w_t^*}{2}, \quad t = 1, 2, \ldots, T.$$

It is following from the complementarity condition for a pair of mutually dual tasks (7)–(9) and (10)–(12) that

$$y_t = \sum_{j=1}^{n(m)} [a_j g_j(\{y_{t-k}\}_{k=1}^m)] \qquad \forall t \notin R^*, \tag{17}$$

$$y_t = \sum_{j=1}^{n(m)} [a_j g_j(\{y_{t-k}\}_{k=1}^m)] + z_t^*, \quad \forall t \in R^* : w_t^* = p_t, \tag{18}$$

$$y_t = \sum_{j=1}^{n(m)} [a_j g_j(\{y_{t-k}\}_{k=1}^m)] - z_t^*, \quad \forall t \in R^* : w_t^* = -p_t. \tag{19}$$

In fact, the solution $(\{a_j^*\}_{j-1}^{n(m)}, z^*)$ of linear algebraic equations system (17)–(19) represents the dual optimal solution of task (13)-(15) and the optimal solution of the task (16), that proves the validity of the following theorem.

Theorem 1. *Let*

- *w^* be the optimal solution of the task (13)-(15),*
- *$(\{a_j^*\}_{j-1}^{n(m)}, z^*)$ be solution of a system of linear algebraic equations (17)-(19).*

Then $\{a_j^*\}_{j-1}^{n(m)}$ is the optimal solution to the task (16).

The main problem with the use of WLDM-estimator is the absence of general formal rules for choosing weight coefficients. Consequently, this approach requires additional research.

Theorem 2. *[17] The sequence$\{(A^{(k)}, z^{(k)})\}_{k=1}^\infty$, constructed by GLDM-estimator Algorithm, converges to the global minimum (a^*, z^*) of the task (2).*

The description of GLDM estimation algorithm shows that its computational complexity is proportional to the computational complexity of the algorithm for solving of primal and/or dual WLDM tasks. Multiply computational experiments show that the average number of iterations of GLDM estimation algorithm is equal to the number of coefficients in the identified equation. If this hypothesis is true then computational complexity in solving practical problems does not exceed

$$O((n(m))^3 T + n(m) \cdot T^2).$$

It is necessary to take into account that the search and finding of the high-order autoregression equation have their own specific conditions. One of these conditions, in particular, is the high sensitivity of the algorithm to rounding errors. To eliminate the possibility of error in the calculations, it is necessary to accurately perform basic arithmetic operations on the field of rational numbers and supplement them with parallelization.

To analyse the quality of the obtained coefficients we use two errors: MAE, and MBE. In statistics, the mean absolute error (MAE) is a measure of the errors between paired observations expressing the same time series:

$$MAE = \frac{\sum_{t=3}^{minFH} \left| y[t] - \overline{y[t]} \right|}{minFH},$$

where $minFH$ be a reasonable forecasting horizon. Mean Bias Error (MBE) is the exact difference between the predicted value and the actual value without any math function like absolute or square root applied to it

$$MBE = \frac{\sum_{t=3}^{minFH} \left(y[t] - \overline{y[t]} \right)}{minFH}.$$

Hence, in terms of the problem of analyzing aerial photographs the developed algorithm allows to get a mathematical model describing the process development for each tile. This is useful if we compare a set of models between each other, and then detect ones corresponding to areas with well-developed crops. Moreover, using the models data from the previous periods we can compare them with data for the current period and forecast the development of the crops in the current vegetation period. As for known approaches, they may work perfectly for some definite cases, and get chaos for some other cases. ARIMA, exponential smoothing and other classical approaches are not suitable for the case when we try to forecast crops for a new period because only a few data is known. Our model needs only 2 first points to apply the obtained model and forecast the development of the process in future. The same is for machine learning approaches. The approach considered in [8] uses the classical differential equation to solve the same task, but the coefficients of their model are defined by Statistica software, so, this approach works for the only case, and should be recalculated for the other objects manually, moreover, the obtained curve does not depend on the initial value.

4   Experimental Results

Let us consider the computational experiment on constructing the solution of Cauchy problem to one quasi-linear difference equation, the identification of this equation, and let us show that the obtained solution shows the high quality of the considered algorithm for the dynamics of NDVI for the following 2 experiments: (1) the dynamics of index for winter wheat sowing in Stavropol region; (2) the dynamics of NDVI for forests of Losiniy island. The data for experiments are taken from dissertations [18,19], these researches are devoted to the other tasks, so there is no way to compare our results with some other models. Although, defining this index is a very important task, there are only a few small free datasets to test the new approaches.

4.1   The Dynamics of NDVI for Winter Wheat Sowing in Stavropol Region

Changes in the vegetative index NDVI in ontogenesis reflect its physiological state and depend both on the phase of development and on growing conditions: temperature and air humidity. Vegetation index NDVI as an optical-biological characteristic of a crop can be used to assess its physiological state. The average dynamics of this indicator for the fields of winter wheat is a peak-shaped non-uniform curve with a maximum at the beginning of the heading phase. We consider the time series for one of the fields examined by the author of dissertation [18] in 2014. The process and the results of modelling are shown in Fig. 6.

This dataset contains the average data for a single field, but if a field is split to several tiles, there is no problem to take average for a single tile if we have the

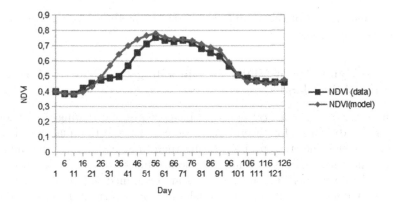

Fig. 6. The observed dynamics of NDVI for one of the fields in Stavropol region in 2014 and the results of modelling

map of a field and its surroundings. The task of splitting the field to the tiles is the subject of a separate research, and the methods of modelling the obtained indexes do not depend on methods of splitting to tiles.

The observing in the considered experiment started in the end of February when the snow goes away and continued until wheat ripening in the beginning of July. The identification results for this experiment allow to obtain the following model coefficients for model (4):

$$y_t = (3.46935 \cdot y_{t-1} - 2.18641 \cdot y_{t-2}) + \left(-5.59237 \cdot y_{t-1}^2 - 2.5635 \cdot y_{t-1}y_{t-2} + 7.72991 \cdot y_{t-2}^2\right).$$

The loss function value is 0.9172588, and the errors are $MAE = 0.02095274$ and $MBE = 0.01885199$.

This experiment shows that the GLDM estimation algorithm allows to get the model coefficients for experimental data obtained in the equal time periods. If the time periods are not equal, for example, if we consider the whole life cycle of the winter wheat developing from September till July (here the period from the middle of November till the end of February is not observed), or we held some measurements of NDVI daily sometimes, the obtained model cannot adequately describe this process. So, the using of arbitrary time periods is the topic of further research. This research is very urgent since it is impossible to measure NDVI in the fixed time periods due to the crop technology and the inability to launch a drone daily due to the high cost of this process for large areas.

Once we obtain the model of NDVI dynamics for several years we can get the average model for a fixed crop in a fixed field. This model can help the farmer to detect the problematic areas (if at one moment the value of NDVI is lower than the model value by an amount exceeding the threshold value) and facilitate monitoring of them, and also count the predicted volume of crops for each considered field.

4.2 The Dynamics of NDVI for Modelling the Growth of Trees in Losiniy Island

One more urgent agricultural and ecological task is monitoring of the forests. Using NDVI dynamics for several years we can observe the quality either of a whole forest or some tree species growing there. Let us for the experiment use consider the time series satisfying the monitoring of forests in Losiniy island from April till October. These time series are examined in [19]. But the author of this work does not consider any model and only describes the observed data. As in the case with crops, once we have a model for one year we can compare it with the same model for the other years, obtain the seasonal model, or the model based on the average coefficients for all the previous years. If we have a model we can analyse the quality of trees and detect the problems with them. As soon as now we are analysing the data for one season, we have very short time series for which GLDM-estimation algorithm gets the following model of the second order (the coefficients of the model (4), loss function, minimal forecasting horizon (in months), and errors are shown in Tables 1). The analysis of Table 1 shows that the developed algorithm works almost perfect for different type of data for different tree types. The errors of modelling are very low for most cases, hence, the obtained models adequately describe the process of NDVI dynamics for various tree types and can be used for analysis of monitoring results. The main advantage of this approach is that all the obtained coefficients may be interpreted in terms of practical task. The use of neural networks for the considered data sets is not justified, since the length of the time series does not allow the efficient use

Table 1. The identification results (coefficients, loss function and errors) for different tree species

Tree	a_1	a_2	a_3	a_4	a_5	Loss fun.	MAE	MBE
Birch	$-3,33289$	4,3374	13,6638	8,41678	$-22,82575$	6,22E$-$015	1,00E$-$012	$-1,00$E$-$012
Elm	$-2,72226$	3,50835	11,0937	6,64245	$-18,05596$	4,00E$-$015	1,42E$-$014	1,42E$-$014
Oak	$-3,04254$	4,35999	13,2885	8,65894	$-23,30116$	4,88E$-$015	1,70E$-$014	$-1,70$E$-$014
Spruce	$-0,55497$	3,03523	12,3839	10,3757	$-26,43966$	7,55E$-$015	4,13E$-$014	$-3,51$E$-$014
Willow	$-2,21953$	2,71921	9,97028	6,33381	$-16,02432$	2,22E$-$015	3,93E$-$014	3,93E$-$014
Maple	$-19,3533$	22,2068	54,9690	30,3626	$-90,5544$	1,95E$-$014	1,71164	$-1,71164$
Linden	$-0,88513$	2,09379	10,0772	9,18861	$-20,48208$	3,11E$-$015	7,26E$-$013	7,26E$-$013
Larch	$-4,45510$	5,07474	13,9688	6,54746	$-20,55857$	4,44E$-$015	3,65776	$-3,65776$
Alder	$-0,94055$	1,94152	9,30427	8,30144	$-18,28108$	3,77E$-$015	1,21342	1,21342
Aspen	$-4,03793$	5,60943	16,2789	10,01083	$-28,17523$	2,66E$-$015	3,07E$-$013	2,53E$-$013
Rowan	$-4,62792$	4,21466	19,5855	12,74890	$-30,28771$	1,15E$-$014	6,22E$-$015	6,22E$-$015
Pine	$-0,96789$	2,93664	11,2552	8,35734	$-22,09748$	1,33E$-$015	7,66E$-$015	$-7,66$E$-$015
Ash	$-0,91149$	1,48240	9,41523	8,47214	$-17,89959$	2,22E$-$015	2,16E$-$014	2,16E$-$014

of neural network methods, moreover, when using these methods, it is impossible to represent the described process in the form of a general equation. The use of a qualitative quasi-linear equation allows the process under consideration to be scaled up for use in the analysis of other crop varieties and other vegetation.

5   Conclusion

The consideration of the relationship mechanisms between the vegetation index, as an objective characteristic of the optical properties of crops, and the formation of its yield is very urgent task. Since photosynthesis is the most essential part of the production process, the study of the influence of the size and duration of the functioning of the assimilation apparatus on the NDVI of plants is an important and relevant area of research in the field of using Earth remote sensing data in biology and agriculture. The model can be used to approximate the missing values of the NDVI vegetation index, estimate the time to reach the maximum value of the index and, therefore, predict the start of harvesting dates.

Speaking about the quality of the considered model itself we can mention that its errors are not worse than ones for neural network approaches or classical statistical models [11] but needs less computational resources. It has one significant advantage in comparison with these models that is in the opportunity to interpret the model coefficients in terms of the research problem. The method considered in the article is another alternative to the construction of digital twins of the production process. Unlike neural networks, this approach makes it possible to explicitly obtain high-quality quasi-linear difference equations (adequately describing the considered process for any type of initial data). Directions of further researches are improving the algorithm for time series using arbitrary time periods, and application of the developed algorithm for multidimensional time series. Also the research of different outer factors such as humidity and temperature influence is the topic of further research.

References

1. Terekhin, E.A.: Analysis of the seasonal dynamics of NDVI index and the reflective properties of corn in the Belgorod region. Mod. Prob. Remote Sens. Earth Space **11**(4), 244–253 (2014). (in Russian)
2. Ahmad, R., Yang, B., Ettlin, G., Berger, A., Rodriguez-Bocca, P.: A machine-learning based ConvLSTM architecture for NDVI forecasting. Int. Trans. Oper. Res. **30**, 2025–2048 (2020). https://doi.org/10.1111/itor.12887
3. Gao, P., Du, W., Lei, Q., Li, J., Zhang, S., Li, N.: NDVI forecasting model based on the combination of time series decomposition and CNN-LSTM. Water Res. Manage. **37**, 1–17 (2023). https://doi.org/10.1007/s11269-022-03419-3
4. Ahmad, R., Yang, B., Rodriguez-Bocca, P.: Deep spatial-temporal graph modeling for efficient NDVI forecasting. Smart Agric. Technol. **4**, 100172 (2023). https://doi.org/10.1016/j.atech.2023.100172
5. Huang, S., Ming, B., Huang, Q., Leng, G., Hou, B.: A case study on a combination NDVI forecasting model based on the entropy weight method. Water Res. Manage. **31**(11), 3667–3681 (2017). https://doi.org/10.1007/s11269-017-1692-8

6. Fernandez-manso, A., Quintano, C., Fernandez-Manso, O.: Forecast of NDVI in coniferous areas using temporal ARIMA analysis and climatic data at a regional scale. Int. J. Remote Sens. **32**, 1595–1617 (2011). https://doi.org/10.1080/01431160903586765

7. Alhamad, M., Stuth, J., Vannucci, M.: Biophysical modelling and NDVI time series to project near-term forage supply: spectral analysis aided by wavelet denoising and ARIMA modelling. Int. J. Remote Sens. **28**, 2513–2548 (2007). https://doi.org/10.1080/01431160600954670

8. Bukhovets, A. G., Semin, E.A., Kostenko, E.I., Yablonovskaya, S.I.: Modelling of the dynamics of the NDVI vegetation index of winter wheat under the conditions of the CFD. Bull. Voronezh State Agrarian Univ. **2**, 186–199 (2018). https://doi.org/10.17238/issn2071-2243.2018.2.186 (in Russian)

9. Greben, A.S., Krasovskaya, I.G.: Analysis of the main methods for forecasting yields using space monitoring data, in relation to grain crops in the steppe zone of Ukraine. Radio Electr. Comput. Syst. **2**(54), 170–180 (2012). (in Russian)

10. Spivak, L.F., Vitkovskaya, I.S., Batyrbayeva, M.Z., Kauazov, A.M.: Analysis of the results of forecasting the yield of spring wheat based on time series of statistical data and integral indices of vegetation. Mod. Prob. Remote Sens. Earth Space **12**(2), 173–182 (2015). (in Russian)

11. Panyukov, A., Makarovskikh, T., Abotaleb, M.: Forecasting with using quasilinear recurrence equation. In: Olenev, N., Evtushenko, Y., Jacimovic, M., Khachay, M., Malkova, V., Pospelov, I. (eds.) OPTIMA 2022. Communications in Computer and Information Science, vol. 1739, pp. 183–195. Springer, Cham (2022). https://doi.org/10.1007/978-3-031-22990-9_13

12. Burmistrov, A.V., Salnikov, I.I.: Method of element-by-element analysis of color images for the formation of distinctive features in the form of linear contours. XXI Century: Results Past Prob. Present Plus **13**(25), 29–34 (2015). (in Russian)

13. Burmistrov, A.V., Salnikov, I.I.: The method of forming linear contours on aerial photographs of rural areas. Mod. Prob. Sci, Educ. **5**, 152–157 (2013). (in Russian)

14. Burmistrov, A.V., Salnikov, I.I.: Information model of the distinguishing features of images on aerial photographs of rural areas. XXI Century: Results Past Prob. Present Plus **3**(19), 41–45 (2014). (in Russian)

15. Panyukov, A., Tyrsin, A.: Stable parametric identification of vibratory diagnostics objects. J. Vibroengineering **10**(2), 142–146 (2008). https://www.extrica.com/article/10181

16. Panyukov, A.V., Mezaal, Y.A.: Stable estimation of autoregressive model parameters with exogenous variables on the basis of the generalized least absolute deviation method. IFAC-PapersOnLine **51**(11), 1666–1669 (2018). https://doi.org/10.1016/j.ifacol.2018.08.217

17. Panyukov, A.V., Mezaal, Y.A.: Improving of the identification algorithm for a quasilinear recurrence equation. In: Olenev, N., Evtushenko, Y., Khachay, M., Malkova, V. (eds.) OPTIMA 2020. CCIS, vol. 1340, pp. 15–26. Springer, Cham (2020). https://doi.org/10.1007/978-3-030-65739-0_2

18. Storchak, I.G.: Winter wheat yield forecast using the NDVI for the conditions of the Stavropol region. Dissertation for a degree Candidate of Agricultural Sciences (2016). http://stgau.ru/science/dis/dis_presto/storchak_2016.pdf. (in Russian)

19. Nur, M.: Development of a methodology for using satellite imagery data for forest monitoring. Dissertation for the degree of candidate of technical sciences (2021). https://www.miigaik.ru/upload/iblock/bb5/bb5fb148785aa10d2c3bae350d853c0b.pdf. (in Russian)

Searching for Distance Graph Embeddings and Optimal Partitions of Compact Sets in Euclidean Space

V. A. Voronov[1,2(✉)] [ID], A.D. Tolmachev[2,3], D.S. Protasov[2,4], and A.M. Neopryatnaya[2,3]

[1] Caucasus Mathematical Center of Adyghe State University, Maikop, Russia
v-vor@yandex.ru
[2] Moscow Institute of Physics and Technology, Dolgoprudniy, Russia
[3] Skolkovo Institute of Science and Technology, Moscow, Russia
[4] Artificial Intelligence Research Institute, Moscow, Russia

Abstract. We consider three problems in combinatorial geometry in which the search for counterexamples and improvement of known estimates is reduced to a finite-dimensional multi-extremal optimization problem with piecewise-smooth constraints. The first problem is to find a distance embedding of some graph into a given surface, i.e. to find a set of points for which a part of pairwise distances and some additional condition are given. The other two problems consist in minimization of some functional computed for partitions of a compact set into a given number of subsets. The solutions found have improved some quantitative estimates in generalizations of the Borsuk hypothesis and variants of the Hadwiger–Nelson–Erdös problem on the chromatic number of space.

Keywords: Borsuk problem · distance graphs · stochastic gradient descent · global optimization

1 Introduction

In discrete and combinatorial geometry there are many famous results established with computer assistance. The most important example is probably the proof of the Kepler hypothesis and its formal verification [11]. In many cases, counterexamples or estimates can be obtained as a solution to some optimization problem. Sometimes even a reasonably good local minimum is of interest.

Some examples are a number of optimization problems on dense packing of balls in a container of given shape and on optimal covering of the set by k balls, cubes etc., for which exact values of the optimum are known only for small k. If the enclosing set is bounded and the number of spheres (or covering sets) is given, then we deal with a finite-dimensional multiextremal optimization problem with piecewise smooth objective function and piecewise smooth constraints.

The first problem considered in this paper, the distance graph embedding problem, is closely related to the Hadwiger–Nelson–Erdös problem on chromatic

© The Author(s), under exclusive license to Springer Nature Switzerland AG 2023
M. Khachay et al. (Eds.): MOTOR 2023, CCIS 1881, pp. 391–403, 2023.
https://doi.org/10.1007/978-3-031-43257-6_29

number space and its variants [2,21]. It is required to estimate the minimum number of non-intersecting subsets into which a whole Euclidean space $\mathbb{R}^n$ (or some set $F \subset \mathbb{R}^n$ with an induced metric) can be partitioned so that none of the subsets contains two points at a unit distance. Lower bounds in most cases are reduced to the construction of a finite graph whose vertices are points of space and whose edges are pairs of points at a unit distance, and then the chromatic number of this graph is calculated or estimated. We consider an approach where at first a k-chromatic graph which presumably has a geometrical realization is constructed and then this realization is calculated as a solution of some global optimization problem.

The second problem is to determine the minimal diameter $d_F(k)$ such that the compact set $F \in \mathbb{R}^n$ can be divided into k parts $F_1, \ldots, F_k$ of diameter at most $d_F(k)$. Equivalently, F can be covered by closures of F_k, and the maximal diameter of the closures is the same. One can say that a question of this kind is a particular case or a quantitative version of the Borsuk problem [20]. In [12] some values of $d_F(k)$ were found for a circle, a triangle and a square. In [8,16,23,24] this problem was studied for so-called universal covers, i.e. sets inside which one can put a copy of any set of diameter 1 in space of a given dimension. In recent years, some results have been achieved in a similar problem for the l_p metric [18,28,31]. Admittedly, problems of this type are computationally more complicated than the much better studied problems on packings of balls or other geometric objects in a container of given form and on covering sets [5,22,25,30]. The most studied particular problem of this kind seems to be the problem on packing of circles in a square [22]. Packing of circles on a torus [19] has also been considered. The paper [17] studied an approach to the problem of covering a shape by circles based on wave propagation.

The third of the problems considered in this paper is to find partitions of a given set into k non-adjoint subsets, none of which contains a pair of points at a distance from the interval $(1, \alpha)$, where α should be maximally possible. In other words, the width of the forbidden distance interval for which the chromatic number of a subset of the metric space does not exceed k is maximized. For the case of subsets of the plane, this problem has previously been studied in [4,6]. In the present paper such partitions are constructed for the case of a two-dimensional sphere.

In the second and third problems the search for a local minimum is reduced to a finite-dimensional optimization problem since it is always possible to identify a finite number of maximal diameters and (in the third problem) minimal distances between sets. Moreover, to compute an optimal value with any given accuracy, it is sufficient to consider a discrete approximation (e.g. points of a rectangular grid) [29]. But reduction to a finite-dimensional problem of sufficiently small dimension requires efficient heuristics for selecting an initial approximation. In this paper, as in the previous work [24], the Voronoi diagram for centres of balls in some package is applied. It should be noted that the final stage of the search for local minimum is actually reduced to the search for an embedding of the distance graph. In contrast to *contact graphs* described in [19], additional geometrical constraints arise when partitioning a given set. For the vertices of a distance graph, which are also vertices of polygons of a partition, Karush–Kuhn–Tucker conditions are usually fulfilled. General results from the theory of combinatorial rigidity of bar and slider networks [13,14] apply to these graphs, which helps to distinguish such a graph before the local minimum is found with high precision.

The purpose of this paper is mostly to list the distance geometry problems for which we have been able to obtain new results using standard optimization methods implemented in open source software. We cannot claim that the methods used are close to the best, but they were good enough to get new results. It seems that both the global search strategy and the numerical method for finding the local minimum can be significantly improved.

The source code is available in the repository [26].

2   Preliminaries

Definition 1. *The chromatic number of a subset of the Euclidean space $F \subseteq \mathbb{R}^n$ for the set of forbidden distances $\Theta \subset \mathbb{R}^+$ is the minimum number of colors needed to color F, so that the points u, v at a distance belonging to the set Θ are colored differently, i.e.*

$$\chi_\Theta(F) = \min\{\chi \ : \ F = F_1 \sqcup F_2 \sqcup \cdots \sqcup F_\chi \quad \forall x, y \in F_i \quad \|x - y\| \notin \Theta\}.$$

In the classical formulation of Hadwiger–Nelson–Erdös problem it is assumed that $\Theta = \{1\}$.

Definition 2. *By an Euclidean distance graph $G = (V, E)$, or simply distance graph we mean a graph whose vertices are points of Euclidean space $\mathbb{R}^n$ and whose edges are connected by points whose distance between them belongs to a given set Θ, i.e.*

$$V \subset \mathbb{R}^n; \quad (x, y) \in E \iff \|x - y\| \in \Theta.$$

According to de Brujin–Erdös theorem [1], if one accepts the axiom of choice or an equivalent statement, then the chromatic number of an infinite graph is reached on a finite subgraph. So, the simplest way to prove the lower extimate $\chi_\Theta(F) \geq m$ is to consider a finite Euclidean distance graph G for which $V \subset F$ and check that $\chi(G) \geq m$.

In addition, in order to obtain such an estimate, it is sometimes reasonable to ask whether a given abstract graph can be realized as a distance graph (see [10]). The particular case of this problem is discussed in the next section.

Definition 3. *For a given integer $k > 0$ we denote by $d_F(k)$ the greatest real number with the property that F can be covered by k sets $F_1, F_2, \ldots, F_k$ whose diameters are at most x, that is,*

$$d_F(k) = \inf\{x \in \mathbb{R}^+ : \exists F_1, \ldots, F_k : F \subseteq F_1 \cup \ldots \cup F_k, \ \forall i \ \mathrm{diam}(F_i) \leqslant x\}.$$

One can observe that this problem is closely related to the previous one. Namely, if a covering of F by k parts of diameter strictly smaller than d exists then

$$\chi_\Theta(F) \le k, \quad \Theta = [d; +\infty).$$

On the other hand, if the chromatic number is larger, $\chi_\Theta(F) \ge k + 1$, and there is a covering by k parts of diameter at most d, *then it is known to be optimal.* As in the previous case, we can argue that any lower bound for the chromatic number is reached on a finite graph. In some cases this graph can be found simply on the vertex set of the computed partition and then it is possible to prove that the minimum is global.

3   Distance Embedding of a Graph

Definition 4. *We call $\phi_r : V \to S^2(r)$ a distance embedding of a graph G in a sphere $S^2(r)$ if the mapping ϕ is injective and for every edge $(u, v) \in E$, $\|\phi_r(u) - \phi_r(v)\| = 1$ is satisfied.*

Denote $v_1, \ldots, v_m$, $v_i \in S^2(r) \subset R^3$ by the coordinates of the vertices of the graph $G = (V, E)$ on the sphere of radius r. Then the distance embedding is defined by the system of algebraic equations

$$\begin{cases} \|v_i\|^2 - r^2 = 0; \\ \|v_i - v_j\|^2 - 1 = 0; \quad (i, j) \in E(G). \end{cases} \tag{1}$$

The problem is supposed to be set correctly (i.e. the number of variables is equal to the number of equations) if the graph G is minimally rigid (i.e. $|E(G)| = 2n - 3$ and for each subgraph $G_1 \subset G$ $|E(G_1)| \le 2|V(G_1)| - 3$ is satisfied) and then some variables are fixed. For example,

$$x_1 = (\sqrt{r^2 - 1/4}, -1/2, 0); \quad x_2 = (\sqrt{r^2 - 1/4}, 1/2, 0); \quad (x_1, x_2) \in E(G).$$

The probabilistic algorithm for finding distance embedding of graph G on n vertices in a sphere is based on generating a random coordinates of vertices in the cube $C = [-1; 1]^3$ and then finding a local minimum of the polynomial using gradient descent:

$$\Phi(G; V) = c \sum_{i=1}^{n} (\|v_i\|^2 - r^2)^2 + \sum_{(i,j) \in E(G)} (\|v_i - v_j\|^2 - 1)^2,$$

where $c > 0$ is a penalty coefficient.

Obviously, not all local minima of $\Phi(G; V)$ will correspond to graph embeddings in the sphere. The value of Φ in a local minimum may be different from zero; moreover, at zero the vertices may coincide. Nevertheless, for a number of examples this algorithm can find a distance embedding with accuracy limited by the implementation of floating point.

Note that the existence of the approximate solution under some additional assumptions will follow from the existence of the exact solution. Namely, for a correct problem of the form (1) the Jacobian $A(x) = \partial F/\partial X$ will be a non-degenerate square matrix. Then it is sufficient to check the conditions of the quantitative version of the implicit function theorem or the convergence theorem of Newton's method [15]. If the convergence conditions are satisfied, then it can also be argued that the solution will be found with positive probability under a random initial approximation.

This approach has been used in [27] to compute distance embeddings of minimally rigid graphs with number of vertices 9 and 10. Here is a previously unpublished result on the embedding of a 17-vertex unit distance graph G_{17} constructed in [7]. In the case of the plane G_{17} is the minimal known graph having the above properties.

In addition, if the solution is found for some value of the radius, we can find it for some interval of values by solving numerically the following system of differential equations.

$$\frac{1}{2}A(x)\frac{\partial x}{\partial t} = b, \quad b_i = \begin{cases} 1, & 1 \le i \le n, \\ 0, & n+1 \le i \le 3n-3 \end{cases}$$

$$\frac{1}{2}A(x) = \begin{pmatrix} x_1 & 0 & 0 & 0 & \cdots & 0 \\ 0 & x_2 & 0 & 0 & \cdots & 0 \\ 0 & 0 & x_3 & 0 & \cdots & 0 \\ \vdots & & & \ddots & & \\ 0 & 0 & 0 & 0 & \cdots & x_n \\ x_1 - x_j & 0 & \cdots & x_j - x_1 & \cdots & 0 \\ \vdots & & & & & \\ 0 & \cdots x_k - x_n & \cdots & 0 & x_n - x_k \end{pmatrix}. \tag{2}$$

In fact, system (2) describes the deformation of the bar and hinge mechanism, in which the hinges are the vertices of the graph, when the radius of the sphere changes continuously (Fig. 1).

Claim. When $0.5522 = r_* \le r \le r^* = 0.577 < \sqrt{3}/3$ there exists a distance embedding of graph G_{17} in sphere S_r^2.

Verifying this statement requires a large amount of one-type computation. Namely, we consider the partitioning of the segment $[r_*, r^*]$ with step $h \le 10^{-4}$. For each point r_i, the distance embedding X_i^* is approximated and the convergence radius of Newton's method is estimated as r changes with the initial approximation X_i^*. If the intervals on which Newton's method converges cover the whole interval $[r_*, r^*]$, then we conclude that the statement is established, otherwise we should reduce h.

Note that when r tends to an upper or lower bound where the distance embedding ceases to exist, the Jacobian matrix $A(x)$ becomes degenerate and the radius of convergence tends to zero. It is not difficult to calculate the embedding

```
k := 0;
while k < k_max do
    for i=1,2, ..., n do
    |   x_i = Random_Point();
    end
    X_0 := x_i, i = 1, 2, ..., n;
    X* = Minimize(Φ(X), X_0);
    if Φ(X*) < ε_1 then
        if ||x_i* - x_j*|| > ε_2, i ≠ j then
        |   (* X* is a distance embedding of the graph G *);
        |   Halt;
        end
    end
    k := k + 1;
end
```

Algorithm 1: Computing the distance embedding

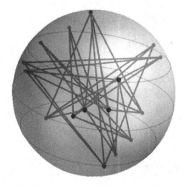

Fig. 1. The distance embedding of G_{17} in $S^2(r)$, $r = 0.57$

at any particular value of r, but this approach does not allow one to justify the existence of the embedding up to the values of r^* and r_* found with high accuracy or analytically.

4   Partitions of the Square and the Equilateral Triangle

Below is an algorithm for finding optimal coverings (partitions). The initial approximation when reduced to a finite-dimensional problem is computed as a Voronoi diagram constructed for some (far from optimal) packing of circles in the polygon F. For a given location of the centers of the circles, the function that gives their maximum radius is written as follows

$$\psi(V) = \min\left\{\min_{p,t}\{(v_p, c_t) + b_t\}, \frac{1}{2}\min_{p,q}\{\|v_p - v_q\|\}\right\} \to \max. \qquad (3)$$

To compute a rough approximation we apply the SGD algorithm.

In the second step, the minimization of the maximum diameter is performed for the finite-dimensional problem of covering the set with polygons initialized with a Voronoi diagram. As in the previous case, we use the implementation of SGD in the **pytorch** library.

$$\varphi(X) = \max_i \max_{p,q \in \mathcal{J}_i} \|x_p - x_q\| \to \min, \tag{4}$$

$$(x_s, c_t) + b_t = 0, \quad (s,t) \in \mathcal{L},$$

where $\mathcal{J}_i$ if the vertex set of the i-th polygon, and $\mathcal{L}$ is the set of edges of auxilary bipartite graph describing the belonging of vertices x_s to the sides of the polygon F.

Note that the function is not formally stochastic, but with a large number of regions we are dealing with the same situation as when minimizing the maximum of a large number of input signals in the neural network training problem. Another motivation for using SGD is the availability of GPU implementations.

Finally, the computation of the optimum with high accuracy is done using classical quasi-Newtonian algorithm BFGS [9], which is possible since we are dealing with a system of polynomial equations in the last step. Note that the partition structure may change if at any optimization step a vertex is found that has points from four or more different subsets of F_i in a small neighborhood.

Let us show the results found by the Algorithm 2. Table 1 shows the found values of diameters and parameters of the distance graph defining the exact value. The bolded estimates were not given in [12]. In addition, one can derive analytical expressions for some of the estimates.

Proposition 1.

$$d_\triangle(6) = d_\triangle(7) = \frac{1}{1 + \sqrt{3}}; \quad d_\triangle(15) \leq \frac{1}{1 + 2\sqrt{3}}.$$

$$d_\square(6) \leq \frac{1}{3\sqrt{3}} \sqrt{7 - \frac{200}{a} + a}; \quad a = \sqrt[3]{2(383 + 129^{3/2})}.$$

An analytic estimate for $d_\square(6)$ can be obtained as the root of the cubic polynomial by minimizing the edge length for the 5-vertex graph with 5 edges (a rhombus with a hanging vertex), see Fig. 3.

Input: F, k
for $1 \leqslant i \leqslant N$ **do**
    $V_0 = \{v_1, \ldots, v_k\}$ are random points distributed uniformly in $[-1, 1]^n$;
    (* Run s steps of SGD for the problem (3) *) **for** $1 \leqslant j < s$ **do**
      | $V_{j+1} \leftarrow \mathrm{SGD}\,(G(V) \rightarrow \max; V_j)$;
    end
    (* Initialize a set of partition vertices with a Voronoi diagram *)
    $X_0 \leftarrow \mathrm{Vor}(F; V_s)$;
    $t \leftarrow 0$;
    (* Find the local minimum for the problem (4) *)
    while *not* $\mathrm{OptCondition}(X)$ **do**
      | $X_{t+1} \leftarrow \mathrm{SGD}\,(F(X) \rightarrow \min; X_t)$;
      | $t \leftarrow t + 1$; **if** *some vertices in X_{t+1} coincide* **then**
      | | rearrange vertices;
      | **end**
    end
    if $F(X_t) < F(X^*)$ **then**
    | $X^* \leftarrow X_t$;
    end
end * Compute the graph of maximal diameters
$H^* \leftarrow \mathrm{MaxDiam}(X^*)$;
(* Find the rigid subgraph H^*_{rigid} in H^* *)
$H^*_{rigid} \leftarrow \mathrm{RigidSubgraph}(H^*)$;
(* Perform a high-precision optimization for H^*_{rigid} *)
$X^*_{prec} \leftarrow \mathrm{BFGS}(X^*, H^*_{rigid})$;
Output: X^*_{prec}.

Algorithm 2: Search for a locally optimal partition

Let us show that with $d_\triangle(7) = d_\triangle(6) = \frac{1}{1+\sqrt{3}}$. Consider a graph with 12 vertices (Fig. 2, left) whose edges are shown by dashed lines (i.e. segments of lenght $d_\triangle(6)$). If we construct on these 12 vertices an auxiliary graph $G_{\triangle,12}$ in which vertices at a distance from the set $[d_\triangle(6); +\infty)$ are connected, then it is easy to see that $\chi(G_{\triangle,12}) = 8$. Therefore it is impossible to divide the triangle into 7 pieces of smaller diameter.   □

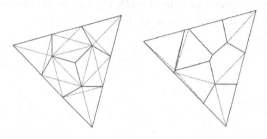

Fig. 2. Optimal partitions of the equilateral triangle, $k \in \{6, 7\}$.

Table 1. Estimates for partitions of □ and △

k	$d_\square(k) \leq$	$d_\triangle(k) \leq$
5	0.64423	0.5
6	**0.59399...**	0.36602
7	0.54858	**0.36602...**
8	**0.51145...**	**0.33581...**
9	**0.45454...**	0.333333
10	**0.43647...**	0.288675
11	**0.41677...**	**0.27123...**
12	**0.39521...**	**0.26795...**
13	**0.38443...**	**0.25242...**
14	**0.36685...**	**0.25**
15	**0.35156...**	**0.22401...**

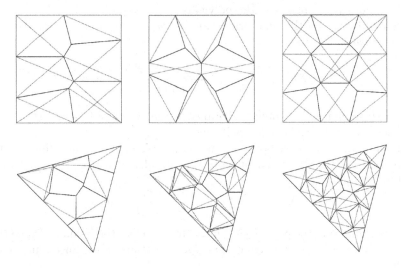

Fig. 3. Optimal partitions of the square, $k \in \{6, 8, 9\}$, and partitions of the equilateral triangle, $k \in \{8, 13, 15\}$.

5 Maximizing the Gap in the Partition of the Sphere

Let us consider a partition the sphere S^2 of unit radius into spherical polygons $F_1, \ldots, F_k$. Denote the partition $\mathcal{F} = \{F_1, \ldots, F_k\}$ and consider the problem

$$\phi(\mathcal{F}) = \max_{1 \leq i \leq k} \operatorname{diam} F_i \to \min. \tag{5}$$

The same algorithm as for polygons is used to find local minima in this problem. The peculiarity is that if the initial approximation is poorly chosen or if the step in the finite-dimensional problem is too large, we may obtain a zero value of the minimum when all vertices of the partition converge to one point of the sphere.

Let us denote by $d_{sph}(k)$ the solution of (5) for a given k. The upper and lower estimates for $d_{sph}(k)$ can be obtained from the results on the density of coverage of the sphere by non-intersecting spherical caps [25]. Table 2 shows the found estimates of the diameter of the sets that manage to cover the unit sphere for a given k.

Now suppose that we are interested in the possibility of coloring polygons in several colors so that some distance interval does not occur between points of the same color. Assuming that the diameters of the polygons are close, this means that neighboring polygons must be colored differently, as well as neighbors of neighbors. That is, for a distance of 1 or 2 in a dual graph, the colors of the polygons must be different. Moreover, we are interested in the fact that distance 3 in the dual graph corresponds to as large a value of the Euclidean distance as possible, or more precisely, we should *maximize the ratio* of this distance to the maximum of the diameters of the polygons.

Denote by $G'(\mathcal{F})$ the dual graph, and by $u_1, \ldots, u_k$ its vertices. We define the objective as follows:

$$\Psi_1(\mathcal{F}) = \frac{\min_{\text{dist}(u_i,u_j)=3} \text{dist}(F_i, F_j)}{\max_{1 \le i \le s} \text{diam } F_i} \to \max, \tag{6}$$

where $\text{dist}(u_i, u_j)$ is the (integer) distance in the dual graph, and $\text{dist}(F_i, F_j)$ is the Euclidean distance between polygons.

Having any given partition one can define some coloring and find the minimum in a finite-dimensional optimization problem with a piecewise smooth function

$$\Psi_2(X) = \frac{\min_{(i,j) \in E_1} \text{dist}(x_i, x_j)}{\max_{(i,j) \in E_2} \text{dist}(x_i, x_j)} \to \max,$$

where the sets of index pairs E_1, E_2 mean respectively the belonging of vertices to one polygon and the belonging of polygons that are at distance 3 in the dual graph.

Finally, the sequence of computations is as follows.

1. Let $V = \{v_1, \ldots, v_s\}$ be a "uniform" arrangement of points on the sphere (e.g., a random local minimum in the Thomson problem).
2. Construct the spherical Voronoi diagram for the set V.
3. Suppose that all the diameters are sides and diagonals of the polygons (i.e. cells of the Voronoi diagram). Then we have a constrainted finite-dimensional optimization problem.
4. Find a local minimum (using the SGD algorithm).
5. Check that the partition is correct, i.e. there is no distances greater than the diagonals in each of the polygons.
6. Compute the objective function.
7. If the specified number of runs is not reached, go to the first step.

The best found estimates for the minimum diameter and maximum ratio are shown in Table 2.

Table 2. Estimates for partitions of the sphere

k	$d_{sph}(k) \leq$	k	$d_{sph}(k) \leq$	k	max ratio $\geq$
2	2.0	12	1.154	12	1.414
3	2.0	20	0.981	13	1.210
4	1.776	30	0.791	14	1.277
5	1.732	40	0.683	15	1.209
6	1.732	50	0.618	16	1.320
7	1.634	100	0.439...	17	1.112
8	1.414	200	0.312...	18	1.298

6 Conclusion

In this paper we considered several problems of combinatorial geometry related to the chromatic number of graphs of diameters or unit distance graphs. Improvements to the algorithm considered in the previous paper [24] are proposed. New upper estimates for cases of square, triangle, and sphere are obtained. In addition, in some cases, by considering a finite set of points, it is possible to prove that the minimum that is obtained is global.

Note that this approach can be applied without any modification to finite dimensional spaces with other metrics, if a particular problem of interest arises.

We list some questions that remain open:

Question 1. Can one find a 4-chromatic triangle-free unit distance graph embedded in the sphere $S^2(r)$, $r > 0.5522$ with less than 17 vertices?

It is possible that the Exoo–Ismailescu graph is the minimal 4-chromatic subgraph for all the radii belonging to the interval $(0.5522, \sqrt{3}/3)$. Howewer, for $r = 0.540...$ there is an embedding of 11-vertex Grötzsch graph [3].

Question 2. Is it true that for $k \to \infty$ the value in problems (5), (6) tends to the limit defined by the hexagonal tiling of the plane?

In other words, is it true that with a sphere radius tending to infinity, irregularities can be evenly distributed?

Question 3. How many local minima can correspond to a given dual graph (i.e. combinatorial structure of the partition)? If this number can be large, what is the asymptotic as $k \to \infty$?

Since the function to be minimised is not convex, we should expect that the number of local minima may be arbitrary large after choosing the partition structure. However, at the moment we are not aware of any examples from which this could be observed.

Acknowledgements. The work was supported by the program "Leading Scientific Schools" under grant NSh-775.2022.1.1.

References

1. de Bruijn, N.G., Erdos, P.: A colour problem for infinite graphs and a problem in the theory of relations. Indigationes Math. **13**, 371–373 (1951)
2. Cherkashin, D., Kulikov, A., Raigorodskii, A.: On the chromatic numbers of small-dimensional Euclidean spaces. Discrete Appl. Math. **243**, 125–131 (2018). https://doi.org/10.1016/j.dam.2018.02.005
3. Cherkashin, D., Voronov, V.: On the chromatic number of 2-dimensional spheres (2022)
4. Chybowska-Sokół, J., Junosza-Szaniawski, K., Węsek, K.: Coloring distance graphs on the plane (2022)
5. Enkhbat, R.: Convex maximization formulation of general sphere packing problem. Bull. Irkutsk State Univ. Ser. Math. **31**, 142–149 (2020). https://doi.org/10.26516/1997-7670.2020.31.142
6. Exoo, G.: ϵ-unit distance graphs. Discrete Comput. Geom. **33**(1), 117–123 (2004). https://doi.org/10.1007/s00454-004-1092-8
7. Exoo, G., Ismailescu, D.: Small order triangle-free 4-chromatic unit distance graphs. Geombinatorics **26**(2), 49–64 (2016)
8. Filimonov, V.P.: Covering sets in $\mathbb{R}^m$. Sbornik: Math. **205**(8), 1160–1200 (2014). https://doi.org/10.1070/sm2014v205n08abeh004414
9. Fletcher, R.: Practical Methods of Optimization. John Wiley & Sons, Hoboken (2013)
10. Frankl, N., Kupavskii, A., Swanepoel, K.J.: Embedding graphs in Euclidean space. J. Comb. Theory Ser. A **171**, 105146 (2020)
11. Hales, T., et al.: A formal proof of the Kepler conjecture. In: Forum of Mathematics, Pi, vol. 5 (2017). https://doi.org/10.1017/fmp.2017.1
12. Heppes, A.: Covering a planar domain with sets of small diameter. Periodica Math. Hung. **53**(1–2), 157–168 (2006). https://doi.org/10.1007/s10998-006-0029-9
13. Jordán, T.: II – combinatorial rigidity: graphs and matroids in the theory of rigid frameworks. In: Mathematical Society of Japan Memoirs, pp. 33–112. The Mathematical Society of Japan (2016). https://doi.org/10.2969/msjmemoirs/03401c020
14. Katoh, N., Tanigawa, S.: On the infinitesimal rigidity of bar-and-slider frameworks. In: Dong, Y., Du, D.-Z., Ibarra, O. (eds.) ISAAC 2009. LNCS, vol. 5878, pp. 524–533. Springer, Heidelberg (2009). https://doi.org/10.1007/978-3-642-10631-6_54
15. Kolmogorov, A.N., Fomin, S.V.: Elements of Function Theory and Functional Analysis [in Russian]. Science publishing (1972)
16. Kupavskii, A.B., Raigorodskii, A.M.: Partition of three-dimensional sets into five parts of smaller diameter. Math. Notes **87**(1–2), 218–229 (2010). https://doi.org/10.1134/s0001434610010281
17. Lempert, A., Kazakov, A., Le, Q.: On reserve and double covering problems fo the sets with non-Euclidean metrics. Yugoslav J. Oper. Res. **29**(1), 69–79 (2018), http://yujor.fon.bg.ac.rs/index.php/yujor/article/view/599

18. Lian, Y., Wu, S.: Partition bounded sets into sets having smaller diameters. Results Math. **76**(3), 1–15 (2021). https://doi.org/10.1007/s00025-021-01425-2

19. Musin, O.R., Nikitenko, A.V.: Optimal packings of congruent circles on a square flat torus. Discrete Comput. Geom. **55**(1), 1–20 (2015). https://doi.org/10.1007/s00454-015-9742-6

20. Raigorodskii, A.M.: Around borsuk's hypothesis. J. Math. Sci. **154**(4), 604–623 (2008). https://doi.org/10.1007/s10958-008-9196-y

21. Soifer, A.: The Mathematical Coloring Book. Springer, New York (2009). https://doi.org/10.1007/978-0-387-74642-5

22. Szabó, P.G., Markót, M.C., Csendes, T.: Global optimization in geometry – circle packing into the square. In: Audet, C., Hansen, P., Savard, G. (eds.) Essays and Surveys in Global Optimization, pp. 233–265. Springer-Verlag, Boston (2005). https://doi.org/10.1007/0-387-25570-2_9

23. Tolmachev, A.D., Protasov, D.S.: Covering planar sets. Doklady Math. **104**(1), 196–199 (2021). https://doi.org/10.1134/s1064562421040141

24. Tolmachev, A., Protasov, D., Voronov, V.: Coverings of planar and three-dimensional sets with subsets of smaller diameter. Discrete Appl. Math. **320**, 270–281 (2022). https://doi.org/10.1016/j.dam.2022.06.016

25. Tóth, G.F.: Packing and Covering. Chapman and Hall/CRC, In Handbook of discrete and computational geometry (2017)

26. Voronov, V., Protasov, D., Tolmachev, A.: Github repository (2023). https://github.com/Vosatorp/Torus/tree/main/triangles_and_squares

27. Voronov, V., Neopryatnaya, A., Dergachev, E.: Constructing 5-chromatic unit distance graphs embedded in the Euclidean plane and two-dimensional spheres. Discrete Math. **345**(12), 113106 (2022). https://doi.org/10.1016/j.disc.2022.113106

28. Wang, J., Zhang, Y.: Borsuk's partition problem in $(\mathbb{R}^n, \ell_p)$. Math. Notes **111**(1–2), 289–296 (2022). https://doi.org/10.1134/s0001434622010321

29. Zong, C.: A quantitative program for Hadwiger's covering conjecture. Sci China Math **53**(9), 2551–2560 (2010). https://doi.org/10.1007/s11425-010-4087-3

30. Zong, C.: Functionals on the spaces of convex bodies. Acta Math. Sinica Engl. Ser. **32**(1), 124–136 (2016). https://doi.org/10.1007/s10114-015-4386-2

31. Zong, C.: Borsuk's partition conjecture. Japan. J. Math. **16**(2), 185–201 (2021). https://doi.org/10.1007/s11537-021-2007-7

Author Index

A

Abotaleb, Mostafa 376
Alekseev, Aleksandr Vladimirovich 324
Alexander, Uspenskii 292
Alkousa, Mohammad 29
Anikin, Anton 92
Aroslankin, Artem 353

B

Berikov, Vladimir 364
Bobrov, Evgeny 203

C

Chentsov, Alexandr G. 218
Chentsov, Pavel A. 218
Cherikbayeva, Lyailya 364
Chukanov, Sergey 92

D

Danik, Yulia 277
Davydov, Ivan 109
Dmitriev, Mikhail 277
Dordzhiev, Adyan 203

E

Ershov, Aleksandr Anatol'evich 324
Ershov, M. A. 67
Ershova, Anna Aleksandrovna 324
Erzin, Adil 122

G

Garbar, Sergey 79
Gasnikov, Alexander 29, 92
Gerontitis, Dimitrios 3
Gnusarev, Alexander 243
Gornov, Alexander 92

I

Il'ev, Victor 134
Il'eva, Svetlana 134

K

Kalmutskiy, Kirill 364
Kalyagin, Valeriy 353
Kazakovtsev, Lev A. 3
Kochetov, Yury 259
Kononov, Alexander 122
Korostil, Alexander V. 146
Krutikov, Vladimir N. 3
Kulachenko, Igor 259

L

Lavlinskii, Sergey 231
Levanova, Tatiana 243
Litvinenko, Alexander 364
Lobanov, Aleksandr 92

M

Makarovskikh, Tatiana 376
Marakulin, Valeriy 308
Melnikov, Andrey 259

N

Nazarenko, Stepan 122
Neopryatnaya, A. M. 391
Nikolaev, Andrei V. 146

P

Panin, Artem 231
Panyukov, Anatoly 376
Pavel, Lebedev 292
Pinyagina, Olga 19
Plyasunov, Alexander 231
Protasov, D. S. 391

R

Ratushnyi, A. 161
Rubtsova, Ekaterina 243

S
Savchuk, Oleg   29
Sharankhaev, Konstantin   122
Shulgina, Oksana   54
Simanchev, R. Yu.   176
Stanimirović, Predrag S.   3
Stonyakin, Fedor   29

T
Tarasyev, Alexander M.   338
Titov, Alexander   29
Tolmachev, A. D.   391

U
Urazova, I.V.   176
Ushakov, Anton V.   109

Ushakov, Vladimir Nikolaevich   324
Usova, Anastasiia A.   338

V
Vasilyev, Igor   109
Voronov, V. A.   391
Voroshilov, A. S.   67
Vyatcheslav, Sigaev   243

Y
Yarullin, Rashid   44, 54
Yuskov, Alexander   259

Z
Zabirova, Rida   29
Zabotin, Igor   44, 54
Zabudsky, Gennady G.   188

Printed in the United States
by Baker & Taylor Publisher Services